INORGANIC CHEMISTRY

INORGANIC CHEMISTRY

An Industrial and Environmental Perspective

T. W. SWADDLE
Professor of Chemistry
University of Calgary

ACADEMIC PRESS
San Diego London Boston New York Sydney Tokyo Toronto

This book is printed on acid-free paper. 

Academic Press
a division of Harcourt Brace & Company
525 B Street, Suite 1900, San Diego, California 92101-4495, USA
http://www.apnet.com

Academic Press Limited
24-28 Oval Road, London NW1 7DX, UK
http://www.hbuk.co.uk/ap/

Library of Congress Cataloging-in-Publication Data
Swaddle, T. W. (Thomas Wilson), date.
Inorganic chemistry : an industrial and environmental perspective / T. W. Swaddle.
p. cm.
Includes index
ISBN 0-12-678550-3 (alk. paper)
1. Chemistry, Inorganic. 2. Environmental chemistry I. Title
QD151.5.S93 1997 97-1552
546--dc21 CIP

PRINTED IN THE UNITED STATES OF AMERICA
97 98 99 00 01 02 EB 9 8 7 6 5 4 3 2 1

This book is dedicated to

May Swaddle

on the occasion of her ninetieth birthday.

Contents

Preface

THIS VOLUME is the outgrowth of my little book *Applied Inorganic Chemistry* (University of Calgary Press, 1990) which aimed to present the essence of inorganic chemistry to second- or third-year undergraduate students of chemistry or chemical engineering in the context of the immense economic, social, and environmental impact of the subject. As the aim of *Applied Inorganic Chemistry* was to present basic chemical principles, it was not intended to be a comprehensive account of the modern inorganic chemical industry and allied subjects or even to be technologically up-to-date. Therefore, it was gratifying to learn that *Applied Inorganic Chemistry* has found favor with educators, practicing engineers, geologists, environmental scientists, and entrepreneurs in several countries as a compact reference book and guide to the title subject.

The present book is intended to provide this extended readership with a concise overview of the applications of inorganic chemistry in a world that is increasingly dominated by technology and its ramifications, beneficial or otherwise. It remains my conviction that this book, like its progenitor, should find a niche either as a textbook for an independent one-semester course in applied inorganic chemistry or as a complement to a conventional academic text in a full-year course on inorganic chemistry. A pedagogical format has therefore been retained and may prove helpful to readers whose background in chemistry is limited or in need of refreshment. The central purposes of this book, however, are to explain the role of inorganic chemistry in the modern world and to provide a sourcebook of readable proportions for scientists and engineers as well as students and the interested public.

A background of basic first-year university or college chemistry, including the rudiments of organic chemistry, is assumed. It has been my experience that modern theories of chemical bonding do not contribute to the understanding of applied inorganic chemistry in proportion to the intellectual effort they require of the student, and indeed they are not necessary for an appreciation of the role of chemistry in the world around us. Except for a brief introduction to band theory, such concepts are not developed here. Simple electron-pair models of covalency, which are covered in almost all high school or first-year university chemistry courses, are adequate for a working understanding of most chemical phenomena.

Many freshman chemistry courses today also present the elements of molecular orbital theory. For students with this background, I have included a few applications of molecular orbital concepts to explain bonding in metal carbonyls and organometallics, but these are not essential to the purpose of the book. Of much greater technological importance is the interplay of thermodynamics and kinetics in choosing strategies for putting chemical knowledge to practical use. A synopsis of useful thermodynamic and kinetic concepts is therefore given in Chapter 2 with the expectation that, for students of chemistry or chemical engineering, it will be supplemented by more rigorous courses or by reference to the bibliography cited. I have found it expedient to take some possibly illicit shortcuts, such as substituting concentrations for activities where convenient, but provide appropriate caveats wherever such simplifications are made.

As an alternative to trying to cover all the required principles at the beginning of the book, I have provided extensive cross-references, both forward and backward, so that a reader seeking information on one particular topic can find supporting information easily. Sources of additional information are given at the end of each chapter; most are monographs or reviews that contain citations of the primary literature, but, in the case of rapidly developing topics, reference is also made to the original sources or to authoritative scientific news articles. I have found *Science*, *Chemical and Engineering News*, *Chemistry in Britain* and *Nature* to be particularly valuable in this respect.

I am deeply grateful to Kim Wagstaff for producing most of the diagrams and for indispensable guidance in preparing the LaTeX manuscript, and to David J. Packer and the staff of Academic Press for bringing this project to fruition. It is also a pleasure to acknowledge the support and technical advice generously given by G. Bélanger (Hydro Québec), G. L. Bolton (Sherritt Gordon), S. Collins (University of Waterloo), S. Didzbalis (Inco), F. M. Doyle (University of California, Berkeley), J. H. Espenson (Iowa State University), A. McAuley (University of Victoria), G. Strathdee (Potash Corporation of Saskatchewan), R. van Eldik (University of Erlangen), and D. R. M. Walton (University of Sussex), as well as fellow Calgarians V. I. Birss, P. M. Boorman, R. M. Butler, T. Chivers, G. M. Gaucher, R. A. Heidemann, J. A. C. Kentfield, J. King, B. Kipling, R. A. Kydd, M. Parvez, W. E. Piers, B. Pruden, W. J. D. Shaw, S. Staub, M. Weir, and H. L. Yeager.

Finally, I thank the Killam General Endowment Fund for a Resident Fellowship to complete this book, and my wife, Shirley, for forbearance and understanding while I was preoccupied with this project.

T. W. Swaddle

Chapter 1

Importance of Inorganic Chemistry

1.1 Historical Overview

INORGANIC CHEMISTRY is a subject that exists by default—it is the part of chemistry that remained when organic chemistry (the chemistry of carbon compounds containing at least some carbon–hydrogen bonds) and physical chemistry (the science of physical measurements as applied to chemical systems) developed as distinct subdisciplines in the nineteenth century. Inorganic chemistry represents the traditional core of chemistry, with a history traceable over thousands of years. Indeed, our word *chemistry* comes from the ancient Greek *khimiya*, meaning the fusion or casting of metals, and to this day the preparation and properties of metals and their compounds remain central themes of inorganic chemistry and of this book.

In the classical era in Europe, the theory and practice of chemistry were pursued mainly by the ancient Greeks, who made many important discoveries in metallurgy in particular and who are also credited with proposing the earliest version of the atomic theory. The Greek chemical tradition declined when mysticism displaced the observational approach in the second century of the Common Era, and subsequently was largely lost in Europe after the fall of Rome in 410 C.E. In the 11th. century C.E., the quasi-science of alchemy returned to Europe via the Arabs, who also introduced Persian, Indian, and Chinese influences.

Alchemy fell into disrepute in Europe in the Middle Ages because of its obscure symbolism, introduction of irrelevant religious ideas and superstitions, and preoccupation with perfectibility that led to belief in the possibility of the transmutation of metals (as opposed to chemical change). The prospect of changing base metals such as lead into gold attracted all manner of charlatans. Furthermore, the ecclesiastically dominated universities of the day declined to teach a subject that seemed to offer, on the one hand, divine powers and, on the other, a wealth of technical knowledge that was considered to be beneath the dignity of academe!

Yet it was the practical utility of chemistry, combined with development of theoretical underpinnings based on accurate observation, experiment and eventually quantitative measurement, that led to the emergence of the modern science. The year 1789 represents a great turning point in world history, not only because of the French Revolution, but also because of the chemical revolution brought about by the publication in France of Antoine Laurent Lavoisier's *Traité Élémentaire de Chimie.* The brief but brilliant book stressed the importance of quantitative weight relationships, introduced the first systematic chemical nomenclature, and drew together key facts and concepts that had been slowly emerging through the work of Black, Cavendish, Priestley, Scheele, and others but had been obscured by misleading preconceptions. In particular, the concept of *oxidation* (which originally meant simply the uptake of oxygen, as in combustion, with consequent gain of weight) and its reverse, *reduction*, replaced the theories of Becher, Stahl, and others, according to which *phlogiston*, a mysterious fluid of apparently negative weight, was released during combustion. Ironically, oxidation of a substance is now identified with loss of *electrons* from it, so that the phlogistonists were not entirely wrong.*

The chemical revolution begun by Lavoisier was completed in 1810 by the Cumbrian schoolteacher and meteorologist John Dalton, who postulated that a chemical element is composed of submicroscopic *atoms* of a single characteristic kind and with a particular *atomic weight* (properly, *atomic mass*). These atoms combine in simple whole-number ratios with those of other elements to form chemical compounds; the whole numbers are the *valences* of the respective elements. The *relative* weights of these atoms are therefore measurable from chemical analyses. With Dalton's theory, the innumerable chemical analyses of substances in the literature suddenly began to make sense in terms of chemical formulas (stoichiometry). Thus, the new framework of meaningful theory supported the accumulated practical chemical knowledge of many centuries and promoted the great upsurge in the chemical industry that accompanied the Industrial Revolution.

In the early 1900s, the work of Volta, Berzelius, Davy, and Faraday on *electrolysis* (i.e., splitting a chemical compound into its constituents by passage of an electrical current, e.g., splitting water into hydrogen and oxygen gases or molten common salt into metallic sodium and chlorine gas)

---

*A further irony is that Lavoisier himself fell victim to the French political revolution, despite his sympathy for its ideals and his contributions to scientific initiatives of the revolutionary government such as establishing the metric system. He was guillotined on May 8, 1794, on absurd charges—perhaps because of his involvement in tax collection for the royalist government and his liaisons with foreign chemists at a time of war.

revealed the *electrical* nature of chemical bonding. Charged atoms (*ions*) that migrate to the negative electrode (*cathode*) are positively charged in the combined state and are called *cations*; those that go to the positive electrode (*anode*) are negatively charged and are named *anions*.

In the early twentieth century, Rutherford showed that an atom consists of a very compact, dense, positively charged *nucleus* around which circulate a number of *electrons*, each with a single unit of negative charge. In a neutral atom, the sum of the negative charges match the charge on the nucleus. Thus, cations are atoms that have lost, and anions are atoms that have gained, one or more electrons, so that compounds can form from matched sets of anions and cations held together by electrostatic attractions. Not all compounds are ionic, however, and G. N. Lewis in particular (1916) recognized that nonionic or *covalent* compounds are held together at the molecular level by the interatomic *sharing* of electrons, usually in pairs. Covalency, then, is also an electrical phenomenon. Quantum mechanical models of the covalent bond were developed around 1930 by L. Pauling (valence bond theory) and R. S. Mulliken (molecular orbital theory), among others; such models, which are covered in introductory chemistry texts, explain covalent chemical bonding as a consequence of the wavelike nature of the electron.

The systematization of inorganic chemistry depends largely on the *periodic table* (see endpapers and Fig. 1.1). In the mid nineteenth century, Döbereiner, de Chancourtois, Odling, Newlands, and especially Lothar Meyer in Germany and Mendeleyev in Russia noted that, if the chemical elements are arranged in order of increasing atomic weight, certain properties such as principal valence recur periodically. The culmination of these observations was the periodic table of Mendeleyev (1869, formally published in 1871), which arranged elements of like properties in vertical columns (*groups*) and showed atomic weights increasing left to right across a given row (*period*). Mendeleyev recognized several gaps in the table, corresponding to then unknown analogs of aluminum, silicon, etc., and he correctly predicted the properties of the missing elements (gallium, germanium, etc.). With the advent of the electronic model of the atom, however, it became apparent that the key parameter is not the atomic weight of an element, but the *nuclear charge number* of its atoms which in almost all cases corresponds to its position (*atomic number*) in the sequence of increasing atomic weights.

Today, it is recognized that an atomic nucleus consists of a number of *protons* (particles of charge number 1+ and mass number approximately 1) and *neutrons* (chargeless particles of mass number approximately 1) bound together by a short-range force known as the *strong force*. The total charge number is then the atomic number, and the total mass number (which is less than the sums of the mass numbers of the free constituent particles by a

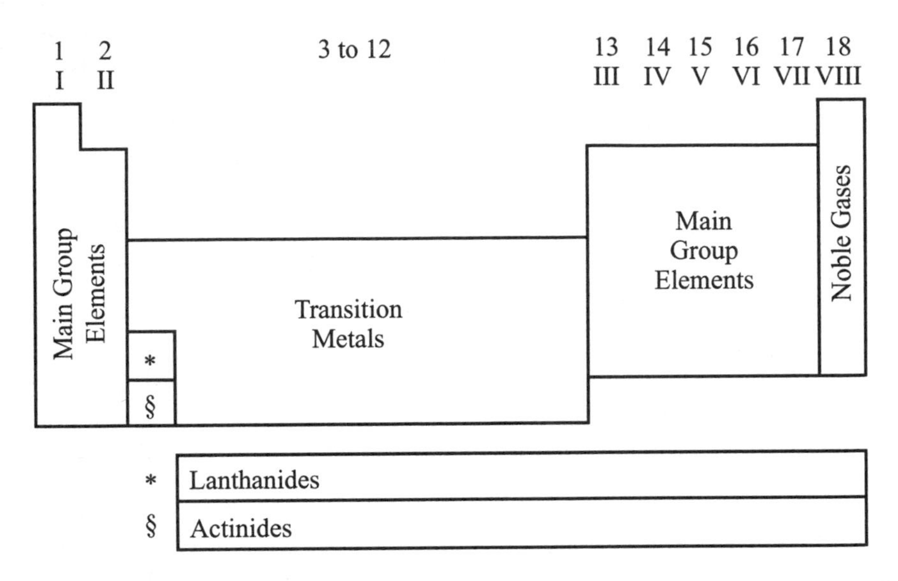

Figure 1.1 Principal features of the periodic table. The International Union of Pure and Applied Chemistry (IUPAC) now recommends Arabic group numbers 1–18 in place of the traditional Roman I–VIII (A and B). Group names include alkali metals (1), alkaline earth metals (2), coinage metals (11), chalcogens (16), and halogens (17). The main groups are often called the *s,p* block, the transition metals the *d* block elements, and the lanthanides and actinides the *f* block elements, reflecting the electronic shell being filled. (See inside front cover for detailed structure of the periodic table.)

small amount corresponding to the mass equivalent of the energy that binds them together) is the atomic number. The atomic number corresponds to the total number of electrons in the atom and so defines its chemical identity. The number of neutrons combined with a particular number of protons may vary, however, giving rise to atoms that have the same chemical identity but different mass numbers; these are called *isotopes*, from Greek words meaning *same place* (in the periodic table). Because naturally occurring elements often consist of mixtures of isotopes, their atomic weights may deviate far from the integral values conceived by Dalton.

The periodicity of chemical properties arises from filling of successive quantum mechanical *shells* of electrons. For example, filling of the *s,p* shells, with capacities of 8 electrons each, and the *d* shells, which can hold up to 10 electrons, is associated with the *main group* and *transition elements*, respectively (Fig. 1.1). Before the advent of quantum theory, two classes of elements were known that seemed not to fit the Mendeleyevian scheme: an uncertain number of "rare earth" elements or *lanthanides*—metallic elements, discovered throughout the 1800s, that form oxides of

TABLE 1.1
The Commonest Elements of the Earth's Crust

Element	*Atom %*	*Weight %*	*Element*	*Atom %*	*Weight %*
O	62.6	46.6	Ti	0.20	0.44
Si	21.2	27.7	F	0.091	0.080
Al	6.47	8.1	P	0.083	0.12
Na	2.64	2.8	C	0.057	0.032
Ca	1.95	3.6	Mn	0.038	0.10
Fe	1.94	5.0	S	0.034	0.052
Mg	1.84	2.1	Cl	0.030	0.048
K	1.42	2.6	Li	0.021	0.007

the type M_2O_3 and tend to occur together in mineral deposits—and the *noble gases* helium, neon, argon, krypton, xenon, and radon, which were discovered in 1895–1903 by Rayleigh, Ramsay, and Rutherford and seemed to have no tendency to form chemical compounds.* The quantum theory of atomic structure makes it clear that the inertness of the noble gases is due to completely filled *s,p* shells, while the lanthanides, and also the radioactive *actinides*, correspond to filling of *f* shells of capacity 14 electrons; thus, there are 15 lanthanide and 15 actinide elements. The short-lived radioactive elements with atomic numbers 104–112, produced in the 1980s and 1990s by nuclear bombardment techniques an atom at a time, appear to have chemical properties characteristic of *d* and *s,p* shell filling.

1.2 Occurrence and Uses of the Commonest Elements

Inorganic chemistry draws its strength from its great practical utility, and this book presents the subject from the standpoint of applications rather than the customary one of quantum mechanical bonding theory. Since the quintessential subject matter is the properties of the 112 known chemical elements and their compounds, we begin with a consideration of the availability of the commonest elements in the Earth's crust (Table 1.1), hydrosphere (i.e., oceans, lakes, rivers, snowfields, ice caps, and glaciers), and atmosphere, along with brief summary of the production and uses of these elements and their compounds.

Oxygen occurs as the free element (O_2) in the atmosphere (21%), from which it is obtained by fractional distillation of liquid air or by membrane technologies, but far greater amounts are found in the Earth's crust in

*Compounds of Kr, Xe and Rn with F and to some extent O and N are now known (e.g., XeF_2, XeF_4, XeO_6^{4-}).

combination with the other elements of Table 1.1 except F and Cl. In the upper atmosphere, a relatively small but crucially important amount of elemental oxygen is present in the form of *ozone* (O_3), and this "ozone layer" protects living things by absorbing much of the damaging short-wavelength ultraviolet radiation of the sun. Evidence that the ozone layer is being destroyed by chlorofluorocarbons and possibly other waste industrial gases has created a crisis of major proportions that must be solved by the chemical industry and the scientific community. Gaseous O_2 is used in large amounts in steelmaking and sewage treatment.

The other major constituent (78%) of the air, *nitrogen*, is of low abundance in crustal rocks and so is absent from Table 1.1. "Fixed" (combined) N, however, is essential for the growth of living things, and nitrogen is therefore separated from air on a large scale to make fertilizers. Other major uses are in the manufacture of explosives and propellants, and in the provision of an inert atmosphere for chemical processing. Nitrogen oxide emissions, on the other hand, are major causes of the atmospheric pollution resulting from certain industrial activities and, particularly, from operation of vehicles with internal combustion engines.

Silicon is the most important constituent of igneous and many sedimentary rocks, occurring in combination with oxygen in feldspars, micas, quartz, sands and shales. The element is used in electronic devices, while silicon in combination with oxygen as silica and silicates finds application in concrete, bricks, pottery, enamels, glasses, optical fibers for telecommunications, and refractory (high-temperature resistant) materials.

Aluminum (properly called *aluminium*, but the former name prevails in North America) is found in combination with Si and O as aluminosilicates in rocks, and as its ore, bauxite. The metal finds use in vehicles, aircraft, packaging, cookware, construction materials, etc., while aluminum salts are used in baking powders, water treatment, and dyeing of textiles. Aluminum oxide is widely used as a refractory and as a support for catalysts. Aluminosilicate catalysts such as zeolites are of key importance in the chemical and petroleum industries.

Sodium occurs extensively in feldspars, clay minerals, etc., but is extracted as NaCl from the oceans, in which Na^+ is the most abundant cation (1050 mg kg^{-1} or parts per million, ppm), and from rock salt deposits. The compounds NaCl, $NaHCO_3$, Na_2CO_3, Na_2SO_4, and NaOH are used in food processing, road deicing, water treatment, glass manufacture, paper making, and the chemical industry.

Calcium compounds, essential for the formation of bones and teeth, are obtainable from limestone ($CaCO_3$), gypsum ($CaSO_4{\cdot}2H_2O$), and fluorite (CaF_2) for use in water treatment, agriculture, construction (concrete), and the chemical industry, for which lime (CaO) is the least expensive source of alkali. Calcium is the third most common cation in seawater (400 mg kg^{-1}).

Iron, which is evidently the main constituent of the Earth's molten metallic core, is widespread in oxidized form in igneous rocks; it is extracted from deposits of hematite (Fe_2O_3), magnetite (Fe_3O_4), or goethite [α-FeO(OH)]. Despite the emergence of newer materials, iron and steels remain the sine qua non of the construction, transportation, energy, manufacturing, and packaging industries.

Magnesium occurs in many igneous rocks and in dolomite. It is usually obtained from seawater (1300 mg kg^{-1}) or from the minerals magnesite ($MgCO_3$) or carnallite ($KCl\cdot MgCl_2\cdot 6H_2O$). The metal is used in lightweight alloys, MgO is employed as a refractory material and as an adsorbent for water treatment, and other Mg compounds find applications in the pharmaceutical and chemical process industries.

Potassium is found in feldspars and micas, and is the fourth most abundant cation in seawater (390 mg kg^{-1}). Potassium compounds are usually obtained from evaporites (i.e., residues from evaporated water) as KCl ("potash") or carnallite, mainly for use in fertilizers.

Titanium is mined as rutile (TiO_2) and ilmenite ($FeTiO_3$). The refined oxide is an important constituent of paints, while the relatively light but highly corrosion-resistant metal is increasingly used structurally and for chemical process equipment despite its high cost.

Fluorine compounds from fluorite (fluorspar, CaF_2) are used in water treatment (to suppress dental caries) and to make fluoropolymers (such as Teflon), lubricants, and refrigerants. Molten cryolite (Na_3AlF_6) is essential as a solvent for Al_2O_3 in the electrolytic production of aluminum metal, while the isotopic enrichment of uranium for nuclear power reactors is usually achieved by diffusion or gas centrifugation of volatile UF_6.

Phosphorus occurs as rock phosphate [$Ca_5(PO_4)_3OH$]. Phosphates are essential to all living things and are therefore important constituents of commercial fertilizers.

Carbon chemistry (organic chemistry and biochemistry) is the basis of all biology and also of the synthetic fibers, plastics, paints, dyes, pharmaceuticals, petrochemicals, and numerous other industries. Since consideration of this enormous field would require a whole book, we confine our attention to elemental carbon and inorganic C compounds. The element is used, in the form of diamond, as a cutting agent and as a protective film; as carbon black, in the tire and printing industries; and, as graphite or coke, for electrodes and as a reductant in extractive metallurgy. Inorganic carbon compounds include carbon dioxide and other gases suspected of causing global warming, as well as toxic carbon monoxide and cyanides. Much carbon occurs combined as carbonate ions in limestone and dolomite, and in the oceans—although, if ranked on the basis of inorganic C content alone (28 mg kg^{-1}), the carbonate ion is only the fourth most abundant anion in seawater, after bromide (65 mg kg^{-1}). Most of the carbon we use comes

TABLE 1.2
Inorganic Chemicals in the Top Fifty, in Terms of Tonnage, Produced in the United States in 1995[a]

Rank	*Product*	*Mt*[b]	*Rank*	*Product*	*Mt*
1	sulfuric acid	40.5	14	ammonium nitrate	8.0
2	nitrogen	30.6	15	urea	7.3
3	oxygen	22.5	21	carbon dioxide	5.0
5	lime	17.4	28	hydrochloric acid	3.0
6	ammonia	17.2	32	ammonium sulfate	2.3
8	sodium hydroxide	11.7	37	carbon black	1.50
9	phosphoric acid	11.5	38	potash	1.42
10	chlorine	11.0	42	titanium dioxide	1.24
11	sodium carbonate	9.3	43	aluminum sulfate	1.04
13	nitric acid	8.0	44	sodium silicate	0.97

[a] Source: *Chemical and Engineering News*, June 26, p. 39 (1995).
[b] Megatonnes = 10^6 metric tons (1 t = 1000 kg).

from petroleum, natural gas, or coal.

Manganese occurs in concentrated form as pyrolusite (MnO_2) and manganite [$MnO(OH)$] deposits, and as manganese nodules on the ocean floor. The metal is used in alloys with iron, while MnO_2 is used in dry cells (flashlight "batteries") and as an oxidant in the chemical industry.

Sulfur occurs native (i.e., as the element), as sulfates such as gypsum, or as sulfides such as pyrite (FeS_2) and galena (PbS), many of which are important sources of nonferrous metals. Sulfate is the second most abundant anion (2700 mg kg^{-1}) in seawater. Sulfuric acid, the most important product of the chemical industry in terms of tonnage (Table 1.2), is the most economical strong acid for a myriad of applications. Among sulfates, alum [$KAl(SO_4)_2 \cdot 12H_2O$] is extensively used to clarify water supplies, ammonium sulfate [$(NH_4)_2(SO_4)$] is an important fertilizer, and sodium sulfate is used to regenerate pulping liquor in the paper industry. Sulfur oxides from the combustion of sulfur-containing fuels, however, are responsible for serious environmental damage through the generation of acid precipitation (i.e., acid rain, snow, and fog).

Chlorine occurs mainly in seawater, in which chloride is the most abundant anion (19350 mg kg^{-1}), and as rock salt (halite, $NaCl$) in evaporites. Elemental chlorine is important in sterilization of water supplies and production of chlorinated plastics such as polyvinyl chloride (PVC), but some chlorine compounds used extensively in the past, such as the insecticide

DDT and polychlorobiphenyl (PCB) transformer fluids, have been phased out in most countries because of their apparent interference with the reproductive cycles of birds and mammals.

Lithium occurs in some igneous rocks, often substituting partly for magnesium; its chief economic source is spodumene ($LiAlSi_2O_6$). Lithium stearate [$CH_3(CH_2)_{16}CO_2Li$] is widely used to thicken oils for use as lubricating greases, and long-lived lithium batteries are now preferred for many dry-cell applications such as powering cameras.

Hydrogen does not rank among the commonest crustal elements (Table 1.1), but, obviously, enormous reserves exist as H_2O in the hydrosphere. Hydrogen gas is in many respects an ideal nonpolluting fuel as well as an important reducing agent for extracting of metals and as a hydrogenating agent in the food and petrochemical industries. There has been much discussion of an environmentally benign, hydrogen-based economy for the future, but extraction of hydrogen from water without recourse to the fossil fuels it is intended to replace presents a formidable challenge to chemists and chemical engineers.

This brief introduction does not even begin to consider the occurrence and uses of such technologically important elements as chromium, copper, tin, lead, boron, bromine and iodine; these are discussed in due course. Nevertheless, it is clear from the foregoing, and from the number of inorganic chemicals appearing among the top fifty (in terms of tonnage) produced in the United States (Table 1.2), that inorganic chemistry plays an indispensable role in the economy of industrialized countries. Our discussion so far also shows, however, that a constructive approach to many of the serious environmental problems besetting the world at this time requires an understanding of inorganic chemistry. These two aspects, economic benefits and environmental protection, are developed in tandem in the following pages.

Finally, we note that many of the most rapidly developing technologies, such as communications, electronics, energy generation and conservation, environmental protection, and aerospace, have generated demands for new materials with unprecedented physical properties or compositional control. Much current research activity in inorganic chemistry is directed toward meeting these needs, as noted throughout this book and especially in Chapter 19.

Exercises

1.1 Tabulate the abundances of elements in seawater, given in the text, as atom % values (include H and O), and compare their rankings with those of the elements in the Earth's crust as given in Table 1.1. Suggest possible reasons for any differences in the rankings.

1.2 Why are elemental hydrogen and helium, the two most abundant elements in the universe, not present in significant amounts in the Earth's atmosphere?

Further reading

1. W. H. Brock, "The Norton History of Chemistry." Norton, New York, 1993.
2. H. M. Leicester, "The Historical Background of Chemistry." Dover, New York, 1971.
3. H. Hartley, "Studies in the History of Chemistry." Oxford Univ. Press, London, 1971.
4. "Chemistry in the Economy." The American Chemical Society, Washington, D.C., 1973.
5. R. Thompson (ed.), "Industrial Inorganic Chemicals: Production and Uses." Royal Society of Chemistry, Cambridge, 1995.
6. J. E. Fergusson, "Inorganic Chemistry and the Earth." Pergamon, Oxford, 1982.
7. P. J. Chenier, "Summary chart of the manufacture of important inorganic chemicals." *J. Chem. Educ.* **60,** 382 (1983).
8. I. Bodek, W. J. Lyman, W. F. Reehl, and D. H. Rosenblatt, "Environmental Inorganic Chemistry." Pergamon, New York, 1988.

Chapter 2

Chemical Energetics

2.1 Kinetics and Thermodynamics

THERE ARE two basic questions that a chemist or chemical engineer must ask concerning a given chemical reaction:

(a) How *far* does it go, if it is allowed to proceed to equilibrium? (Indeed, does it go in the direction of interest at all?)

(b) How *fast* does it progress?

Question (b) is a matter of chemical *kinetics* and reduces to the need to know the *rate equation* and the *rate constants* (customarily designated k) for the various steps involved in the reaction mechanism. Note that the rate equation for a particular reaction is not necessarily obtainable by inspection of the stoichiometry of the reaction, unless the mechanism is a one-step process—and this is something that usually has to be determined by experiment. Chemical reaction time scales range from fractions of a nanosecond to millions of years or more. Thus, even if the answer to question (a) is that the reaction is expected to go to essential completion, the reaction may be so slow as to be totally impractical in engineering terms. A brief review of some basic principles of chemical kinetics is given in Section 2.5.

Question (a) is in the province of chemical *thermodynamics*[1] and amounts to evaluating the *equilibrium constant* (K). Unlike the rate equation, the equilibrium expression for a typical reaction

$$a\mathrm{A} + b\mathrm{B} \rightleftharpoons c\mathrm{C} + d\mathrm{D} \tag{2.1}$$

can be written down by inspection of the stoichiometry:

$$K^\circ = \frac{\{\mathrm{C}\}^c\{\mathrm{D}\}^d}{\{\mathrm{A}\}^a\{\mathrm{B}\}^b} \tag{2.2}$$

where the braces represent the *activities* of the chemical species A, B, C, and D. In simple terms, *activity* is a thermodynamically effective concentration and is related to the stoichiometric concentration (moles per liter of solution, *molar*, M; or moles per kg of solvent, *molal*, m) by the activity coefficient γ:

$$\{A\} = \gamma[A] \tag{2.3}$$

where the square brackets denote stoichiometric concentration. For ideal systems (in practice, for gaseous reactions at low total pressures or solution reactions at high dilution), γ is unity. For simplicity in this book, we assume in general that this is the case for nonelectrolytes and thus equate activity with concentration.

The superscript $^\circ$ in Eq. 2.2 indicates a true thermodynamic equilibrium constant. We use plain K when concentrations replace activities or, for electrolyte solutions, when K refers to a nonzero ionic strength (see Section 2.2).

2.2 Activities in Electrolyte Solutions

For solutions of ions, departures from ideality can be large even in quite dilute solutions because of the strong electrostatic attractions or repulsions between the ions. Furthermore, the simple definition of activity coefficient given in Eq. 2.3 fails for electrolytes because we can never measure the activity of, say, a cation M^{m+} without anions X^{x-} being present at the same time; instead, we usually define a *mean* ionic activity $a_\pm$ and coefficient $\gamma_\pm$ as

$$a_\pm = \gamma_\pm c(m^m x^x)^{1/(m+x)} \tag{2.4}$$

where c is the molal concentration of the electrolyte M_xX_m. Although it is not possible to *measure* single-ion activity coefficients γ_i for the ith kind of ion in the solution, they may be estimated theoretically. The *Debye–Hückel* approach[2–4] relates γ_i to a quantity called the *ionic strength*, I, given by $0.5\sum c_i z_i^2$ where c_i is the concentration and z_i the charge of ions of the ith kind. These i kinds include *all* the ions in the solution, not just M^{m+} and X^{x-}. The *limiting* Debye-Hückel equation

$$\ln \gamma_i = -Az_i^2\sqrt{I}, \tag{2.5}$$

in which A is a constant calculable from the theory for a particular temperature and pressure (1.172 at 25.0 °C and 0.1 MPa, or 0.509 if $\log \gamma_i$ is required), is valid only up to $I \approx 0.005\ m$. Equation 2.5 is actually a less rigorous form of the expression for $\gamma_\pm$:

$$\ln \gamma_\pm = -Az_+z_-\sqrt{I} \tag{2.6}$$

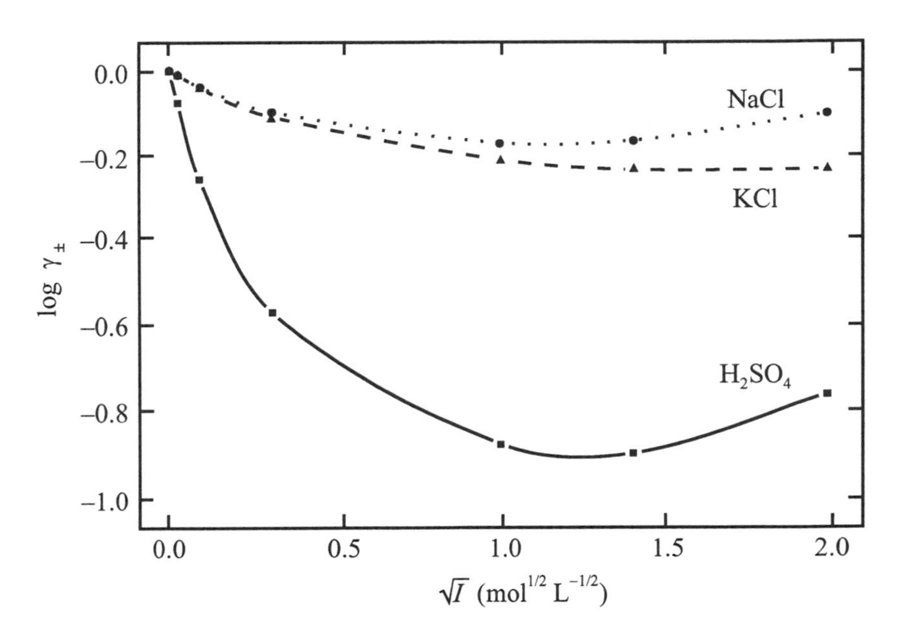

Figure 2.1 Dependence of $\log \gamma_{\pm}$ on the square root of the ionic strength I.

The *extended* form of Eq. 2.5,

$$\ln \gamma_i = -Az_i^2 \frac{\sqrt{I}}{1 + Ba_i\sqrt{I}}, \tag{2.7}$$

is satisfactory up to $I \approx 0.1\ m$, but contains, along with the theoretical constant B ($3.286 \times 10^9\ m^{-1}$ at 25.0 °C and 0.1 MPa), a parameter a_i that represents the distance of closest approach of anions to cations but is not clearly defined (there may be several kinds of ions in the solution) or reliably calculable. Güntelberg found that the product Ba_i can be conveniently set to 1 for many electrolyte solutions, with little loss of accuracy. Davies showed that introduction of an empirical correction CI to Eq. 2.7 extended its ceiling to $I \approx 0.5\ m$:

$$\ln\gamma_i = -Az_i^2 \left(\frac{\sqrt{I}}{1 + Ba_i\sqrt{I}} - CI \right). \tag{2.8}$$

The value of C has to be determined experimentally for a particular system but is typically 0.2–0.3. Pitzer[5] presented equations that give activity coefficients for both binary solutions and mixed electrolytes for I up to 6 m, but there are several parameters that must be determined empirically.

Figure 2.1 shows how experimental $\log \gamma_{\pm}$ values vary with ionic strength for some simple aqueous electrolytes. The dependence on $\sqrt{I}$

is linear at very low I, with the same slope for electrolytes of the same charge type (as the limiting Debye–Hückel theory predicts), but the Ba_i and Davies C factors, which are different for different electrolytes, exert their influence above $I \approx 0.005$ mol L^{-1}.

Rather than attempting to allow for the concentration dependence of activity coefficients of ions M^{m+} and X^{x-} when we calculate an equilibrium constant K, we can vary $[M^{m+}]$ and $[X^{x-}]$ *with constant activity coefficients* if we maintain an effectively constant ionic strength I with a swamping concentration of some inert electrolyte. Usually a perchlorate (ClO_4^-) or trifluoromethanesulfonate ($CF_3SO_3^-$, "triflate") salt is used because, of all the commoner anions, these are the least basic and have the least tendency to form complexes with M^{m+} (Chapter 13). Thus, for practical purposes, equilibrium constants for reactions between ions are usually quoted with reference to some particular ionic strength (K), rather than for the ideal case of "infinite dilution" ($K°$). Figure 2.1 shows that $\gamma_{\pm}$ for typical ionic solutes varies only slowly with I in the range 0.1–2.0 mol L^{-1}, so that swamping electrolyte concentrations in that range are particularly suitable.

2.3 Equilibrium and Energy

A rigorous treatment of chemical thermodynamics[1] is beyond the scope of this book. However, there are several thermodynamic relationships that can provide important insights, even if we resort to a few oversimplifications of thermodynamic concepts. In an overview of inorganic chemistry and its applications, it is more important to appreciate what thermodynamics can tell us than to worry about its rigor or theoretical significance.

Perhaps the most important equation relates the *thermodynamic equilibrium constant* $K°$ to the *standard free energy change* $\Delta G°$ of the reaction:

$$\Delta G° = -RT \ln K° \tag{2.9}$$

where R is the gas constant (8.3143 J K^{-1} mol^{-1}) and T is the temperature in kelvins (degrees Celsius + 273.15). The *standard Gibbs free energy* $G°$ represents the capacity of the system for doing external work at constant pressure; thus, the thermodynamic driving force of a chemical reaction is the tendency to minimize this free energy. When $G°$ is minimized, the reaction ceases, that is, has reached equilibrium.

Activities, and hence $K°$ values, are necessarily dimensionless quantities and are defined with reference to a convenient *standard state*. The standard state now universally adopted in the Système International d'Unités (SI) is

(*a*) pressure (P) = 100 kPa (0.1 MPa, 1 bar),

(*b*) temperature (T) = 298.15 K (25.00 °C),

(*c*) solute concentrations hypothetically 1 molal,

(*d*) ideal gas behavior (i.e., as if $P \to 0$), and

(*e*) ideal solution behavior (i.e., as if infinitely dilute).

The last restriction explains the ***hypothetical*** standard concentration of 1 molal: obviously, 1 mol kg^{-1} is quite concentrated, and the standard state conditions are extrapolated from high dilution. Gas concentrations are conveniently expressed as partial pressures:

$$\text{concentration} = \frac{n \text{ mol}}{\text{volume } V} = \frac{P}{RT}(\text{ideally}) \propto P, \text{at } 25\,°\text{C} \qquad (2.10)$$

so the standard state for a gaseous reactant is 1 bar partial pressure with (extrapolated) ideal behavior.

Thus, if we start with our reactants and products under standard conditions and allow the reaction to proceed to equilibrium, an amount of energy $\Delta G°$ becomes available for external work. In the context of doing external electrical work, an oxidation–reduction reaction can generate a ***standard electromotive force*** $\Delta E°$ given by

$$\Delta G° = -nF\Delta E° \qquad (2.11)$$

where n is the number of moles of electrons transferred in the reaction and F is the charge of one mole of electrons (the ***Faraday constant***, 96,485 A s mol^{-1}). However, the change in standard heat content, or ***enthalpy change*** ($\Delta H°$), associated with the reaction is not the same as $\Delta G°$, since some of the heat content can never be extracted. We speak of the ***standard entropy*** of a substance $S°$ as being its isothermally unavailable heat content per kelvin:

$$\Delta G° = \Delta H° - T\Delta S°. \qquad (2.12)$$

The enthalpy change of the reaction, $\Delta H°$, can be calculated by subtracting the heat contents of the reactants from those of the products (Hess's law):

$$\Delta H° = \sum H°(\text{products}) - \sum H°(\text{reactants}). \qquad (2.13)$$

However, the absolute heat contents of individual substances are generally not available. The arbitrary assumption is therefore made that $H°$ is zero for the chemical elements in their most stable form at 25 °C and 1 bar. The heat contents of chemical compounds are then defined as the ***standard heat of formation*** $\Delta H_f°$ from the elements at 25 °C and 1 bar. For the formation of liquid water,

$$H_2(g) + \tfrac{1}{2}O_2(g) \xrightarrow[1\text{ bar}]{25\,°\text{C}} H_2O(l), \qquad (2.14)$$

$\Delta H_f°$ is −285.830 kJ mol^{-1}; for water vapor, $\Delta H_f°$ is −241.818 kJ mol^{-1}, the difference being simply the heat of evaporation of 1 mole of water under standard conditions. For the formation of carbon monoxide,

$$C(s) + \tfrac{1}{2}O_2(g) \rightleftharpoons CO(g), \tag{2.15}$$

ΔH_f° is -110.525 kJ mol^{-1}, relative to $\Delta H_f^\circ = 0$ for graphite. (Diamond, the other familiar form of elemental carbon, is actually less stable than graphite at 25 °C and 1 bar; $\Delta H_f^\circ = 1.895$ kJ mol^{-1}.) We can therefore calculate a standard heat of reaction for the water–gas reaction,

$$C(s) + H_2O(g) \rightleftharpoons H_2(g) + CO(g), \tag{2.16}$$

to be $(-110.525) - (-241.818) = +131.293$ kJ mol^{-1}.

Note that reaction 2.16 is *endothermic* (the plus sign for ΔH° means that heat is taken into the system), whereas reactions 2.14 and 2.15 are *exothermic* (the reactions give out heat to the surroundings). Many heats of formation or of reaction can be measured by calorimetry (i.e., by recording the temperature rise of a thermally insulated apparatus of known heat capacity when the reaction of interest is carried out in it) or can be obtained from other ΔH_f° data, as shown for the water–gas reaction. If we know ΔH° and also know the *standard entropy change* (ΔS°) for a given reaction, we can calculate its equilibrium constant (K°) from a combination of Eqs. 2.9 and 2.12:

$$\ln K^\circ = \frac{\Delta S^\circ}{R} - \frac{\Delta H^\circ}{RT}. \tag{2.17}$$

Now S°, the standard entropy of a single substance, unlike H°, *can* be calculated absolutely if we know the *standard heat capacity* C_P° (at constant pressure) of that substance as a function of temperature from zero kelvin:

$$S^\circ = \int_0^T \frac{C_P^\circ}{T}\, dT = \int_0^T C_P^\circ \; d(\ln T). \tag{2.18}$$

Entropies are thus determinable from statistical mechanics or from calorimetry. They are listed along with ΔH_f° values for many substances in Appendix C and in various reference books.[6,7] Values of S° for $H_2(g)$, $O_2(g)$, and $H_2O(l)$ are 130.684, 205.138, and 69.91 J K^{-1} mol^{-1}, respectively.

It is important to note that the entropies of gases are larger than those of liquids or solids. This is because entropy is a function of the degree of randomness or disorder at the molecular level. Customarily, S° values are given in *joules* per kelvin per mole, whereas ΔG° and ΔH° are given in *kilojoules*; remember to multiply the latter by 1000 when doing calculations. The entropy of formation of liquid H_2O (reaction 2.14) is then

$$\Delta S_{f\ H_2O(l)}^\circ = S_{H_2O(l)}^\circ - S_{H_2(g)}^\circ - \tfrac{1}{2}S_{O_2(g)}^\circ = -163.34 \text{ J K}^{-1} \text{ mol}^{-1}. \tag{2.19}$$

The free energy of formation of liquid water at 25 °C is

$$\Delta G_f^\circ = \Delta H_f^\circ - T\Delta S_f^\circ = -237.13 \text{ kJ mol}^{-1}, \tag{2.20}$$

and the thermodynamic equilibrium constant for reaction 2.14 is:

$$K^\circ = \frac{\{H_2O(l)\}}{\{H_2(g)\}\{O_2(g)\}^{1/2}}. \tag{2.21}$$

The activity of the pure water phase, which is separate from the gaseous reactants, can be set to unity, and the gas activities may be approximated as equal to the partial pressures in bars:

$$K = \frac{1}{P_{H_2} {P_{O_2}}^{1/2}} \text{ bar}^{-3/2} = \exp \frac{-\Delta G^\circ}{RT} = 3.5 \times 10^{41} \text{ bar}^{-3/2}. \tag{2.22}$$

Thus, the equilibrium pressure of hydrogen created by dissociation of air-free liquid water would be only 2×10^{-28} bar, or one molecule in 200 m^3. In effect, then, reaction 2.14 goes to total completion because it is highly *exergonic*, that is, because ΔG_f° is so strongly negative, and liquid water simply does not dissociate detectably at 25 °C (even if this were kinetically favored).

Usually, ΔH_f° is the dominant term in Eq. 2.17. For example, the formation of nitric oxide,

$$\tfrac{1}{2}N_2(g) + \tfrac{1}{2}O_2(g) \rightleftharpoons NO(g), \tag{2.23}$$

is endothermic ($\Delta H_f^\circ = +90.25$ kJ mol^{-1}) and also *endergonic*, that is, ΔG_f° is positive ($+86.55$ kJ mol^{-1}) corresponding to $K^\circ \ll 1$. In other words, equilibrium 2.23 lies very far to the left at 25 °C. Thus, thermodynamics predicts that NO should decompose almost completely to nitrogen and oxygen at room temperature. In fact, it does not, because the reaction is so slow (unless catalyzed).

Note that reactions 2.14, 2.15, and 2.23 involve fractional stoichiometric coefficients on the left-hand sides. This is because we wanted to define conventional *enthalpies of formation* (etc.) of *one mole* of each of the respective products. However, if we are not concerned about the conventional thermodynamic quantities of formation, we can get rid of fractional coefficients by multiplying throughout by the appropriate factor. For example, reaction 2.14 could be doubled, whereupon ΔG° becomes $2\Delta G_f^\circ$, $\Delta H^\circ = 2\Delta H_f^\circ$, and $\Delta S^\circ = 2\Delta S_f^\circ$, and the right-hand sides of Eqs. 2.21 and 2.22 must be squared so that the new equilibrium constant $K' = K^2 = 1.23 \times 10^{83}$ bar^{-3}. Thus, whenever we give a numerical value for an equilibrium constant or an associated thermodynamic quantity, we must make clear how we chose to define the equilibrium. The concentrations we calculate from an equilibrium constant will, of course, be the same, no matter how it was defined. Sometimes, as in Eq. 2.22, the units given for K will imply the definition, but in certain cases such as reaction 2.23 K is dimensionless.

2.4 Temperature and Pressure Effects on Equilibrium

2.4.1 Temperature Effects

Equation 2.17 shows that, *if* ΔS° and ΔH° for a reaction are known and can be taken to be independent of temperature, then in principle we can calculate the equilibrium constant K_T at any temperature T:

$$\ln K_T = a - bT^{-1}. \tag{2.24}$$

This approximation is good enough to allow extrapolations of equilibrium constants over modest temperature ranges for many commonly encountered reactions. The reason is that the molar heat capacities C_P° of the reactants and the products tend to cancel, giving a standard heat capacity change of reaction ΔC_P° which is often negligible:

$$\Delta C_P^\circ = \sum C_P^\circ(\text{products}) - \sum C_P^\circ(\text{reactants}). \tag{2.25}$$

Reaction 2.23 is a case in point. We have two molecules (O_2, N_2) of similar heat capacities reacting to give two others (2NO) that have thermodynamic properties intermediate between O_2 and N_2; so $\Delta C_P^\circ \approx 0$. Consequently, application of the equations

$$\Delta H_T = \Delta H^\circ + \int_{298}^{T} \Delta C_P^\circ \, dT \tag{2.26}$$

$$\Delta S_T = \Delta S^\circ + \int_{298}^{T} \Delta C_P^\circ \, d(\ln T) \tag{2.27}$$

tells us that ΔH_T and ΔS_T will be essentially independent of temperature. If, however, ΔC_P° is *not* negligible, we must resort to evaluation of Eqs. 2.26 and 2.27. If ΔC_P° is significant but approximately constant over the temperature range chosen, Eq. 2.28, derived by combining Eqs. 2.26 and 2.27, is better than Eq. 2.24. The ΔC_P° for any reaction can be calculated from tables of standard (25 °C) thermodynamic properties of the reactants and products.[6]

$$\ln K_T = a - bT^{-1} + c \ln T \tag{2.28}$$

2.4.2 Pressure Effects

Pressure effects on equilibria in liquids or solids are generally less spectacular than temperature effects, at least at the pressures normally encountered in chemical engineering (a few tens of megapascals) or in the environment (hydrostatic pressures in the ocean trenches exceed 100 MPa, but about 40 MPa would be more typical of the ocean floors). Higher lithostatic pressures are, of course, found beneath the Earth's surface, reaching 370 GPa (0.37

TPa, 3.7 Mbar) near the center of the Earth. Laboratory measurements at up to 550 GPa are now possible with samples confined between tiny opposed diamond "anvils,"[8,9] but *in situ* measurements on such samples are hard to make and interpret, as shown by the controversy over reports that hydrogen becomes metallic above 140 GPa.[10] While pressures of several gigapascals are used in producing some solid materials, for example, in synthesizing diamonds[8] or for compacting powders by isostatic pressing, most applications of high pressures involve liquids or solutions,[11,12] and this limits the pressure to 1–2 GPa, as most liquids freeze at higher pressures (~1 GPa for water at ambient temperature).

The pressure dependence of an equilibrium constant $K°$ is determined by the standard *reaction volume* $\Delta V°$, which is the change in molar volume of the system as the reaction goes from the initial to the final state, that is, the difference in molar volumes between the products and the reactants:

$$\Delta V° = \sum V°(\text{products}) - \sum V°(\text{reactants}). \tag{2.29}$$

The pressure dependence of $K°$ is given by

$$(\partial \ln K°/\partial P)_T = -\Delta V°/RT. \tag{2.30}$$

Over relatively small pressure ranges, for example, 0.1–100 MPa, $\Delta V°$ may be taken to be independent of pressure, so that (dropping the superscripts) $\ln K$ becomes a linear function of pressure. Thus, on going from atmospheric ("zero") pressure to a pressure P, we have

$$\ln K_P = \ln K_0 - P\Delta V/RT. \tag{2.31}$$

If P is in megapascals, the matching unit for ΔV will be cubic centimeters per mole. For a typical reaction volume of -10 cm^3 mol^{-1}, K increases by only 50% on going from atmospheric pressure to 100 MPa. Because of the logarithmic form of Eq. 2.31, however, pressure effects on K become important for larger ΔV values or higher pressures. Of course, for reactions of gases, pressure effects are very large, as the partial pressures of the reactants take the place of concentrations (Eq. 2.10).

One important effect of applied pressure is to raise the boiling point of a liquid, so that as little as 1 MPa of externally applied pressure can extend the liquid range of a solvent enormously. Thus, a pressure of 1.56 MPa raises the boiling point of water to 200 °C, doubling its liquid range, while at 8.6 MPa water remains liquid to 300 °C. For water, the *critical point*, that is, the point beyond which there is no longer any difference between liquid and vapor, is 374 °C and 22.13 MPa, at which the density of the fluid is 320 kg m^{-3} (cf. 997.0 kg m^{-3} at 25 °C). Supercritical water (i.e., water at temperatures higher than 374 °C and liquid-like densities; see Fig. 2.2) is a promising medium for the destruction of hazardous organic wastes:

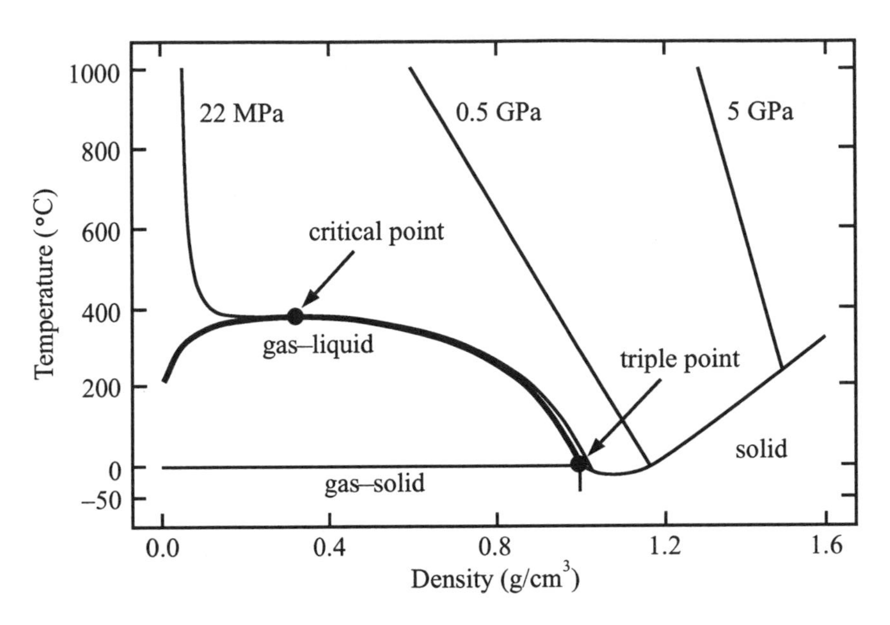

Figure 2.2 Temperature, pressure, and density relations for water substance. The heavy curve corresponds to the saturated vapor pressure of liquid water. Note the extreme sensitivity of density to pressure near the critical temperature.

the solvent power of supercritical water increases with increasing density (i.e., pressure), for both organic and inorganic substances, and at such high temperatures kinetic limitations on slow hydrolysis or oxidation reactions are largely overcome.[13]

2.4.3 Hydrothermal Chemistry

High temperature aqueous or *hydrothermal* chemistry is important in many contexts, including electrical power generation from nuclear or fossil fuels, exploitation of geothermal energy, growth of quartz and other crystals for the electronics industry, oil recovery from tar sands or oil shales by steam injection, mineral processing, and study of geological phenomena such as the mechanism of formation of hydrothermal mineral deposits.[14–21] Spectacular examples of this last phenomenon are the active hydrothermal vents on mid-ocean ridges that spew superheated water at temperatures up to 400 °C into the cold, pressurized water of the deep sea. The hot water is usually acidic and often contains substantial concentrations of metal ions and hydrogen sulfide, in which case the sudden chilling and fall in acidity can cause immediate precipitation of a dense cloud of dark-colored metal sulfide particles and formation of chimneys of deposited minerals around the vent ("black smokers"). The deposited solids include potentially valuable resources of minerals such as copper sulfides.

Associated with the vents are colonies of exotic life forms such as thermophilic bacteria and giant tube worms that have evolved to exploit the energy and sulfide content of the hydrothermal fluids, much as more familiar living things have evolved to use energy coming directly or indirectly from sunlight.[22,23] Hydrogen sulfide, though extremely toxic to higher animals, provides much energy when it is oxidized by dissolved oxygen to sulfate ion, and vent bacteria can use this oxidation as a source of metabolic energy to synthesize organic compounds from dissolved carbon dioxide. Vent ecosystems are self-sufficient except for requirement for dissolved oxygen, which cannot exist in the vent fluids and must come by convection from the rest of the ocean. Thermophilic bacteria may prove useful industrially for rapid fermentation procedures and for speedier bacterial leaching of metal sulfide ores (Chapter 17).

For any serious study of hydrothermal chemistry, one needs thermodynamic data at the relevant temperatures and pressures. Obtaining precise physicochemical measurements on aqueous solutions at high temperatures is difficult and time-consuming,[18,19,20,24–27] so several workers have sought methods of extrapolating the abundant thermodynamic data for 25 °C to the hydrothermal regime. It follows from Eqs. 2.17, 2.26, and 2.27 that the key quantities are the heat capacities C_P° of the reacting solute species. If C_P° is approximately constant between 298 K and the temperature of interest, the extrapolation is straightforward. Criss and Cobble[27] found that, for a variety of ions in aqueous solution, there exists a simple empirical relationship between their entropies S_T° at an elevated temperature T and the corresponding entropies S° at 298 K:

$$S_T^\circ = a_T + b_T S^\circ \tag{2.32}$$

where a_T and b_T are constants for the particular temperature T, with one pair of values for all cations, a different pair for monatomic anions including OH^-, and a third pair for oxoanions. This *correspondence principle* allows us to estimate not only unknown ionic entropies at elevated temperatures T, but also an average value for the ionic heat capacity over the range 298 K to T:

$$< C_P^{\text{average}} >]_{298}^{T} = \frac{S_T^\circ - S_{298}^\circ}{\ln(T - 298)}. \tag{2.33}$$

The Criss–Cobble correspondence principle is useful for aqueous solutions to about 200 °C. At higher temperatures, the heat capacities C_P° of ionic solutes such as NaCl[26] *at constant pressure* rapidly become strongly negative and appear to be headed toward infinite negative values on approaching the critical temperature (which, incidentally, is somewhat higher for aqueous electrolyte solutions than for pure water). If, however, we examine the heat capacities C_V° of aqueous electrolytes *at constant volume*, this bizarre phenomenon disappears: the heat capacities increase, but only slightly, with rising temperature. This is because dissolved ions exert a large *electrostrictive* effect on the solvent, that is, they draw in solvent to *solvate* themselves (Section 13.1), so reducing the volume of the system; at constant volume these solute-solvent interactions change little with rising temperature, but at constant pressure the thermal expansion of the bulk solvent causes the electrostrictive effect to loom large. Close to the the critical temperature, the hypersensitivity of solution density to pressure shown in Fig. 2.2 leads to extreme values of constant-pressure parameters.

A further problem is that *ion association*, that is, the tendency of oppositely charged ions to form pairs or larger aggregates in solution, becomes increasingly important as the temperature rises unless the density is kept constant; this is because ion association is inversely related to the *dielectric constant* (relative permittivity) of the medium, which is correlated with density for a given solvent. Helgeson and co-workers have attacked these problems theoretically for aqueous solutions up to 1000 °C.[28] For our purposes, it is enough to note that quantitative treatment of ionic reactions in sub- and supercritical aqueous solutions is extremely difficult at present, and likely to remain so for some time.

As an illustration of the effects of temperature, pressure, and ionic strength on ionic equilibria in solution, we can consider the self-ionization (autoprotolysis) of water:

$$H_2O(l) \rightleftharpoons H^+(aq) + OH^-(aq). \tag{2.34}$$

The thermodynamic equilibrium constant K_w° is given by the activity quotient $\{H^+\}\{OH^-\}/\{H_2O\}$, and pK_w° $(= -\log K_w^\circ)$ has the value 14.00 (more precisely, 13.998) at 25.0 °C. In practice, $\{H_2O\}$ is taken to be unity for dilute solutions, and the *concentration* product $K_w = [H^+][OH^-]$ is used. The common assumption that pK_w is 14.00, however, is correct *only for very dilute solutions at 25.0° C and 0.1 MPa.* For seawater at this temperature and pressure, pK_w is 13.76 because the ionic strength is about 0.71 mol L^{-1}. In pure water at 0.0 °C and 0.1 MPa, pK_w is 14.94 because the ionization of water is endothermic; ΔH° is 55.8 kJ mol^{-1} at 25 °C but is temperature dependent, so that Eqs. 2.26–2.28 must be invoked to extrapolate K_w with respect to temperature (see Exercise 2.1). Since ΔV_w° for the ionization is −21 cm^3 mol^{-1}, Eq. 2.31 predicts pK_w = 13.63 at 100 MPa and 25.0 °C if ΔV_w° is pressure independent. At the *saturated vapor pressure*, pK_w passes through a broad minimum of about 11.36 around 240 °C as the temperature is increased, then rises to 16.63 at 374 °C and 22.1 MPa, the critical point. For supercritical water, pK_w decreases (ionization increases) with increasing pressure (density) at a given temperature; at the critical temperature, for example, pK_w falls to only 5.70 at 7.1 GPa. Thus, as Hawkes has remarked,[29] pK_w almost never has the traditional value of 14.00.

2.5 Chemical Kinetics: Basic Principles

Chemical kinetics[30] (i.e., reaction rates) and equilibria are interrelated in that chemical equilibria are *dynamic*—that is, reaction continues in both the forward and reverse directions at equilibrium, but the *rates* in the opposing directions are exactly equal. It follows that the energetic factors considered above also influence reaction rates. Kinetic phenomena, however, are most in evidence far from equilibrium, and they are more difficult to treat theoretically than are equilibria. Thus, although thermodynamics can tell us whether a particular reaction *should* proceed from reactants to specific products under given conditions of temperature, pressure, concentration, etc., it tells us nothing about how the reaction could take place at the molecular level (the reaction mechanism) nor how *fast* equilibrium will be approached. The reaction may be "infinitely slow" for lack of a favorable mechanism. Reaction kinetics and mechanism are intimately related; the rate of a reaction is largely determined by (*a*) how the reacting molecules organize themselves in space, ready to set up new bonds and break old ones, (*b*) how many of the potential reactant molecules have enough energy to get over the bond making/breaking energy barrier (activation energy), and (*c*) (for reactions in solution) how the solvent molecules rearrange to facilitate the activation process.

The rate of a reaction is expressed as the increase in the amount of a particular product, or the decrease in the amount of a reactant, per second at a time t. For reactions in a fluid phase, it is usual to substitute concentrations, or for gases partial pressures, for amounts. (See Section 5.6 for treatment of the kinetics of some solid-state reactions.) Note that the numerical value of the rate may depend on how we choose to define it. For example, in the oxidation of aqueous iodide ion by arsenic acid to give triiodide ion and arsenious acid,

$$H_3AsO_4 + 3I^- + 2H^+ \rightleftharpoons H_3AsO_3 + I_3{}^- + H_2O, \qquad (2.35)$$

the rate may be defined as $d[I_3{}^-]/dt$, or $-d[H_3AsO_4]/dt$, or $-d[I^-]/dt$, etc., but the stoichiometric coefficients in reaction 2.35 tell us that the last rate is numerically three times either of the other two. Thus, we must state clearly what we mean by "reaction rate" (such as the rate of disappearance of iodide ion, or the rate of appearance of arsenious acid) wherever ambiguity may arise.

2.5.1 Relation of Rate Equation to Mechanism

The *rate equation*, sometimes called the "rate law," relates the rate at time t to the activities (less rigorously, the concentrations) of the reactants remaining at that instant. Unlike the equilibrium expression (cf. Eq. 2.2), however, the form of the rate equation generally *cannot* be obtained merely by inspection of the stoichiometric equation unless the reaction is known to proceed to completion in a single step. The rate equation must be determined experimentally, and this includes finding out whether the reaction is or is not a single-step process. The activities (in practice, concentrations or, for gases, partial pressures) and the powers to which they are raised (reaction *orders*) in the experimentally determined rate equation tell us which molecules, or fragments thereof, are actually involved in the rate-determining step of the reaction. Any reactant particles that are not involved in this bottleneck configuration or *transition state* will not affect the observed reaction rate but will be consumed relatively rapidly in subsequent steps. Thus, it is found that the rate of the forward reaction in Eq. 2.35 is *first order* with respect to each of arsenic acid, iodide ion, and hydrogen ion (and so *third order overall*). We cannot determine the order with respect to water, as this is the solvent and its activity is essentially unity at all times.

$$\frac{d[I_3^-]}{dt} = k_f[H_3AsO_4][I^-][H^+] \tag{2.36}$$

In Eq. 2.36, k_f is the forward rate constant, and the composition of the transition state is $\{H_4AsO_4I\}$, although it could contain additionally (or be short of) the elements of one or more water molecules, since we cannot determine the order with respect to the solvent. Equation 2.36 cannot be arrived at from reaction 2.35, but consideration of the concentration factors in the two equations tells us at once the rate law for the reverse reaction (Eq. 2.38, rate constant k_r), since, according to reaction 2.35, the equilibrium expression has to be

$$K = \frac{[H_3AsO_3][I_3^-]}{[H_3AsO_4][I^-]^3[H^+]^2} \tag{2.37}$$

and, at equilibrium, the forward and reverse rate must be equal:

$$-\frac{d[I_3^-]}{dt} = \frac{k_r[H_3AsO_3][I_3^-]}{[I^-]^2[H^+]}. \tag{2.38}$$

It follows that the equilibrium constant K is given by k_f/k_r. The reverse reaction is *inverse* second order in iodide, and inverse first-order in H^+. This means that the transition state for the reverse reaction contains the elements of arsenious acid and triiodide ion *less* two iodides and one hydrogen ion, namely, $\{H_2AsO_3I\}$. This is the same as that for the forward reaction, except for the elements of one molecule of water, the solvent, the participation of which cannot be determined experimentally. The concept of a common transition state for the forward and reverse reactions is called the *principle of microscopic reversibility.*

However, more than one reaction pathway may exist, in which case the rate equation will contain sums of terms representing the competing reaction pathways. For example, one of the oxidation reactions that convert the atmospheric pollutant sulfur dioxide to sulfuric acid (a component of acid rain) in water droplets in clouds involves dissolved ozone, O_3 (see Sections 8.3 and 8.5):

$$SO_2(g) + H_2O(l) \rightleftharpoons H_2SO_3(aq) \rightleftharpoons HSO_3^-(aq) + H^+(aq) \tag{2.39}$$

$$HSO_3^- + O_3 \rightarrow H^+ + SO_4^{2-} + O_2. \tag{2.40}$$

The rate equation for 2.40 turns out to be

$$-\frac{d[HSO_3^-]}{dt} = \left(k_1 + \frac{k_2}{[H^+]}\right)[HSO_3^-][O_3], \tag{2.41}$$

which implies two parallel pathways: an $[H^+]$-independent one with a transition state composition $\{HSO_6\}^-$ and rate constant k_1, and an $[H^+]$-dependent one with transition state $\{SO_6\}^{2-}$ and rate constant k_2. (Actually, $[H^+]$ also affects the overall aqueous ozone oxidation process through the solubility equilibrium of SO_2, reaction 2.39.)

Rate equations of considerable complexity can result from *chain reactions*, such as the reaction of bromine with hydrogen in the gas phase between 200 and 300 °C to form hydrogen bromide. These are reactions in which a chain carrier is created in an initiation step (here, a Br· atom from dissociation of Br_2) and goes on to create more carriers ($Br\cdot + H_2 \rightarrow HBr + H\cdot$, followed by $H\cdot + Br_2 \rightarrow HBr + Br\cdot$, and so on) until a recombination step ends the chain. The rate equation for HBr formation has been shown to be:

$$\frac{d[HBr]}{dt} = \frac{k_1[H_2][Br_2]^{1/2}}{1 + k_2[HBr]/[Br_2]}. \tag{2.42}$$

Chain reactions are important in certain polymerizations, organic halogenation reactions, combustion processes, and explosions. Usually they involve a *radical* (a molecule or atom with an odd valence electron, e.g., Br·), since these will create another radical each time they react with an ordinary molecule having only paired electrons. Detailed discussions are available in standard texts.[30]

Differential rate equations such as Eq. 2.38 are often used in integrated form, so that the rate constant can be evaluated simply by measuring concentrations rather than rates at selected times t. Integration of the rate equation, however, may be difficult or may give an unwieldy result. In this book, we are concerned primarily with reaction rates as such, and the reader is therefore referred to texts devoted to chemical kinetics[30] for information on integrated rate equations.

2.5.2 Temperature Effects on Rates

The temperature dependence of a rate constant is usually given empirically by the Arrhenius equation:

$$k = A \exp \frac{-E_a}{RT} \tag{2.43}$$

in which E_a is the Arrhenius activation energy (typically on the order of 50 to 100 kJ mol^{-1}) and the pre-exponential factor A includes such factors as the frequency of collision between the reactant molecules and the probability of their mutual orientation being favorable for reaction. Thus, $\ln k$ is a linear function of the reciprocal Kelvin temperature, so that if we know k at any two temperatures, or at one temperature when E_a is known, we can use Eq. 2.43 to calculate k at any other temperature. This does not apply for reactions proceeding by two or more parallel pathways of different E_a (unless one can evaluate Eq. 2.43 for the various pathways separately, e.g., if one can solve Eq. 2.41 for k_1 and k_2 at two temperatures) or for multistep reactions if a different step becomes rate determining as the temperature changes. In either case, a curved plot of $\ln k$ against T^{-1} will result. Simple Arrhenius treatments work well in a great many cases.

An alternative to Eq. 2.43 that is popular with modern kineticists is the *Eyring equation* (Eq. 2.44), which derives from the notion that the transition state is in (very unfavorable) equilibrium with the reactants but decays with a *universal frequency* given by $k_B T/h$, where k_B is Boltzmann's constant and h is Planck's constant. Then, by analogy with Eqs. 2.9 and 2.12, we have

$$k = \frac{k_B T}{h} \exp \frac{-\Delta G^{\ddagger}}{RT} = 2.083 \times 10^{10}\, T \exp \frac{-(\Delta H^{\ddagger} - T\Delta S^{\ddagger})}{RT} \tag{2.44}$$

where $\Delta G^{\ddagger}$, $\Delta H^{\ddagger}$, and $\Delta S^{\ddagger}$ are the *free energy, enthalpy*, and *entropy of activation*, respectively. Thus, $\ln(k/T)$ should be a linear function of T^{-1}. This is slightly different from the Arrhenius expectation, but in practice data usually fit either Eq. 2.43 or Eq. 2.44 equally well within the inevitable experimental errors. For reactions in solution, $\Delta H^{\ddagger} = E_a - RT$.

2.5.3 Pressure Effects on Rates

Pressure effects on k can be accommodated in terms of a *volume of activation*, $\Delta V^{\ddagger}$, which is the pressure derivative of $\Delta G^{\ddagger}$ (cf. Eqs. 2.30 and 2.44):

$$(\partial \ln k/\partial P)_T = -\Delta V^{\ddagger}/RT. \tag{2.45}$$

If $\Delta V^{\ddagger}$ can be taken to be constant over the experimental pressure range (this approximation is often a poor one for ranges of more than 100 MPa), then $\ln k$ becomes a linear function of pressure (cf. Eq. 2.31):[11]

$$\ln k_P = \ln k_0 - P\Delta V^{\ddagger}/RT \tag{2.46}$$

Values of $\Delta V^{\ddagger}$ are frequently on the order of ± 10 cm^3 mol^{-1} for simple reactions in solution, in which case k would be retarded or accelerated by a factor of only about 1.5 on going from ambient P and T to 100 MPa. If, however, $\Delta V^{\ddagger}$ is around -30 cm^3 mol^{-1} for a reaction that is impracticably slow at room temperature and pressure, and a 2 GPa press is available, the reaction can in principle be accelerated by a factor of 3×10^{10} without heating. In practice, the acceleration will almost certainly be somewhat less than this, as $\Delta V^{\ddagger}$ is likely to become numerically smaller (less negative) as the pressure increases over this wide range.

The Eyring approach has the advantage that the pseudothermodynamic activation parameters can be readily related to the true thermodynamic quantities that govern the equilibrium of the reaction. The Arrhenius equation, on the other hand, is easier to use for simple interpolations or extrapolations of rate data.

2.6 Ionization Potential and Electron Affinity

Chemical combination entails the transfer or sharing of electrons between atoms. Consequently, the energetics of bonding between atoms can be related to the energy changes associated with the loss or gain of electrons by the isolated (i.e., gaseous) atoms. These spectroscopically measurable energy changes are called *ionization potentials* and *electron affinities*, respectively.

The nth *ionization potential* (IP) is the energy required to remove an electron from an atom $M^{(n-1)+}$ in the gaseous state. Thus, for aluminum, we have

$$\begin{array}{rl}
Al(g) \rightleftharpoons Al^{+}(g) + e^- & \text{First } IP\text{: } 577\,\text{kJ mol}^{-1} \\
Al^{+}(g) \rightleftharpoons Al^{2+}(g) + e^- & \text{Second } IP\text{: } 1816 \\
Al^{2+}(g) \rightleftharpoons Al^{3+}(g) + e^- & \text{Third } IP\text{: } 2764 \\
\hline
Al(g) \rightleftharpoons Al^{3+}(g) + 3\,e^- & 5157\,\text{kJ mol}^{-1}.
\end{array}$$

In general, ionization potentials *decrease* as we descend the periodic table within a given group. This is as we might expect, since the atoms increase in size. The first IP is 899 kJ mol^{-1} for beryllium and falls to 503 kJ mol^{-1} for barium, down Group 2. As we cross the periodic table from left to right, IP values tend to rise:

	Al	Si	P	S	Cl	Ar	
First IP:	577	786	1061	1000	1255	1521	kJ mol^{-1}.

According to Slater, this is because electrons in the same quantum shell (here, the $3p$ orbitals) screen one another's "view" of the nuclear charge by only ~0.35 unit. Thus, going from Al to Si, the nuclear charge increases by +1.00, but the added electron screens only +0.35 of this. Electrons in lower shells screen the nuclear charge by essentially +1.00 unit, as seen by the outermost electrons. This same effect explains the *lanthanide contraction*—the steady shrinking of lanthanide(III) ion radii from 103 to 86 pm as we fill the $4f$ quantum shell from La^{3+} ($4f^0$) to Lu^{3+} ($4f^{14}$).

There is a large increase in IP as we pass from the $n(s, p)$ to the $(n-1)(s, p)$ quantum shell. For example, the first IP of Na (removal of the lone $3s$ electron) is 496 kJ mol^{-1}, as compared to 2081 kJ mol^{-1} for Ne (removal of a $2p$ electron from a complete shell).

The nth *electron affinity* (EA) of an atom $X^{(n-1)-}$ was traditionally defined as the energy *released* when an electron is added to a gaseous atom. This, however, goes against the thermodynamic convention of defining energy changes as positive when energy is *gained* by the system. In this book, we always use the thermodynamic convention instead. Confusion can arise because electron affinities can be either positive or negative (unlike IP values, which are always positive as normally—and thermodynamically correctly—defined). If, in a table, the first EA values of the halogens are negative, then we know that the modern thermodynamically correct convention is being used.

Trends in EA are less clear-cut than those in IP, but in general EA values are most negative for elements at the upper right of the periodic table, notably the halogens:

	F	Cl	Br	I	
First EA:	−322	−349	−325	−295	kJ mol^{-1}.

In general, electron affinities become positive for $n > 1$ and as we start a new quantum shell, for example, on going from F (adding the last electron to the $2p$ shell) to Ne (starting the $3s$ shell). Slater's concept of the effective nuclear charge "seen" by the electrons makes it clear why this is so.

	O	F	Ne
First EA:	−141	−322	+29 kJ mol^{-1}
Second EA:	+710		
$O \rightarrow O^{2-}$	+569 kJ mol^{-1}		

Taking the trends in IP and EA together, we see that elements at the left of the periodic table (especially the lower left), including the transition elements and the lanthanides and actinides, have the greatest tendency to form cations, while the elements at the right (especially O, F, and Cl) tend to form anions. Those in the upper middle region tend to form compounds by sharing pairs of electrons, that is, by covalent rather than ionic bonding.

It can be anticipated that, even in covalent bonds, the electrons will often be unequally shared by the bound atoms, giving rise to a certain degree of *ionic character* or *polarity* to the bond, whenever the *electronegativities* (χ) of the atoms are different.

2.7 Electronegativity and Bond Energies

Electronegativity (χ) was defined by Linus Pauling in 1932 as a measure of the power of an atom *in a molecule* to draw electrons to itself. The stressed words are necessary to distinguish electronegativity from electron affinity. Pauling devised an electronegativity scale on the basis of thermochemical measurements of bond energies. His definition of electronegativity, however, implies two essential components of χ: an atom in a molecule may attract electrons from other atoms (a function of the negative of EA, presumably) but must also retain its own electrons against the attractions of the other atoms (a function related to IP). Accordingly, Mulliken, in 1934, proposed an alternative electronegativity scale that averaged $-EA$ and IP:

$$\chi_{\mathrm{M}} = \frac{IP - EA}{2}. \tag{2.47}$$

Here, IP and EA refer to the atom in the appropriate valence state. Encouragingly, values of Mulliken's χ_{M} are roughly proportional to those of Pauling's χ, the former being larger by a factor of about 2.8.

Although nowadays electronegativity tends to be regarded as an old idea, it has received renewed interest.[31] In 1989, Allen[32] presented a fresh approach in which he redefined electronegativity as the average one-electron energy of the valence shell electrons in the free atoms. He suggested that the electronegativities of the elements so calculated are sufficiently important that they should be included graphically in the periodic table as a third dimension. Parr, Pearson, and others have considered calculated *electron densities* in molecules as the basis for electronegativity scales.[31–36] The key property in this approach is the *electronic chemical potential* μ:

$$\mu = (\partial E/\partial N)_v \tag{2.48}$$

where E is the electronic energy, N is the number of electrons, and v is the external potential (due mainly to the atomic nuclei) acting on the electrons. It turns out that $-\mu$ approximately equals the Mulliken electronegativity $\chi_{\mathrm{M}} = (IP - EA)/2$; it has been termed the *absolute electronegativity*. This approach applies equally well to molecules, radicals (fragments of molecules), and free atoms, rather than just to individual atoms in molecules. As Pearson has remarked,[33] the various electronegativity scales come into their own in different applications. One particular virtue of μ is that, when two entities of different μ interact, electron density flows between them until μ is equalized—this is the essence of the *electronegativity equalization principle*, proposed in 1951 by Sanderson.

Pauling's original scale, however, relates rather directly to bond polarities and bond strengths, and so it remains the one most widely used. Actually, the other scales parallel it reasonably closely. The Pauling scale is empirical, being derived from thermochemical data that were subsequently refined by Allred.[34] Although the Pauling scale should be regarded as only semi-quantitative, it is instructive in that it relates bond structure (ionic character) in molecular substances directly to thermochemically measured bond energies and hence, ultimately, to the bulk-matter thermodynamic properties discussed above.

Unfortunately, there is sometimes confusion over what exactly we mean by "bond energy." We could define a *mean bond energy* E as the contribution of an A—B pair to the total energy of the molecule AB_n. Thus, for the gas methane (CH_4),

$$E = \frac{\text{heat of atomization}}{\text{number of bonds}} = \frac{1657}{4} = 414 \text{ kJ mol}^{-1}.$$

However, if we take the CH_4 molecule apart stepwise, we get four different results that we may call the *bond dissociation energies* D (the nomenclature is arbitrary):

$$\begin{array}{llll} CH_4 \rightarrow CH_3 & + & H & D = 431\,\text{kJ mol}^{-1} \\ CH_3 \rightarrow CH_2 & + & H & 364 \\ CH_2 \rightarrow CH & + & H & 523 \\ CH \rightarrow C & + & H & 339 \\ \hline CH_4 \rightarrow C & + & 4H & 1657\,\text{kJ mol}^{-1}\ (= 4E) \end{array}$$

Clearly, one should not use tabulated bond energy data for polyatomic molecules without knowing whether they are mean bond energies or bond dissociation energies. In any event, the relationship with the heat of formation of CH_4 can be illustrated by an enthalpy cycle:

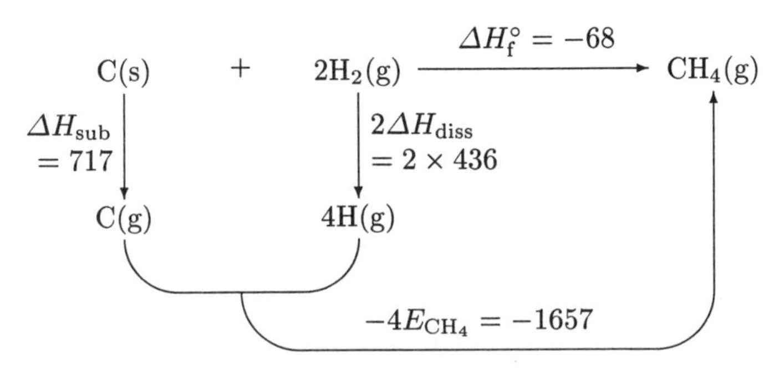

TABLE 2.1
Typical Single-Bond Dissociation Energies (kJ mol^{-1})

Homoatomic		*Heteroatomic*	
H_2	436	HF	565
F_2	159	HCl	431
Cl_2	243	HBr	368
Br_2	192	HI	297
I_2	150	$H_3C—H$	431
C—C	347		
Si—Si	222		

Here, ΔH_{sub} is the heat of sublimation (direct vaporization) of graphite, ΔH_{diss} is the bond energy of molecular hydrogen, and the energies are quoted in kilojoules per mole.

$$\Delta H_{\text{f}} = \Delta H_{\text{sub}} + 2\Delta H_{\text{diss}} - 4E_{\text{CH}_4} = -68 \text{ kJ mol}^{-1}$$

Table 2.1 lists a few typical single-bond energies for homoatomic (A—A or B—B) and heteroatomic (A—B) bonds.

One might naively expect the energy D_{AB} of an A—B bond to be about the average of the A—A and B—B bond energies, but heteroatomic bonds are nearly always stronger than this average. Pauling maintained that this was due to an ionic contribution Δ_{AB} to D_{AB}:

$$\Delta_{\text{AB}} = D_{\text{AB}} - \frac{D_{\text{AA}} + D_{\text{BB}}}{2} \tag{2.49a}$$

or

$$\Delta_{\text{AB}} = D_{\text{AB}} - \sqrt{D_{\text{AA}} \times D_{\text{BB}}}. \tag{2.49b}$$

In Eq. 2.49a, the arithmetic mean of D_{AA} and D_{AB} is taken, as is usual, but some authors use the geometric mean as in Eq. 2.49b to avoid negative Δ_{AB} values in a few cases involving very weak bonds. Either way, the Pauling electronegativity *difference* between A and B is defined by

$$\begin{aligned} \Delta_{\text{AB}} &= (\chi_{\text{A}} - \chi_{\text{B}})^2 \text{ in electron volts} \\ &= 96.5(\chi_{\text{A}} - \chi_{\text{B}})^2 \text{ in kJ mol}^{-1}. \end{aligned} \tag{2.50}$$

The conversion factor of 96.5 is Faraday's constant (divided by 1000 for kilojoules). Since the electronegativity difference is squared, the ionic contribution Δ_{AB} represents a positive contribution to the A—B bond energy whether $\chi_{\text{A}} > \chi_{\text{B}}$ or vice versa. If we set the value of χ for any one element, we can, in principle, use Eqs. 2.49 and 2.51 to calculate χ for any

TABLE 2.2
Allred–Pauling Thermochemical Electronegativities

H	2.20						
Li	0.98	Na	0.93	K	0.82	Rb	0.82
Be	1.57	Mg	1.31	Ca	1.00	Sr	0.95
B	2.04	Al	1.61	Ga	1.81	In	1.78
C	2.55	Si	1.90	Ge	2.01	Sn	1.96
N	3.04	P	2.19	As	2.18	Sb	2.05
O	3.44	S	2.58	Se	2.55	Te	—
F	3.98	Cl	3.16	Br	2.96	I	2.66

other element from thermochemical data:

$$|\chi_{\mathrm{A}} - \chi_{\mathrm{B}}| = 0.102\sqrt{\Delta_{\mathrm{AB}}} \quad (\Delta_{\mathrm{AB}} \text{ in kJ mol}^{-1}). \tag{2.51}$$

The *thermochemical* electronegativities are obviously only semiquantitative; as we noted earlier, D_{AB} varies considerably depending on what other atoms are attached to A or B as well as the valence states of A and B, and Eq. 2.49a gives somewhat different D_{AB} values from Eq. 2.49b. Values of Allred–Pauling thermochemical electronegativities for main group elements in their principal valence states are given in Table 2.2. The concept is less useful for the transition elements, for which χ ranges between 1.36 and 1.91 in the first transition period, and for the lanthanides ($\chi \approx 1.1$–1.2).

2.8 Electronegativity and Chemical Properties

An obvious use of an electronegativity scale is to predict the direction of electrical polarity of a covalent bond with ionic character. Table 2.2 tells us that the C—H bond in alkanes (C_nH_{2n+2}) is polar in the same sense as the O—H bonds in water, although to a much lesser degree:

$$R_3C^{\delta-}\text{—}H^{\delta+} \qquad {}^{2\delta-}O(H^{\delta+})_2.$$

On the other hand, the polarity of the Si—H bond in silanes (silicon analogs of alkanes) is reversed:

$$R_3Si^{\delta+}\text{—}H^{\delta-}.$$

Thus, the hydrogens in silanes (or organosilanes, if R is an organic group) behave more like hydride ions (H^-) than like protons (H^+). Whereas triphenylmethane (R = C_6H_5) is unreactive toward water, triphenylsilane *hydrolyzes* (reacts with water) to give triphenylsilanol [$(C_6H_5)_3Si—OH$] and hydrogen gas (cf. $H^- + H_2O \rightarrow OH^- + H_2$):

$$\begin{matrix} (C_6H_5)_3Si^{\delta+} \text{ - } \vdots \text{ - } H^{\delta-} \\ HO^{\delta-} \text{ - } \vdots \text{ - } H^{\delta+} \end{matrix} \longrightarrow (C_6H_5)_3SiOH + H_2.$$

Similarly, it comes as no surprise to learn that Si—Cl bonds are much more susceptible to hydrolysis than C—Cl bonds, since Si—Cl has the greater electronegativity difference. These naive predictions can be misleading, however. The $R_3C—F$ bond should be very polar, and hydrolysis might be expected to give $R_3C—OH$ and HF. In fact, however, fluorocarbons are usually extremely inert to hydrolysis, even by aqueous alkali, because the *overall* energy change on replacing the C—F and HO—H bonds by C—OH and H—F bonds is, on balance, unfavorable (see Exercise 12.5).

Essentially ionic compounds, consisting of a crystalline lattice of M^{m+} and X^{x-} ions held together by simple electrostatic attraction with negligible covalent bonding, can be expected for M of very low χ (e.g., most metals but especially the heavier members of Groups 1 and 2) and X of high χ (notably F, O, and Cl). Simple electrostatic considerations, confirmed by X-ray crystallography, indicate that highly charged anions X^{x-} (and also H^-) will be large, with easily polarized electron clouds. Conversely, small and/or highly charged cations M^{m+} have high electrical potentials at their surfaces and will tend to draw the electron clouds of nearby anions or molecules toward themselves. Consequently, such M^{m+} or X^{x-} ions tend to indulge in electron-pair *sharing* (covalency) to at least some degree, rather than to form truly ionic compounds (*Fajans' rules*). Examples are as follows:

NaF, NaCl, MgF_2, $MgCl_2$: ionic, high-melting solids
AlF_3: ionic, high-melting solid
$AlCl_3$: solid which is easily sublimed; exists as covalent Al_2Cl_6 units in vapor state
SiF_4, $SiCl_4$: covalent gas, covalent liquid

Similarly, very few true ionic hydrides, borides, nitrides, and carbides exist (see Chapter 5).

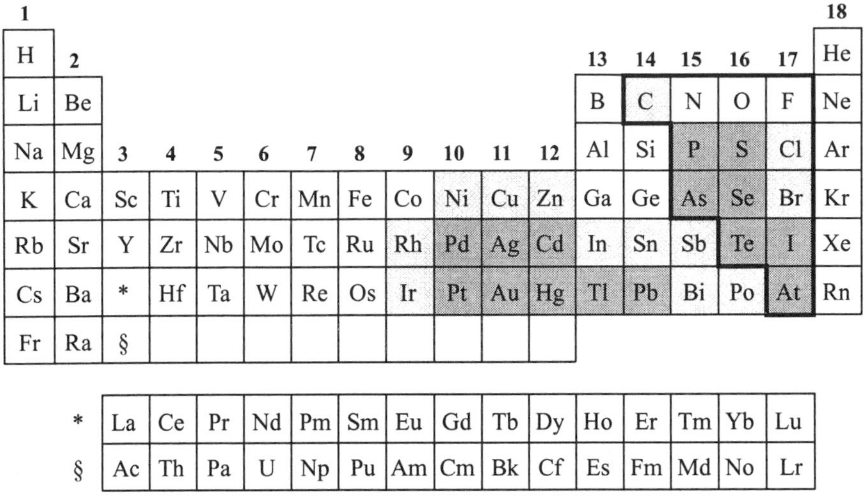

1	2	3	4	5	6	7	8	9	10	11	12	13	14	15	16	17	18
H																	He
Li	Be											B	C	N	O	F	Ne
Na	Mg											Al	Si	P	S	Cl	Ar
K	Ca	Sc	Ti	V	Cr	Mn	Fe	Co	Ni	Cu	Zn	Ga	Ge	As	Se	Br	Kr
Rb	Sr	Y	Zr	Nb	Mo	Tc	Ru	Rh	Pd	Ag	Cd	In	Sn	Sb	Te	I	Xe
Cs	Ba	*	Hf	Ta	W	Re	Os	Ir	Pt	Au	Hg	Tl	Pb	Bi	Po	At	Rn
Fr	Ra	§															

*	La	Ce	Pr	Nd	Pm	Sm	Eu	Gd	Tb	Dy	Ho	Er	Tm	Yb	Lu
§	Ac	Th	Pa	U	Np	Pu	Am	Cm	Bk	Cf	Es	Fm	Md	No	Lr

Figure 2.3 Elements displaying soft acid or soft base behavior as donor or acceptor atoms in their compounds (heavy shading). Elements shown in light shading are borderline soft; carbon is soft in CN^-, CO, and related donor molecules. Unshaded elements are hard. Heavy line encloses bases.

2.9 Hard and Soft Acids and Bases

Pearson, following up on an idea of Chatt, Ahrland, and Davies, divided the chemical elements into groups according to the acid–base (electron pair acceptor–donor) properties of their atoms or common ions:

(*a*) *soft acids*: relatively large, easily polarizable cations, usually of low charge (e.g., Tl^+, Hg^{2+}, Pt^{2+});

(*b*) *soft bases*: easily oxidized, highly polarizable donor atoms of elements of low electronegativity, typically with low-lying empty orbitals (e.g., I^-, S^{2-});

(*c*) *hard acids*: small and/or highly charged cations with no easily polarizable electrons (e.g., H^+, cations of Groups 1 and 2, B, Al, the lanthanides and actinides, and the left side of the transition series);

(*d*) *hard bases*: electron donor atoms of high electronegativity, low polarizability, and relatively small size (e.g., F, O, and N); and

(*e*) borderline cases.

This classification is summarized by element in Fig. 2.3. Its value lies in the generalization that soft acids interact more strongly with (i.e., accept electron pairs more readily from) soft bases, whereas hard acids bond more strongly to hard bases. This very simple *principle of hard and soft acids and bases* (HSAB) allows us to make immediate chemical predictions (or rationalizations of observations) without going into detail. For example, mercury binds strongly with sulfur atoms (hence the name *mercaptan* for organic sulfides of the type RSH), but weakly with O or F. Similarly, "hard" Al and Si occur naturally as oxides but not as sulfides, whereas "soft" gold is often found as its telluride AuTe but never as an oxide or fluoride. Not surprisingly, there are many borderline cases, and some elements such as copper show soft behavior in low oxidation states (Cu^+ binds strongly with I^-, but solid CuF is unknown) but hard behavior in more oxidized states (CuF_2 is a very stable solid, but CuI_2 does not exist—indeed, aqueous Cu^{2+} oxidizes I^- to iodine, forming solid CuI).

In thermodynamic terms, soft–soft interactions in solution can be described as enthalpy driven; heats of reaction tend to be quite strongly negative, reflecting the formation of strong covalent bonds. Conversely, hard–hard reactions between ions in solution are typically entropy driven; they tend to have near-zero enthalpy changes but markedly positive entropy and volume changes, reflecting the release of electrostricted solvent as the oppositely charged ions come together and cancel the charges.

2.9.1 Principle of Maximum Hardness

Pearson[35, 36a] and Parr and co-workers[36b, c] developed the *principle of maximum hardness*, which states that reacting molecules will arrange their electrons so as to be as hard as possible. Chemical equilibrium, then, is the state of maximum hardness. Soft donors prefer soft acceptors because *both* partners can increase their hardness by reacting with one another—the shared electrons flow to become less polarizable. To implement this theory quantitatively, Pearson *et al.* introduced scales of "absolute" hardness η and its reciprocal, softness σ:

$$2\eta = (\partial\mu/\partial N)_v = 2/\sigma \tag{2.52}$$

where μ is defined by Eq. 2.48. Operational definitions, using the thermodynamic convention for the sign of EA, are

$$-\mu = (IP - EA)/2 = \chi_M \tag{2.53}$$

$$\eta = (IP + EA)/2. \tag{2.54}$$

In these relationships, IP and EA refer to free atoms or ions, whole molecules or radicals, rather than atoms within molecules, and so $-\mu$ bears little relationship to Pauling's χ. In terms of molecular orbital (MO) theory, IP is the negative of the energy of the highest occupied MO ($-\epsilon_{\mathrm{HOMO}}$), and EA is the energy of the lowest unoccupied MO (ϵ_{LUMO}). Thus, *hard molecules are those with a large HOMO–LUMO gap, and soft molecules are those with a small HOMO–LUMO gap.* Pearson[35] has tabulated values of η for various molecules and ions; it is sufficient for our purposes to note that these correspond to the qualitative expectations of the naive HSAB theory.

2.10 Multiple Bonding and Its Chemical Consequences

Multiple bonding is not universal across the periodic table, but is important for carbon and its neighbor nitrogen. One might expect the energies of double and triple bonds to be twice and three times those of comparable single bonds, but Table 2.3 shows that this is only roughly true. The N—N bond energies have important chemical consequences. The N—N single bond is very weak, but the N≡N bond is very strong. Thus, the toxic liquid hydrazine ($H_2N—NH_2$) is very reactive and easily oxidized to N_2. It finds application as a rocket fuel. In industry, hydrazine is used as a reducing agent for such things as corrosion control (Section 16.7), electroless plating of metals onto metallic or nonmetallic surfaces (e.g., copper onto glass or plastics for circuit boards), and removal of toxic chromates from waste water as in

$$4CrO_4{}^{2-} + 3N_2H_4 + 4H_2O \rightarrow 4Cr(OH)_3(s)_\downarrow + 3N_2 + 8OH^-. \qquad (2.55)$$

The extraordinary strength of the triple bond in N_2 means that many reactions of nitrogen compounds (e.g., reaction 2.55) tend to produce N_2 gas, often with a large release of energy. If this occurs rapidly, we have the sudden production of a hot, expanding gas—in other words, the makings of a dangerous explosion. Many nitrogen compounds such as lead azide ($Pb(N_3)_2$) or ammonium nitrate (NH_4NO_3) are therefore only *kinetically* stable; that is, they exist only because of substantial activation energy barriers to their violent decomposition to N_2 and other products. Once such a reaction is initiated, the heat of reaction ensures that the activation barrier

TABLE 2.3
Typical Multiple-Bond Dissociation Energies (kJ mol^{-1})

	Bond order		
	1	2	3
C—C	347	611	837
C—N	293	615	891
N—N	159	418	946

is easily overcome, and an explosion results. Explosions are the ultimate manifestation of the energetics of chemical reactions and are therefore considered separately in Section 2.11. The other side of the coin is that the reverse process, the conversion of inert atmospheric N_2 into needed fertilizers (NH_3, ammonium salts, or nitrates) is very difficult. This problem is discussed in more detail in Chapter 9.

We must not, however, suppose that *all* multiple bonds are necessarily inert. In fact, the data of Table 2.3 imply that this is a special feature of *triply* bonded dinitrogen, since the N=N bond is rather weak. Acetylene, for example, reacts readily with hydrogen gas (especially if catalyzed) to form ethane because the energy required to reduce the triple C≡C bond to a single C—C bond and to break two H—H bonds is more than compensated by the formation of four new *heteroatomic* bonds:

$$\begin{array}{llrl}
\mathrm{H{-}C \equiv C{-}H} & \text{put in: } & 837 - 347 = & 490\,\mathrm{kJ\,mol^{-1}} \\
\quad + & & & \\
\quad \mathrm{2H_2} & \text{put in: } & 2 \times 436 = & 872 \\
\quad \downarrow \; \Delta H & & & \\
\mathrm{H_3C - CH_3} & \text{get out: } & 4E_{\mathrm{CH}} = & \underline{-1657} \\
 & & \Delta H = & -295\,\mathrm{kJ\,mol^{-1}}.
\end{array}$$

Further information on issues raised in Sections 2.6 to 2.10 may be obtained from standard texts.[37]

2.11 Explosives and Propellants

The popular view of practical chemistry as having to do with dense fumes, spectacular colors, vile stinks, and rattling good explosions is not entirely without foundation, and every chemist or chemical engineer should have at least some appreciation of the nature of explosives[38–41] if for no other reason than to protect him/herself and others from mishaps in the line of professional practice. Furthermore, disregarding the more sinister activities associated with explosives and propellants, it is hard to imagine life in a civilized society without these products of the chemical industry: explosives are essential for blasting in civil engineering projects (highways, dams, waterways, pile and bolt placement in rock, and demolition), mining, and quarrying, as well as for a host of less familiar applications. These include explosive riveting; forming of metals that are otherwise difficult to work; welding of dissimilar metals; propellants, explosive bolts and nuts and other key products for the aerospace industry; icebreaking and avalanche control; and safety equipment such as airbags for automobiles or ejector seats and escape chutes for aircraft.

For an explosion to occur, the first requirement is that the decomposition or combustion of the explosive—a substance or mixture of substances that is only *kinetically* stable at ambient temperature and pressure—be highly exothermic, so that, as a few molecules react, the temperature of neighboring molecules is raised beyond the point where the reaction becomes extremely rapid and a detonation wave spreads throughout the sample. A second requirement is that the hot reaction products be gaseous; it is their violent expansion that propagates the explosion and provides its destructive force. A shock wave can itself raise the local temperature to very high values. Thus, in practice, when a solid or liquid explosive is struck sufficiently sharply or is ignited, a nonreactive shock spreads through it at several thousand meters per second (much faster than in gas-phase explosions) and is followed closely by the reaction zone where gases are produced at pressures on the order of tens of gigapascals. The exact conditions of the explosion will depend on both the identity of the explosive and its density (e.g., in the case of a powder, whether it is loosely or tightly packed). For example, in the detonation of the commercial explosive RDX (Fig. 2.4) of density 1600 kg m^{-3}, the decomposition wave travels at 8200 m s^{-1}, so that it would traverse a 1 m long charge in about 0.1 ms; since the hot gaseous products cannot dissipate in this time, they generate a peak pressure of about 28 GPa.

2.11.1 Nitrogen-Containing Explosives

As noted in Section 2.10, the extraordinarily high bond energy of the triple bond in N_2 endows many nitrogen compounds with the tendency to decompose highly exothermically to dinitrogen gas and often other gases, and all commonly used explosives contain large percentages of N combined in some manner. For this reason, since 1989, some airport security units have used neutron activation analysis to detect explosives concealed in luggage: any material with high nitrogen content is revealed by the characteristic γ radiation of neutron-activated N. Most commercial high explosives are organic compounds that contain N in the form of nitro groups ($-NO_2$) which can, in the explosive reaction, oxidize the carbon content to CO or CO_2 gas and the hydrogen content to water vapor, thus increasing both the exothermicity of the explosion and the volume of gas generated. Examples of some commercial high explosives are given in Fig. 2.4.

In general, the order of decreasing sensitivity to detonation of organic nitro explosives is *O*-nitro > *N*-nitro > *C*-nitro compounds. Examples of O-nitro explosives are nitroglycerin (more accurately called glyceryl trinitrate), cellulose nitrates, and pentaerythritol tetranitrate (PETN). Nitroglycerin is a notoriously shock- and temperature-sensitive oily liquid; traditionally, workers entrusted with monitoring its manufacture were required to sit on one-legged stools to prevent their dozing off and possibly failing to note a rise in temperature of the reaction mixture. Alfred Nobel's "dynamite" consisted of nitroglycerin absorbed onto kieselguhr (a diatomaceous

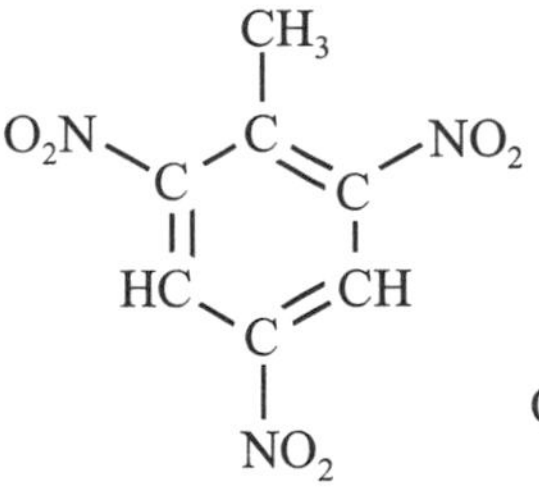

TNT 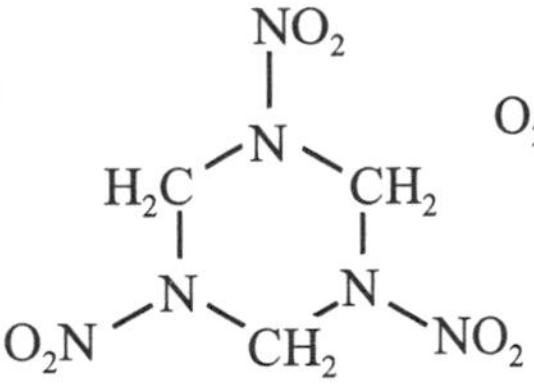 RDX (cyclonite) 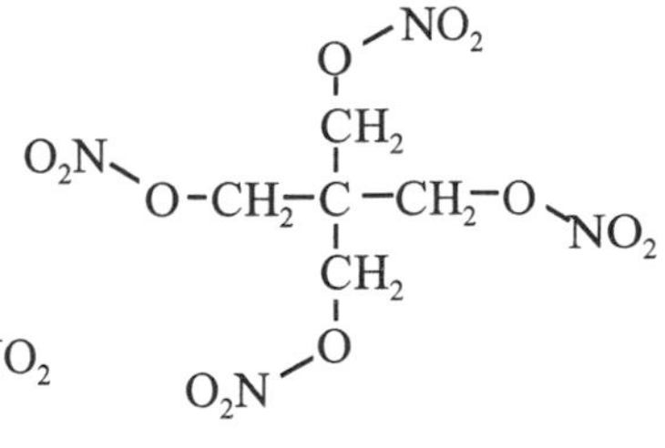 PETN

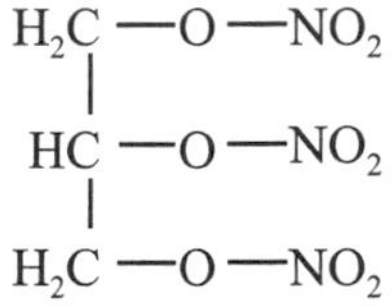

nitroglycerin 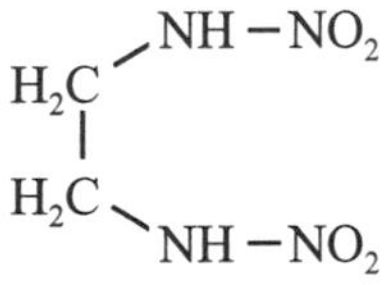 EDNA (haleite) 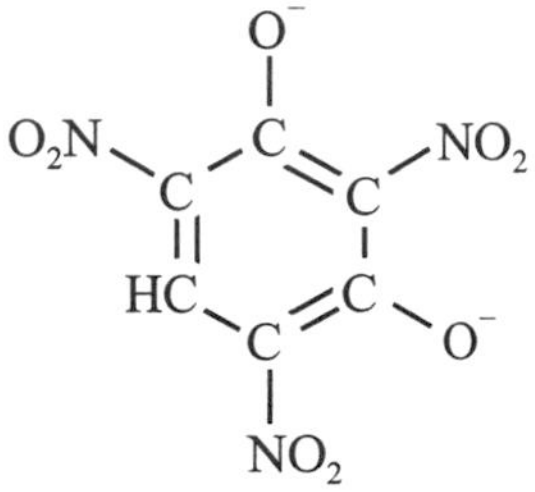 styphnate ion

Figure 2.4 Common commercial explosives. The styphnate ion is used as its lead salt in detonators.

earth) to render it safe to handle; nowadays, nitroglycerin is used on a combustible solid phase such as nitrocellulose, which is soluble to some extent in nitroglycerin (giving preparations known variously as cordite, gelignite, or ballistite), or even wood pulp impregnated with sodium nitrate (modern dynamite). Explosives in which the nitro group is attached to nitrogen include ethylene dinitramine (EDNA, an exceptionally strong high explosive with a low ignition temperature but rather low sensitivity to impact) and RDX, whereas the commonest *C*-nitro high explosive is trinitrotoluene (TNT). Often, mixtures of explosives are made to optimize their characteristics; for example, the notorious plastic explosive Semtex is a mixture of RDX, PETN, and oil in a poly(butadiene-styrene) support.

2.11.2 Ammonium Nitrate as an Explosive

The most commonly used blasting explosive, however, is not an organic compound but ammonium nitrate, which is produced in large tonnages primarily for use as "34–0–0" fertilizer (Section 9.4). Ammonium nitrate is stable enough in normal handling, and *gentle* heating to its melting point (170 °C) leads to the smooth evolution of nitrous oxide (laughing gas):

$$NH_4NO_3 \xrightarrow{\text{heat}} N_2O + 2H_2O. \tag{2.56}$$

However, reaction 2.56 is exothermic, and if the temperature rises above 250 °C or if the solid is strongly shocked, violent decomposition to N_2 rather than to N_2O (which has a positive free energy of formation) may result:

$$NH_4NO_3 \rightarrow N_2 + 2H_2O + \tfrac{1}{2}O_2. \qquad (2.57)$$

Ammonium nitrate can be used as a blasting explosive, even in the form of a slurry in its own saturated aqueous solution (such "aquagels" are usually thickened with guar gum, polyacrylamides, or gelatin). Conversely, it presents a potential explosion hazard during manufacture even in the concentrated solution stage (although normally an initiating explosive or exceptional circumstances would be required to cause it to explode). Producers and shippers of ammonium nitrate must therefore be aware of the potential hazard posed by reaction 2.57. In 1921, an attempt was made at a plant in Oppau, Germany, to break up a stockpile of caked NH_4NO_3 by dynamiting it. Over 600 people died, 1500 were injured, and 7000 lost their homes in the resulting explosion. In 1947, a fire broke out on a freighter carrying 1400 tonnes of NH_4NO_3 and led to an explosion that killed 561 people and destroyed several ships, refineries and a large rubber plant in Texas City, Texas. More recently, on December 13, 1994, an explosion at a Yazoo City, Mississippi, ammonium nitrate plant killed four employees, injured a dozen more, and destroyed the plant. Nor is the problem confined to ammonium nitrate; other oxidizing agents besides nitrate ion can oxidize NH_4^+ explosively to N_2. For example, on May 4, 1988, near Henderson, Nevada, a plant that produced ammonium perchlorate as a rocket fuel oxidant was destroyed by a series of massive explosions involving 4000 metric tons of NH_4ClO_4. Two senior company executives were killed, while 350 local residents were injured and thousands of buildings, some over 30 km away, were damaged.

It will be seen from Eq. 2.57 that oxygen is among the decomposition products of exploding ammonium nitrate, so that not all the oxidizing potential of the nitrate ion is used up on the ammonium ion. It is therefore usual, when ammonium nitrate is to be used deliberately as an explosive, to mix it with an oxidizable substance such as fuel oil. Although such mixtures have great utility as relative safe and inexpensive explosives for legitimate blasting purposes, their ready availability has been widely exploited by terrorists. In the most notorious case, a U. S. federal government building in Oklahoma City was destroyed on April 19, 1995, killing 166 people and injuring 400, by a bomb containing an estimated 2000 kg of an ammonium nitrate–fuel oil mixture. Various means have been suggested to make fertilizer-grade NH_4NO_3 unexplodable.[42] These include addition of calcium carbonate (used with limited success in Northern Ireland) or ammonium hydrogen phosphates [$(NH_4)_2HPO_4$ or $NH_4H_2PO_4$]; it is not impossible to remove these additives, although terrorists seem disinclined to try to do so.

2.11.3 Initiators

Since safe handling is obviously a desirable characteristic of commercial explosives, they are designed to be rather difficult to detonate. A primary explosive (initiator) is therefore required to produce a sufficiently intense shock to detonate the main charge. Today, the most commonly used initiators are lead styphnate (Fig. 2.4) and lead azide; the latter has almost completely displaced the mercury fulminate [$Hg(ONC)_2$)] formerly used universally in detonators. Lead azide is actually quite stable to heat; it explodes at 390 °C if heated quickly, but on slow heating decomposition sets in at about 200 °C and the explosive properties are lost. It may be made to explode in a flame by coating it with lead styphnate—a weaker explosive with greater heat sensitivity—and indeed it is usually employed together with lead styphnate to ensure sensitivity enough to be detonated by a firing pin. Alternatively, PETN, an *O*-nitro compound and therefore very sensitive to percussion (though less so than nitroglycerin), may be used in conjunction with lead azide as a detonator in blasting caps or in explosive cords. RDX can be similarly used. Heavy metal azides in general are highly shock-sensitive explosives, and great care must be taken not to leave azides or azide solutions in contact with heavy metals and particularly with copper; accordingly, lead azide detonators are made with aluminum or zinc casings.

2.11.4 Other Explosives and Propellants

Other explosives of interest include the ancient *black powder*, the composition of which was described by Roger Bacon in 1249 C.E. as 41% saltpeter (KNO_3), 29.5% charcoal (carbon, which makes the powder black), and 29.5% elemental sulfur. These ingredients, if not the proportions, have remained the same through the centuries—the first recorded ascent of a major peak in the New World was that of the Mexican volcano Popocatépetl in 1519 by Cortez' troops looking for sulfur to replenish their ammunition. It is now recognized that one can omit the sulfur if 87% KNO_3 and 13% C are used. The modern formula is 75% KNO_3, 15% C, and 10% S ("strong blasting powder," "poudre forte"). One problem with the former military use of black powder was the smoke produced; nitrocellulose (guncotton), diluted with waxes or glues to slow its excessively fast burn, replaced black powder, and it in turn was replaced by the more modern explosives described above. Black powder is still used for time fuses, fireworks, and some sporting ammunition. Potassium chlorate ($KClO_3$) can be used in place of KNO_3 as the oxidant for C and S; however, this practice is to be strongly discouraged, as the mixture is very sensitive to detonation by friction or impact. Solid alkali-metal chlorates, and perchlorates such as $KClO_4$, are not explosive in the pure state, but they can form dangerous, unpredictably explosive mixtures with oxidizable material such as organic matter and must be handled accordingly.

In the laboratory, sulfur–nitrogen compounds provide a very rich and interesting chemistry but present some risk of explosion, as the large N content would suggest. Thus, the orange-yellow solid S_4N_4, which is rather easily made by reaction of S_2Cl_2 with ammonia in a solvent such benzene, may explode if struck or heated. Similarly, the infamous "nitrogen triiodide" (probably $NI_3{\cdot}NH_3$), beloved of teenage pranksters, forms as a black residue from the reaction of aqueous ammonia with iodine which, on drying out, detonates with an impressive crackle at the slightest touch (practical jokers should, however, be aware that, explosion hazards apart, the purple fumes of elemental iodine formed in the explosion can burn the skin). Numerous other explosion hazards exist in the laboratory; there can be no hard-and-fast rules to cover all eventualities, but it will be clear that one should *never mix strong oxidants with oxidizable material whenever there is the possibility of reaction to produce gas.* In particular, *nitrogen compounds must be handled with respect.*

The foregoing discourse stressed the sudden generation of large amounts of hot gases in confinement as being the essence of explosive behavior. It does not follow, however, that an explosive mixture should be designed to maximize the *amount* of gas; sometimes, the strongly exothermic formation of a solid product such as Al_2O_3 gives energy enough to produce a higher gas pressure than would otherwise be realized. Thus, if aluminum metal powder is mixed with TNT, less CO or CO_2 is formed, but the greater release of energy makes for an exceptionally powerful explosion. Aluminum powder (16–18%) is therefore often mixed with ammonium nitrate (65–72%) and TNT to give *ammonal* (not to be confused with *amatol*, which is a mixture of ammonium nitrate with TNT alone).

Finally, we note that propellants for rockets and the like are often explosive mixtures of liquids: 100% hydrogen peroxide (H_2O_2), liquid oxygen, and even liquid fluorine have been used as oxidants in rockets for light substances such as hydrazine (N_2H_4) or unsymmetrical dimethylhydrazine [$(CH_3)_2N{-}NH_2$], while torpedoes and submarines have been powered with diesel fuel oxidized by hydrogen peroxide. Indeed, had the fast Walter H_2O_2-turbine U-boats come into service earlier than December 1944, the Second World War might have taken a rather different turn.

Exercises

2.1 At any given temperature T, the thermodynamic equilibrium constant K° for a given reaction is related to the standard enthalpy of reaction ΔH_T° and the entropy of reaction ΔS_T° by

$$-RT \ln K_T^\circ = \Delta G_T^\circ = \Delta H_T^\circ - T\Delta S_T^\circ \tag{1}$$

which at the standard reference temperature of 298.15 K becomes

$$\ln K_{298}^\circ = \frac{\Delta S_{298}^\circ}{R} - \frac{\Delta H_{298}^\circ}{298.15R}. \tag{2}$$

If the standard heat capacity change of reaction is ΔC_P°, we can write

$$\Delta H_T^\circ = \Delta H_{298}^\circ + \int_{298}^{T} \Delta C_P^\circ \, dT \tag{3}$$

$$\Delta S_T^\circ = \Delta S_{298}^\circ + \int_{298}^{T} \frac{\Delta C_P^\circ}{T} \, dT. \tag{4}$$

(a) Assuming that ΔC_P° is independent of temperature (an acceptable simplification, if not strictly correct), show that

$$\ln K_T^\circ = \ln K_{298} + \left(\frac{\Delta H_{298}^\circ}{298.15} - \Delta C_P^\circ\right)\left(\frac{T - 298.15}{RT}\right) + \frac{\Delta C_P^\circ}{R} \ln \frac{T}{298.15} \tag{5}$$

which has the general form

$$\ln K_T = a - \frac{b}{T} + c \ln T \tag{6}$$

where a, b, and c are constants.

(b) For the self-ionization of water,

$$H_2O(l) \rightleftharpoons H^+(aq) + OH^-(aq), \tag{7}$$

Olofsson and Hepler report $K_{298}^\circ = 1.002 \times 10^{-14}$, $\Delta H_{298}^\circ = 55.815$ kJ mol^{-1}, and $\Delta C_{P298}^\circ = -224.6$ J K^{-1} mol^{-1}. Calculate K° for 0 °C, on the basis of a constant ΔC_P° (Eq. 5). We customarily associate the neutral point of water with pH 7.00; this is valid at 25 °C, but what is the corresponding pH at 0 °C?

(c) Repeat calculation (b) with the assumption that ΔC_P° is negligible (i.e., that ΔH° and ΔS° can be regarded as constants).

(d) Calculations (b) and (c) show that heat capacity effects can be neglected for approximate calculations of K° over modest temperature ranges. Show, however, that the data given in (b) predict that K° for the ionization of water will pass through a *maximum* at about 274 °C, if ΔC_P° and the pressure remain constant. [*Hint*: What is the slope of a $\ln K$ versus T^{-1} plot at a maximum?] What would the neutral pH be at this maximum? What factor has been tacitly disregarded in the data supplied above, and what qualitative effect would its inclusion be expected to have?

[*Answers:* (b) 1.15×10^{-15}, pH 7.47; (d) pH 5.67.]

2.2 (a) Suppose a chemical reaction has two simultaneous (parallel) pathways of comparable importance in the middle of a particular temperature range, but different activation energies. Sketch the plot of $\ln k$ against T^{-1} over that range.

(b) Repeat exercise (a) for a single-path, multistep reaction, to show how the identity of the rate-determining step may change as the temperature is increased.

(c) Catalysts are often said to work by providing a reaction path of lower activation energy. If so, would you expect catalysts to be proportionally more effective at high temperatures or low?

2.3 Given the following mean bond energies in kJ mol^{-1}

H—H	436	Cl—Cl	243	F—F	159	C—C	347
H—Cl	431	C—H	414	C—Cl	331		

and assuming the electronegativity χ_H of hydrogen to be 2.1, calculate values for the electronegativities of carbon and chlorine. If the electronegativity of fluorine is 4.0, estimate the C—F bond energy. (The measured value is 490 kJ mol^{-1}.) Note that some minor inconsistencies may arise when these calculations are carried out by different routes; remember that electronegativity is only a semiquantitative concept. Furthermore, if one calculates, say, $|\chi_{Cl} - \chi_H|$, it is not clear whether χ_{Cl} is larger or smaller than χ_H; remember, however, that Cl tends to form Cl^- and H prefers to form H^+.

2.4 Compare the estimated enthalpy change for the reaction of acetylene with hydrogen to give ethane

$$\mathrm{H{-}C{\equiv}C{-}H + 2H_2 \rightarrow H_3C{-}CH_3}$$

with that of nitrogen to give hydrazine

$$\mathrm{N{\equiv}N + 2H_2 \rightarrow H_2N{-}NH_2}$$

given the mean bond energies $E_{C-H} = 414$, $E_{C-C} = 347$, $E_{C\equiv C} = 837$, $E_{H-H} = 436$, $E_{N\equiv N} = 946$, $E_{N-N} = 159$, and $E_{N-H} = 389$ kJ mol^{-1} (all negative, in the thermodynamic sense). Note that high triple bond strength does not necessarily imply lack of reactivity.

2.5 Lead azide is an explosive solid that can be detonated by shock (or by heating to 350 °C), while sodium azide exists as stable white crystals that decompose smoothly on heating (unless allowed to react with extraneous material). Why are heavy metal azides so explosive? Why are lead azide detonators not sheathed in copper or brass?

2.6 Hydrothermal vent and other sulfur bacteria derive their metabolic energy from the oxidation of H_2S. Use Appendix C to calculate the standard Gibbs free energy change per mole of dissolved HS^- when it is oxidized to HSO_4^- at 25 °C. Some thermophilic bacteria can live at 150 °C; if ΔC_P° can be neglected, what would the free energy change be at that temperature?
[*Answers:* −768 and −725 kJ mol^{-1}, respectively.]

2.7 Show that the rate of a reaction for which the volume of activation is −30 cm^3 mol^{-1} can be accelerated by a factor of 3.1×10^{10} if the pressure is raised from 0.1 MPa to 2 GPa at 25 °C.

References

1. G. N. Lewis and M. Randall, "Thermodynamics." McGraw-Hill, Toronto, 1961 (revised by K. S. Pitzer and L. Brewer); P. A. Rock, "Chemical Thermodynamics." University Science Books, Mill Valley, California, 1984; B. N. Roy, "Principles of Modern Thermodynamics." Institute of Physics Publishing, Bristol, U.K., 1995; I. M. Klotz and R. M. Rosenberg, "Chemical Thermodynamics." Benjamin/Cummings, Menlo Park, California, 1986; P. W. Atkins, "Physical Chemistry," 5th Ed. Freeman, New York, 1994; R. H. Parker, "An Introduction to Chemical Metallurgy," 2nd Ed. Chapters 1–4 (thermodynamic and kinetic concepts relevant to inorganic chemistry). Pergamon, Oxford, 1978.

2. J. O'M. Bockris and A. K. N. Reddy, "Modern Electrochemistry," Vol. 1. Plenum, New York, 1973.

3. R. A. Robinson and R. H. Stokes, "Electrolyte Solutions," 2nd Ed. (revised). Butterworth, London, 1965.

4. P. H. Rieger, "Electrochemistry," 2nd Ed. Chapman & Hall, New York, 1994.

5. K. S. Pitzer, Ion interaction approach. *In* "Activity Coefficients in Electrolyte Solutions" (R. M. Pytkowicz, Ed.), pp. 157–208. CRC Press, Boca Raton, Florida, 1979; K. S. Pitzer and J. J. Kim, Thermodynamics of electrolytes. IV. Activity and osmotic coefficients for mixed electrolytes. *J. Am. Chem. Soc.* **96**, 5701–5707 (1974) and earlier articles cited.

6. D. D. Wagman, W. H. Evans, V. B. Parker, R. H. Schumm, I. Halow, S. M. Bailey, K. L. Churney, and R. L. Nuttall, "The NBS Tables of Chemical Thermodynamic Properties." American Chemical Society, Washington, D.C., 1982 [also published as *J. Phys. Chem. Ref. Data*, **11**, Supplement No. 2 (1982)].

7. R. C. Weast and M. J. Astle (eds.), "CRC Handbook of Chemistry and Physics." CRC Press, Boca Raton, Florida, 1996 (revised annually).

8. R. M. Hazen, "The New Alchemists: Breaking through the Barriers of High Pressure." Times Books (Random House), New York, 1993.

9. J. A. Xu, H. K. Mao, and P. M. Bell, High-pressure ruby and diamond fluorescence: Observations at 0.21 and 0.55 terapascal. *Science* **232**, 1404–1406 (1986).

10. H. K. Mao and R. J. Hemley, Optical studies of hydrogen above 200 gigapascals: Evidence for metallization by band overlap. *Science* **244**, 1462–1645 (1989); I. F. Silvera, Evidence for band overlap metallization of hydrogen. *Science* **247**, 863–864 (1990); R. Pool, The chase continues for metallic hydrogen. *Science* **247**, 1545–1546 (1990); R. A. Kerr, Shock forges pieces of Jovian interior. *Science* **271**, 1667–1668 (1996); F. Hensel and P. P. Edwards, Hydrogen: the first metallic element. *Science* **271**, 1692 (1996); S. T. Weir, A. C. Mitchell, and W. J. Nellis, Metallization of fluid hydrogen at 140 GPa (1.4 Mbar). *Phys. Rev. Lett.* **76**, 1860–1863 (1996).

11. N. S. Isaacs, "Liquid Phase High Pressure Chemistry." Wiley, New York, 1981.

12. R. van Eldik and J. Jonas (eds.), "High Pressure Chemistry and Biochemistry." Reidel, Dordrecht, The Netherlands, 1987.

13. R. W. Shaw, T. B. Brill, A. A. Clifford, C. A. Eckert, and E. U. Franck, Supercritical water: A medium for chemistry. *Chem. Eng. News* December 23, 26–39 (1991).

14. H. L. Barnes (ed.), "Geochemistry of Hydrothermal Ore Deposits," 2nd Ed. Wiley, New York, 1979.

15. D. T. Rickard and F. E. Wickman, "Chemistry and Geochemistry of Solutions at High Temperatures and Pressures." Pergamon, New York, 1981.

16. R. A. Laudise, Hydrothermal synthesis of crystals. *Chem. Eng. News* September 28, 30–43 (1987).

17. A. Rabenau, The role of hydrothermal synthesis in preparative chemistry. *Angew. Chem. Int. Ed. Engl.* **24**, 1026–1040 (1985).

18. Symposium: Chemistry in high temperature aqueous solutions. *J. Solution Chem.* **21**(8), 711–932 (1992).

19. T. W. Swaddle, High temperature aqueous chemistry. *Chem. Can.* January, 21–23 (1980)

20. G. C. Ulmer and H. L. Barnes, "Hydrothermal Experimental Techniques." Wiley, New York, 1987.

21. P. R. Tremaine, E. E. Isaacs, and J. A. Boon, Hydrothermal chemistry applied to *in situ* bitumen recovery. *Chem. Can.* April, 29–33 (1983).

22. V. Tunnicliffe, Hydrothermal vent communities of the deep sea. *Am. Sci.* **80**, 336–349 (1992).

23. R. M. Kelly, J. A. Baross and M. W. W. Adams, Life in boiling water. *Chem. Br.* **30**, 555–558 (1994); M. W. W. Adams and R. M. Kelly, Enzymes from microorganisms in extreme environments. *Chem. Eng. News* December 18, 32–41 (1995).

24. R. M. Izatt, S. E. Gillespie, X. Chen and J. L. Oscarson, Thermodynamics of chemical interactions in aqueous solutions at elevated temperatures and interaction with protons and metal ions in aqueous solutions at high temperatures. *Chem. Rev.* **94**, 467–517 (1994).

25. L. F. Silvester and K. S. Pitzer, Thermodynamics of electrolytes. X. Enthalpy and the effect of temperature on the activity coefficients. *J. Solution Chem.* **7**, 327–337, (1978) and earlier work cited therein.

26. R. H. Wood, D. Smith-Magowan, K. S. Pitzer, and P. S. Z. Rogers, Comparison of experimental values of $\overline{V}^{\circ}, \overline{C}_P^{\circ}$, and $\overline{C}_V^{\circ}$ for aqueous NaCl with predictions using the Born equation at temperatures from 300 to 573.15 K at 17.7 MPa. *J. Phys. Chem.*, **87**, 3297–3300 (1983).

27. J. W. Cobble, High-temperature aqueous solutions. *Science*, **152**, 1479–1485 (1966); C. M. Criss and J. W. Cobble, The thermodynamic properties of high temperature aqueous solutions. IV. Entropies of the ions up to 200 °C and the correspondence principle. V. The calculation

of ionic heat capacities up to 200 °C. Entropies and heat capacities above 200 °C. *J. Am. Chem. Soc.*, **86**, 5385–5401 (1964).

28. E. L. Shock, E. H. Oelkers, J. W. Johnson, D. A. Sverjensky and H. C. Helgeson, Calculation of the thermodynamic properties of aqueous species at high temperatures and pressures. *J. Chem. Soc. Faraday Trans.* **88**, 803-826 (1992).

29. S. J. Hawkes, pK_w is almost never 14.0. *J. Chem. Educ.* *72*, 799 (1995).

30. A. A. Frost and R. G. Pearson, "Kinetics and Mechanism." Wiley, New York, 1953; K. J. Laidler, "Reaction Kinetics", Vols. 1 and 2. Pergamon, Oxford, 1963; J. H. Espenson, "Chemical Kinetics and Reaction Mechanisms." McGraw-Hill, New York, 1981.

31. S. A. Borman, Three-dimensional periodic table proposed. *Chem. Eng. News* January 1, 18–21 (1990).

32. L. C. Allen, Electronegativity is the average one-electron energy of the valence-shell electrons in ground-state free atoms. *J. Am. Chem. Soc.*, **111**, 9003–9014 (1989); L. C. Allen, Electronegativity scales. *Acc. Chem. Res.*, **23**, 175 (1990).

33. R. G. Pearson, Electronegativity scales. *Acc. Chem. Res.* **23**, 1–2 (1990).

34. A. L. Allred, Electronegativity values from thermochemical data. *J. Inorg. Nucl. Chem.* **17**, 215–221 (1990).

35. R. G. Pearson, Hard and soft acids and bases—the evolution of a concept. *Coord. Chem. Rev.* **100**, 403–425 (1990); R. G. Pearson, Absolute electronegativity and hardness. Application to inorganic chemistry. *Chem. Br.* **31**, 444-447 (1991); R. G. Pearson, Recent advances in the HSAB concept. *J. Chem. Educ.* **64**, 561 (1987).

36. (*a*) R. G. Pearson, The principle of maximum hardness. *Acc. Chem. Res.* **26**, 250–255 (1993); (*b*) R. G. Parr and P. K. Chattaraj, Principle of maximum hardness. *J. Am. Chem. Soc.* **113**, 1854–1855 (1991); (*c*) P. K. Chattaraj, H. Lee and R. G. Parr, HSAB principle. *J. Am. Chem. Soc.* **113**, 1855–1856 (1991).

37. J. E. Huheey, E. A. Keiter and R. L. Keiter, "Inorganic Chemistry: Principles of Structure and Reactivity," 4th Ed. Harper Collins, New York, 1993; D. F. Shriver, P. Atkins and C. H. Langford, "Inorganic Chemistry," 2nd Ed. Freeman, New York, 1994; F. A. Cotton and

G. Wilkinson, "Advanced Inorganic Chemistry," 5th Ed. Wiley (Interscience), New York, 1988; N. N. Greenwood and A. Earnshaw, "Chemistry of the Elements." Pergamon, Oxford, 1984.

38. T. Urbański, "Chemistry and Technology of Explosives," Vols. 1 and 2 (1964), Vol.3 (1965), and Vol. 4 (1984). Pergamon, New York, 1964.

39. J. Köhler and R. Meyer, "Explosives," 4th. Ed. VCH Publ., New York, 1993.

40. K. O. Brauer, "Handbook of Pyrotechnics." Chemical Publ. Co., New York, 1974.

41. J. E. Dolan, Molecular energy—the development of explosives. *Chem. Br.* **21**, 732–737 (1985).

42. A. M. Rouhi, Government, industry efforts yield array of tools to combat terrorism. *Chem. Eng. News* July 24, 10-19 (1995).

Chapter 3

Catenation: Inorganic Macromolecules

3.1 Factors Favoring Catenation

THE TENDENCY of atoms of certain elements to form chains with themselves (homoatomic catenation) or in alternation with other atoms (heteroatomic catenation) is of extreme importance in chemistry. The immense subject of organic chemistry and, indeed, life as we know it depend on the special ability of carbon to catenate; from the chemical engineering standpoint, catenation and the associated ability to form molecular rings and cages provide opportunities to make materials of desired mechanical, electrical, thermal, chemical, or catalytic properties.

Which elements can be expected to homocatenate? They must have electronegativities large enough to attract electrons for covalent bonding but not so large as to attract them *too* strongly. In other words, they must be "willing" to share electrons. Thus we expect elements with mid-range values of χ, namely, elements in the upper-central region of the main groups of the periodic table, to exhibit homocatenation. From Table 2.2, we see that carbon and sulfur are the prime candidates.

When atoms of two different elements catenate in alternation, the chain gains strength from the disparity in electronegativity between them (Eqs. 2.49 and 2.50). The elements must, however, have little tendency to form discrete ions, otherwise an ionic lattice results instead of a covalent chain. An important example is the *siloxane* (Si–O–Si) link, in which there is a large electronegativity difference (1.54, according to Table 2.2) but no possibility of the Si atoms forming discrete Si^{4+} ions—such ions would be so small and highly charged as to polarize any neighboring anion into covalency (Fajans' rules, Section 2.8). Thus, siloxane chains, branched chains, rings, and cages form an enormous variety of structures in minerals, synthetic silicates, and organosiloxanes.

In general, however, elements other than carbon offer only limited prospects for the synthesis of robust molecular polymers. Homonuclear catenation of heavier (larger) atoms gives relatively long and therefore weak bonds, while the same bond polarity that gives added bond strength to heteronuclear chains also increases the mechanistic vulnerability of the polymeric chain to attack by water (hydrolysis) and other polar reagents (see Section 2.8). Further, one of the most important mechanisms for forming organic polymers—the *addition* of molecules with double bonds one to another—is not widely available to inorganic molecules, in which double bonds are uncommon (Section 2.10):

$$n\,[R_2C{=}CR_2] \longrightarrow [-R_2C{-}CR_2-]_n\,. \tag{3.1}$$

For example, although some organodisilenes $R_2Si{=}SiR_2$ have been characterized in the photolysis (light-induced breakup) of cyclic organosilanes in hydrocarbon glasses at liquid nitrogen temperatures, the barrier to Si–Si catenation is so low that organodisilenes form polymers or cyclic oligomers immediately on warming unless R is a very bulky group such as mesityl.

Condensation polymerization, in which some very stable molecule XZ is eliminated from equimolar amounts of species $R_xE{-}X$ and $R_xE{-}Z$ (E being the element that catenates and R some side group), is more widely applicable:

$$R_xE{-}X + Z{-}ER_x \longrightarrow R_xE{-}R_xE + XZ. \tag{3.2}$$

The success of reaction 3.2 in producing long-chain polymers depends on having precisely equimolar amounts of the two pure reactants, and essentially complete reaction; otherwise, small polymeric molecules (*oligomers*) will be produced instead. Accordingly, a rather more effective variant of reaction 3.2 has X and Z in the same molecule:

$$n[R_xEX{-}ZER_x] \longrightarrow [-R_xE{-}R_xE-]_n + nXZ. \tag{3.3}$$

Finally, ring opening of small cyclic oligomers to form short open chains of transient existence can lead to polymers of high molecular weight through chain growth processes. This is particularly effective where the cyclic molecules have strained geometries, as with the cyclophosphazenes discussed in Section 3.6.

3.2 Homocatenation of Carbon

3.2.1 Diamond and Graphite

Carbon, having a normal valence of four, readily forms elaborate branched chains, rings, and networks (organic molecules). In the limiting case, the

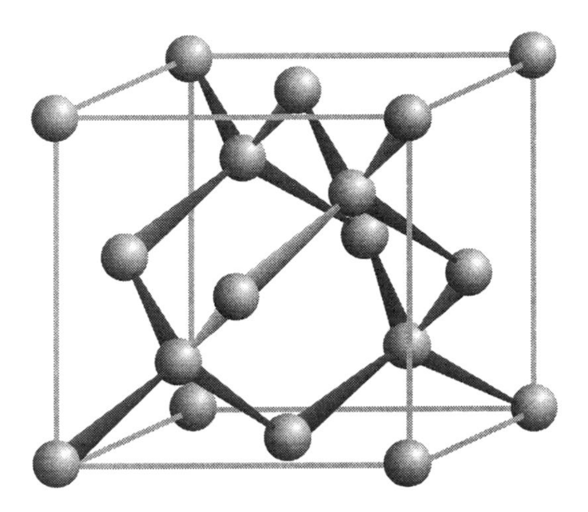

Figure 3.1 Structure of diamond.

three-dimensional network of C–C single bonds gives diamond, the hardest known substance. The structure shown in Fig. 3.1 is the normal *cubic* form of diamond; a *hexagonal* form has also been made in the laboratory. The distinction between these cubic and hexagonal structures is precisely that made in Section 4.4 (Fig. 4.13) between the zinc blende and wurtzite forms of ZnS. Interestingly, the HSAB principle (Section 2.9) seems to apply to mechanical as well as chemical hardness,[1] and tetravalent carbon is expected to be very "hard" according to the HSAB concept.

Diamonds are essential in industry for cutting, polishing, and drilling hard solids, but natural diamonds, whether of gemstone or industrial quality, are rare—90% of the world's supply currently comes from South Africa and most of the remainder from Russia. Consequently, since the 1890s, there have been efforts to make diamonds from the much more plentiful elemental modification or *allotrope* of carbon, graphite, the structure of which is shown in Fig. 3.2. The problem is that, thermodynamically, graphite is the most stable form of carbon at ambient temperature and pressure. Diamonds may be "forever," as advertisements would have us believe, but that is only because the *kinetics* of the transformation to graphite is extremely slow. As the phase diagram for carbon shows (Fig. 3.3), diamond, free of metastable graphite, can be expected to form only at very high pressures and, for kinetic reasons, high temperatures—with all the associated technical problems, notably that of devising suitable equipment for containment of the reactants.

Nevertheless, synthetic diamonds have been made by the high temperature/high pressure route, first by Allmänna Svenska Elektriska Aktiebolaget

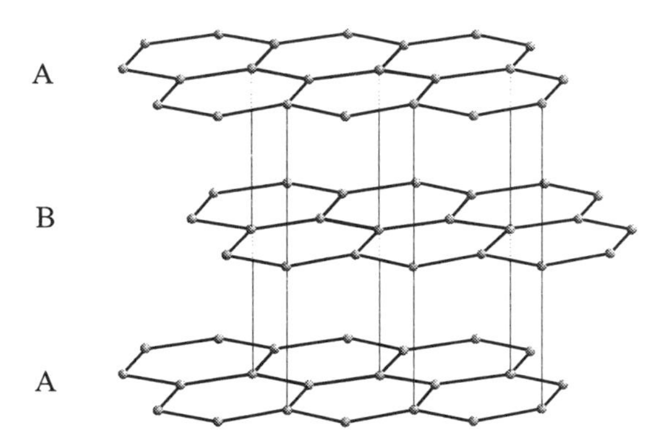

Figure 3.2 Structure of graphite (hexagonal form).

(ASEA) in Sweden in 1953, then in 1954 by General Electric in the United States, and now routinely to the extent of some 40 metric tons per year worldwide (about five times the annual production of natural diamonds). These scientific adventures in diamond synthesis are the main subject of a popular book.[2] A molten transition metal (Ni, Co, Mn, Cr, or Ta) is generally used as a solvent and as a catalyst. The diamonds are usually black because of graphite inclusions, but are suitable for use as polishing grit. Diamonds of *gem quality* are made by Sumitomo Electric for use as heat sinks for small electronic components, because one of diamond's other special properties is an extraordinarily high thermal conductivity, which can be up to five times that of silver at 25 °C if the nitrogen impurity content is kept low (as in the Sumitomo product).[3] When N substitutes for some C atoms in the diamond lattice, the crystal is yellow; when boron replaces a small number of C atoms, the crystal is blue and also semiconducting, so that B-doped diamonds find specialized applications in high power, high frequency electronic devices.

In view of the extreme pressures required by the high-T/high-P method, it is astonishing that *thin films* of diamond can be formed on suitable substrates by the low pressure thermal decomposition of C-containing gases such as methane.[3,4] The key is to inhibit the formation of graphite by having hydrogen gas present; graphite reacts much faster with hydrogen (particularly atomic H) than does diamond and so is removed as fast as it is deposited, while carbon deposited in the diamond form accumulates. In a simple application, a mixture of 1% methane in H_2 at 5 kPa pressure is passed over a wire at 2200 °C, and the atomic H and C so produced impinge on the heated substrate to form a diamond film.

Alternatively, a plasma (ionized gas) of H and C can be created at the substrate surface with guided microwaves and allowed to react with the

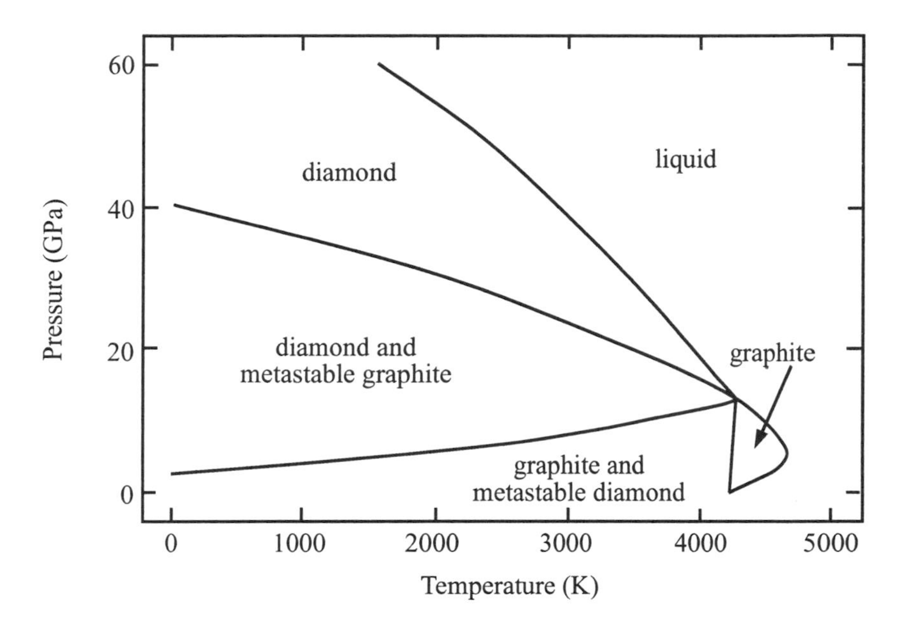

Figure 3.3 Phase diagram for carbon.

substrate. The substrates can be insulators such as silica glass, metals such as copper, hard materials such as silicon carbide, and so on, and the diamond coating may confer semiconducting properties for electronics, high heat conduction for use as a heat sink, or (in the case of diamond-coated tweeter diaphragms for loudspeakers) improved high frequency acoustic performance. The formation of these films is one example of the general technique of *chemical vapor deposition* (CVD),[3b] which is discussed in more detail in Chapter 19.

Researchers are now seeking practical ways to fluorinate the surfaces of diamond films, as the resulting surfaces are expected to have very low coefficients of friction (cf. "non stick" polytetrafluoroethylene or Teflon, Section 12.3) and hence have applications in low-friction tools. Direct fluorination with elemental fluorine is impractical, but photodecomposition of fluoroalkyl iodides chemically absorbed on the diamond surface looks promising.[4]

In graphite, three ordinary electron-pair bonds link each C to its neighbors in a quasi-infinite tessellated hexagonal or "chicken wire" pattern, and the remaining valence electrons, one from each C, are delocalized in a π-orbital system extending across the entire chicken wire sheet, above and below. The three-dimensional structure consists of stacks of these sheets. The structure of the sheets is, in effect, one huge collection of fused benzene rings. The "wall-to-wall" delocalized π electron system endows graphite with its black color and also with high electrical conductivity in directions parallel to the carbon atom planes. Graphite is therefore used as an electrode in many industrial processes as well as in flashlight cells. Graphite cleaves easily in planes parallel to the carbon sheets, giving it a greasy feel; thus, it is therefore used as a dry lubricant and (mixed with various proportions of clay for greater hardness) as "lead" for pencils. The planar sheets, however, are extremely resistant to reorganization, and this is why conversion of graphite to diamond requires very high temperatures as well as a catalyst even when the pressure criterion is amply met.

3.2.2 Fullerenes and Carbon Nanotubes

Because each ring of carbons in the sheets in graphite is a regular hexagon, the sheets are precisely planar. One of the more sensational developments in chemistry in the 1980s was the identification in certain soots of a series of carbon modifications in which C_5 pentagons replace some of the C_6 hexagons in graphite like C sheets, with the result that the sheets become curved. The curvature may result in the formation of a closed surface, usually spheroidal in shape as in C_{70} or, in the specific case of C_{60} (*Science* magazine's Molecule of the Year for 1991[5]), perfectly spherical, with the C atoms tracing the pattern of panels on a soccer ball (Fig. 3.4).[6] Such curved structures generated in this way are associated with the architectural concepts of Buckminster Fuller, and so these forms of carbon are referred to as *fullerenes* or, jocularly, as buckyballs (C_{60} has been referred to as "soccerene" and "footballene," but buckminsterfullerene is the preferred name).[5]

Soots containing abundant fullerenes can be made by striking an electric arc between two graphite electrodes or by controlled combustion of hydro-

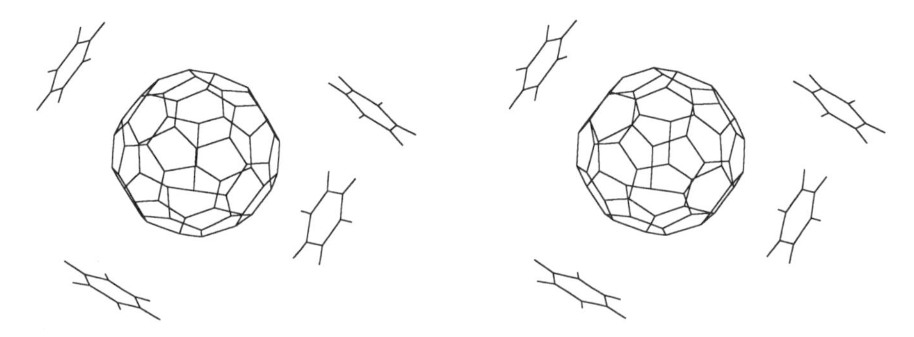

Figure 3.4 Structure of $C_{60}{\cdot}4C_6H_6$ (stereoscopic pair). Reproduced with permission from the data of Ref. 6.

carbons; mixed, variously colored fullerenes can be extracted from the soot with a suitable solvent (e.g., toluene) and separated by chromatographic methods. Although yields of fullerenes at first were small, efficient gram-scale preparative methods are now available.[7] As might be expected, the formation of C_{60} is endothermic (ΔH_f° = 2.42 and 2.66 kJ mol^{-1} for the crystalline and vapor phases, respectively) with respect to graphite, the thermochemical benchmark for carbon.[7c] Since ΔH_f° for diamond is just 1.895 kJ mol^{-1} (Appendix C) the possibility of using C_{60} as an alternative to graphite as a feedstock for diamond production suggests itself (see later in this section).

Fullerenes are rather easily oxidized, which explains the fact that, despite their commonplace occurrence in soots, they had escaped detection for so long. Fullerenes have now been found in Precambrian carbonaceous rocks from Karelia, Russia; in breccias associated with the 1.85-billion-year-old Sudbury impact structure in Canada; and in a sooty layer (believed to be due to fires from the asteroid impact that is thought to have killed off the dinosaurs) marking the Cretaceous–Tertiary boundary in New Zealand.[8]

Fullerenes can be derivatized by various means. For example, reaction with fluorine gas proceeds stepwise to the formation of colorless $C_{60}F_{60}$, which, according to the ^{19}F nuclear magnetic resonance (NMR) spectrum, contains just one type of F site and so evidently retains a high degree of symmetry.[9] In view of the low adhesion typical of fluorocarbons, this spherical molecule is expected to have extraordinary lubricant properties. Curiously, bromination of C_{60} is reversible on heating; otherwise, the reactions of fullerenes resemble those of alkenes or arenes (aromatic hydrocarbons).

Generation of soot in the presence of alkali metal vapors leads to the formation of "fullerides," compounds such as K_3C_{60} which, remarkably, are superconducting (Chapter 19) up to 18 K and, in many cases, higher. In these fullerides, the alkali metal atoms occupy sites in the crystal lattice outside the fullerene cage. However, since the cavities within fullerenes are large on the scale of atoms, it is possible, by doping the carbon electrodes used to produce the soot with a salt of an appropriate metal, to produce fullerenes with one or more metal atoms encarcerated inside them, for example, La@C_{82} and Sc_2@C_{80} (where @ signifies encapsulation). In La@C_{82}, the lanthanum is present as the 3+ ion, and the C_{82} cage carries a 3− charge. In the case of buckminsterfullerene, discrete anions $C_{60}{}^{n-}$, where n = 1–6, have been generated chemically or electrochemically in nonaqueous solvents such as dimethylformamide (DMF).[10] Fullerenes are insoluble in water unless they are derivatized with a solubilizing group such as sulfonate (—$SO_3{}^-$) or are themselves encapsulated inside a cyclodextrin "barrel."[11]

At present, the technological future of fullerenes is unclear, although uses in batteries, as drug delivery vehicles, in polymeric materials, and as catalysts have been proposed. It has been shown[12] that C_{60} can be converted to amber-colored diamond crystals by anisotropic pressures of 15–25 GPa *at ambient temperature*, so avoiding the problems of simultaneous containment of high temperatures and high pressures involved in the production of industrial diamonds from graphite. In the short term, however, it seems likely that the subsequently discovered[13] carbon *nanotubes*—structures in which a graphite like carbon sheet is rolled into a long tube rather than a fullerene spheroid—will find more applications. A Swiss–Brazilian team[13c] has developed a very small but highly effective electron gun, operating on field emission of electrons from bundles of carbon nanotubes, that can be used to activate a phosphorescent display screen; the device may provide an alternative to liquid-crystal and thin-film-transistor display screens for portable computers and similar applications. Immobilization of small protein molecules such as the biological redox reagent cytochrome c_3 inside such carbon nanotubes[13d] also holds much potential for the development of catalysts and sensors. Many structures based on curved carbon sheets are theoretically possible,[14] and one can anticipate major developments in this field of chemistry in the near future.

In 1996, the Nobel Prize for chemistry was awarded to R. F. Curl, H. W. Kroto, and R. E. Smalley, in recognition of their discovery of the fullerenes.

3.3 Boron Nitride

Boron nitride[15] (BN) is *isoelectronic* with elemental carbon, that is, $(BN)_n$ has the same number of electrons, coming from the same atomic orbitals, as C_{2n}. Consequently, BN behaves very much like carbon, even to the extent of having a phase diagram almost identical to Fig. 3.3 (the solid-liquid boundaries are shifted some 500 °C toward lower temperatures for BN relative to C). Thus we have a graphitic BN (*g*-BN), which has a structure similar to that of graphite except that the B_3N_3 rings are stacked one above another in successive layers (B and N alternating) whereas in graphite the C_6 rings are offset from one layer to the next.

Like graphite, *g*-BN has a greasy feel because of the easy cleavage of the solid parallel to the BN sheets, but there the similarities end: *g*-BN is white, an electrical insulator, and slowly hydrolyzed by water to ammonia and boric acid, $B(OH)_3$. Evidently, the strong alternation of electronegativities between the B and N atoms (Table 2.2) disrupts the crystal-wide π-electron system that is responsible for the electrical conductivity and blackness of graphite. Of the several rather difficult methods for making *g*-BN, the commonest involves fusion of borax, $Na_2[B_4O_5(OH)_4]\cdot 8H_2O$, with ammonium chloride.

At high temperatures (1800 °C) and pressures (8.6 GPa), *g*-BN goes over to cubic *c*-BN, also known as borazon, which has the normal diamond structure and, like diamond, is kinetically stable under ambient conditions. As might be expected, *c*-BN is very hard—not as hard as diamond, but harder than either boron carbide (B_4C) or the widely used abrasive silicon carbide (carborundum, SiC). A hexagonal form, *w*-BN, which has the wurtzite structure (Section 4.4), is also known and corresponds to hexagonal diamond; indeed, *g*-BN is converted by intense shock compression (which produces transient high temperatures and pressures) into a mixture of *c*-BN, *w*-BN, and amorphous BN.[16]

Finally, a report[17] that pure BN nanotubes are formed in an electric discharge between a BN-packed tungsten electrode and a cooled copper electrode extends the list of formal similarities between BN and carbon, and implies that much interesting BN chemistry remains to be discovered.

3.4 Homocatenation of Sulfur

Sulfur is usually only divalent, unless oxidized, and its catenation chemistry is therefore more limited and less familiar than that of carbon. Solid α-sulfur contains very stable eight-membered crown like rings, S_8,

[Structure: crown-shaped S_8 ring]

but many other ring sizes and open chain lengths are possible. Thus, in addition to the familiar hydrogen sulfide H_2S (a very toxic gas with the well-known "rotten eggs" smell), there exist numerous *sulfanes* (ostensibly analogs of alkanes and silanes), $H—S_x—H$, which occur in natural gas wells and may block the production path if they decompose to deposit solid S_8 on depressurization.

Much important aqueous sulfur chemistry involves oxosulfur species of the type $^-O_3S—S_x—SO_3^-$ ($x = 0$ to 20 or more), such as the tetrathionate ion ($x = 2$) which results when thiosulfate ion ($S_2O_3{}^{2-}$) is oxidized by, for example, iodine (the basis of the classical analytical technique known as *iodometry*):

$$\mathrm{O{=}S(O^-)({=}O)—S^-} + I_2 + \mathrm{{}^-S—S(O^-)(=O){=}O} \longrightarrow \mathrm{O{=}S(O^-)(=O)—S—S—S(O^-)(=O){=}O} + 2I^-. \quad (3.4)$$

The catenation of sulfur is discussed in more detail in Section 10.1.

3.5 Catenation of Silicon

3.5.1 Silanes and Organosilanes

Silicon has electronegativity 1.9, the silicon atom is larger than carbon, and the Si—Si bond is weaker than C—C (Table 2.1). Furthermore, Si is somewhat softer (in HSAB parlance) than C. Consequently, although Si is usually tetracovalent like carbon and elemental Si has the diamond structure, solid Si is less hard than diamond, and the reluctance of Si to form homonuclear double bonds would seem to preclude the formation of graphite- or fullerene-like analogs (it does, however, have important electrical properties which form the basis of the present-day electronics industry, as discussed in Sections 5.3 and 19.2). For the same reasons, the *silanes* Si_xH_{2x+2} are less diverse, and markedly more reactive, than the analogous alkanes.[18–21] For example, the Si–Si bond in silanes and many organosilanes [hexamethyldisilane, $(CH_3)_3Si—Si(CH_3)_3$, is a notable exception] is cleaved by aqueous or aqueous–alcoholic alkali with evolution of hydrogen gas,

$$R_3Si—SiR_3 + 2H_2O \longrightarrow 2R_3SiOH + H_2, \tag{3.5}$$

whereas an analogous C–C bond would be unaffected by alkali. Silanes with x up to 6 were first prepared by Alfred Stock from the reaction of metal silicides such as Mg_2Si with hydrochloric acid, followed by fractional distillation of the product, excluding air throughout because of the sensitivity of the silanes to molecular oxygen (a classic study that introduced vacuum line techniques to inorganic chemistry).[21] Organosilanes are readily made by chloride abstraction from organochlorosilanes (described below) with molten sodium, usually in a suitable solvent such as xylene:

$$2R_3SiCl + 2Na(l) \longrightarrow R_3Si—SiR_3 + 2NaCl. \tag{3.6}$$

Organopolysilanes $(R_2Si)_n$ can be similarly made by abstraction of Cl by alkali metals from dichlorodi(organo)silanes. Initially, the observation that $(CH_3)_2SiCl_2$ reacts with sodium to give an insoluble crystalline polymer that decomposes at 250 °C without melting diverted interest from organopolysilanes, and Eugene G. Rochow, who was a prime mover in the founding of the organosiloxane ("silicone") polymer industry in the years following the Second World War,[22] was quite dismissive of the potential utility of organopolysilanes. Subsequently, however, it was recognized that decomposition of this material to give silicon carbide fibers of high tensile strength or other SiC objects had genuine commercial value. Furthermore, West and co-workers found that, if phenyl groups replace some of the methyls, a meltable thermoplastic polymer results that is quite soluble in organic solvents and resistant to thermal decomposition and hydrolysis:

$$n(CH_3)_2SiCl_2 + n(C_6H_5)(CH_3)SiCl_2 + 4n\text{Na(l)} \longrightarrow [-(CH_3)_2Si-Si(CH_3)(C_6H_5)-]_n + 4n\text{NaCl}. \quad (3.7)$$

Unlike polymers based on carbon chains, however, organosilane polymers display significant mobility of the electrons in the Si chain (cf. elemental Si itself) and also remarkable sensitivity to photolysis, particularly by ultraviolet light. The latter property finds application in the use of organopolysilanes as ***photoresists*** for *microlithography* (*photolithography*) in the manufacture of microchips in the electronics industry. In photolithography, which is described in more detail in Chapter 19, a very thin layer of a photoresist is coated on top of a photo-insensitive polymer film on a silicon wafer and then exposed to a patterned image in visible (or, for diffraction-limited images, ultraviolet) light, for example, from a patterned mask. The image on the photoresist is then developed if necessary with a suitable solvent to remove the irradiated part of the photoresist, and an oxygen plasma is then applied to burn away the exposed polymer underneath to expose the silicon chip surface. Organosilanes have special advantages as photoresists in that they are especially sensitive to ultraviolet light and, in addition, that the oxygen plasma converts them to a protective film of silica on the raised surface of the surviving polymer. In this way, microcircuit patterns can be inscribed on the silicon wafer.

3.5.2 Siloxanes and Organosiloxanes

As noted in Section 3.1, Si—O bonds gain high strength from the large electronegativity difference between Si and O, while remaining covalent and thus directional. Consequently, —Si—O—Si—O— (polysiloxane) chains are very stable indeed. Furthermore, since Si is almost always tetravalent, branched Si—O chains and, indeed, sheets and three-dimensional networks are common both in naturally occurring minerals and in synthetic silicates, many of which are of primary economic importance (Chapter 7).

The propensity of silicon to form branched Si—O structures does not mean that siloxane chemistry is directly analogous to carbon (organic) chemistry. Carbon readily forms multiple bonds, whether to itself, as in ethene ($H_2C{=}CH_2$, "ethylene") and ethyne ($HC{\equiv}CH$, "acetylene"), or to another atom, such as oxygen in $R_2C{=}O$ (a ketone). The multiple-bonding tendency of Si is negligible; Si=Si double bonds, for example, are stable only in compounds of the form $R_2Si{=}SiR_2$, where R is a very bulky organic group that prevents polymerization to give Si—Si chains. Thus, attempts to make ketone analogs like $R_2Si{=}O$, where R is some organic group, invariably lead to polymeric substances $[R_2SiO]_n$ commonly called *silicones* (by analogy with *ketones*), although the preferred name is organopolysiloxanes.

Siloxane polymers can take the form of oils, greases, waxes, waxy solids, and rubbers, with several very useful and adjustable properties: high thermal stability and resistance to oxidation, excellent electrical insulation, remarkable water repellency, good biocompatibility (i.e., they can be used for surgical implants*), and low chemical reactivity (though aqueous OH^- or HF can break the Si—O—Si links). The liquids have surprisingly low viscosities for such large molecules (and low temperature coefficients of viscosity) , and they are widely used as very thermally stable high temperature heat transfer fluids. The solids are much more permeable to gases than are most plastics, and they are used as membranes for gas separations. Solid polydimethylsiloxane (which is transparent) is used for soft contact lenses, as the cornea of the eye must have access to the oxygen in air for its metabolic processes. Siloxane polymers also serve as the outer backing for the protein layer of the artificial skin used for the treatment of burns—the silicone layer keeps bacteria out but allows the wound to "breathe," so removing excess water.

The physical properties of siloxane polymers can be "tuned" by appropriate choices of the side groups R, which need not all be the same. However, stereoregularity of the kind possible with carbon polymers of the type $[\text{—CRR}'\text{—}]_n$ (isotactic and syndiotactic polymers, Section 18.4) has not been realized to date with $[\text{—SiRR}'\text{—O—}]_n$. The polysiloxane chain is inherently more flexible than an analogous hydrocarbon chain because the greater length of Si—O—Si (2 × 164 pm) relative to C—C (154 pm) reduces the interaction of the side groups R on successive Si atoms substantially. Furthermore, the Si—O—Si bond angle is only about 143° (cf. the tetrahedral C—C—C angle of 109.5°), so reducing the congestion still further. Thus, silicone polymers are generally more flexible, and remain so over a wider temperature range, than hydrocarbon polymers. In practice, however, polysiloxane polymers are usually not used in the pure form but are prepared with fillers such as finely divided silica (preferably *silica fume* from the high temperature oxidation of $SiCl_4$, Section 7.5) or with chain cross-linking agents to improve the mechanical and other properties.

Silicones are made by hydrolysis of organochlorosilanes R_nSiCl_{4-n}, which are produced from elemental silicon (obtainable from the carbon reduction of silica, i.e., sand, Section 17.7) by the *Rochow process*:

$$\text{RCl(g)} + \text{Si(s)} \xrightarrow[\text{Cu catalyst}]{\text{heat}} \text{R}_n\text{SiCl}_{4-n} \tag{3.8}$$

*Many women who have received silicone-fluid-filled breast implants have claimed to suffer from implant-related diseases, apparently caused by small leakages of the fluid into the body tissue. Successful lawsuits on those grounds in the United States have driven Dow Corning, the principal manufacturer of the implants, to file for bankruptcy protection. Food and Drug Administration Commissioner David A. Kessler, however, has reported to a congressional subcommittee that, as of July, 1995, there was no scientific evidence that silicone implants are a significant cause of disease. See M. S. Reisch and R. L. Rawls, Silicone gel breast implants. *Chem. Eng. News*, December 11, 10–17 (1995).

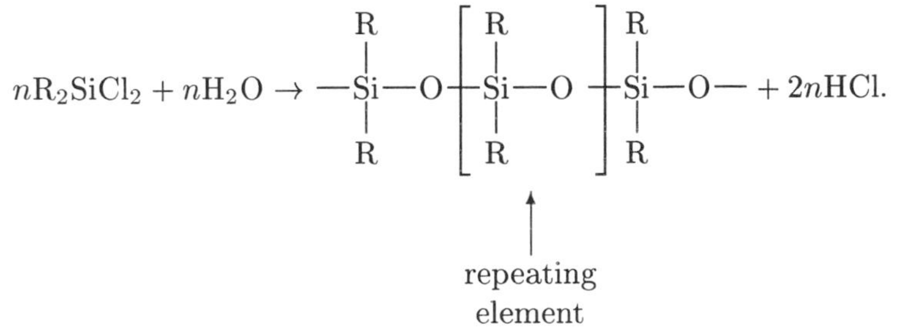

$$nR_2SiCl_2 + nH_2O \rightarrow -\underset{R}{\overset{R}{\underset{|}{\overset{|}{Si}}}}-O\left[-\underset{R}{\overset{R}{\underset{|}{\overset{|}{Si}}}}-O-\right]\underset{R}{\overset{R}{\underset{|}{\overset{|}{Si}}}}-O- + 2nHCl. \qquad (3.9)$$

↑ repeating element

Inclusion of $RSiCl_3$ will lead to chain branching

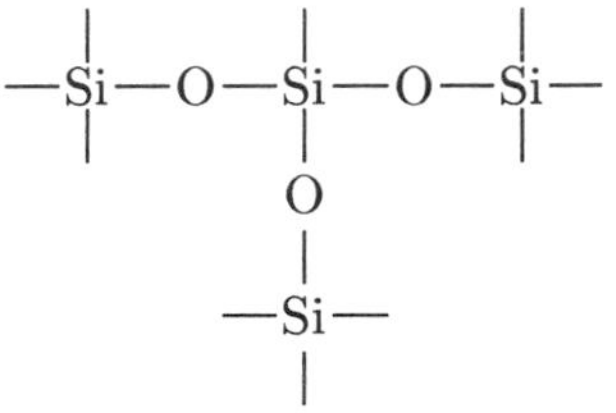

$$-\underset{|}{\overset{|}{Si}}-O-\underset{\underset{\underset{-\underset{|}{Si}-}{|}}{O}}{\overset{|}{Si}}-O-\underset{|}{\overset{|}{Si}}-$$

while R_3SiCl will cause chain termination:

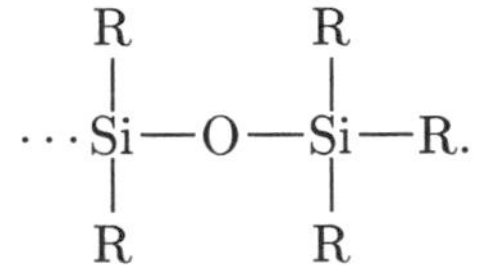

$$\cdots\underset{R}{\overset{R}{\underset{|}{\overset{|}{Si}}}}-O-\underset{R}{\overset{R}{\underset{|}{\overset{|}{Si}}}}-R.$$

Siloxane polymerization differs mechanistically from the formation of hydrocarbon polymers in that it is essentially an acid–base process, as might be expected from the strong alternation of electronegativites along the heteroatomic chain, and the radical initiators that catalyze the homocatenation of alkenes do not work for siloxanes. Long, unbranched polysiloxane chains are favored by higher condensation reaction temperatures and basic catalysts such as alkali metal hydroxides. Acidic condensation catalysts tend to produce polymers of lower molar mass, or cyclic oligomers.

3.6 Phosphazenes

The largest group of inorganic heteroatomic chain compounds with considerable industrial potential is the ***phosphazenes***,[18, 23] which are based on a phosphorus–nitrogen skeleton. The electronegativity difference here ($\chi_P = 2.19$, $\chi_N = 3.04$) is less than in the siloxane link, but some multiple bonding is involved between N and P, which strengthens the framework. The bonding in these compounds is often represented as $\cdots$—N=PR$_2$—$\cdots$, but because all P—N bond lengths in the skeleton are of about the same length, the P—N bond order is evidently about 1.5 throughout the chain.

Polydichlorophosphazenes can be made by a variety of methods, the most familiar of which is the reaction of phosphorus pentachloride with ammonium chloride in an appropriate medium such as chlorobenzene:

$$n\mathrm{PCl_5} + n\mathrm{NH_4Cl} \rightarrow \left[-\mathrm{N}{=}\underset{\displaystyle\mathrm{Cl}}{\overset{\displaystyle\mathrm{Cl}}{\underset{|}{\overset{|}{\mathrm{P}}}}}- \right]_n + 4n\mathrm{HCl}. \tag{3.10}$$

Typical chain lengths n run around 15,000 (molecular weights up to about 2×10^6). Traditional methods of making dichlorophosphazenes, however, involve high temperatures and give little control over the molecular weight of the product. An alternative procedure that gives high yields of polydichlorophosphazenes with good control over molecular weights involves the PCl_5-catalyzed elimination of $(CH_3)_3SiCl$ from $Cl_3P{=}NSi(CH_3)_3$.[24] The chlorine atoms in a dichlorophosphazene polymer can be then replaced with organic groups R by reaction with RM, where M is an alkali metal (e.g., LiC_6H_5, phenyllithium; see Chapter 18), to give polymers of very high thermal stability, although the N—P links may tend to hydrolyze slowly.

In fact, some P—N cleavage tends to occurs during reaction of the dichlorophosphazene with RM; this can be avoided if fluorophosphazenes $[-NPF_2-]_n$ are used in place of chlorophosphazenes. Reaction of P—Cl bonds with alkali metal alkoxides MOR (e.g., M = Na, R = $-C_2H_5$) gives $[-NP(OR)_2-]_n$, while chlorophosphazenes react with amines RNH_2 to give $[-NP(NHR)_2-]_n$ and $[RNH_3]Cl$. Since these reactions are more readily effected with small molecules than with polymers, it is often preferable to derivatize a small oligomer of a dihalophosphazene, such as $(NPCl_2)_3$, and then polymerize the product (ring-opening polymerization); several P—Cl bonds must still be available, however, as the polymerization mechanism is ionic and apparently involves separation of a chloride ion as an initial step.

Perfluoroalkoxy side chains ($R = C_xF_{2x+1}O-$) confer high resistance to hydrolysis. Phosphazenes of this type may be used to coat fabrics as water repellents and flame retardants, while the flexibility at low temperature (i.e., low glass transition temperatures) and excellent solvent resistance of phosphazene plastics hold high promise for special applications such as fuel hoses for use in cold environments. Lightly cross-linked phosphazene polymers are used as noncrystalline solid electrolytes in rechargeable lithium batteries, where they act as membranes for passage to and fro of the ions of Li salts (typically, lithium triflate, CF_3SO_3Li). The cost of such materials, however, is still prohibitively high for large-scale applications in the manner of polyethylene or nylon.

The tendency of the P—N skeleton of simple phosphazene polymers to hydrolyze slowly can be put to good use. The products of the hydrolyses are generally nontoxic, and so phosphazene capsules can be used as "bioerodible" surgical implants for targeted drug delivery.

Exercises

3.1 Phosphorus and arsenic vapors consist of molecules P_4 and As_4, respectively, in which the four atoms occupy the vertices of a regular tetrahedron. Suggest reasons for the difference between these structures and that of gaseous nitrogen (which is also in Group 15).

3.2 What factors conspire to limit commercial exploitation of fullerenes?

3.3 Given that $\Delta H_f^\circ = -254.4$ kJ mol^{-1} and $S^\circ = 14.81$ J K^{-1} mol^{-1} for hexagonal solid boron nitride, use the data of Appendix C to calculate the free energy change for the hydrolysis of one mole of BN(s) by liquid water at 25 °C. Ignore the weak acid–base interaction of the products and the very slight solubility of $B(OH)_3$(s).
[*Answer:* -55.45 kJ mol^{-1}.]

3.4 Phosphazenes of the type $(NPCl_2)_n$ have high thermal stability but hydrolyze slowly on exposure to water or moist air. What would you expect the hydrolysis products to be? (Consider Table 2.2.)

3.5 Looking ahead to Section 18.4, explain why isotacticity is not achievable by standard methods of organosiloxane polymer production.

References

1. R. G. Pearson, Hard and soft acids and bases—the evolution of a concept. *Coord. Chem. Rev.* **100**, 403–425 (1990).

2. R. M. Hazen, "The New Alchemists: Breaking Through the Barriers of High Pressure." Times Books (Random House), New York, 1993.

3. (*a*) P. K. Bachmann and R. Messler, Emerging technology of diamond thin films. *Chem. Eng. News*, May 15, 24–39 (1989); (*b*) M. N. R. Ashfold, P. W. May, C. A. Rego, and N. M. Everitt, Thin film diamond by chemical vapour deposition methods. *Chem. Soc. Rev.* **23**, 21–30 (1994).

4. E. Wilson, Fluorine adds easily to diamond films. *Chem. Eng. News*, January 15, 6–7 (1996).

5. D. Koshland, Molecule of the year. *Science* **254**, 1705 (1991); E. Culotta and D. Koshland, Buckyballs: Wide open playing field for chemists. *Science* **254**, 1706–1707 (1991); G. S. Hammond and

V. S. Kuck (eds.), "Fullerenes: Synthesis, Properties, and Chemistry of Large Carbon Clusters," ACS Symp. Ser. 481, American Chemical Society, Washington, D.C., 1992; H. W. Kroto and D. R. M. Walton (eds.), "Fullerenes: New Horizons for the Chemistry, Physics, and Astrophysics of Carbon." Cambridge Univ. Press, Cambridge, 1993; H. W. Kroto (ed.), "Fullerenes." Pergamon/Elsevier, London, 1993; A. Hirsch, "The Chemistry of the Fullerenes." Thieme Verlag, Stuttgart, 1993; H. Aldersley-Williams, "The Most Beautiful Molecule: Discovery of the Buckyball." Wiley, New York, 1995.

6. M. F. Meidine, P. B. Hitchcock, H. W. Kroto, R. Taylor, and D. R. M. Walton, Single crystal structure of benzene-solvated C_{60}. *J. Chem. Soc. Chem. Commun.* 1534 (1992).

7. (*a*) P. Bhyrappa, A. Penicaud, M. Kawamoto, and C. A. Reed, Improved chromatographic separation and purification of C_{60} and C_{70} fullerenes. *J. Chem. Soc. Chem. Commun.* 936–937 (1992); (*b*) A. M. Vassallo, A. J. Palmisano, L. S. K. Pang, and M. A. Wilson, Improved separation of fullerenes–60 and –70. *J. Chem. Soc. Chem. Commun.* 60–61 (1992); (*c*) W. V. Steele, R. D. Chirico, N. K. Smith, W. E. Billips, P. R. Elmore, and A. E. Wheeler, Standard enthalpy of formation of buckminsterfullerene. *J. Phys. Chem.* **96**, 4731–4733 (1992); (*d*) W. A. Scrivens, P. V. Bedworth, and J. M. Tour, Purification of gram quantities of C_{60}: a new inexpensive and facile method. *J. Am. Chem. Soc.* **114**, 7917-7919 (1992).

8. L. Becker, J. L. Bada, R. E. Winans, J. E. Hunt, T. E. Bunch, and B. M. French, Fullerenes in the 1.85-billion-year-old Sudbury impact structure. *Science* **265**, 642-645 (1994); D. Heymann, L. P. F. Chibante, R. R. Brooks, W. S. Wolbach, and R. E. Smalley, Fullerenes in the Cretaceous-Tertiary boundary layer. *Science*, **265**, 645-647 (1994).

9. J. H. Holloway, E. G. Hope, R. Taylor, G. J. Langley, A. G. Avent, T. J. Dennis, J. P. Hare, H. W. Kroto, and D. R. M. Walton, Fluorination of buckminsterfullerene. *J. Chem. Soc. Chem. Commun.* 966–969 (1991).

10. R. Subramanian, P. Boulas, M. N. Vijayashree, F. D'Souza, M. T. Jones, and K. M. Kadish, A facile and selective method for the solution-phase generation of C_{60}^{-} and C_{60}^{2-}. *J. Chem. Soc. Chem. Commun.* 1847–1848 (1994).

11. T. Andersson, K. Nilsson, M. Sundahl, G. Westman, and O. Wennerström, C_{60} embedded in γ-cyclodextrin: A water-soluble fullerene. *J. Chem. Soc. Chem. Commun.* 604-606 (1994).

12. M. N. Regueiro, P. Monceau, and J.-L. Hodeau, Crushing C_{60} to diamond at room temperature. *Nature (London)* **355**, 237–239 (1992).

13. (*a*) S. Iijima, Helical microtubules of graphitic carbon. *Nature* **354**, 56–58 (1991); (*b*) P. Ball, New horizons in inner space. *Nature (London)* **361**, 257 (1993); (*c*) W. A. de Heer, A. Châtelain, and D. Ugarte, A carbon nanotube field emission electron source. *Science* **270**, 1179–1180 (1995) (with commentary by R. F. Service, *ibid.*, 1119); (*d*) S. C. Tsang, J. J. Davis, M. L. H. Green, H. A. O. Hill, Y. C. Leung, and P. J. Sadler, Immobilization of small proteins in carbon nanotubes. *J. Chem. Soc. Chem. Commun.* 1803-1804 (1995).

14. H. Terrones, M. Terrones, and W. K. Hsu, Beyond C_{60}: Graphite structures for the future. *Chem. Soc. Rev.* **24**, 341–350 (1995).

15. A. Earnshaw and N. N. Greenwood, "Chemistry of the Elements." Pergamon, London, 1984.

16. A. B. Sawaoka, New sintering processing of high-density boron nitride and diamond. *In* "High Temperature Ceramics" (G. Kostorz, ed.), pp. 41-58. Academic Press, London, 1989.

17. N. G. Chopra, R. J. Luyken, K. Cherrey, V. H. Crespi, M. L. Cohen, S. G. Louie, and A. Zettl, Boron nitride nanotubes. *Science* **269**, 966–967 (1995).

18. J. E. Mark, H. R. Allcock, and R. West, "Inorganic Polymers." Prentice–Hall, Englewood Cliffs, New Jersey, 1992.

19. S. Patai and Z. Rappoport (eds.), "The Chemistry of Organosilicon Compounds." Wiley, New York, 1989.

20. J. M. Ziegler and F. W. G. Fearon (eds.), "Silicon-Based Polymer Science." Advances in Chemistry Series 224. American Chemical Society, Washington, D.C., 1990.

21. A. Stock, "Hydrides of Boron and Silicon." Cornell Univ. Press: Ithaca, New York, 1933 (reissued in 1957).

22. E. G. Rochow, "An Introduction to the Chemistry of the Silicones," 2nd Ed. Wiley, New York, 1951.

23. H. R. Allcock, "Inorganic macromolecules." *Chem. Eng. News* March 18, 22–36 (1985).

24. C. H. Honeyman, I. Manners, C. T. Morrissey, and H. R. Allcock, Ambient temperature synthesis of poly(dichlorophosphazene) with molecular weight control. *J. Am. Chem. Soc.* **117**, 7035–7036 (1995).

Chapter 4

Crystalline Solids

IN A PERFECT crystal, the constituent atoms, ions, or molecules are packed together in a regular array (the *crystal lattice*), the pattern of which is repeated periodically *ad infinitum*. Thus, regularly repeating planes of atoms are formed. The smallest complete repeating three-dimensional unit is called the *unit cell*, and the crystallographer's primary objective is to determine the dimensions and geometry of the unit cell, as well as the precise deployment of the atoms within it.[1–6]

4.1 Determination of Crystal Structure

Just as the rulings on a diffraction grating create colored interference patterns in the reflected light, so layers of atoms in a crystal give rise to diffraction patterns in incident radiation of the appropriate wavelength—in this case, monochromatic (i.e., single wavelength) X-rays or beams of electrons or neutrons (which also have wave like properties). X-Ray diffraction is caused by interaction of an incoming X-ray photon with the *electron density* in the crystal. Since atoms of very low atomic number such as hydrogen have relatively low electron density, they do not show up strongly in X-ray diffraction patterns, and so H atoms in particular are often missing from crystal structures determined by X-ray diffraction. To locate the hydrogens in such cases, we can resort to diffraction of a beam of neutrons of a single, known velocity (since the wavelength of a material particle is inversely proportional to its momentum).

When X-rays of wavelength λ are reflected from parallel planes of atoms of spacing d, they will reinforce one another if rays from successive planes arrive at the detector a distance λ apart (or $n\lambda$, where n is a positive integer); otherwise, they will tend to cancel. As Fig. 4.1 shows, rays reflected from successive planes at an angle θ will each travel $2d \sin\theta$ further than

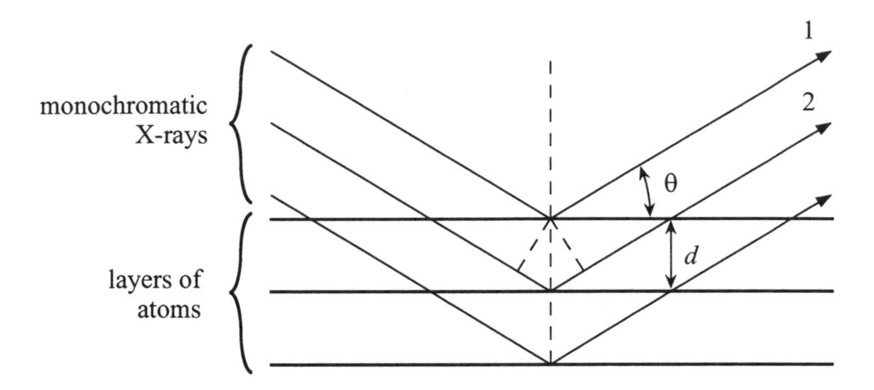

Figure 4.1 Diffraction of X-rays by layers of atoms. The path of ray 2 to the detector is longer than the path of ray 1 by $2(d \sin \theta)$.

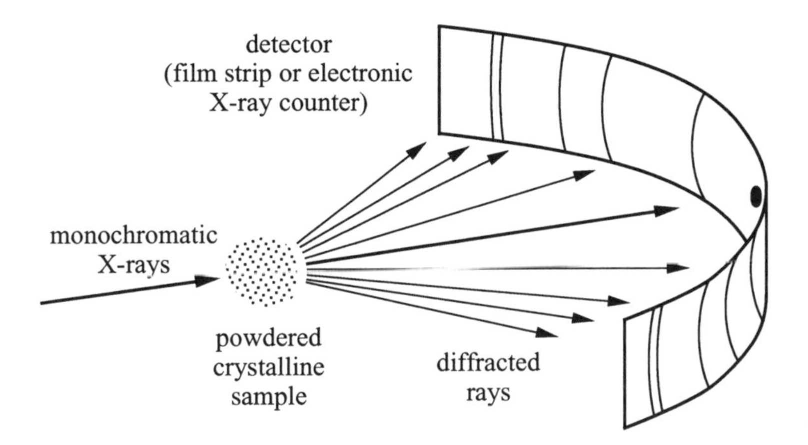

Figure 4.2 Characterization of a powdered solid by its X-ray diffraction pattern.

their immediate predecessors to reach the detector. Thus, when reinforced X-rays are recorded at the detector, Eq. 4.1 (the *Bragg equation*) must hold, and, knowing λ and measuring θ, we can obtain *d:*

$$n\lambda = 2d \sin \theta. \tag{4.1}$$

If a *single crystal* is rotated in a monochromatic X-ray beam, a pattern of spots of reinforced X-rays can be recorded, traditionally on a photographic film placed behind the crystal perpendicular to the primary beam (giving the so-called *Laue photographs*). Nowadays, X-ray diffractometers use electronic photon counters as detectors. Since, as noted above, different atoms have different X-ray scattering powers, both the positions *and the intensities* of these spots are important in working out the structure of the crystal in terms of the planes of atoms present. In the case of crystals containing molecular units, these molecular structures will also show up in the analysis of the diffraction pattern. With the advent of powerful digital computers, such determinations of structure have become routine in modern research in synthetic chemistry.

Alternatively, when a *powdered* crystalline solid diffracts monochromatic X-radiation, the diffraction pattern will be a series of concentric rings, rather than spots, because of the random orientation of the crystals in the sample (Fig. 4.2). The structural information in this pattern is limited; however, because even solid compounds that have the same structure but different composition will almost inevitably have different d values, each individual solid chemical compound will have its own characteristic *powder diffraction pattern.*

X-Ray powder diffraction patterns are catalogued in the JCPDS data file,[7] and can be used to identify crystalline solids, either as pure phases or as mixtures. Again, both the positions and the relative intensities of the features are important in interpretation of powder diffraction patterns, although it should be borne in mind that diffraction peak heights in the readout from the photon counter are somewhat dependent on particle size. For example, a solid deposit accumulating in a heat exchanger can be quickly identified from its X-ray powder diffraction pattern, and its source or mechanism of formation may be deduced—for instance, is it a corrosion product (if so, what is it, and where does it come from) or a contaminant introduced with the feedwater?

4.2 Bonding in Solids

Bonding in solids takes several forms. Some elements such as carbon or compounds such as silica (SiO_2 in its various forms—see Section 7.5) can form quasi-infinite networks of covalent bonds, as discussed in Section 3.2; such crystalline solids are typically very high melting (quartz has mp 1610 °C). On the other hand, small, discrete molecules like dihydrogen (H_2) or sulfur (S_8, Section 3.4) interact only weakly with one another through van der Waals forces (owing to electric dipoles induced by the electrons and nuclei of one molecule in the electron cloud of a neighbor and vice versa) and form low melting crystals (H_2 has mp -259 °C; α-S melts at 113 °C).

When a metal M of low electronegativity (χ) combines with a nonmetal X of high χ, the product is likely to be a high-melting solid consisting of ions M^{m+} and X^{x-}, held together in a regular pattern (the crystal lattice) by electrostatic forces rather than electron-sharing bonding (covalency). The energy of these electrostatic interactions—called the *lattice energy*, U—makes formation of the ionic solid possible by compensating for the energy inputs, such as ionization potential needed to form the ions, and is clearly dependent to some degree on the structure of the crystal at the atomic or molecular level (Section 4.7).

Bonding in metals involves delocalization of electrons over the whole metal crystal, rather like the π electrons in graphite (Section 3.2) except that the delocalization, and hence also the high electrical conductivity, is three dimensional rather than two dimensional. Metallic bonding is best described in terms of *band theory*, which is in essence an extension of molecular orbital (MO) theory (widely used to represent bonding in small molecules) to arrays of atoms of quasi-infinite extent.

Molecular orbital theory is explained at length in almost all introductory chemistry textbooks, and only a brief summary is given here. As the simplest possible case, we consider interactions between two free hydrogen atoms, A and B, each with a single electron, the time-average spatial distributions of which are described mathematically as wave functions or *orbitals* (1s orbitals, if the atoms are in their lowest energy states) ϕ_A and ϕ_B. When A and B approach one another closely enough to interact, the two *atomic* orbitals combine mathematically to give two *molecular* orbitals, one (say, $\phi_{MO} = \phi_A + \phi_B$, if we assume the combination of atomic orbitals to be linear) lower in energy than the original atomic orbitals and the other ($\phi^*_{MO} = \phi_A - \phi_B$) higher in energy. When the difference in energy between the two molecular orbitals is sufficient to overcome the spin pairing energy of the two electrons, the two electrons will both occupy ϕ_{MO}, with a net energetic stabilization. At the $A - B$ approach distance where the increasing mutual repulsion of the atomic nuclei balances this stabilization, we have a stable H_2 molecule. The occupied molecular orbital ϕ_{MO} is therefore known as a *bonding orbital*, and ϕ^*_{MO} (electronic occupation of which would *destabilize* the system) is known as an *antibonding orbital.*

If we now have, say, four H atoms in a row, there would be four molecular orbitals of different energies, two bonding and two antibonding; eight atoms give a stack of eight molecular orbitals; and so on (Fig. 4.3). As the number of participating atoms increases and we move from a one-dimensional row to a three-dimensional array of atoms, the range in energies between the lowest bonding and highest antibonding molecular orbitals levels out to asymptotic limits, giving eventually a *band* of bonding and antibonding orbitals, very closely spaced in energy. At the absolute zero of temperature, only the lower half of this band (the bonding orbitals) would be filled with electrons; the highest occupied energy level is known as the *Fermi level*, after the Italian-born U.S. physicist Enrico Fermi. For all accessible temperatures, a small fraction of the electrons will be excited thermally into energy levels higher than the Fermi level, leaving some depleted levels below it.

Because the band is partially filled and extends throughout the crystal, electrons can move freely through it, with the number flowing in any one

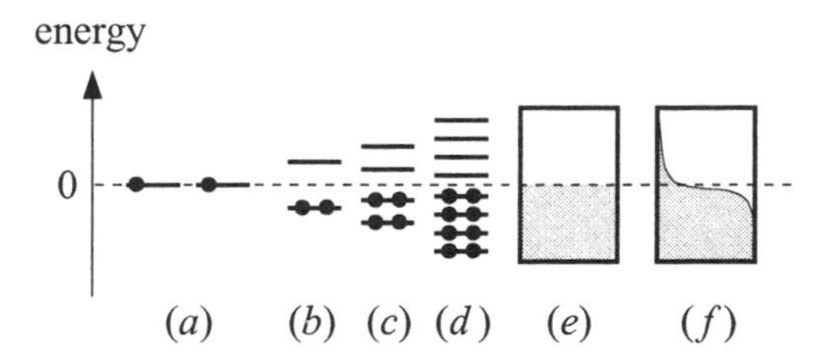

Figure 4.3 Formation of electronic bands in a hypothetical array of hydrogen atoms. (*a*) Two H atoms, infinitely far apart. (*b*) Two H atoms interacting (as in the actual H_2 molecule). (*c*) Four, (*d*) eight, and (*e*) a very large number of H atoms interacting. The dots or (in the band) shading represent the occupancy of the energy levels by the electrons, in the absence of thermal excitation. (*f*) Percentage distribution of the electron population in a band at a nonzero temperature.

direction being ordinarily balanced by an equal number coming the other way. If, however, an electric potential is applied across the solid, the band energy levels will be depressed in energy near the positive connection, while near the negative end they will be elevated. Consequently, there will be a net flow of electrons through the band to occupy the lower energy levels preferentially. This amounts to conduction of electricity through a partly filled electronic band and is a characteristic property of *metals*. The free flow of electrons is limited by scattering by the atoms, which are in effect present as cations. As the temperature increases, the amplitudes of the vibrations of the metal ions about their mean positions in the lattice increase, resulting in increased scattering of the electrons as they flow; thus, the electrical conductivity of a metal *decreases* with increasing temperature.

In our illustrative example, hydrogen was chosen for simplicity, even though it is well known that solid hydrogen ordinarily consists of an array of discrete H_2 molecules as in Fig. 4.3b and is therefore not normally a conductor of electricity. It is expected, however, that pressures on the order of several hundred gigapascals will force the H atoms in solid hydrogen into sufficiently close proximity that a band structure will form, as in Fig. 4.3f—in other words, hydrogen will metallize. Such metallic hydrogen probably exists in the core of the planet Jupiter, which is composed largely of hydrogen. Demonstration of such metallization of hydrogen in the laboratory, however, has proved elusive (see Section 2.4).

For hydrogen, only the 1s orbital is energetically accessible for band formation. For elements of the first row of the periodic table, the 2s and, at somewhat higher energy, the three 2p orbitals are available, and, depending on the ways in which the atomic orbitals align with the crystal structure, these may form either a continuous s, p band or a pair of bands with the

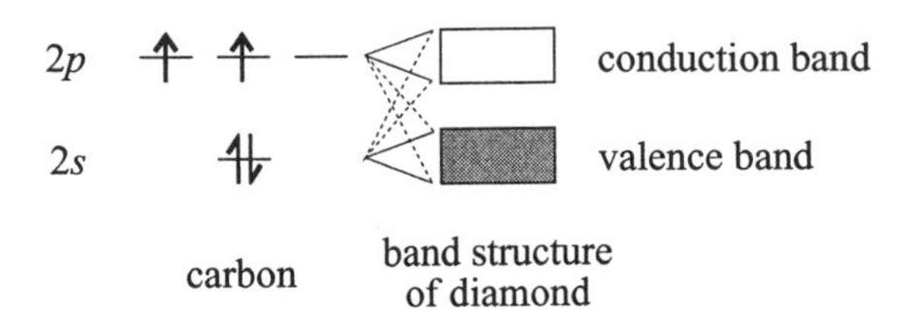

Figure 4.4 Band structure associated with the diamond structure in Group 14 elements.

same number of energy levels in each and with an energy gap (*band gap*) between the bands. The latter is the case for carbon with the diamond structure (Fig. 4.4); since four electrons are available per C atom, they fill the lower band (the *valence band*) completely, so that electronic conduction within this band is not possible, while the upper band (the *conduction band*) contains no electrons. The Fermi level is located midway between the two bands. For diamond, the band gap is wide enough to prevent any population of the conduction band from the valence band by thermal excitation of electrons from the latter, and so diamond is an electrical insulator. As we descend the periodic table to Si and Ge, however, the band gap narrows, and a small fraction of the electrons in the valence band can be thermally excited into the otherwise empty conduction band, giving rise to a limited degree of electrical conductivity that *increases* with rising temperature (the opposite of the temperature dependence of electrical conductivity in metals). Such materials are known as *intrinsic semiconductors.*

In certain solids such as titanium dioxide or cadmium sulfide, the energy of the band gap corresponds to that of light (visible, ultraviolet, or infrared), with the result that the solid, when illuminated, may become electrically conducting or acquire potent chemical redox characteristics because of the promotion of electrons to the conduction band (which is normally unoccupied). These properties have obvious practical significance and are considered at length in Chapter 19.

4.3 The Close Packing Concept

Figure 4.5 shows the manner of close packing of identical atoms—assumed to be spheres of equal radii—in a single plane, A. If a second layer B of the same atoms is close packed on top of layer A (Fig. 4.6), it will be seen that each B atom rests on three A atoms that are in mutual contact, so enclosing a void. The centers of the four atoms describe a regular tetrahedron about the void, which is therefore called a *tetrahedral interstice* or *T-hole.* A second kind of interstice, bounded by six atoms (three from each

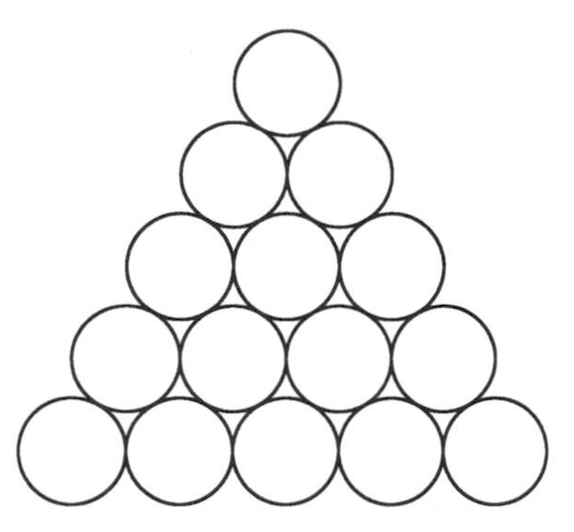

Figure 4.5 Close packing of spheres in a single layer.

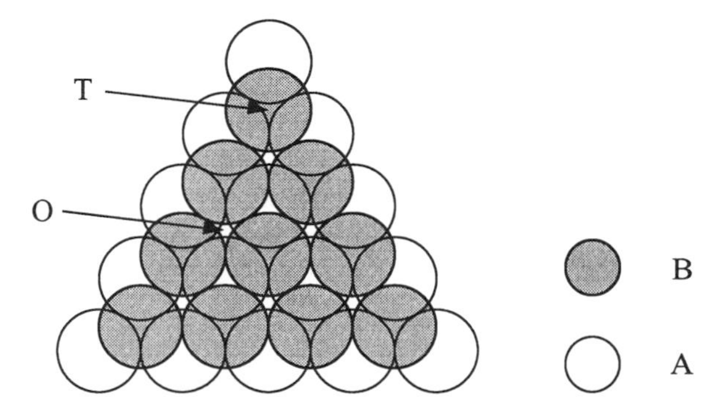

Figure 4.6 Placement of a layer B of close-packed spheres on top of a layer A, generating octahedral (O) and tetrahedral (T) interstices between the layers.

layer), is also generated, and these are called *octahedral interstices* or *O-holes* (Fig. 4.6). A different (smaller) kind of atom could be accommodated in the T-holes between the A and B layers. As we shall see in Section 4.4, the crystal structures of many ionic compounds can be represented in terms of the systematic filling of O- and/or T-holes in a close-packed array of ions X by smaller ions M (usually cations). The X sublattice may be somewhat expanded from close packed to accommodate M, but the point is that the essential geometric features of closest packing are frequently present in ionic crystal structures.

When we add a third layer C of atoms on top of the two layers, we find there are two close packed possibilities; this can be tested with small disks on Fig. 4.6. Each atom of layer C must rest on three of layer B. One possibility is to place atoms of layer C directly above atoms of layer A. Thus, we create a new layer just like A, and further layers are added to

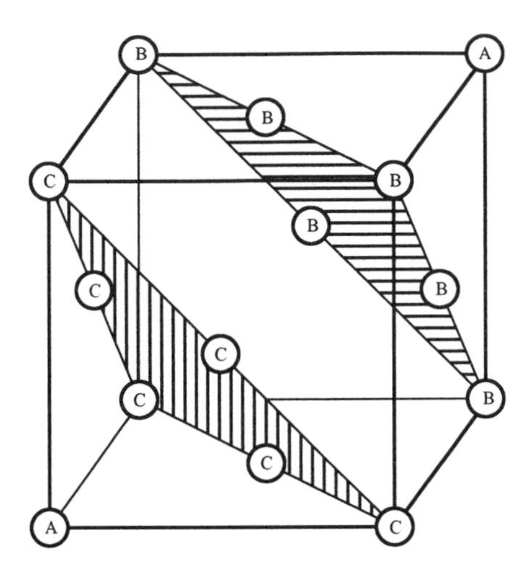

Figure 4.7 Cubic close-packed layers A, B, and C within a face-centered cubic unit cell.

give a sequence ABABAB.... This is known as *hexagonal close packing* (hcp). If, however, we place the C atoms in positions directly above the octahedral holes that exist between A and B (such as the one marked "O" in Fig. 4.6), we have a new arrangement. The fourth layer would go above the A atoms (if the same packing sequence is adhered to), so the layer order would be ABCABC.... This is called *cubic close packing* (ccp) because, as Fig. 4.7 shows, it generates a unit cell that has cubic symmetry. More specifically, it is a *face-centered cubic* (fcc) unit cell, so called because there is an atom at the center of each face of a cube *in addition to* one at every corner. In contrast, a *simple cubic* unit cell is one in which only the corner atoms are present. Simple cubic unit cells, however, are rarely encountered in practice.

4.3.1 Structures of Metals

The concept of close packing is particularly useful in describing the crystal structures of metals, most of which fall into one of three classes: hexagonal close packed, cubic close packed (i.e., fcc), and body-centered cubic (bcc). The bcc unit cell is shown in Fig. 4.8; its structure is not close packed. The stablest structures of metals under ambient conditions are summarized in Table 4.1. Notable omissions from Table 4.1, such as aluminum, tin, and manganese, reflect structures that are not so conveniently classified. The artificially produced radioactive element americium is interesting in that the close-packed sequence is ABAC..., while one form of polonium has

TABLE 4.1
Structures of Some Metallic Elements at Ambient Temperature and Pressure

Body-centered cubic	*Cubic close packed*	*Hexagonal close packed*
alkali metals	Cu, Ag, Au	Be, Mg
V, Nb, Ta	Rh, Ir	Zn, Cd, In
Cr, Mo, W	Ni, Pd, Pt	Tc, Re, Ru, Os
U, Np	Pb, Th	most lanthanides

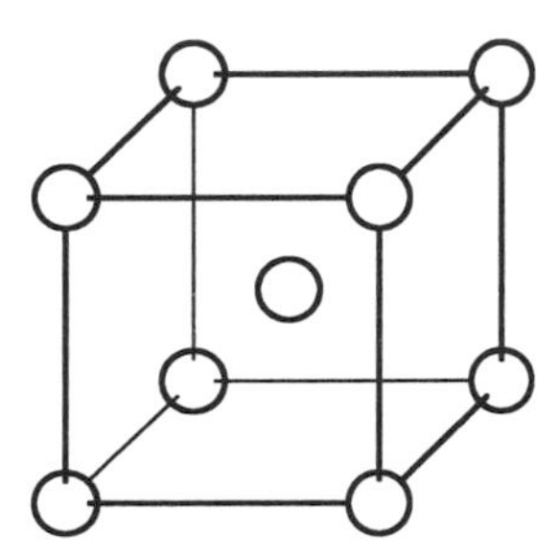

Figure 4.8 The body-centered cubic (bcc) unit cell.

the rare simple cubic structure. Cobalt has an essentially close-packed structure, but the layer sequence is not regular.

Metals are often *polymorphic*, that is, they may exhibit alternative structures, particularly at other temperatures and pressures. An important example is iron, which has a body-centered structure (α-iron) at room temperature, but, on heating, goes over to a face-centered cubic form (γ-iron) at 906 °C and returns to another body-centered cubic structure (δ-iron) above 1401 °C. The usual expectation, though, is that close-packed structures will be favored by low temperatures and body-centered by high, since the increased lattice vibrations at high temperatures will work against close packing. The polymorphism of tin can be exasperating in cold climates. The familiar "white tin" (β-Sn) has a dense, complicated structure that slowly goes over to "gray tin" (α-Sn) with the more open diamond structure (Fig. 3.1) on prolonged exposure to temperatures significantly below the transition temperature of 14.2 °C. Tin represents one of the relatively rare cases in which the low temperature form is the *less* dense (by 21%). The effect is to make tin sheets that have been exposed to the cold for extended periods appear to have contracted a terrible skin disease.

Alloys are metals made by combining two or more elements. Two structural types may be identified: ***substitutional alloys***, in which atoms of one kind of metal partially replace those of another in the normal lattice sites, and *interstitial compounds*, in which the intruding atoms (usually H, B, C, or N) reside in some, but usually not all, of the interstices (O- or T-holes) in the lattice of a metal. For example, copper and gold form a range of substitutional alloys with the ccp structure, as expected from Table 4.1. The Cu and Au positions in the 1:1 alloy CuAu can be ordered (with alternating layers of Cu and Au parallel to the floor of the unit cell of Fig. 4.7) or disordered (with random placement of Au and Cu throughout the lattice). In the alloy Cu_3Au, when ordered, the Cu atoms occupy all the face-center sites and the Au all the corners of the cubic unit cell, but disordered structures are again possible. Interstitial compounds are not alloys in the usual sense of a combination of metals, but they do retain a metallic appearance and electrical conductivity. They are considered at length in Chapter 5.

4.3.2 Metallic Glasses

Metals are not necessarily crystalline, and there is much interest currently in the preparation and properties of *metallic glasses*.[8] Glasses in general are solids that lack long-range structure at the atomic level; often, they are described as extremely viscous, supercooled liquids (i.e., liquids far below their normal freezing temperatures), but there is no evidence that they have any more tendency to flow than do crystalline solids. They are, however, thermodynamically unstable with respect to slow crystallization (*devitrification*; see Section 7.5). Metallic glasses are usually made by melt-spinning certain alloys at forced cooling rates high enough (10^5 to 10^6 K s^{-1}) to avert crystallization, but they can also be produced by vapor deposition or chemical precipitation. In the absence of any opportunity for crystal slip mechanisms such as characterize crystalline solids, metallic glasses often have very high tensile strength and wear resistance. Their properties are typically metallic, but, as might be inferred from Section 4.2, their electrical resistivities can be relatively low. Examples of compositions of glass-forming alloys are $Fe_{100-x}B_x$ (x = 12–25), $Al_{100-x}La_x$ (x = 10, 50–80), $U_{100-x}Co_x$ (x = 24–40), and $Be_{40}Zr_{10}Ti_{50}$.

Paradoxically, one of the most technologically promising properties of metallic glasses is the partial recrystallization under controlled annealing to form metals with extremely fine, uniform, microcrystalline structures.[8] Such devitrification is extensively practiced with oxide glasses to produce tough microcrystalline ceramics (Section 7.5), and we now have the prospect of producing devitrified metal glasses with extraordinary mechanical properties. For example, samples of devitrified $Al_{88}Ni_9Ce_2Fe$ have been prepared with tensile strengths as high as 1560 MPa without brittleness, and devitrified $Fe_{14}Nd_2B$ has found application as a very hard magnetic material.

4.4 Binary Ionic Solids: Common Structural Types

Among binary (i.e., two-element) ionic compounds, six simple types of unit cell structures are commonly encountered, although many more exist:

1:1 (MX) types: halite (NaCl), cesium chloride (CsCl), zinc blende (ZnS), and wurtzite (also ZnS);

2:1 (MX_2) types: fluorite (CaF_2) and rutile (TiO_2).

Sodium chloride (halite) structure (Fig. 4.9). Sodium and chloride ions alternate in three directions at right angles. As Fig. 4.9 is drawn, the shaded ions are Na^+ and obviously form a *face-centered cubic* (fcc) array. However, the unshaded ions (nominally Cl^-) also form an fcc array, as can be seen by stacking another unit cell on top of the one shown. Thus, the Na^+ and Cl^- sublattices are interchangeable, and we could just as well have specified the shaded ion to be chloride. The structure could be described as interpenetrating fcc arrays of anions and cations of equal charge. The *coordination number* (number of nearest neighbors) is six for both Na^+ and Cl^-. The centers of these six nearest neighbors trace out a *regular octahedron* (Fig. 4.10), and we can therefore also describe the structure as consisting of an fcc array of Cl^- (or Na^+) with a Na^+ (or Cl^-) ion in all octahedral "holes" (interstices) in that array. These octahedral holes are entirely analogous to the O-holes described in Section 4.3 (and illustrated in Fig. 4.6).

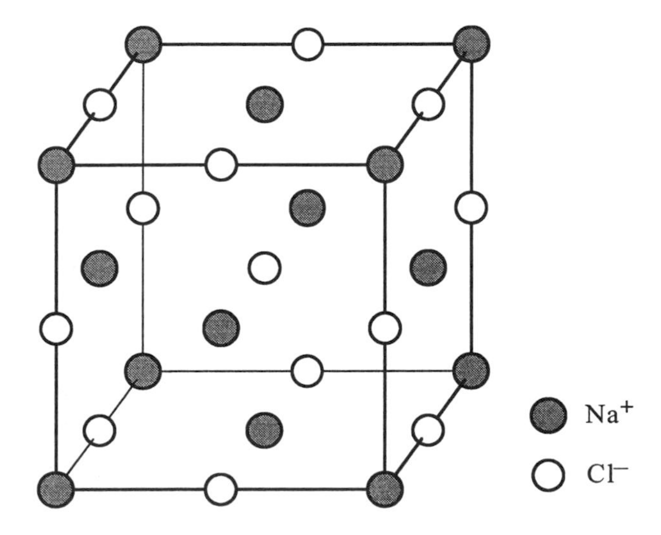

Figure 4.9 Sodium chloride unit cell.

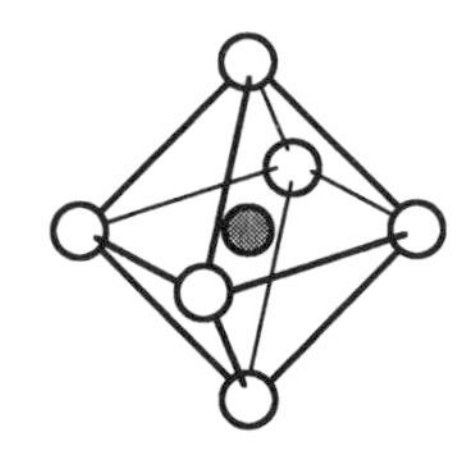

Figure 4.10 Octahedral coordination.

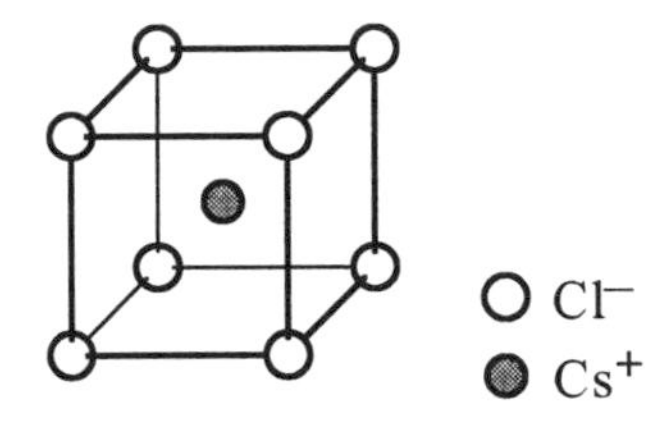

Figure 4.11 Cesium chloride unit cell.

The contents of the unit cell shown in Fig. 4.9 are

$$\left.\begin{array}{l} 8 \times \text{one-eighth } Na^+ \text{ at each corner} \\ 6 \times \text{one-half } Na^+ \text{ on each face} \end{array}\right\} \text{4 whole } Na^+$$

$$\left.\begin{array}{l} 12 \times \text{one-quarter } Cl^- \text{ on each edge} \\ 1 \times \text{a whole } Cl^- \text{ at the body center of the cube.} \end{array}\right\} \text{4 whole } Cl^-$$

Each unit has a total of four NaCl. This is because, with the exception of the Cl^- at the body center, parts of each ion lie in adjacent unit cells.

The NaCl structure is very common; examples include the halides of Li, Na, K, and Rb; AgCl and AgBr; NiO; MgO; CaO; and PbS.

Cesium chloride (CsCl) structure (Fig. 4.11). The CsCl structure can be described as interpenetrating *simple cubic* arrays of Cs^+ and Cl^-. Again, the Cs^+ and Cl^- positions are fully interchangeable. The structure is sometimes wrongly called *body-centered cubic* (bcc). The terminology is appropriate only when the shaded and unshaded atoms of Fig. 4.11 are identical, as in Fig. 4.8. In any case, the coordination number is eight for any atom. The unit cell of CsCl contains one net CsCl unit.

Zinc blende structure (Fig. 4.12). Zinc blende or *sphalerite* (ZnS) has a cubic structure in which we again have interchangeable, interpenetrating fcc arrays of Zn^{2+} and S^{2-}. If, as in Fig. 4.12, we call the shaded ions zinc and divide the unit cell into eight subcubes, we see that the zinc ions occupy the body centers of every *alternate* subcube. Furthermore, the coordination number of each ion is now four, and the nuclei of the four nearest neighbors

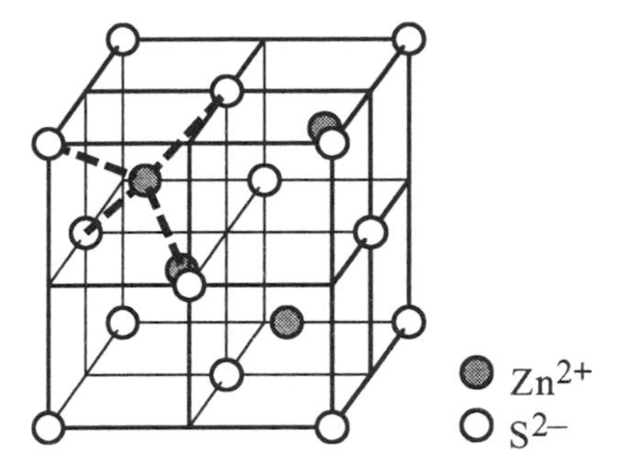

Figure 4.12 Zinc blende unit cell.

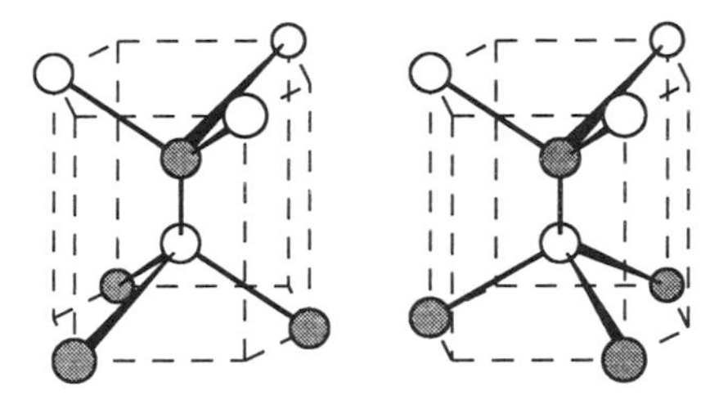

Figure 4.13 Illustration of the difference between zinc blende (left) and wurtzite (right) structures.

trace out a *regular tetrahedron;* we say that the Zn^{2+} (or the S^{2-}) ions are *tetrahedrally coordinated.* (Compare this with tetrahedral bonding in organic compounds such as CH_4. Indeed, if we make all the shaded and unshaded atoms in Fig. 4.12 the same, we have the diamond structure, Fig. 3.1.) Thus, the Zn^{2+} ions occupy one-half of the *tetrahedral holes* in an fcc array of S^{2-} ions, and vice versa. These tetrahedral holes are the same as those described in terms of close packing in Section 4.3. Note that there are *two* tetrahedral (T) holes and *one* octahedral (O) hole per fcc sublattice atom; in zinc blende, the O-holes are all empty.

Wurtzite structure. Zinc sulfide can also crystallize in a hexagonal form called *wurtzite* that is formed slightly less exothermically than the cubic zinc blende (sphalerite) modification ($\Delta H_f = -192.6$ and -206.0 kJ mol^{-1}, respectively) and hence is a high temperature *polymorph* of ZnS. The relationship between the two structures is best described in terms of close packing (Section 4.3): in zinc blende, the anions (or cations) form a cubic close-packed array, whereas in wurtzite they form hexagonal close-packed arrays. This relationship is illustrated in Fig. 4.13; note, however, that this does not represent the actual unit cell of either form.

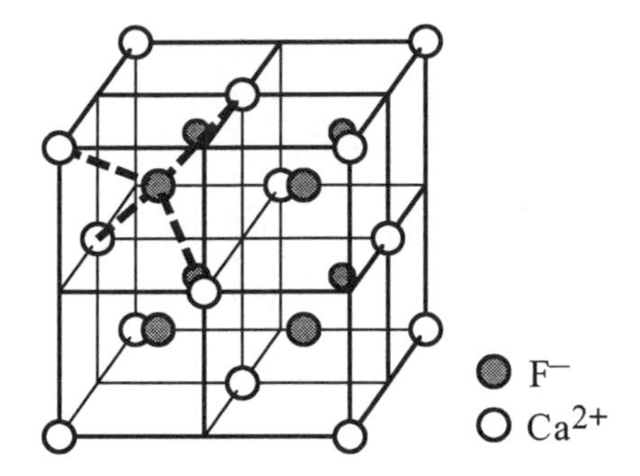

Figure 4.14 Fluorite unit cell.

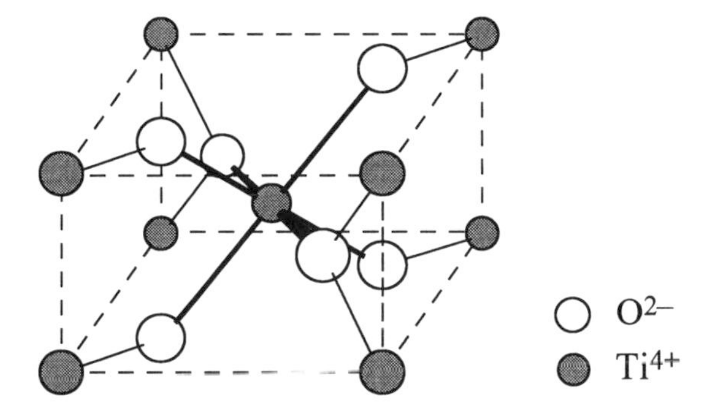

Figure 4.15 Rutile unit cell (tetragonal).

Fluorite (CaF_2) structure (Fig. 4.14). Comparison of the fluorite structure (Fig. 4.14) with Fig. 4.12 shows that fluorite can be described as an fcc array of Ca^{2+} with F^- ions in *all* the tetrahedral holes (forming a simple cubic sublattice of fluorides). In this case, the Ca^{2+} and F^- sites are *not* interchangeable. This is to be expected, since we have twice as many F^- as Ca^{2+}; as noted earlier, there are indeed twice as many T-holes as lattice atoms. The coordination numbers of Ca^{2+} and F^- are eight and four, respectively. Other solids with this structure include the nuclear fuel UO_2.

If we make the shaded spheres in Fig. 4.14 the *cations*, and the unshaded ones the *anions*, we have the *antifluorite* structure, which is typified by lithium oxide (an fcc array of O^{2-} with Li^+ in every T-hole).

Rutile structure (Fig. 4.15). Titanium dioxide occurs naturally as anatase, brookite, and rutile, all of which contain octahedral ${TiO_6}^{8-}$ units. The coordination number of the central Ti^{4+} is very obviously six, and a little thought confirms that the same is true of the Ti^{4+} ions at the corners. That the coordination number of the O^{2-} ions is three is seen from the nearest oxygen at the right. The extended structure can be regarded as chains of edge-sharing TiO_6 units. The anatase structure is like NaCl with every other cation missing and the anions accordingly displaced somewhat. The heats of formation of anatase, brookite, and rutile are −939.7, −941.8, and −944.7 kJ mol^{-1}, respectively, and ΔG_f° values follow this trend, so that rutile is the stable low-temperature form. The rutile structure is shared by MnO_2, SnO_2, and most divalent transition metal fluorides.

4.5 Radius Ratio Rules

Consider a set of nearest neighbor anions surrounding a cation (which will almost inevitably be smaller) in a crystal. The maximum electrostatic attractions will result when we have as many anions as possible surrounding the cation and in contact with it but not *quite* touching each other (since like charges repel one another). The sketch on the left-hand side of Fig. 4.16 shows a case in which anion–anion repulsions would be maximal and anion–cation attractions low. This system will seek to reduce its coordination number. Conversely, in the central sketch in Fig. 4.16, more anions could be accommodated to advantage, so the coordination number tends to increase. The limiting case, with all seven ions just in contact, is represented on the right-hand side of Fig. 4.16. (A further anion above and one more below the plane of the paper have been omitted for clarity.) This limiting case occurs when the sum $(r_+ + r_-)$ of the radii of the cation and anion is equal to $r_-\sqrt{2}$ (from Pythagoras' theorem), that is, when the radius ratio r_+/r_- equals $(\sqrt{2} - 1) = 0.414$. For a radius ratio less than this, we expect the coordination number to be reduced to four. Proceeding in the same way, we can calculate the ion radius ratios that set the limits between eight and six coordination, and (less importantly) four and three coordination, and so make it possible to anticipate from tables of ionic radii (e.g., Appendix F)[9,10] which of the basic structures described in Section 4.4 a given binary ionic compound is likely to adopt (Table 4.2).

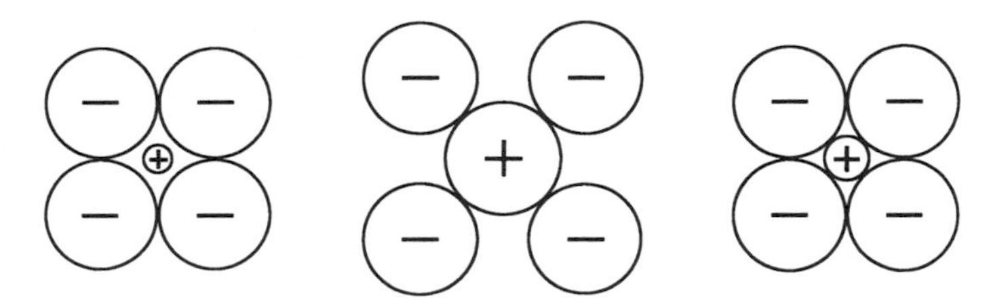

Figure 4.16 Effects of relative sizes of anions and cations in octahedral (six) coordination. Two additional anions, positioned above and below the plane, have been omitted for clarity.

TABLE 4.2
Predicted Dependence of Structure Type on Cation/Anion Radius Ratio for Binary Ionic Compounds

Coordination	r_+/r_-	*Structure type*	
number of M^{m+}	*(minimum)*	*MX*	MX_2
8	0.73	CsCl	CaF_2
6	0.41	NaCl	TiO_2
4	0.22	ZnS	
3	0.15		

For example, the ionic radius of Mg^{2+} is usually quoted as 72 pm, and for magnesium oxide, sulfide, and telluride we have

X^{2-}:	O^{2-}	Te^{2-}	S^{2-}
r_-:	140 pm	221 pm	184 pm
r_+/r_-:	0.52	0.33	0.39
Predicted structure:	NaCl	ZnS	ZnS
Observed structure:	NaCl	ZnS	NaCl.

This example was chosen to show that the radius ratio structure predictions work quite well *except* near the limiting radius ratio values. Here, r_+/r_- is borderline for MgS, and we predicted the wrong structure. This is not surprising, since the concept of fixed ionic radii is not well founded.

First of all, we cannot measure r_+ or r_- directly. As explained in Section 4.1, X-ray diffraction measurements give us internuclear spacings corresponding to $(r_+ + r_-)$ and some noncontact separations, rather than r_+ or r_- as such. If, however, we can guess any one r_+ or r_- on the basis of some theory or other, then we can estimate all the others.

Further, ions are *not* hard, billiard ball like spheres. Since the wave functions that describe the electronic distribution in an atom or ion do not suddenly drop to zero amplitude at some particular radius, we must consider the surfaces of our supposedly spherical ions to be somewhat "fuzzy."

A more subtle complication is that the apparent radius of an ion increases (typically by some 6 pm for each increment) whenever the coordination number increases. Shannon[10] has compiled a comprehensive set of ionic radii that take this into account. Selected Shannon-type ionic radii are given in Appendix F; these are based on a radius for O^{2-} of 140 pm for six coordination, which is close to the traditionally accepted value, whereas Shannon takes the reference value as 126 pm on the grounds that it gives more realistic ionic sizes. For most purposes, this distinction does not matter; the two scales simply differ consistently by 14 pm.

Finally, changes in coordination number can be induced by high pressures. Thus, in the Earth's mantle, four coordination of silicon by oxygens in enstatite ($Mg_{0.9}Fe_{0.1}SiO_3$) gives way to six coordination (a change from pyroxene-like to perovskite-type structures; see Sections 4.6 and 7.1) when the overburden of rock exceeds 670 km, at which depth the pressure is 17.5 GPa. Sodium chloride itself goes over to the CsCl structure at sufficiently high pressures.

4.6 Ionic Solids and Close Packing

Many common ionic crystal structures can be conveniently described as a ccp or hcp array of one kind of ion with one or more other kinds occupying some or all of the T- or O-holes, usually in a regular way. Sodium chloride can be regarded as a ccp array of Na^+ with Cl^- ions in all the O-holes (or vice versa), and fluorite as a ccp array of Ca^{2+} with F^- ions in all the T-holes (Section 4.4). This device comes into its own with more complicated structures for which unit cell structures may be hard to represent on paper. For example, the α-Al_2O_3 (corundum) structure is an hcp array of O^{2-} with Al^{3+} ions occupying two-thirds of the O-holes; in other words, as we move from O-hole to O-hole in a given direction, every third hole is vacant. The familiar red oxide of iron, α-Fe_2O_3 (hematite), has a similar structure. Table 4.3 summarizes standard examples. The close-packing approach affords a simple way to describe some important *ternary* (three-element) structures. Thus, the titanium ore ilmenite, $FeTiO_3$, may be regarded as an hcp array of O^{2-} with Fe^{2+} in one-third of the O-holes, Ti^{4+} in a second third, and the remaining O-holes vacant. (Ilmenite can therefore be described as hematite in which half the Fe^{3+} are replaced by Ti^{4+} and the rest by Fe^{2+}.)

4.6.1 Perovskites

The ilmenite structure is common among ternary oxides of the general form ABO_3 when the ions A and B are of roughly similar radii. However, if one ion B much smaller than A, and A is not too much smaller than O^{2-}, the *perovskite structure* (Fig. 4.17) is adopted.[11] This comprises a ccp array of A^{n+} and O^{2-}, while $B^{(6-n)+}$ occupies one-fourth of the O-holes and the rest remain vacant. The perovskite structure can be expected to be associated with elements B that tend to high oxidation states, such as Ti^{IV} in the mineral perovskite itself ($CaTiO_3$) or niobium(V) in $NaNbO_3$. It is also common among compounds of the type ABX_3, where X is F or (less commonly) Cl, Br, or I (e.g., $KNiF_3$). We noted in Section 4.5 that the normal pyroxene structure of enstatite ($MgSiO_3$) gives way to a perovskite

TABLE 4.3
Binary Crystal Structures in Terms of Close Packing

Structure type	*Close-packed array*	*Filling of interstices*
zinc blende	ccp Zn^{2+}	S^{2-} in half of the T-holes
wurtzite	hcp Zn^{2+}	S^{2-} in half of the T-holes
fluorite	ccp Ca^{2+}	F^- in every T-hole
halite	ccp Cl^-	Na^+ in every O-hole
nickel arsenide	hcp As^{3-}	Ni^{3+} in every O-hole
rutile	hcp O^{2-}	Ti^{4+} in half of the O-holes
anatase	ccp[a] O^{2-}	Ti^{4+} in half of the O-holes
corundum	hcp O^{2-}	Al^{3+} in two-thirds of the O-holes

[a]Distorted.

arrangement at around 17.5 GPa pressure.

Several perovskite-type metal oxides, such as $BaTiO_3$ and $Pb(Zr,Ti)O_3$ (lead zirconate titanate or PZT), have *ferroelectric* properties, which in practical terms means that a high external electrical field can cause *poling* (i.e., induction of a permanent dipole moment) in polycrystalline ceramics made from them. Field gradients of several thousand volts per centimeter are required to produce this poling, but fabrication of such materials as thin films permits recording of low-voltage electrical signals; these films have important applications in the electronics industry.[3, 11–13] Finally, the non-stoichiometric (Section 5.4) compounds related to $YBa_2Cu_3O_{7-x}$, which Müller and Bednorz showed in 1986 to be superconducting at unprecedentedly high temperatures for certain ranges of x, have a complicated structure that is nevertheless derived from the perovskite type.[13–17] We revisit these topics in Chapter 19.

4.6.2 Spinels

A large group of ternary oxides AB_2O_4 have structures related to that of *spinel*, $MgAl_2O_4$: a ccp array of O^{2-} with A in one-eighth of the T-holes and B in one-half of the O-holes. Recalling that there are two T-holes and one O-hole per close-packed atom (O^{2-}), we see that this does indeed correspond to AB_2O_4. An example of an economic mineral that has this *normal* spinel structure is the chromium ore *chromite*, $FeCr_2O_4$, where A is Fe^{2+} and B is Cr^{3+}. In Nature, however, minerals rarely occur as pure phases because ions of similar size can usually substitute for one another. In particular, Mg^{2+} frequently replaces Fe^{2+}, and Fe^{3+} or Al^{3+} replaces

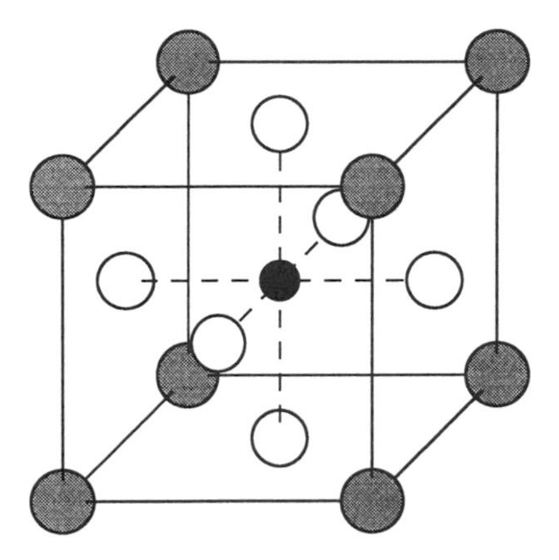

Figure 4.17 Perovskite structure. Open spheres, anion; shaded spheres, larger cation; filled sphere, smaller cation.

Cr^{3+}. Consequently, natural chromite is more realistically formulated as $(Mg,\mathbf{Fe^{II}})(Al,Fe^{III},\mathbf{Cr})_2O_4$; $FeCr_2O_4$ is its idealized or *end-member* composition. The substituent ions replace Fe^{2+} and Cr^{3+} in a random way, much as solute molecules displace solvent molecules in a liquid solution. Consequently, chromite samples could be described as *solid solutions* of $MgAl_2O_4$ and $MgFe^{III}_2O_4$ in $Fe^{II}Cr_2O_4$.

A closely related group of AB_2O_4 oxides has the *inverse* spinel structure. Here, again, there is a ccp array of O^{2-}, but the B atoms are equally divided between T- and O-sites, and all of the A ions appear in O-, not T-, holes. Thus, we have $B(AB)O_4$ with B in one-eighth of the T-holes, A in one-quarter of the O-holes, and B in one-quarter of the O-holes.

An important example of an oxide with the inverse spinel structure is *magnetite*, a black, ferromagnetic oxide of iron (red hematite is effectively nonmagnetic) containing both iron(II) and iron(III): Fe_3O_4 or $Fe^{II}Fe^{III}_2O_4$, better written as $Fe^{III}(Fe^{II}Fe^{III})O_4$. Magnetite is the usual product of corrosion of iron at elevated temperatures with a limited oxygen supply and is also a valuable though uncommon mineral of iron, as it has a high percentage Fe content and is readily located and concentrated by virtue of its ferromagnetism. The brown ferromagnetic oxide used in Type I recording tape is *maghemite*, which contains no Fe^{2+} but has some H^+ as well as Fe^{3+} randomly distributed through the O- and T-holes of a ccp array of O^{2-}. It is often formulated as "γ-Fe_2O_3"; in reality, however, its H^+ content is variable, and the composition ranges up to HFe_5O_8, sometimes written as $5Fe_2O_3{\cdot}H_2O$. Since iron(II) oxide (wüstite, nominally FeO) also has a ccp O^{2-} array with Fe^{2+} in the O-holes (NaCl structure), the oxidation sequence at elevated temperatures

$$\text{iron} \rightarrow \text{wüstite} \rightarrow \text{magnetite } (\rightarrow \text{maghemite}) \rightarrow \text{hematite}$$

involves movement of Fe^{2+} and Fe^{3+} ions through a common cubic oxide lattice structure. The final conversion to hematite, however, requires a changeover from ccp to hcp O^{2-} (γ to α structure type).

4.6.3 Layer Structures

As a final example of the use of the close packing idea, we note the *layer structures* typified by cadmium chloride and iodide. The iodide ions in CdI_2 form an hcp array (ABABAB...), and the Cd^{2+} ions occupy *alternate* layers of O-holes completely, thus giving a 1:2 stoichiometry but leaving every other layer of O-holes empty. The structure may be regarded as a stack of single-decker sandwiches with iodide "bread" (layers A and B) and a cadmium filling, but nothing between sandwiches. (The food analogy should not be pressed too far, as Cd is very toxic!) The crystal will obviously have planes of easy cleavage parallel to the layers. This structure is exhibited by many 1:2 ionic solids, including $M(OH)_2$ (if the hydrogens are ignored), where M may be Ca, Mg, Fe, Ni, or Cd. The $CdCl_2$ structure is also of this sandwich type, except that the anions form a ccp rather than an hcp array, with Cd^{2+} in alternating layers of O-holes.

Molybdenum sulfide (MoS_2) has an unusual layer structure in which sulfide atoms in a given upper layer sandwiching the Mo atoms are located directly above the S atoms in the lower layer, but alternate MoS_2 layers are offset as in hcp. Thus, the structure is A(Mo)A, B(Mo)B, A(Mo)A, B(Mo)B.... There is little cohesion between successive MoS_2 sheets, resulting in a greasy consistency. Molybdenum sulfide is therefore widely used as a solid lubricant, particularly for high-temperature applications (cf. graphite).

4.7 Energetics of Ionic Compounds

4.7.1 Lattice Energies

To calculate the lattice energy U of an ionic crystal, consider first the potential energy E of a particular ion, say, a sodium ion in NaCl (Fig. 4.18), in terms of the coulombic attractions (of Cl^-) and repulsions (of other Na^+) of the other ions in the lattice. In NaCl, the nearest neighbors of Na^+ are 6 Cl^- at a distance r (i.e., $r\sqrt{1}$), 12 Na^+ at a distance $r\sqrt{2}$, 8 Cl^- at a distance $r\sqrt{3}$, 6 Na^+ at a distance $2r$ (i.e., $r\sqrt{4}$), and so on. (For reasons of clarity, Fig. 4.18 shows only a few of these ions.) Then, for ions of charge $z+$ and $z-$ in a lattice with the NaCl structure ($z = 1$ for NaCl itself, 2 for

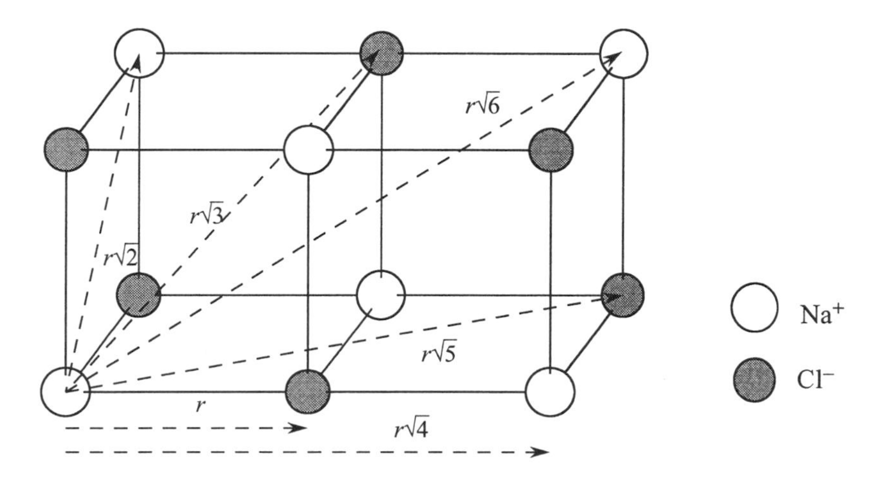

Figure 4.18 Internuclear distances in a crystal of NaCl.

MgO, etc.) where the unit cell side is $2r$, we have

$$E = -\frac{z^2e^2}{4\pi\varepsilon_0 r}\left(\frac{6}{\sqrt{1}} - \frac{12}{\sqrt{2}} + \frac{8}{\sqrt{3}} - \frac{6}{\sqrt{4}} + \frac{24}{\sqrt{5}} - \cdots\right). \tag{4.2}$$

Here (as usual in the SI), the electronic charge e is 1.6022×10^{-19} A s and the permittivity of a vacuum ε_0 is 8.854×10^{-12} A^2 s^4 kg^{-1} m^{-3}. The infinite series in parentheses is convergent, and its sum (1.748) is called the *Madelung constant*, M, for the NaCl structure. Values of M are characteristic of the lattice structure but actually are not greatly different from one structure to another (e.g., M for the CsCl lattice is 1.763).

The net coulombic attractions of the ions for one another are counterbalanced at their equilibrium separation r_0 by the short-range repulsions due to interpenetration of the electronic clouds of ions "in contact." This short-range repulsion falls off sharply with increasing r. Max Born suggested that it could be represented by B/r^n, where B is a constant and n can be estimated from the compressibility of the crystal. As a rule of thumb, for ions having the electron configurations of Ne, Ar, Kr, and Xe, $n \approx 7$, 9, 10, and 12, respectively,

$$E = \frac{1}{4\pi\varepsilon_0}\left(-\frac{Mz^2e^2}{r} + \frac{B}{r^n}\right). \tag{4.3}$$

At the equilibrium separation, $r = r_0$ and $dE/dr = 0$, so that

$$4\pi\varepsilon_0\frac{dE}{dr} = \frac{Mz^2e^2}{{r_0}^2} - \frac{nB}{{r_0}^{n+1}} = 0 \tag{4.4}$$

whence

$$B = \frac{Mz^2e^2{r_0}^{n-1}}{n}, \tag{4.5}$$

and B can be eliminated from Eq. 4.3:

$$E = \frac{1}{4\pi\varepsilon_0}\left(-\frac{Mz^2e^2}{r_0} + \frac{Mz^2e^2}{nr_0}\right). \tag{4.6}$$

The lattice energy U is defined as the energy *released* (U is therefore negative by thermodynamic convention) when a mole of the requisite free gaseous ions comes together from infinite interionic separation to make up the crystal. If N is Avogadro's number (6.0221×10^{23}), we have the *Born–Landé formula:*

$$U = -NE = \frac{NMz^2e^2}{4\pi\varepsilon_0 r_0}\left(1 - \frac{1}{n}\right). \tag{4.7}$$

Equation 4.7 can clearly be improved by taking a more sophisticated approach to ionic interactions, but it gives satisfactory values for U.

Lattice energies can be related to the heats of formation ΔH_f° of ionic solids through the *Born–Haber cycle*, which is the counterpart of the thermochemical cycle for covalent compounds given in Section 2.7.

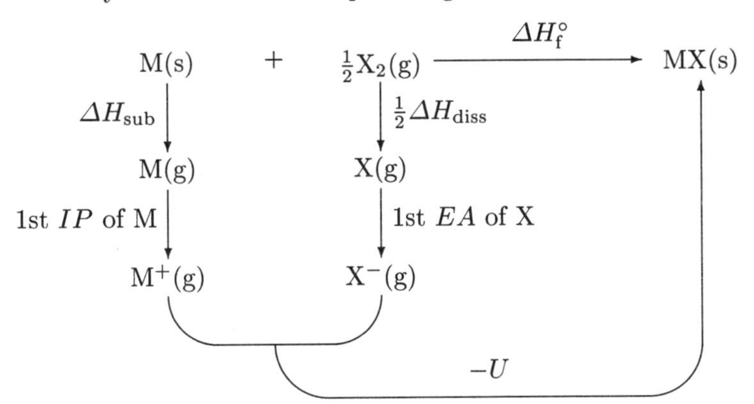

We assume that we have a solid metal M which reacts with a diatomic, gaseous nonmetal X_2 (e.g., Cl_2, F_2, O_2). Similar cycles can be written for solid elements such as sulfur as the nonmetal. In either case, before we can connect U with ΔH_f° we must form *gaseous ions* of M and X. We need not only the relevant ionization potentials (IP) and electron affinities (EA), but also the heats of atomization of solid M and gaseous X_2. These atomization energies are traditionally referred to as heats of sublimation ΔH_{sub} of M(s)* and of dissociation ΔH_{diss} of X_2. For NaCl itself, we have

***Sublimation* is an ancient alchemical term meaning direct passage from solid to vapor.

(in kJ mol^{-1}): $\Delta H_f = -411$; $IP = 496$; $\Delta H_{sub} = 108$; $EA = -349$; and $\Delta H_{diss} = 243$. So, bearing in mind that U represents energy *given out* from the system,

$$U = -(-411) + 496 + (-349) + 108 + \tfrac{1}{2}(243) = 788 \text{ kJ mol}^{-1}. \qquad (4.8)$$

Clearly, U is the biggest number in the cycle and is the main driving force for the formation of ionic compounds. Nevertheless, the other factors can tip the balance one way or another. For example, ΔH_{sub} is particularly large for the transition metals niobium, tantalum, molybdenum, tungsten, and rhenium, with the result that, in their lower oxidation states, they do not form simple ionic compounds such as $ReCl_3$ but rather form compounds that contain *clusters* of bonded metal atoms (in this example, Re_3 clusters are involved, so the formula is better written Re_3Cl_9).

4.7.2 Predicting Stabilities of Ionic Compounds

In principle, we can use the Born–Haber cycle to predict whether a particular ionic compound should be thermodynamically stable, on the basis of calculated values of U, and so proceed to explain all of the chemistry of ionic solids. The relevant quantity is actually the *free energy of formation*, ΔG_f°, and this is calculable if an entropy cycle is set up to complement the Born–Haber enthalpy cycle. However, in practice ΔH_f° dominates the energetics of formation of ionic compounds.

The pitfall in making predictions of this kind is that there may be more stable compounds of the elements involved than the one considered. One can ask, for example, why calcium does not form a stable solid monofluoride, CaF, as well as CaF_2, since the first ionization potential for Ca is only 596 kJ mol^{-1} as against 1748 kJ mol^{-1} for the sum of the first and second IP values. To answer this question, we first make an educated guess of the unknown ionic radius of Ca^+; about 122 pm seems reasonable, by comparison with other monovalent ions (Appendix F), and the radius ratio for CaF is therefore 122/133. Since this is greater than 0.73 (see Table 4.2), the crystal structure should be of the CsCl type. The Madelung constant for the CsCl structure is 1.763, and Eq. 4.7 gives $U = 865$ kJ mol^{-1}. This last value can be plugged into the Born–Haber cycle to give $\Delta H_f^\circ = -332$ kJ mol^{-1} for CaF(s); it would appear that CaF(s) should be comfortably stable, since ΔH_f° for NaCl is only 79 kJ mol^{-1} more negative than this! The heat of formation of CaF_2(s), however, has been measured to be -1220 kJ mol^{-1}. Thus we have

$$\begin{array}{lll} \text{Ca(s)} + \text{F}_2\text{(g)} \rightarrow \text{CaF}_2\text{(s)} & \Delta H = -1220 \text{ kJ mol}^{-1} & (4.9) \\ 2\text{CaF(s)} \rightarrow 2\text{Ca(s)} + \text{F}_2\text{(g)} & \Delta H = 2(+332) \text{ kJ mol}^{-1} & (4.10) \\ \hline \text{Add: } 2\text{CaF(s)} \rightarrow \text{Ca(s)} + \text{CaF}_2\text{(s)} & \Delta H = -556 \text{ kJ mol}^{-1}. & (4.11) \end{array}$$

The *disproportionation* (self-oxidation/reduction) of CaF(s) is therefore strongly favored thermodynamically, and so any attempt to make solid CaF from the correct amounts of calcium and fluorine will simply give CaF_2 and unreacted calcium. In this particular example, the conclusion is the same one reached by hand-waving arguments about the special stability of the argon like, filled-quantum-shell electronic configuration of Ca^{2+}, but we see now that many other factors are involved; indeed, in the transition metals, lanthanides, and actinides the "filled-shell" approach is of little use. The lanthanide elements, for example, are *all* preferentially trivalent in their ionic compounds. This is because the trends in (first + second + third) IP, ΔH_{sub}, and U are influenced in a compensatory manner by the progressive shrinking (*lanthanide contraction*, Section 2.6) of the metal atoms as we go from lanthanum to lutetium. Similarly, the most stable oxidation states of the metals of the first transition period are scandium(III), titanium(IV), vanadium(V), chromium(III), manganese(II), iron(II) or iron(III), cobalt(II), nickel(II), copper(II), and zinc(II). Only the first three and the last one would be predicted from the full/empty shell argument.

Exercises

4.1 Count the effective numbers of anions and cations contained within the zinc blende, fluorite, and rutile unit cells.

4.2 With the aid of the data of Appendix F, predict the crystal structure of magnesium fluoride. (The observed structure is of the rutile type.)

4.3 The structure of the sodium selenide crystal may be described as a cubic close-packed array of Se^{2-} ions, with an Na^+ ion in every tetrahedral hole.

(*a*) Sketch the unit cell of this compound.

(*b*) Name the structure type.

4.4 Estimate the lattice energy of sodium chloride, using the appropriate data from Appendix F and assuming that the Born exponent is 9. Compare your result with the value given in Eq. 4.8.

4.5 Set up a Born–Haber cycle for the formation of hematite (α-Fe_2O_3) from the elements.

4.6 Calculate the heat of formation of fluorite, given the following information (units are kJ mol^{-1}):

Electron affinity of fluorine	=	−334.4
First ionization potential of calcium	=	596.1
Second ionization potential of calcium	=	1151.7
Lattice energy of fluorite	=	2635
Heat of sublimation of calcium	=	178.2
Bond energy of fluorine	=	158.0.

[*Answer:* see Section 4.7.2.]

4.7 Calculate the lattice energy of AlF_3 from the data of Appendix C. The heat of formation of F^-(g) (for example) incorporates $\frac{1}{2}\Delta H_{\text{diss}}$ for F_2 and the first electron affinity of F(g), so your task is simplified. [*Answer:* 6221 kJ mol^{-1}.]

4.8 Use the ionic radii r of Appendix F to illustrate the following generalities:

(*a*) r increases as we descend a group in the periodic table;

(*b*) r decreases as the ionic charge becomes more positive, for cations and anions with the same electronic configuration;

(*c*) for a given element, r decreases with increasing oxidation state;

(*d*) for a given cation, the ionic radius increases with increasing coordination number.

References

1. A. F. Wells, "Structural Inorganic Chemistry," 5th Ed. Oxford Univ. Press, London, 1984.

2. D. McKie and C. McKie, "Crystalline Solids." Nelson, London, 1974.

3. L. Smart and E. Moore, "Solid State Chemistry: An Introduction." Chapman & Hall, London, 1992.

4. M. F. C. Ladd and R. A. Palmer, "Structure Determination by X-Ray Crystallography," 3rd Ed. Plenum, New York, 1993.

5. G. H. Stout and L. H. Jensen, "X-Ray Structure Determination: A Practical Guide," 2nd Ed. Wiley, New York, 1989.

6. J. P. Glusker and K. N. Trueblood, "Crystal Structure Analysis: A Primer," 2nd. Ed. Oxford Univ. Press, New York, 1985.

7. "The Mineral Powder Diffraction File." Joint Committee for Powder Diffraction Standards (JCPDS), International Center for Diffraction Data, Swarthmore, Pennsylvania (updated continuously).

8. A. L. Greer, Metallic glasses. *Science* **267**, 1947–1953 (1995).

9. R. C. Weast and M. J. Astle (eds.), "CRC Handbook of Chemistry and Physics." CRC Press, Boca Raton, Florida, 1996 (revised annually).

10. R. D. Shannon, Revised effective ionic radii and systematic studies of interatomic distances in halides and chalcogenides. *Acta Crystallogr.* **A32**, 751–767 (1976).

11. M. Sayer and K. Sreenivas, Ceramic thin films: fabrication and applications. *Science* **247**, 1056–1060 (1990).

12. A. Reller and T. Williams, Perovskites—chemical chameleons. *Chem. Br.* **25**, 1227–1230 (1989).

13. A. Wold and K. Dwight, "Solid State Chemistry." Chapman & Hall, New York, 1993.

14. K. A. Müller and J. G. Bednorz, The discovery of a class of high-temperature superconductors. *Science* **237**, 1133–1139 (1987).

15. A. S. Sleight, Chemistry of high-temperature superconductors. *Science* **242**, 1519–1527 (1988).

16. F. J. Adrian and D. O. Cowan, The new superconductors. *Chem. Eng. News*, December 21, 24–41 (1992).

17. P. P. Edwards and C. N. R. Rao, A new era for chemical superconductors. *Chem. Br.* **30**, 722–726 (1994); A. R. Armstrong and W. I. F. David, Pinpointing atomic positions. *Chem. Br.* **30**, 727-729 (1994); C. Greaves, Cementing relations. *Chem. Br.* **30**, 743-745 (1994).

Chapter 5

The Defect Solid State

5.1 Inevitability of Crystal Defects

IN THE PRECEDING chapter, it was tacitly assumed that crystalline solids were perfect, that is, that all of the sites characteristic of a particular structure would be occupied, that the sites that should be vacant in the ideal structure would indeed be unoccupied, and that the atoms or ions making up the lattice were all of the specified kind. In practice, thermodynamics tells us that no crystal can ever be structurally perfect. The equilibrium state of the crystal will be one in which free energy is minimized, and this would seem to favor a perfect crystal lattice since misplaced or foreign ions will lead to a reduced lattice energy and hence to a less negative heat of formation. Disorder at the atomic level, however, will be reflected in a more positive *entropy* term, and the increase in the product $T\Delta S^\circ$ will tend to compensate for the loss of lattice energy due to disorder, with increasing effectiveness as the temperature rises:[1–4]

$$\Delta G^\circ = \Delta H^\circ - T\Delta S^\circ. \tag{5.1}$$

Consequently, although hypothetical perfect crystals can be said to exist at the unattainable absolute zero of temperature, real crystals contain defects that increase in number (thermodynamics) and mobility (kinetics—atoms can move from one site to another on surmounting an Arrhenius-type activation energy barrier) as the temperature rises. Eventually, these defects (e.g., thermal vibrations of atoms around their equilibrium positions, dislocations of atoms or planes of atoms from their ideal sites, creation of vacant sites) become severe enough that long-range atomic ordering breaks down, and the crystal melts. Even in the liquid some transient short-range ordering may persist, especially in ionic melts and in hydrogen-bonded liquids like water. The local ice like structures that form and decay continually in liquid water have been referred to as *flickering clusters.*

Wherever there is a defect in a crystal lattice, interatomic forces will remain unbalanced and the free energy will be less negative than elsewhere in the crystal, although generally the lattice will deform locally to smooth this out. Nevertheless, defect sites (especially of the extended variety) tend to be more chemically reactive than the bulk crystal and tend to be active sites for crystal growth, dissolution, corrosion, and catalytic activity.

5.2 Main Types of Crystal Defects

Point defects. Point defects (Fig. 5.1) are limited to a single point in the lattice, although the lattice will buckle locally so that the influence of point defects may spread quite far. A *Frenkel defect* consists of a misplaced interstitial atom and a lattice vacancy (the site the atom should have occupied). For example, silver bromide, which has the NaCl structure, has substantial numbers of Ag^+ ions in tetrahedral holes in the ccp Br^- array, instead of in the expected octahedral holes. Frenkel defects are especially common in salts containing large, polarizable anions like bromide or iodide.

Defects in which both a cation and sufficient anions to balance the charge (or vice versa) are completely missing from the lattice are called *Schottky defects.* Schottky defects result in a density that is lower than that calculated on the basis of unit cell dimensions, whereas Frenkel defects do not affect this density. Titanium(II) oxide, for example, also has the NaCl structure, but, even when its composition is $TiO_{1.00}$ (which it rarely is; see Section 5.4), about one-sixth of the Ti^{2+} and O^{2-} sites are vacant.

The existence of Schottky or Frenkel defects, or both, within an ionic solid provides a mechanism for significant electrical conductance through ion migration from site to empty site (leaving, of course, a fresh empty site behind).[4] Solid β-AgI provides a classic example of a nonmetallic solid with substantial electrical conductivity at elevated temperatures; at 147 °C, it undergoes a transition to α-AgI in which the silver ion sublattice is disordered and consequently allows for relatively free movement of Ag^+ and

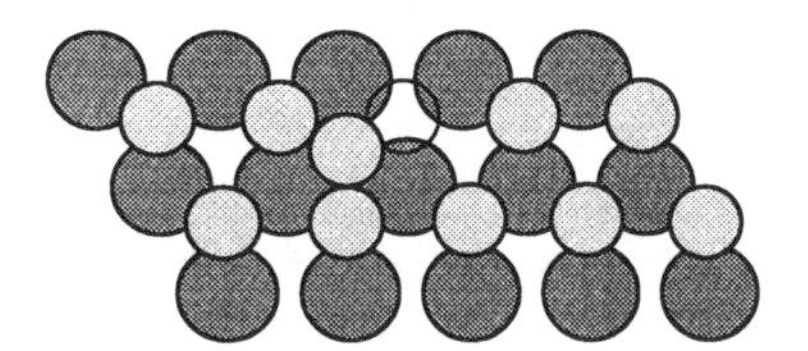

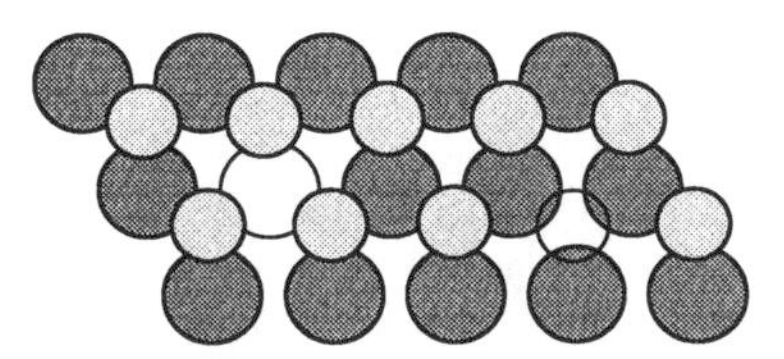

Figure 5.1 Frenkel and Schottky defects. In the Frenkel case (left), a member of the lightly shaded ion sublattice is found in the wrong kind of interstice. In the Schottky defect (right), an ion and a counter-ion (here presumed to have equal but opposite charges) are completely missing. Unshaded circles represent the vacant sites.

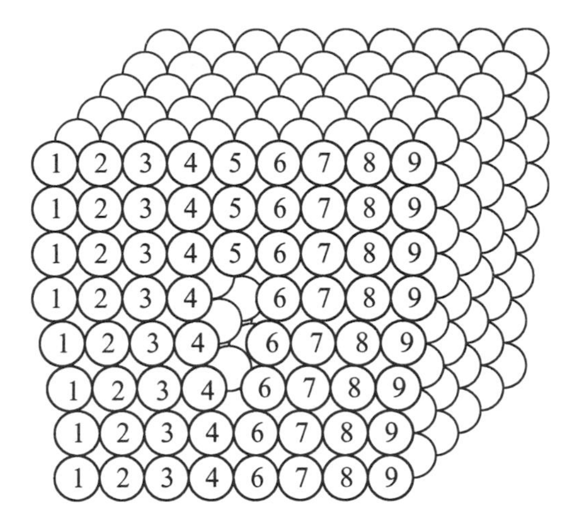

Figure 5.2 Edge dislocation. Layer 5 is incomplete (all atoms are the same).

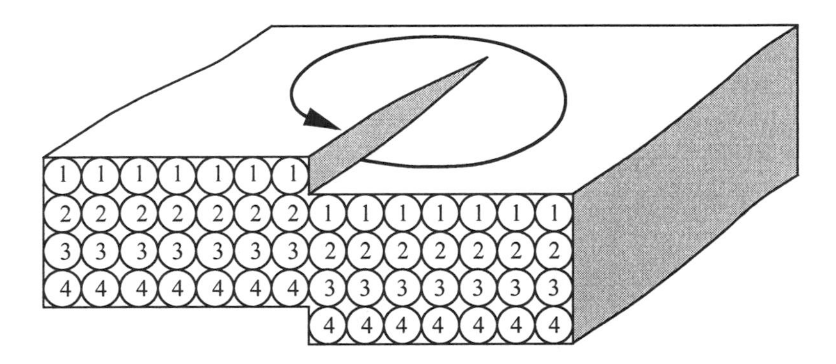

Figure 5.3 Screw dislocation. All the atoms are the same, but layer 1 has become mismatched with layer 2 (etc.). In effect, all the layers become one continuous helicoid surface.

hence a rather high conductivity of approximately 100 S m^{-1}. The related solid $RbAg_4I_5$ has a conductivity of 25 S m^{-1} even at 25 °C. The phenomenon of high temperature conductance in ionic solids has been known since Faraday's observations on silver sulfide and lead(II) fluoride in 1838, but interest in it has become intensive only relatively recently in the search for solid electrolytes suitable for use in fuel cells or advanced electrical storage batteries. Most practical solid electrolytes developed to date, however, have been of the nonstoichiometric type (Section 5.4).

Line defects. Line defects extend in one dimension and may originate in an incomplete layer of atoms (an *edge dislocation*, Fig. 5.2) or from

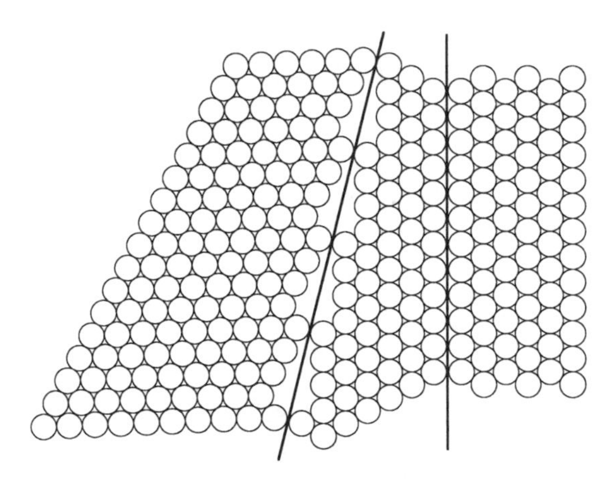

Figure 5.4 Plane defects.

mismatching of layers (a *screw dislocation*, Fig. 5.3). The latter converts the whole crystal lattice into a single helicoid surface (cf. a multistorey parking structure), but it is a line defect in the sense that the linear axis of the helix defines the defect. Line defects create lines of weakness in the crystals that may initiate fracture under stress. Where they emerge at the surface of the crystal, they form sites for the initiation of crystal dissolution (revealed as *etch pits* under the microscope) or further crystal growth (e.g., continuing the helicoid structure). These defects, like others, are mobile to some degree, by concerted slippage of layers of atoms.

Plane defects. Plane defects are, in effect, grain boundaries. Large crystalline specimens are usually made up of microcrystals, or grains, the lattices of which do not match precisely with those of their neighbors (Fig. 5.4). Impurities tend to be concentrated in the regions of mismatch. Large crystals may also contain sizable inclusions of the solution from which they were grown. Consequently, when precipitating solids from solution, chemists and X-ray crystallographers usually adjust solution conditions such as temperature and concentration to give small, well-formed crystals. On the other hand, if large crystals tend to form readily from solution, it usually means that levels of impurities in the system are quite low, so that regular crystal growth is not inhibited. Advances in transmission electron microscopy have made it possible to see grain boundaries and other defects with individual atoms resolved.[2]

Crystal surfaces. Crystal surfaces may be viewed as vast defects inasmuch as the lattice forces are incompletely balanced. The effects of this imbalance are partially offset by distortion of the crystal lattice near the surface, but crystal surfaces still show a strong tendency to adsorb other molecules. Surface areas of powders on the order of 1 to 100 m^2 g^{-1} are customarily measured by the amount of nitrogen gas they adsorb at low temperatures, and release at high temperatures, in forming a multimolecular covering of N_2 (the *Brunauer–Emmett–Teller* or *BET* method). The adsorbed molecules may be activated chemically, as well as held in juxtaposition with other reactants in locally high concentrations, so the surfaces of solids are often active in *heterogeneous catalysis* (Chapter 6).

Stacking faults. If, for example, we have a crystal structure based on cubic close packing, the sequence ABCABCABC... of layers may contain occasional errors such as ABCABABC.... Such stacking faults are of minor chemical significance.

5.3 Impurity Defects and Semiconduction

It is inevitable that some foreign atoms will be present in any macroscopic crystal, usually substituting for the normal lattice atoms (cf. natural chromite, Section 4.6.2). Few analytical reagent-grade chemicals are much better than 99.9% pure, which means 0.1% impurities; thus, if the atomic or molecular weights are comparable, there will on average be a foreign atom or molecule in any cubic portion of the lattice of 10 atoms or molecules along the edge. This can have a profound effect on the bulk properties of the crystal, such as electrical conductivity, whether because of ion migration as in point defects or electrons moving through conduction bands (Section 4.2).[3,4]

For example, the conductivity of elemental germanium (an intrinsic semiconductor, Section 4.2), which is tetracovalent and has the diamond structure, is greatly increased by *doping* the crystal with small amounts of neighboring elements in the periodic table, such as gallium, which has only three valence electrons, or arsenic, which has five. A Ga atom substituting for a Ge therefore leaves the lattice-wide covalent bonding structure (valence band) one electron short, creating a so-called *positive hole* in the otherwise filled $4s$,$4p$ valence band. This hole can move around the band under the influence of an applied electrical field. Gallium-doped Ge is therefore a *p-type semiconductor*. Another way of representing this phenomenon is to regard the effect of doping as the creation of a narrow band of *acceptor levels* just above the valence band in energy, so that the Fermi level (defined in Section 4.2) now falls between it and the valence band (Fig. 5.5); since valence-band electrons can be easily promoted thermally into the readily accessible empty acceptor band, the conductivity increases markedly with rising temperature (contrast with metallic conduction). This *extrinsic* semiconduction mechanism augments the *intrinsic* semiconductor properties of Ge that are due to thermal promotion of valence electrons into the main conduction band (Section 4.2). Clearly, with judicious doping, chemists can

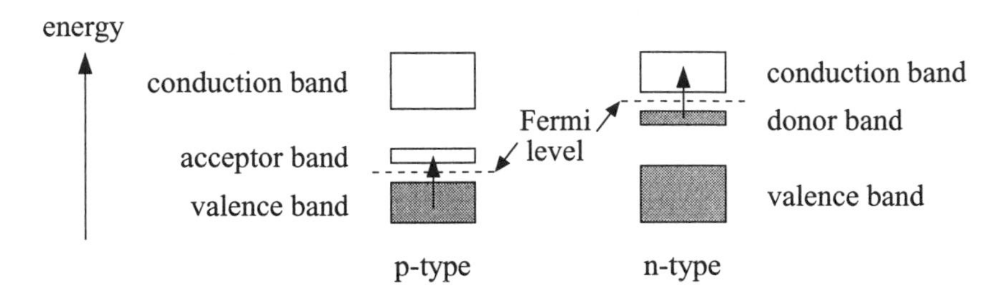

Figure 5.5 Band model of the mechanisms of p-type (left) and n-type semiconduction. Arrows represent thermal excitation of electrons.

fine-tune the electrical properties of semiconductors such as germanium and silicon.

Conversely, doping Ge with As introduces an extra electron that cannot be accommodated in the tetracovalent network (valence band), and this creates a narrow band of occupied *donor* levels, just below the conduction band in energy. The Fermi level is now located between the donor band and the conduction band, and electrons in the donor band can be readily excited thermally into the conduction band (Fig. 5.5). Thus, a negative or *n-type semiconductor* is created. Semiconductors can exhibit electrical conductivities in the range 10^{-3} to 10^{4} S m^{-1}, as compared to 10^{3} to 10^{7} S m^{-1} for metals.

If we place n- and p-type semiconducting crystals in contact (a p-n junction), we create a device that conducts electricity preferentially in one direction; this is the basis of action of the semiconductor diodes used in the electronics industry, although specially refined silicon (Section 17.8.2) is usually employed rather than Ge. Transistors and electronic "chips" are designed using similar basic principles—typically with n–p–n or p–n–p junctions. We consider chemical aspects of electronic devices in more detail in Chapter 19.

5.4 Nonstoichiometry

In introductory chemistry courses, much emphasis is necessarily placed on the concept of stoichiometry, that is, the fact that elements combine in certain definite proportions by weight, proportions that reflect their valences and atomic masses. For much of the chemistry of the main group elements and organic compounds, this concept works extremely well, but in transition metal chemistry in particular it is common for ions of more than one oxidation state to form with comparable ease, and sometimes to occur together in the same ionic solid. The presence of more highly oxidized cations in a metal ion sublattice is counterbalanced by vacancies in that sublattice.

Conversely, the presence of some metal ions of lower oxidation state in the metal ion sublattice requires vacant anion sites to balance the charge. In some cases, the charge imbalance is caused by ions of some other element or, rarely, by multiple valence of the anions. In any event, the empirical formula of a recognizable solid transition metal compound may be variable over a certain range, with nonintegral atomic proportions. Such *nonstoichiometric* compounds may be regarded as providing extreme examples of impurity defects.

For example, iron(II) sulfide can exist as stoichiometric $Fe^{2+}_{1.000}S^{2-}$, which has the nickel arsenide structure (Table 4.3) and is known as troilite, but it is more commonly encountered with up to 10% of the Fe ions missing. Of the Fe ions that *are* present, there are two in the trivalent state for every Fe^{2+} ion that is missing, so balancing the charges. The S^{2-} array is essentially complete. Thus, although one may come across reports of phases with compositions such as "Fe_8S_9," the formula would be better written as $Fe_{1-x}S$ with $x = 0.111$ (no special significance being attached to 0.111), or

$$\underset{1-3x}{Fe^{2+}_{0.667}}\ \underset{2x}{Fe^{3+}_{0.222}}\ \underset{x}{\square_{0.111}}\ S$$

where the square symbol represents vacant sites in the iron ion sub-lattice. In general, compositions such as $Fe^{2+}_{0.634}Fe^{3+}_{0.244}\square_{0.122}S$ are just as likely for the product of any given preparation, although the tendency of the vacant sites to become ordered in the lattice does give special stability to certain compositions such as Fe_7S_8. The *iron-deficient* sulfides are known as *pyrrhotite;* an iron-rich material $Fe_{1+x}S$ or *mackinawite* is also known. There is also the familiar *pyrite* or *fool's gold* (FeS_2), but this is usually stoichiometric, the anion being $S_2{}^{2-}$ (rather like the peroxide ion, $O_2{}^{2-}$).

The various iron sulfide phases are important in connection with the mechanism of corrosion of steels by aqueous hydrogen sulfide, H_2S. Corrosion by $H_2S(aq)$ is a limiting factor in the applicability of the *GS (Girdler sulfide) process* for the production of heavy water (D_2O)*. Heavy water is used as the moderator and/or coolant in certain nuclear power reactors, notably the Canadian CANDU (Canada deuterium uranium) and U.S. PHW (pressurized heavy water) designs, in which D_2O is necessary because, unlike ordinary light water (H_2O), it has a low tendency to capture the neutrons that maintain the ^{235}U fission chain reaction. In the GS process, the deuterium content of water is built up by repeated equilibrations of the D distribution between water and H_2S at successively high and low temperatures. Several often ill-defined Fe sulfides are involved, presenting a complex chemical problem to the corrosion engineer.

*D stands for deuterium (i.e., ^2H, the isotope of hydrogen of mass 2).

Iron(II) oxide (wüstite, Section 4.6) also ranges in composition, from $Fe_{0.88}O$ to $Fe_{0.95}O$; stoichiometric FeO is not encountered in practice. As an extreme case, δ-TiO, an NaCl-type solid previously mentioned as exhibiting gross Schottky defects even when stoichiometric, ranges widely in composition:

$TiO_{0.69}$	96% Ti sites, 66% O^{2-} sites occupied
$TiO_{1.00}$	85% Ti sites, 85% O^{2-} sites occupied
$TiO_{1.33}$	74% Ti sites, 98% O^{2-} sites occupied.

5.5 Metal Oxides and Sulfides as Extrinsic Semiconductors

Many metal oxides and sulfides exhibit semiconductor properties by virtue of their nonstoichiometry. In Section 5.3, we distinguished positive (p-type) and negative (n-type) semiconductors.

p-Type oxides gain oxygen on heating in air, for example, NiO:

$$\underset{\text{green}}{NiO} + \tfrac{x}{2}O_2 \rightarrow \underset{\text{black}}{NiO_{1+x}}. \tag{5.2}$$

The black phase persists on cooling. It contains $2Ni^{3+}$ in place of $2Ni^{2+}$, plus a vacancy in the Ni^{2+} sublattice, for every additional O^{2-}. The Ni^{3+} ions, being smaller and more highly charged than Ni^{2+}, act as charge carriers by skipping through vacant sites under an applied external electrical potential.

n-Type oxides typically lose O_2 reversibly on heating in air, for example, zinc oxide:

$$\underset{\text{white}}{ZnO(s)} \rightleftharpoons \underset{\text{yellow}}{ZnO_{1-x}(s)} + \tfrac{x}{2}O_2(g). \tag{5.3}$$

The excess Zn^{2+} ions in the nonstoichiometric high temperature phase are counterbalanced by free electrons centered upon the empty O^{2-} sites (*F* or *color* centers, F standing for *Farben*, German for color; excitation of these electrons by light causes the yellow color). The electrons act as charge carriers under an applied electrical field. Similarly, ZnS, on heating to 500 °C, loses sulfur vapor to give ZnS_{1-x}, which fluoresces strongly in ultraviolet light.

It is possible to make nonstoichiometric solids that have *ionic* conductivities as high as 0.1–1000 S m^{-1} (essentially the same as for liquid electrolytes) yet negligible *electronic* conductances. Such solid electrolytes are needed for high energy density electrical cells, fuel cells, and advanced batteries (Chapter 15), in which mass transport of ions between electrodes is necessary but internal leakage of electrons intended for the external circuit must be minimized. One such material is *calcia-stabilized zirconia* (calcia means CaO; zirconia is ZrO_2), which has a fluorite like structure and is stable over a composition range $Zr_{0.87}Ca_{0.26}O_2\square_{0.26}$ to $Zr_{0.78}Ca_{0.44}O_2\square_{0.44}$ at temperatures up to at least 2200 °C. Here, the square symbol represents vacancies in the oxide array of the fluorite like lattice (one for every Ca^{2+}) and transport of oxide ions through these *anion* vacancies gives the solid remarkable electrical conductivity.

Conversely, the so-called *β-alumina* electrolytes permit transport of *cations*—sodium ions—in layers with incompletely occupied cation sites separating blocks of aluminum oxide lattice having a spinel like structure. In effect, there is a close-packed array of O^{2-} ions, but in every fifth layer up to three-quarters of the O^{2-} are missing, with sodium ions helping to balance the charges of those that are present (the term "alumina," implying Al_2O_3, is clearly a misnomer). The limiting composition of β-alumina is therefore $NaAl_{11}O_{17}$, but it is usually nonstoichiometric: $Na_{1+2x}Al_{11}O_{17+x}$. In the nomenclature of Section 4.3, the sequence of close-packed layers in β-alumina is

$$\ldots ABCA(B)ABCA(B)ABCA(B)ABCA(B)\ldots$$

where the parentheses denote the oxide-deficient, sodium-containing layers. A modification known as β″-alumina has a generally higher Na content and a different oxide layer sequence

$$\ldots ABCA(B)CABC(A)BCAB(C)ABCA(B)\ldots$$

but requires some Mg^{2+} or Li^{+} in the spinel like blocks to stabilize the structure.

5.6 Mechanism of Scaling of Metals

Most metals acquire an oxide film or *scale* on their surfaces on exposure to oxygen or air, especially at elevated temperatures.[5–7] The kinetics and mechanism of formation of such films provide examples of applications of the concepts of the defect solid state outlined above, and indeed of solid-state kinetics in general.[8]

Oxide films are often protective in the sense of hindering further oxidation, but this is not always the case. Pilling and Bedworth made an early attempt (1923) to rationalize the protective behavior of oxide films on the basis of the volume occupied by the oxide relative to the volume of metal from which it was formed. If the molar volume $V°$ of oxide per mole of metal is less than the molar volume of the metal, the scale will be under tension as it forms and will tend to crack and so be nonprotective. An example would be magnesium:

$$V^\circ \text{ of Mg} = \frac{\text{atomic mass}}{\text{density}} = \frac{24.31}{1.74} = 14.0 \text{ cm}^3 \text{ mol}^{-1},$$

$$V^\circ \text{ of MgO} = \frac{\text{molar mass}}{\text{density}} = \frac{40.31}{3.58} = 11.3 \text{ cm}^3 \text{ mol}^{-1}.$$

If V° of the oxide per mole of metal content is greater than V° for the metal, the oxide film forms under compression and may be protective, as in the case of nickel:

$$V^\circ \text{ of Ni} = 6.6, \quad V^\circ \text{ of NiO} = 11.2 \text{ cm}^3 \text{ mol}^{-1}.$$

However, the Pilling–Bedworth approach is of limited applicability, as is shown by the behavior of copper:

$$\frac{V^\circ \text{ of CuO}}{V^\circ \text{ of Cu}} = \frac{12.5}{7.1}$$

$$\frac{\frac{1}{2}V^\circ \text{ of Cu}_2\text{O}}{V^\circ \text{ of Cu}} = \frac{11.9}{7.1}.$$

Copper would therefore be expected to form a protective film, be it of copper(I) or copper(II) oxide. (Note, however, that the picture becomes confused where two oxide stoichiometries are involved.) Indeed, oxidation of Cu at 600 to 800 °C proceeds according to the *parabolic law:*

$$\frac{dy}{dt} = \frac{k}{y} \tag{5.4}$$

or

$$y^2 = 2kt + c \tag{5.5}$$

where y is the thickness of scale (or weight gain per unit surface area) at time t, and k and c are constants. The parabolic law (Fig. 5.6a) implies protection by the thickening, coherent, oxide film. At 500 °C, however, the film is apparently insufficiently plastic to support the Pilling–Bedworth compressional stress, and it cracks intermittently to expose fresh metal (Fig. 5.6b). In the limit of frequent film cracking (as with sodium), Fig. 5.6b reduces to a straight line (*rectilinear rate law*, Fig. 5.6c). The Pilling–Bedworth concept also fails to explain why some metal oxidation processes follow *logarithmic* or *inverse logarithmic rate laws* (Fig. 5.6d, and Eqs. 5.13 through 5.16), which lead to much sharper slowing of oxidation than the intuitively reasonable *parabolic rate law*.

In 1933, C. Wagner rationalized metal oxidation rates in terms of metal oxide lattice defects. The oxidizing surface was regarded as an electrochemical cell (Fig. 5.7):

$$\text{M} \rightarrow \text{M}^{n+} + n\,e^- \qquad \text{at anode (metal surface)} \tag{5.6}$$

$$\text{O}_2 + 4e^- \rightarrow 2\text{O}^{2-} \qquad \text{at cathode (air oxide interface)} \tag{5.7}$$

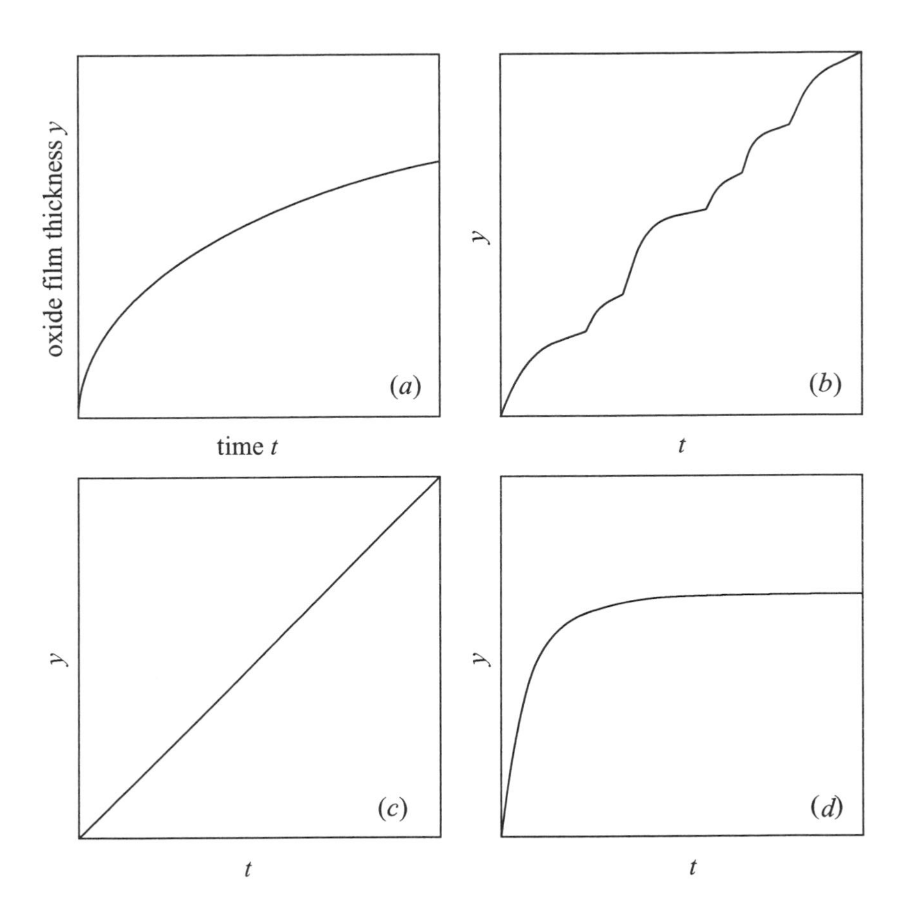

Figure 5.6 Rate laws for formation of oxide films. (*a*) Parabolic rate law. (*b*) Effect of film cracking (successive parabolic segments). (*c*) Limiting case of (*b*). (*d*) Logarithmic rate law.

in which the electrolytic medium is the film (of thickness y) of defective metal oxide. This places "dry" oxidation on a similar basis as "wet" corrosion (Section 16.2). Electron transfer across the film is usually relatively rapid, and, since M^{n+} is usually smaller than O^{2-}, the film grows mainly by diffusion of M^{n+} outward to meet the O^{2-} that are being added to the oxide sublattice at the gas–oxide interface. The M^{n+} ions move through Schottky and/or Frenkel defects under the influence of electric field and concentration gradients (these are equivalent, according to Einstein), which will be inversely proportional to the separation y of the "anode" and "cathode."

The probability that an ion will jump to a neighboring vacancy, in the absence of an external electrostatic field, is $\exp(-E_a/k_BT)$, where E_a

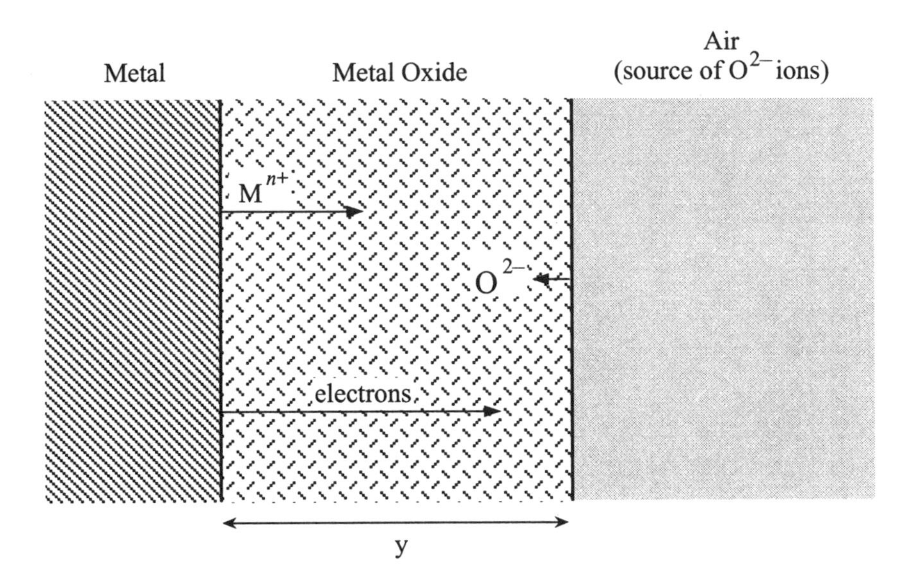

Figure 5.7 Oxidation of a metal by gaseous oxygen, viewed as an electrochemical process.

is the activation energy and k_B is Boltzmann's constant. In the presence of an electric field gradient ($\propto y^{-1}$), the probability of an up-field jump is $\exp[(-E_a - K/y)/k_BT]$ and the probability of a down-field jump is $\exp[(-E_a + K/y)/k_BT]$, where K is a constant. Thus,

$$\frac{dy}{dt} = AK(e^{\theta} - e^{-\theta}) = 2AK \sinh\theta \tag{5.8}$$

(the "sinh equation"), where $A = \exp(-E_a/k_BT)$ and $\theta = K/yk_BT$. Since

$$e^{\theta} = 1 + \theta + \frac{\theta^2}{2!} + \frac{\theta^3}{3!} + \cdots \tag{5.9}$$

we can expand Eq. 5.8 to

$$\frac{dy}{dt} = AK\left(2\theta + \frac{\theta^3}{3} + \cdots\right). \tag{5.10}$$

If the temperature T is *high* and the oxide film is *thick* (T and y both being large), θ can become much less than unity, so

$$\frac{dy}{dt} \approx \frac{(\text{constant})}{y} \tag{5.11}$$

which is the *parabolic law* (Eq. 5.4). On the other hand, when T is *low* and the film is *thin*, θ may be much larger than unity, so that the term in $\exp(-\theta)$ becomes negligible, whence

$$\frac{dy}{dt} = AK \exp\theta \tag{5.12}$$

which, after further approximations,[5] gives an expression equivalent to the *inverse logarithmic law*

$$y_0^{-1} - y_t^{-1} = b\ln(k't + 1) \tag{5.13}$$

or, for a *small* increase of weight w per unit area,

$$w^{-1} = c - k''\ln t \tag{5.14}$$

where b, c, k', and k'' are constants. Inverse logarithmic oxidation rates are exhibited by iron at room temperature: the oxide film grows on a freshly exposed metal surface to a thickness of 1.6 nm after one day, and 3.5 nm after one year—a drastic retardation.

If, contrary to our assumption, electron transfer is slower than M^{n+} diffusion and thus controls the oxidation rate, it can be shown that the *direct logarithmic law* applies:

$$y_t - y_0 = a\ln(k^*t + 1) \tag{5.15}$$

or

$$w = a\ln(k^*t + 1) \tag{5.16}$$

where a and k^* are constants. Again, this is expected to apply to fairly thin (<100 nm) films at relatively low temperatures, such as in the oxidation of cobalt at 320 to 520 °C, but in practice it is hard to distinguish direct from inverse logarithmic kinetics.

Finally, there are two special cases in which the *rectilinear law* is observed: when the rate-controlling factor is the rate of supply of O_2 and when the metal oxide is volatile at the temperature of oxidation. The latter case occurs in the high temperature oxidation of molybdenum, since MoO_3 is quite volatile, and in this case dw/dt is negative.

The mechanism of the scaling of iron is so complex as to require special mention. Above 570 °C, wüstite ($Fe_{1-x}O$) is thermodynamically stable and forms the relatively thick basal layer in the oxide film. This is followed by a magnetite (Fe_3O_4) layer which is followed by a final layer of Fe_2O_3. Magnetite itself tends to become nonstoichiometric under oxidizing conditions, with excess Fe^{3+}, so that its composition and color can vary from $Fe_{3.000}O_4$ (black) toward cubic $Fe_{2.667}O_4$ (i.e., γ-Fe_2O_3, chocolate brown). Thus, as outlined in Section 4.6, the oxidation of iron above 570 °C involves mainly the migration of Fe^{2+} and Fe^{3+} outward through a ccp sublattice of O^{2-} in a largely nonstoichiometric oxide film. At high temperatures, the final Fe_2O_3 layer must be α-Fe_2O_3, since γ-Fe_2O_3 is unstable with respect to α-Fe_2O_3 (and seems to require the elements of water to stabilize it; Section 4.6.2), but this relatively thin layer has an hcp O^{2-} array and grows by diffusion of O^{2-} inward. The growth kinetics are further complicated by the tendency for cavities to form within the film. At *low* temperatures, wüstite is unstable with respect to disproportion to Fe and Fe_3O_4, and so the oxide film is usually Fe_3O_4 (possibly nonstoichiometric to some degree) and/or γ-Fe_2O_3 (if the stabilizing traces of water are present).

5.7 Interstitial Compounds

When discussing metal alloys (Section 4.3), we saw that atoms of nonmetallic elements such as H, B, C, and N can be inserted into the interstices (tetrahedral and octahedral holes) of a lattice of metal atoms to form metal-like compounds that are usually nonstoichiometric and have considerable technological importance. These *interstitial compounds* are commonly referred to as metal hydrides, borides, carbides, or nitrides, but the implication that they contain the anions H^-, B^{3-}, C^{4-}, or N^{3-} is misleading. To clarify this point, we consider first the properties of truly ionic hydrides, carbides, and nitrides.

As noted in Section 2.8, anions that have large, easily polarized electron clouds will tend not to form stable ionic crystals but, instead, will become involved in covalent bonding. If they exist at all, they will be found in compounds only with large, low-charged cations of the least electronegative elements. The large, squashy hydride ion H^-, with only two electrons and an ill-defined ionic radius of 140 to 200 pm, is known in the white solid salts NaH and CaH_2. The presence of hydride ions in these salts is confirmed by their vigorous reaction with water and other potential sources of H^+ to give H_2 gas; consequently, they are sometimes used as powerful reducing agents. The organic chemists' reductants, *lithium aluminum hydride* and *sodium borohydride*, contain the complex ions (Section 13.2) AlH_4^- and BH_4^- rather than free H^-. Likewise, true salts containing the highly charged monatomic anions boride (B^{3-}), carbide (C^{4-}), and nitride (N^{3-}) are rarely encountered. The familiar *calcium carbide*, once widely used to fuel lamps and still favored for this purpose by cavers, is actually an acetylide, that is, a salt of the $^-C{\equiv}C^-$ ion; it is made by reduction of lime with coke in an electric furnace and is hydrolyzed by water to give acetylene (ethyne) gas, which burns in air with a luminous flame:

$$CaO + 3C \xrightarrow{2100\,^\circ C} CaC_2 + CO \tag{5.17}$$

$$CaC_2(s) + 2H_2O(l) \longrightarrow Ca(OH)_2(s) + HC{\equiv}CH(g). \tag{5.18}$$

In interstitial compounds, however, the nonmetal is conveniently regarded as *neutral* atoms inserted into the interstices of the expanded lattice of the elemental metal. Obviously, this is an oversimplification, as the electrons of the nonmetal atoms must interact with the modified valence and conduction bands of the metal host, but this crude picture is adequate for our purposes. On this basis, Hägg made the empirical observation that insertion is possible when the *atomic* radius of the nonmetal is not greater than 0.59 times the *atomic* radius of the host metal—there is no simple geometrical justification for this, however, as the metal lattice is concomitantly expanded by an unknown amount. These interstitial compounds are sometimes called *Hägg compounds*.[9,10] They are, in effect, *interstitial solid solutions* of the nonmetal in the metal (as distinct from *substitutional* solid solutions, in which actual lattice atoms are replaced, as in the case of gold–copper and other alloys; Section 4.3).

Atoms of C or N can enter the O-holes of a close packed metal lattice to give, in the limit, one nonmetal for each metal atom (as in the case of TiC and TiN, which have the NaCl structure). Usually, however, there are fewer interstitial than lattice atoms, as in the case of $PdH_{0.6}$, the formation of which is presumably responsible for the ease with which hydrogen gas diffuses through palladium metal. If the nonmetal atoms are small enough, as with hydrogen, they may enter the T-holes instead, in which case a limiting composition of two nonmetal atoms per lattice atom is theoretically possible, although not usually attained. Titanium and zirconium form hydrides TiH_x and ZrH_x with a distorted fluorite structure (Fig. 4.14), where x ranges up to 1.73 and 1.92, respectively. The tendency of Ti and Zr to form hydrides, so becoming embrittled, presents a serious technical problem wherever these otherwise tough and highly corrosion-resistant metals are used in contact with hydrogen gas, for example, in pressurized heavy water nuclear power reactors in which the fuel elements are usually sheathed in zircaloy (a Zr alloy).

As we go from left to right across the transition metals in the periodic table, the metal atoms become smaller, much as in the *lanthanide contraction* (Section 2.6). Furthermore, the atoms of elements of the first transition series are smaller than those of corresponding members of the second and third. Consequently, interstitial carbides are particularly important for metals toward the lower left of the series, as with TiC, ZrC, TaC, and the extremely hard tungsten carbide WC, which is used industrially as an abrasive or cutting material of almost diamond like hardness. The parallel with trends in chemisorption (Section 6.1) will be apparent.

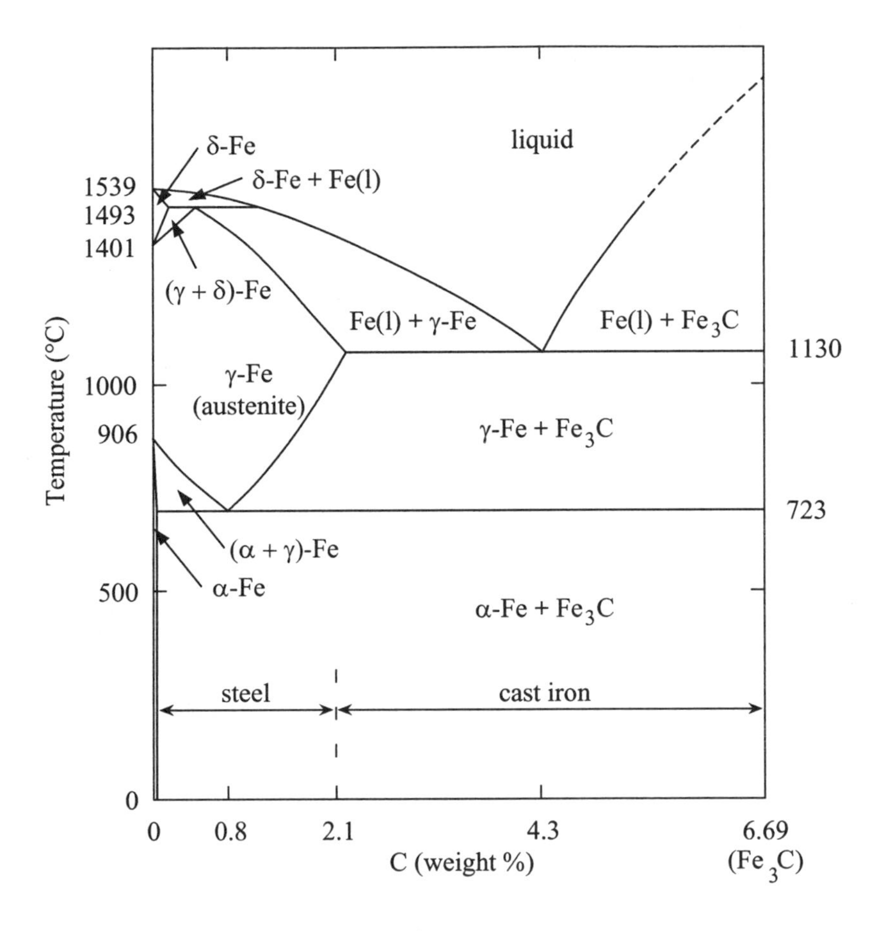

Figure 5.8 The iron–carbon phase diagram.

5.7.1 Carbon Steels

The carbon–metal atomic radius ratio for the room temperature form of iron, α-Fe, is 0.60, just in excess of the Hägg limit. Consequently, the ability of α-Fe to accept interstitial C is marginal: only 0.022 weight % or 0.06 atom % C can be accommodated in the random solid solution known as *ferrite*.

There is, however, a well-defined solid iron carbide phase known as *cementite*, Fe_3C (6.69 weight % C). Further, as the temperature is increased toward the melting point of 1539 °C, the crystal structure of pure iron changes as follows:

α-Fe (ferromagnetic): up to 766 °C; bcc
β-Fe (nonmagnetic): 766 to 906 °C; bcc
γ-Fe (nonmagnetic): 906 to 1401 °C; fcc
δ-Fe (nonmagnetic): 1401 to 1539 °C; bcc.

The γ-Fe can accommodate up to 2.11 weight % C in the O-holes of the fcc structure. This solid solution is called *austenite*. The Fe–C phase diagram[7, 11, 12] is shown in outline in Fig. 5.8.

If austenite is cooled *slowly* toward ambient temperature, the dissolved carbon in excess of 0.022 weight % comes out of solid solution as *cementite*, either in continuous layers of Fe_3C (*pearlite*) or as layers of separated Fe_3C grains (*bainite*). In either case, the iron is soft and grainy, as with cast iron. If, on the other hand, the hot austenite is cooled *quickly* (i.e., *quenched*), the γ-Fe structure goes over to the α-Fe form without crystallization of the interstitial carbon as cementite, and we obtain a hard but brittle steel known as *martensite* in which the C atoms are still randomly distributed through the interstices of a strained α-Fe lattice. Martensite is kinetically stable below 150 °C; above this temperature, crystallization of Fe_3C occurs in time.

If martensite is reheated to between 200 and 300 °C for an appropriate time and is then requenched, a *partial* crystallization of Fe_3C occurs, and a tough steel, *sorbite*, with properties intermediate between martensite and pearlite or bainite, is obtained. This process is known as the *tempering* of steel.[12]

High-carbon austenitic structures can be preserved at ambient temperatures if the iron is alloyed with sufficient nickel or manganese, since these metals form solid solutions with γ-Fe but not with α-Fe. If over 11% chromium is also present, we have a typical *austenitic stainless steel.* Such steels are corrosion resistant, nonmagnetic, and of satisfactory hardness, but, because the $\alpha\text{-Fe} \rightleftharpoons \gamma\text{-Fe}$ transition is no longer possible, they cannot be hardened further by heat treatment. Figure 5.9 summarizes these observations.

5.7.2 Nitriding

Although nitrogen gas is normally considered to be unreactive, it forms an interstitial nitride with titanium sufficiently easily that the production of titanium metal has to be carried out in an argon atmosphere (the *Kroll process*, Section 17.8). Similarly, oxygen is used in preference to air in modern steelmaking, in part because some nitriding of the steel may otherwise occur, leading to a more brittle product (Section 17.7.3). Embrittlement of low-alloy steels by nitriding is now recognized as a serious problem in ammonia plants after 10 years or more of operation.

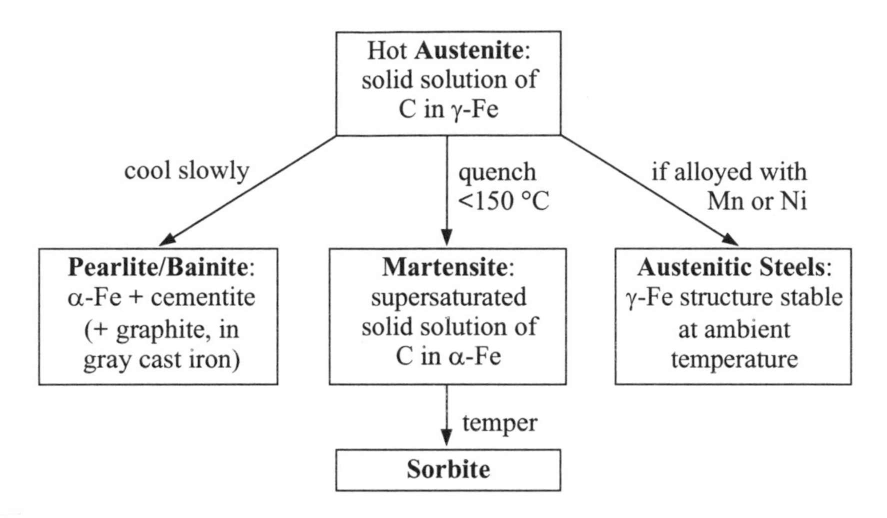

Figure 5.9 Heat treatment of solid iron containing carbon.

On the other hand, surface nitriding or *case hardening* of a steel specimen can improve its durability. Both N and C atoms are introduced interstitially to the Fe lattice on the surface of the steel by immersing it in, for example, a solution of sodium cyanide (NaCN) in a molten $Na_2CO_3/NaCl$ mixture at 870 °C.

Exercises

5.1 Stoichiometric titanium(II) oxide and magnesium oxide have densities of 4.93 and 3.58 g cm^{-3}, respectively. They both have the NaCl structure, with unit cells of edge 416.2 and 421.2 pm, respectively. Show that only about 84% of the lattice sites in TiO, but essentially all in MgO, are filled.

5.2 On which of the following metals would the oxide film be expected to be protective, according to the Pilling–Bedworth principle?

Metal	*Density (g cm^{-3})*	*Oxide*	*Density (g cm^{-3})*
Be	1.85	BeO	3.01
Ca	1.54	CaO	3.25–3.38
Al	2.70	α-Al_2O_3	3.97
Ti	4.50	rutile	4.26

[*Answer:* All except calcium.]

5.3 The oxidation of iron by (dry) air at ambient temperature proceeds according to the *inverse logarithmic rate law*, which may be written as:

$$\frac{1}{y} = c - k \ln t$$

where the oxide film thickness y is 1.6 nm after time $t = 1$ day and 3.5 nm after 1 year; c and k are empirical constants. How much longer will a doubling of the oxide film, from 3.5 to 7.0 nm, require? [*Answer:* 10.5 more years.]

5.4 In a much-publicized study in 1989, Pons and Fleischmann claimed to have observed *cold fusion* of nuclei of deuterium (heavy hydrogen, D) within palladium electrodes that were being used to electrolyze D_2O. Had this been the case, what other electrode materials might also have shown the same phenomenon?

5.5 Hägg found that metals can accommodate interstitial nonmetal atoms of radius up to 59% of that of the metal atoms. Show that, in this limiting case, accommodation of the nonmetal atoms in the octahedral holes of a face-centered cubic metal lattice should result in an expansion of the unit cell dimension by 12.4%. [*Hint*: Review the radius ratio rules in Section 4.5.]

References

1. N. N. Greenwood, "Ionic Crystals, Lattice Defects, and Non-Stoichiometry." Chemical Publ. Co., New York, 1970.

2. A. L. Robinson, Spotting the atoms in grain boundaries. *Science* **233**, 842–844 (1986); M. M. McGibbon, N. D. Browning, M. F. Chisholm, A. J. McGibbon, S. J. Pennycook, V. Ravikumar, and V. P. David, Direct determination of grain boundary structures in $SrTiO_3$. *Science* **266**, 102–104 (1994); M. Eberhart, Computational metallurgy. *Science* **265**, 332-333 (1994).

3. A. Wold and K. Dwight, "Solid State Chemistry." Chapman & Hall, New York, 1993; L. Smart and E. Moore, "Solid State Chemistry." Chapman & Hall, New York, 1992; A. Navrotsky, "Physics and Chemistry of Earth Materials." Cambridge Univ. Press, New York, 1994.

4. P. G. Bruce (ed.), "Solid State Electrochemistry." Cambridge Univ. Press, Cambridge, 1995.

5. U. R. Evans, "The Corrosion and Oxidation of Metals." Arnold, London, 1960, and Supplements 1 (1968) and 2 (1978).

6. U. R. Evans, "An Introduction to Metallic Corrosion," 3rd Ed., Chapter 1. Arnold, London, 1981.

7. M. G. Fontana, "Corrosion Engineering," 3rd Ed. McGraw-Hill, New York, 1986.

8. H. Schmalzried, "Solid State Reactions." Academic Press, New York, 1974; H. Schmalzried, "Chemical Kinetics of Solids." VCH, Weinheim, 1995.

9. F. A. Cotton and G. Wilkinson, "Advanced Inorganic Chemistry," 5th Ed. Wiley (Interscience), New York, 1988.

10. A. F. Wells, "Structural Inorganic Chemistry," 5th Ed. Oxford University Press: London, 1984.

11. R. A. Heidemann, A. A. Jeje, and M. F. Mohtadi, "An Introduction to the Properties of Fluids and Solids." pp. 106–113. Univ. of Calgary Press, Calgary, Alberta, 1984.

12. G. J. Long and H. P. Leighly, Jr., The iron–iron carbide phase diagram. *J. Chem. Educ.* **59**, 948–953 (1982).

Chapter 6

Inorganic Solids as Heterogeneous Catalysts

6.1 Heterogeneous Catalysis

A CATALYST is a substance that increases the rate at which a chemical reaction approaches equilibrium, while not being consumed in the process. Thus, a catalyst affects the *kinetics* of a reaction, through provision of an alternative reaction mechanism of lower activation energy, but cannot influence the *thermodynamic* constraints governing its equilibrium.

Use of an insoluble, nonvolatile solid as catalyst for a fluid-phase reaction has major advantages: loss of the catalyst is minimized, it does not significantly contaminate the reaction products, and it stays physically in place in the reaction chamber. Because such *heterogeneous catalysts* often must operate at high temperatures, most are refractory (i.e., high melting) inorganic materials themselves, or else require a refractory support material such as a metal oxide, often alumina (Al_2O_3). In special cases, the actual active catalyst is a liquid at the reaction temperature and must be used on a refractory support. An important example is the *vanadium pentoxide catalyst* used in the contact process (Section 10.2) for the oxidation of SO_2 to SO_3 at 400 to 600 °C. In practice, the V_2O_5 is used with a K_2SO_4 *promoter* in the pores of an inert support material, and the mixture is molten under the reaction conditions. (V_2O_5 melts at 690 °C, but the fusion point of the mixture is lower.) However, such cases are exceptional.

Although the general definition of a catalyst given above emphasizes the acceleration of the approach to equilibrium, the *selectivity* of a catalyst is often of more importance than its overall catalytic activity. An unselective catalyst may accelerate undesirable reaction pathways as well as, or more than, the desired one. A commercially important example of selective catalysis is the oxidation of ammonia by air over platinum metal to produce nitric oxide (NO) (used to make nitric acid, Chapter 9) rather than the thermodynamically favored product, N_2, from which the ammonia was originally made!

Heterogeneous catalysis of a reaction between two molecules can occur only when *both* molecules are adsorbed on to a solid surface. As noted in Section 5.2, the ability of solids to adsorb molecules is due to the presence of unbalanced electrostatic attractions or unsatisfied covalence at the surface of a crystalline solid. Many new and powerful techniques to characterize the surfaces of solids and their adsorbates at the atomic level are available. These include *low-energy electron diffraction* (LEED), *X-ray photoelectron spectroscopy* (XPS, which measures energies of electrons emitted from surface atoms excited by X-rays), *infrared spectroscopy* (IR), *extended X-ray absorption fine structure analysis* (EXAFS), *thermal desorption spectroscopy* (TDS), *transmission electron microscopy* (TEM), *scanning tunneling microscopy* (STM, which registers the "tunneling" of electrons from a sharply pointed probe traveling parallel to a conducting surface), and *atomic force microscopy* (AFM, which measures the attractive force between a finely pointed probe and features on a surface as the probe traverses it). The last three techniques are capable of *imaging individual molecules*, even atoms, and furthermore the STM probe can be used to move molecules at will across the surface on which they are adsorbed. Although the cost of sophisticated instruments, the complexity of surface phenomena, and the difficulty of relating laboratory observations under clean conditions to what goes on in the "dirty" environment of an industrial catalytic reactor have conspired to hinder progress in placing heterogeneous catalysis on a truly scientific footing, it is no longer the "black art" of a few years ago.[1–5]

6.1.1 Physical Adsorption and Chemisorption

The presence of unbalanced attractions at the surface of a solid—say, a metal such as nickel—means that small molecules will tend to become rather loosely attached to the surface in one or (more likely) several molecular layers with an exothermic adsorption energy ranging to about -20 kJ mol^{-1} for nonpolar molecules. (The term *adsorption* is used to denote surface sorption without penetration of the bulk solid, which would be called *absorption.*) No chemical bonds are formed or broken. This state is usually called *physical adsorption* or *physisorption.* If, however, the adsorbate forms chemical bonds with the surface atoms, the adsorption process is called *chemisorption.* Chemisorption can be quite strongly exothermic (-40 to -800 kJ mol^{-1}) but involves only the first monomolecular layer of adsorbate.

As Fig. 6.1 shows, with reference to the adsorption of hydrogen on nickel,

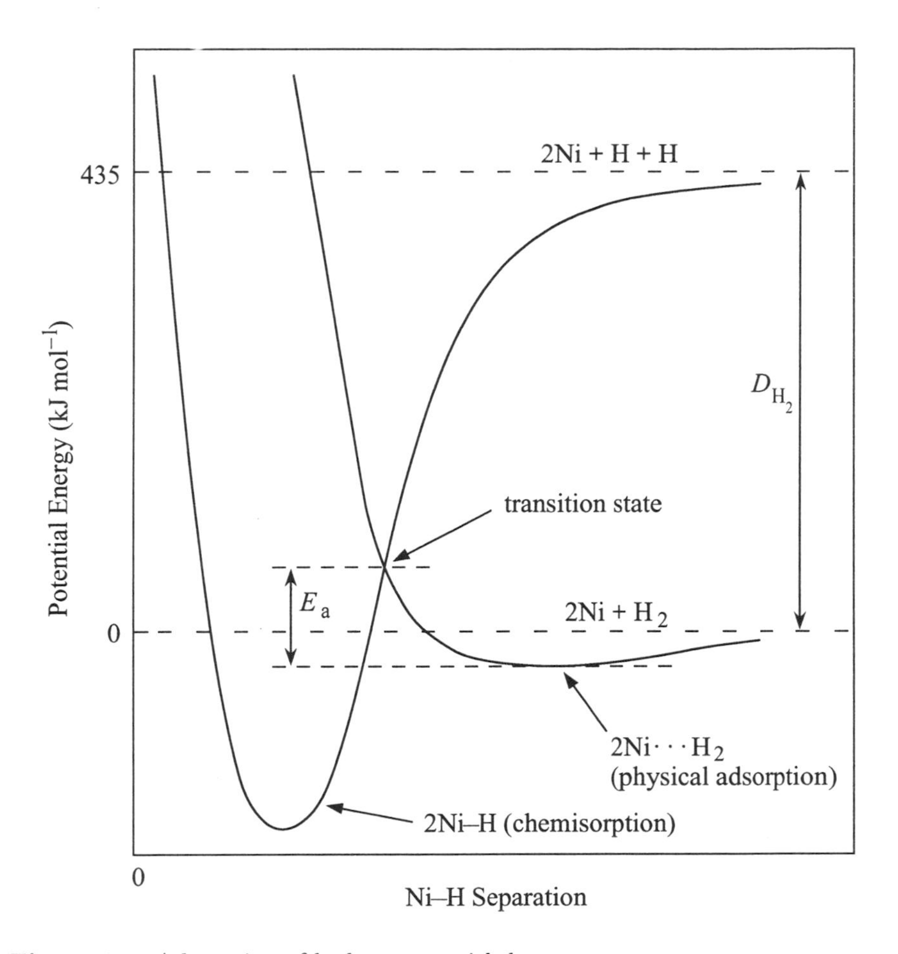

Figure 6.1 Adsorption of hydrogen on nickel.

there is no activation energy for physical adsorption, but the transition to the chemisorbed state (in which the H—H bond is broken in favor of two Ni—H) bonds can be made by surmounting an activation energy barrier E_a. This barrier to the breaking of the H—H bond is seen to be small compared with the atomization energy of 435 kJ mol^{-1} for gaseous H_2. This is the key to the activation of H_2 by nickel surfaces—the strong H—H bond is broken for further reactions with only a small investment of energy:

$$\underset{\text{metal}}{2\text{Ni}} + \underset{\text{gas}}{\text{H}_2} \rightleftharpoons \underset{\text{physical adsorption}}{2\text{Ni}\cdots\text{H}_2} \rightleftharpoons \underset{\text{(surface only) chemisorption}}{2\text{NiH}}. \tag{6.1}$$

In the activation of O_2 by a metal M,[6] the double O=O bond may be

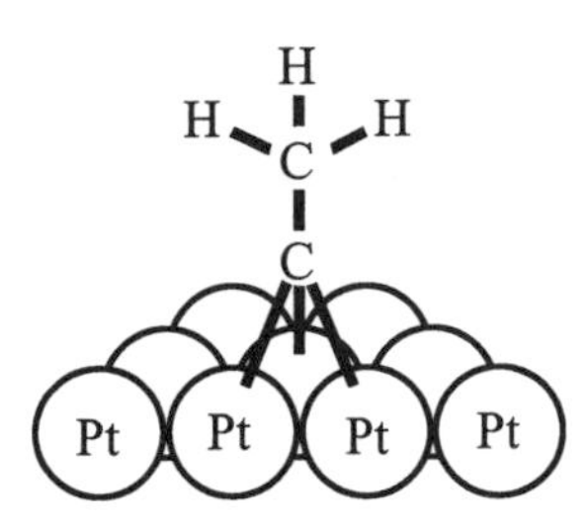

Figure 6.2 Chemisorption of ethylene on platinum.

weakened to a single bond or else broken altogether:

$$\begin{array}{ccccc} \mathrm{O{=}O} & & \mathrm{O{-}O} & & \mathrm{O\quad O} \\ \vdots\quad\vdots & & |\quad | & & \|\quad \| \\ \mathrm{-M{-}M-} & \rightleftharpoons & \mathrm{-M{-}M-} & \rightleftharpoons & \mathrm{-M\quad M-}\,. \\ |\quad | & & |\quad | & & |\quad | \end{array} \tag{6.2}$$

In fact, the most catalytically active transient oxygen species on surfaces is often the singly charged atom O^-.

If the energy of chemisorption is *too* strongly negative, the adsorbed species will become *un*reactive toward others and will "poison" the catalyst by virtue of using up the reactive sites. The reactant molecules must, however, be sufficiently strongly adsorbed to allow surface reactions to take place at all. Thus, given that chemisorption of the reactants *does* occur, catalytic activity will be *inversely* related to their (negative) energy of adsorption. For a solid to be an effective catalyst for a given reaction, it must interact with the reactants just (but *only* just) firmly enough to cause chemisorption with essentially complete surface coverage. Since it is unlikely that both reagents in a bimolecular reaction will interact equally strongly with the surface, the critical question is whether the solid can just barely chemisorb the molecule that is the less reactive toward surfaces.[4]

With the advent of sophisticated experimental techniques for studying surfaces, it is becoming apparent that the structure of chemisorbed species may be very different from our intuitive expectations.[1c] For example, ethylene (ethene, $H_2C{=}CH_2$) chemisorbs on platinum, palladium, or rhodium as the *ethylidyne* radical, $CH_3{-}C{\equiv}$ (Fig. 6.2). The carbon with no hydrogens is bound symmetrically to a triangle of three metal atoms of a close-packed layer [known as the (111) plane of the metal crystal]; the three carbon–metal bonds form angles close to the tetrahedral value that is typical of aliphatic hydrocarbons. The missing H atom is chemisorbed separately. Further H atoms can be provided by chemisorption of H_2, and facile reaction of the metal-bound C atom with three chemisorbed H atoms diffusing across the catalyst surface results in the formation of ethane, C_2H_6; in other words, the ethylene becomes hydrogenated via chemisorption.

Laszlo[7] pointed out that solids of fractal dimension D near 2.0 are usually more efficient catalysts than those of D near 3.0.* An adsorbed species diffusing across the surface of a catalyst will find a target much more quickly (i.e., catalysis will be more efficient) in space of dimension near 2 than of D near 3. We find, for example, that catalytic activities of variously prepared activated charcoals increase as we go from $D = 3.0$ to $D = 1.9$.

6.2 Transition Metals as Catalysts

Metal catalysts are normally associated with *hydrogenation* (or *dehydrogenation*) reactions such as the Fe-catalyzed Haber process (Section 9.3) or the nickel-catalyzed "hardening" (hydrogenation) of edible vegetable oils to make margarine and similar products. They also feature in oxidation–reduction processes such as oxidation of ammonia to nitric oxide (Section 9.4) and in cleanup of automobile exhausts (Section 8.4); in both cases, platinum or platinum alloys are used. For all metals except gold, the chemisorption bond strength sequence for common gaseous reagents is usually

$$O_2 > \text{alkynes} > \text{alkenes} > CO > H_2 > CO_2 > N_2. \tag{6.3}$$

The electronic nature of the interaction between a transition metal and one of these molecules in chemisorption is not easily explained, but it seems that the valence orbitals (mainly d orbitals) of the metal should have sufficient electronic vacancies to permit interaction with the electron-rich adsorbate molecules, but not so many as to allow the interaction to become too strong (cf. poisoning). In any event, the strength of the interactions with a given molecule rises as we go from right to left across the transition metals in the periodic table. The result is that metals of the Sc, Ti, V, Cr, and Mn groups generally interact so strongly, even with N_2, that they are inevitably poisoned by adsorbates. Going further to the right in the periodic table, we find that

(*a*) iron, ruthenium, and osmium *barely* chemisorb N_2 (and, implicitly, the rest of series 6.3 are held more firmly);

(*b*) cobalt and nickel *barely* chemisorb CO_2 and H_2;

(*c*) the Co and Ni analogs rhodium, palladium, iridium, and platinum *barely* chemisorb H_2 but *not* CO_2; and

---

*The fractal dimension D of an ideal solid is 3.00, of a perfectly flat surface 2.00, and of a straight line 1.00—the familiar geometric dimensions. Porous solids and rough surfaces, however, have effective dimensions corresponding to fractional values of D.

(*d*) copper, silver, and gold *barely* chemisorb CO and ethylene. (Gold, however, does not usually chemisorb oxygen at low temperatures, apparently for kinetic reasons.)

These relative chemisorption strengths enable us to make some simple predictions regarding suitable metal catalysts for specific reactions. For example, a catalyst for the Haber process must chemisorb both N_2 and H_2, but not too strongly. Since N_2 is the less readily bound, we choose Fe, Ru, or Os. The latter two are expensive, so our best choice is iron—usually finely divided, on a suitable refractory support.

Similarly, we can ask what would be the best catalyst for hydrogenating an olefin such as ethylene. Since olefins (alkenes) are more strongly chemisorbed than hydrogen, we choose a metal that just barely chemisorbs H_2—this means Co, Rh, Ir, Ni, Pd, or Pt. In practice, nickel is the least expensive choice. Again, it should be finely divided (maximum surface area) for greatest catalytic efficiency and be dispersed on the internal surfaces of a porous support such as alumina with surface area on the order of $200\ m^2\ g^{-1}$.[8]

It must be stressed, however, that the foregoing is only a rough guide to the catalytic properties of metals, and that experiment is the ultimate arbiter in the choice of a metal catalyst for a particular reaction, especially where selectivity is a primary concern. For example, both Pt and Rh can be used to catalyze the formation of *synthesis gas* ("syngas," a mixture of H_2 and CO used for the production of methanol, CH_3OH) by the partial oxidation of methane or ethane by air. Schmidt and co-workers[9] showed that Rh catalyzes the partial oxidation of methane with a slight stoichiometric deficiency of oxygen (65% methane, 35% O_2) very selectively to H_2 (60%) and CO (30%) with only 2% of the by-product CO_2 and 5% H_2O; the contact time can be as short as 1 ms, the reaction rate is mass transport limited, and the reaction is exothermic enough that no process heat need be supplied to maintain the reaction temperature of about 1000 °C. Platinum, on the other hand, gives poorer selectivity, producing more H_2O than H_2. For the oxidation of ethane to syngas, Rh again favors the formation of CO and H_2, but Pt produces mainly ethylene with 70% selectivity. It seems that the activation energy for the reaction of chemisorbed H and O atoms to form OH is much lower on Pt than on Rh; thus, OH radicals form readily on Pt and go on to make H_2O by acquiring another H either from adsorbed ethyl groups where present (in the ethane oxidation, by facile β-H elimination), resulting in ethylene formation, or from further adsorbed H (in the methane case). On hot Rh, O atoms survive long enough to attack the chemisorbed hydrocarbon fragments and give CO. Palladium is of no use in either reaction because of coking (carbon formation). Coking is also a problem with Pt or Rh catalysts of low fractal dimensions; metal films of low surface area on ceramic supports are therefore preferred.

Much current interest centers on the catalytic properties of supported *metal clusters*, that is, metal particles having more metal atoms than discrete molecules such as $Fe_3(CO)_{12}$ (which chemists have studied exhaustively in solution, often in connection with *homogeneous* catalysis), and yet not enough to behave as a bulk metal.[10] For example, in platinum clusters of just six atoms, forming an octahedron of edge 0.55 nm, *all* the metal atoms are on the surface of the cluster and so have potential catalytic activity, while in a cluster of 146 (six atoms forming an edge of 1.65 nm) 70% of the Pt are surface atoms. Metal particles in the size range 1–5 nm already show interesting changes in catalytic properties from bulk metals, but even smaller particles show markedly improved selectivity over conventional metal catalysts. Obviously, such small clusters must be supported on a suitable solid such as γ-Al_2O_3 or within the pores of nanoporous materials such as zeolites (Section 7.3). Platinum clusters supported in zeolite pores are used commercially for the highly selective catalysis of alkane dehydrocyclization.

6.3 Defect Oxides and Sulfides in Catalysis

Nonstoichiometric metal oxides are effective catalysts for a variety of oxidation–reduction reactions (as might be expected) since the variable valence of the constituent ions enables the oxide to act as a sort of "electron bank." Nonstoichiometric metal oxides resemble metals in that they can also catalyze hydrogenation and alkene isomerization reactions. However, on zinc oxide, for instance, these two processes are independent, whereas hydrogen must be present for isomerization to occur on metals.

The mode of action of semiconducting metal oxides can be illustrated with reference to catalysis of the oxidation of carbon monoxide:

$$CO + \tfrac{1}{2}O_2 \rightarrow CO_2. \qquad (6.4)$$

p-Type metal oxides, such as $Ni_{1-x}O$, can accommodate excess positive charge in the lattice and can thus be expected to adsorb oxygen to form anions such as $O_2{}^-$, O^-, or O^{2-} on their surfaces. It turns out that O^-(ads) is the most active adsorbed oxygen species in terms of catalytic activity. It is not, of course, a commonly encountered anion under circumstances other than the special environment of a crystal surface. Within the lattice, the electrons are provided by

$$M^{n+} \rightarrow M^{(n+1)+} + e^-. \qquad (6.5)$$

Thus, for NiO, we have

$$\tfrac{1}{2}O_2(g) + Ni^{2+} \rightarrow O^-(ads) + Ni^{3+} \qquad (6.6)$$

$$O^-(ads) + CO(ads) \rightarrow CO_2(g) + e^- \qquad (6.7)$$

followed by the regeneration of Ni^{2+}:

$$e^- + Ni^{3+} \rightarrow Ni^{2+}. \tag{6.8}$$

n-Type metal oxides, such as ZnO, tend to give up oxygen and accommodate excess electrons in their lattice defects:

$$CO(g) + 2O^{2-}(\text{lattice}) \rightarrow {CO_3}^{2-}(\text{lattice}) + 2e^- \tag{6.9}$$

$$\tfrac{1}{2}O_2(g) + 2e^- \rightarrow O^{2-} \tag{6.10}$$

$${CO_3}^{2-}(\text{lattice}) \rightarrow CO_2(g) + O^{2-}(\text{lattice}). \tag{6.11}$$

Thus, the two oxides consumed in reaction 6.9 are regenerated in reactions 6.10 and 6.11.

Metal sulfides play an important role in catalyzing a wide variety of hydrogenations (e.g., of fats, coal, or olefins) and also desulfurization reactions, which are used in pretreatment of fossil fuels to reduce the emission of sulfur oxides during combustion (Section 8.5). *Molybdenum disulfide*, an important defect catalyst, can be made to function as an n-type ($Mo_{1+x}S_2$) or p-type ($Mo_{1-x}S_2$) semiconductor by exposure to an appropriate mixture of H_2S and hydrogen at temperatures on the order of 600 °C. The equilibrium

$$H_2S(g) \rightleftharpoons H_2(g) + \tfrac{1}{2}S_2(g) \tag{6.12}$$

then *buffers* the S_2 pressure and hence controls the composition of the solid MoS_2. Alternatively, the concentrations of positive or negative defects can be controlled by *doping* MoS_2 with cobalt(II) or antimony(V), respectively.

Molybdenum disulfide catalysts are usually employed with hydrogen gas, as in the *hydrodesulfurization* (HDS) of the cyclic molecule thiophene (C_4H_4S), which, along with benzo- and dibenzothiophenes, is one of the common sulfur-bearing contaminants of petroleum fuels. The use of Co-doped MoS_2 gives mainly alkanes rather than unsaturated HDS products because p-type MoS_2 phases are excellent catalysts for hydrogenations in general.

$$C_4H_4S + 4H_2 \xrightarrow[MoS_2]{\text{p-type}} CH_3CH_2CH_2CH_3 + H_2S. \tag{6.13}$$

A typical commercial hydrodesulfurization catalyst might contain 14% MoO_3 and 3% CoO, on an alumina support. The oxides are converted to Co-doped MoS_2 by exposure to H_2S/H_2 under carefully controlled conditions of temperature and partial pressures.

6.4 Catalysis by Stoichiometric Oxides

6.4.1 Acidic Oxides

Several *insulator* oxides, such as Al_2O_3 and SiO_2, can hydrate reversibly to give hydroxides of the type $Al(OH)_3$, $AlO(OH)$, or $Si(OH)_4$, and, not surprisingly, can catalyze dehydration reactions such as the conversion of alcohols to olefins at elevated temperatures:

$$R{-}CH_2{-}CH_2OH(g) \xrightarrow[Al_2O_3]{350\text{–}400\,^\circ C} R{-}CH{=}CH_2(g) + H_2O(g). \quad (6.14)$$

The Al_2O_3 in reaction 6.14 probably functions as a Lewis *acid*,[11] since alcohols usually dehydrate via formation of a carbocation, $R{-}CH_2{-}CH_2^+$. The role of Al_2O_3 as a catalyst for the Claus process (Section 10.1) may be similarly viewed.

Zeolites, which are aluminosilicates that can be regarded as being derived from Al_2O_3 and SiO_2, function as acidic catalysts in much the same way (Section 7.3). In addition, they catalyze isomerization, cracking, alkylation, and other organic reactions. A structurally related class of microporous materials based on aluminum phosphate ($AlPO_4$) has also been developed (Section 7.7); like zeolites, they have cavities and channels at the molecular level and can function as shape-selective catalysts.

6.4.2 Basic Oxides

Basic oxides such as MgO and ZrO_2 favor *anionic* reaction mechanisms on their surfaces.[12] The lattice oxide ions on the surface can abstract protons from adsorbates, forming OH^-. For example, in the production of acrylonitrile ($CH_2{=}CH{-}CN$) for the synthetic fibers industry, acetonitrile (CH_3CN) and methanol (CH_3OH) adsorbed on MgO each lose H^+ to form $^-CH_2CN$ and CH_3O^-; then, attack of the anionic end of the former on the C atom of the latter leads to $CH_2{=}CH{-}CN$ and, when the protons are redistributed, H_2, H_2O, and unchanged MgO. It may be noted that the catalyst is acting in part as a dehydrating agent, just as do acidic oxides. In much the same way, zirconia (zirconium dioxide, ZrO_2) is used industrially to catalyze the direct hydrogenation of aromatic carboxylic acids (e.g., benzoic acid, C_6H_5COOH) to the corresponding aldehydes (in this case, C_6H_5CHO) by a combination of hydrogenation and dehydration.

Basic oxides such as La_2O_3 and mixtures like Li_2O(5% by weight)–MgO have important potential as catalysts in the oxidative coupling of methane.[13] Large reserves of methane are available in the form of natural gas and could serve as feedstock for production of many organic chemicals rather than as a mere fuel, but the CH_4 molecule is exasperatingly unreactive in most circumstances. With basic metal oxide catalysts at 600–900 °C, however, methane will react with the oxygen of the air to form higher hydrocarbon products such as ethane and particularly ethene (i.e., ethylene, the largest tonnage organic chemical) together with lesser amounts of CO and CO_2. Industrial application of these processes will require improvement of the ethene yields (in effect, minimization of ethene losses through oxidation) and development of appropriate technology to separate the products.

6.5 Photocatalysis by Inorganic Solids

Much effort has gone into development of catalysts for photochemical reactions, initially with the objective of converting solar energy into storable fuels (typically H_2 from the photolysis of water) but, more recently, mainly for the destruction of noxious pollutants such as chlorocarbons. There are two ways in which a catalyst may be involved in a photochemical reaction: it may simply provide a surface on which the reactants can be adsorbed, so that, when a molecule of one reactant is activated by absorption of light, a molecule of the other is held in close proximity to facilitate reaction (a *catalyzed photoreaction*); or it may itself be excited by the absorption of light and then activate the adsorbed molecules (a *sensitized photoreaction*). The latter mode is the more relevant to the theme of this chapter, and is exemplified by the photocatalytic properties of titanium dioxide, TiO_2.[14,15]

6.5.1 Photocatalytic Splitting of Water

Titanium dioxide is an insulator, having a completely filled valence band and a completely empty conduction band. The band gap, however, is about 3.2 eV, which corresponds to the energy of photons of light of wavelength 400 nm—just outside the visible region, in the near-ultraviolet (near-UV). Thus, the UV portion of sunlight can excite electrons from the valence band of TiO_2 into the conduction band, leaving positive holes in the valence band. In this way, TiO_2 in sunlight can provide very energetic electrons from the conduction band to reduce an adsorbed substance, while positive holes from the valence band can act as electron scavengers (i.e., as oxidants). Small band-gap semiconductors such as CdS could conceivably make use of the entire visible region of light, but they tend to photodegrade. In contrast, TiO_2 is very stable thermally and photochemically, and towards water over the range pH 1-14; it is also non-toxic and inexpensive, and provides a highly energetic electron/hole couple.

In principle, the 3.2 eV (309 kJ mol^{-1}) electron donor/electron acceptor pairs in TiO_2 should have more than enough energy to decompose water into hydrogen and oxygen (1.23 V), but the evolution of both O_2 and H_2 on TiO_2 surfaces is hindered by very high *overpotentials*. The phenomenon of overpotential is considered at length in Section 15.4, but for present purposes we can regard it as a mechanistic and hence *kinetic* limitation. What is needed, then, are surfaces in electrical contact with the TiO_2 that catalyze the formation of gaseous H_2 and O_2. Two substances with these respective properties are platinum metal and ruthenium(IV) oxide (RuO_2), and chemists have prepared TiO_2 powders in which the particles are partially coated with RuO_2 to serve as anodes for O_2 evolution and with Pt metal for cathodes where H_2 can be evolved. The RuO_2 traps holes from the valence band and oxidizes the water to oxygen, while Pt traps the electrons from the conduction band and reduces the water to hydrogen. Alternatively, a photoactivated TiO_2 electrode can be connected to a separate Pt electrode; here, O_2 evolution must take place on the TiO_2, and a small voltage may have to be impressed externally to facilitate this. Another possibility is to scavenge the holes with a "sacrificial donor" such as methanol, which would be oxidized in water to CO_2.

6.5.2 Photocatalysis and Environmental Protection

With the current ready availability of inexpensive petroleum fuels, interest in splitting water for the hydrogen content has waned; at the same time, however, concern over possible toxic effects of chlorocarbons released into the environment has intensified, and much effort has gone into the development of TiO_2-photocatalyzed oxidation of organic pollutants at low concentrations in water. Whether or not O_2 is ultimately produced when water is exposed to UV-irradiated TiO_2, the usual immediate effect is oxidation of water by the holes to produce hydroxyl radicals $\cdot OH$, and these are very effective at oxidizing organic matter, notably environmental pollutants. Alternatively, organic matter may be oxidized directly by reaction with the holes. The electrons that were promoted to the conduction band to form the holes can be scavenged by O_2, so that the net reaction is oxidation of the pollutant with consumption of O_2.

For example, trichloroethylene, a widely used industrial solvent, can be completely oxidized by air to inorganic products ("mineralized") over UV-irradiated TiO_2, giving CO, CO_2, HCl, and Cl_2. Similarly, chloroform ($CHCl_3$), a potentially carcinogenic (cancer-causing) contaminant present in some municipal water supplies as a result of chlorination of organic matter (see Chapters 12 and 14), can be mineralized by TiO_2:

$$2CHCl_3(aq) + O_2(g) + 2H_2O(l) \xrightarrow{UV,\ TiO_2} 2CO_2(g) + 6HCl(aq). \quad (6.15)$$

Photooxidations by O_2 over TiO_2 can also be carried out in the gas phase; however, the presence of water still appears to be necessary to assist in the removal of intermediates or products from the solid surface.

Commercial TiO_2 (e.g., Degussa P25), which is usually mainly the anatase form, typically has a particle size of 10–70 nm. Work by Hoffmann's group,[15] however, has shown that TiO_2 particles a mere 2–4 nm in diameter generate even more powerfully oxidizing holes and more strongly reducing electrons than the commercial material on UV irradiation because, with far fewer atoms to form the valence and conduction bands, the bands are narrower in energy and hence the gap between them is wider (see Figs. 4.3 and 4.4). The small TiO_2 particles form transparent *sols* (colloidal dispersions, Section 14.2) that admit light more freely and have markedly higher quantum yields (i.e., photochemical efficiencies) than suspensions of commercial TiO_2.

Titanium dioxide powder is widely used as a white pigment or filler in paints. Its intense whiteness, and hence high covering power, arises from its high index of refraction. Unlike the lead(II) carbonate pigment formerly used, it is nontoxic and chemically inert. As noted earlier, however, TiO_2 is capable of photocatalyzing oxidation of organic constituents of the paint when exposed to UV light in moist air, and this can lead to discoloration (typically yellowing) of the paint over time. An antioxidant should therefore be added to the paint at the time of manufacture.

Exercises

6.1 Japanese chemists succeeded in obtaining good yields of methane by reaction of H_2 with a mixture of carbon monoxide *and carbon dioxide*, at temperatures as low as 270 °C, by use of a special mixed catalyst containing nickel as the most important metallic constituent. Why is nickel used? In the same vein, why is platinum or platinum–rhodium alloy (but *not* nickel) used in catalytic converters for automobile exhausts? (See also Section 17.4.)

6.2 Two important ways in which heterogeneously catalyzed reactions differ from homogeneous counterparts are the definition of the rate constant k' and the form of its dependence on temperature T. The heterogeneous rate equation relates the rate of decline of the concentration (or partial pressure) c of a reactant to the fraction f of the catalytic surface area that it covers when adsorbed. Thus, for a first-order reaction,

$$\text{rate} = -\frac{dc}{dt} = k'f$$

if the products are not adsorbed. At low enough temperatures, $f \approx 1$, and the usual Arrhenius equation gives the true activation energy for the heterogeneous reaction. Adsorption, however, is invariably exothermic, so that, as T is increased, f will eventually decrease and, in the high temperature limit, will approach zero. What will the apparent activation energy E_{app} be, in the high temperature limit? Sketch the dependence of ln(rate) on $1/T$.

6.3 Show that a substance that is an effective catalyst for a particular reaction will also catalyze the *reverse* reaction (cf. Section 2.5 and Exercise 2.2).

6.4 It is possible, by progressively "dealuminating" certain zeolites, to make solids that approach SiO_2 in composition while still retaining the cavities and channels typical of zeolite architecture (Section 7.3). What technological value might such materials have, and why? [*Note:* One such substance, the Socony–Mobil catalyst ZSM-5, catalyzes the conversion of methanol to gasoline-type alkanes; the first step is apparently the formation of dimethyl ether.]

References

1. (*a*) T. H. Maugh II, Catalysis; no longer a black art. *Science* **219**, 474–477 (1983); (*b*) S. T. Ceyer, New mechanisms for chemistry at surfaces. *Science* **249**, 133–139 (1990); (*c*) J. T. Yates Jr., Surface chemistry. *Chem. Eng. News*, March 30, 22–35 (1992); (*d*) H. Cai, A. C. Hillier, K. R. Franklin, C. C. Nunn, and M. D. Ward, Nanoscale imaging of molecular adsorption. *Science* **266**, 1551–1555 (1994); (*e*) D. W. Goodman, Model studies in catalysis using surface science probes. *Chem. Rev.* **95**, 523–536 (1995); (*f*) J. A. Schwarz, C. Contescu, and A. Contescu, Methods for the preparation of catalyst materials. *Chem. Rev.* **95**, 477–510 (1995).

2. G. C. Bond, "Heterogeneous Catalysis: Principles and Applications," 2nd Ed. Oxford Univ. Press (Clarendon), Oxford, 1987; G. C. Bond, *In* "Insights into Speciality Inorganic Chemicals" (D. Thompson, ed.), pp. 63–104. Royal Society of Chemistry, Cambridge, 1995.

3. B. C. Gates, "Catalytic Chemistry." Wiley, New York, 1992.

4. M. V. Twigg and M. S. Spencer (series eds.), "Fundamental and Applied Catalysis." Plenum, New York. This series includes J. T. Richardson, "Principles of Catalyst Development," 1989; V. P. Zhdanov, "Elementary Physicochemical Processes on Solid Surfaces," 1991; J. R. Jennings (ed.), "Catalytic Ammonia Synthesis," 1991.

5. M. V. Twigg (ed.), "Catalyst Handbook." Wolfe, London, 1989.

6. M. W. Roberts, Chemisorption and reaction pathways at metal surfaces: The role of surface oxygen. *Chem. Soc. Rev.* **18**, 451–476 (1989).

7. P. Laszlo, Catalysis of organic reactions by inorganic solids. *Acc. Chem. Res.* **19**, 121–127 (1986).

8. G. C. Bond, Supported metal catalysts—Some unsolved problems. *Chem. Soc. Rev.* **20**, 441–475 (1991).

9. L. D. Schmidt, M. Huff, and S. S. Bharadwaj, Catalytic partial oxidation reactions and reactors. *Chem. Eng. Sci.* **49**, 3981-3994 (1994); S. S. Bharadwaj and L. D. Schmidt, Catalytic partial odixation of natural gas to syngas. *Fuel Process. Technol.* **42**, 109-127 (1995).

10. B. C. Gates, Supported metal clusters: synthesis, structure, and catalysis. *Chem. Rev.* **95**, 511–522 (1995).

11. A. Corma, Inorganic solid acids and their use in acid-catalyzed hydrocarbon reactions. *Chem. Rev.* **95**, 559–614 (1995).

12. H. Hattori, Heterogeneous base catalysis. *Chem. Rev.* **95**, 537–558 (1995).

13. J. H. Lunsford, The catalytic oxidative coupling of methane. *Angew. Chem. Int. Ed. Engl.* **34**, 970–980 (1995).

14. A. L. Linsebigler, G. Lu, and J. T. Yates, Jr., Photocatalysis on TiO_2 surfaces: Principles, mechanisms, and selected results. *Chem. Rev.* **95**, 735–758 (1995).

15. M. R. Hoffmann, S. T. Martin, W. Choi, and D. W. Bahnemann, Environmental applications of semiconductor photocatalysis. *Chem. Rev.* **95**, 69–96 (1995).

Chapter 7

Silicates, Aluminates, and Phosphates

NEXT TO oxygen, silicon and aluminum are by far the most abundant constituents of the crust of the Earth (Table 1.1),[1] where they occur as silica (SiO_2), alumina (Al_2O_3), and a great variety of solid silicates and aluminosilicates. The diversity of silicates originates in the propensity of Si to form strong Si—O—Si links as noted in Section 3.5, leading to very stable chains, rings, sheets, and networks based on oxygen sharing (corner sharing) between $SiO_4{}^{4-}$ tetrahedra. The Al^{3+} ion is capable of substituting for "Si^{4+}" in some of the tetrahedra, giving rise to aluminosilicate frameworks. In all silicates and aluminosilicates, cations of various kinds must be present to counterbalance the negative charge of the anionic units or frameworks. Thus, four of the five volumes of Deer, Howie, and Zussman's invaluable reference work on mineralogy[2] are taken up with silicates of one kind or another, and these materials constitute increasingly important economic resources.[3]

7.1 Silicate Structures

To simplify the essential features of complicated silicate structures, we represent the $SiO_4{}^{4-}$ (or aluminate $AlO_4{}^{5-}$) tetrahedron will be represented by the triangle shown in Fig. 7.1 (left), which is essentially the tetrahedron as viewed down one of the Si—O bonds. Triangle corner sharing therefore implies O^{2-} sharing between linked $SiO_4{}^{4-}$ units. Aluminum has the ability (shared by Si, but only at very high pressures; see Section 4.5) to be six as well as four coordinate and to link with $SiO_4{}^{4-}$ units by corner sharing of the $AlO_6{}^{9-}$ octahedron. We represent this octahedron by the double-triangle symbol in Fig. 7.1, showing an octahedron as viewed down

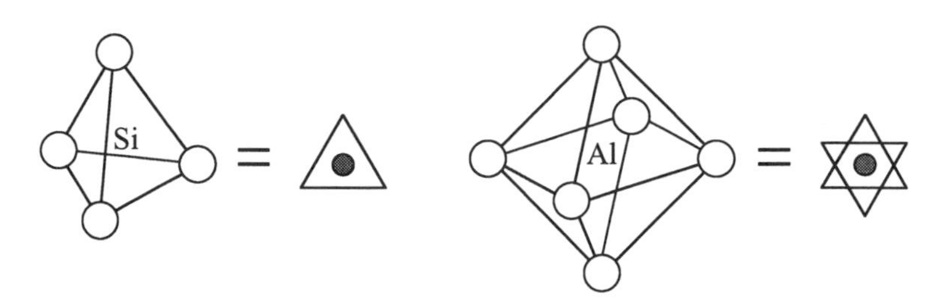

Figure 7.1 Representation of SiO_4^{4-} tetrahedron (left) and AlO_6^{9-} octahedron in terms of the view down threefold axes.

a threefold axis (perpendicular to one of the triangular faces). Typical silicate frameworks are summarized in Fig. 7.2.[2–5]

(*a*) *Orthosilicates.* Structures containing isolated SiO_4^{4-} units are called *orthosilicates.* Examples include *phenacite* ($Be_2[SiO_4]$) and *willemite* ($Zn_2[SiO_4]$) in which the beryllium and zinc cations are also tetrahedrally surrounded by the oxygens of the silicate anions. A more important case is *olivine,* $(Fe,Mg)_2[SiO_4]$, which can also be regarded as a rumpled hcp array of O^{2-}, with "Si^{4+}" in one-eighth of the T-holes and the divalent cations in one-half of the O-holes. [Compare and contrast this with the spinel structure (Section 4.6) in which the O^{2-} sub-lattice is ccp.] The end-member compositions of olivine are known as *forsterite* (Mg_2SiO_4) and *fayalite* (Fe_2SiO_4).

(*b*) *Pyrosilicates.* The *pyrosilicate* ion, $Si_2O_7^{6-}$, is not very common in solid minerals. *Thortveitite* ($Sc_2[Si_2O_7]$) is a scandium mineral with this anion. Note that the charge on the ion is given by four times the number of silicons minus twice the number of oxygens.

(*c*) *Three-silicate rings.* Three-silicate rings, $Si_3O_9^{6-}$, occur in a few minerals such as *benitoite* ($BaTi[Si_3O_9]$). The ring consists of three silicons linked through three oxygens, so it is actually six membered and not particularly strained.

(*d*) *Six-silicate rings.* Six-silicate rings, $Si_6O_{18}^{12-}$, are known in the precious stone *beryl* ($Be_3Al_2[Si_6O_{18}]$).

(*e*) *Single-chain silicates.* Single-chain silicates, $(SiO_3)_n^{2n-}$, are called *pyroxenes.* (Note that n is not intended to mean the chain length, which is "infinite.") Examples are *enstatite* ($Mg[SiO_3]$) and *diopside,* $MgCa[(SiO_3)_2]$.

(*f*) *Double-chain silicates.* Double-chain silicates, $(Si_4O_{11})_n^{6n-}$, are known as *amphiboles,* such as *tremolite,* $Ca_2Mg_5(OH)_2[(Si_4O_{11})_2]$. These include the true asbestoses, such as *crocidolite* or *blue asbestos,*

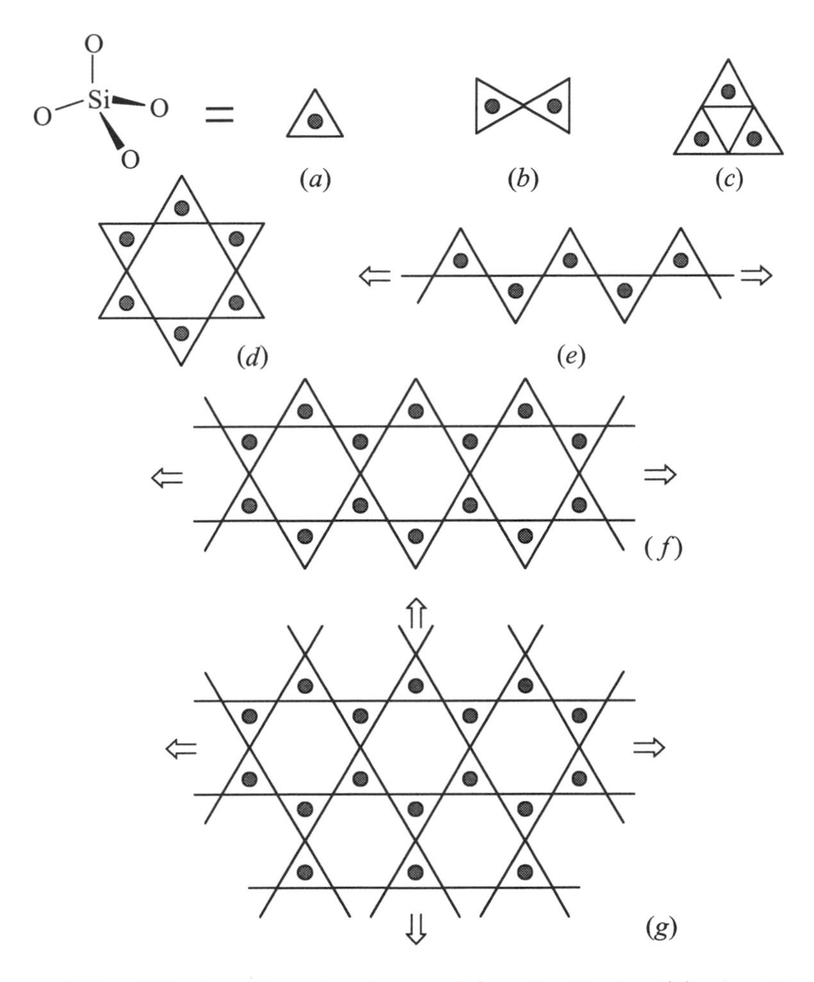

Figure 7.2 Silicate anion structures: (*a*) orthosilicate, (*b*) pyrosilicate, (*c*) three-silicate ring, (*d*) six-silicate ring, (*e*) pyroxene, (*f*) amphibole, and (*g*) phyllosilicate.

which owe their fibrous character and heat-insulating properties to the silicate chain structures (crocidolite requires milling to separate the fibers). Most commercial asbestos, however, is *chrysotile*, which is actually a layer silicate that forms tubular fibers like a rolled carpet [see (*g*)].

(*g*) *Sheet silicates.* Sheet silicates, $(Si_2O_5)_n{}^{2n-}$, are called *phyllosilicates* (phyllo means leaflike) and are characterized by easy cleavage parallel to the sheets of silicate (where these are flat). This is typified by *talc*,

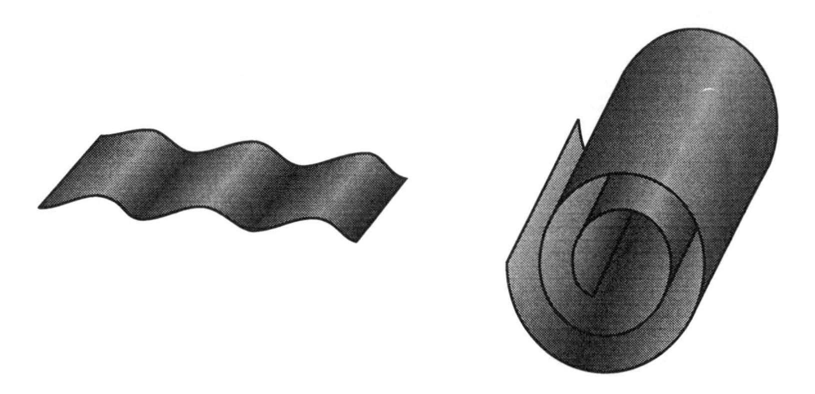

Figure 7.3 Serpentine sheet structures: antigorite (left) and chrysotile.

$Mg_3(OH)_2[Si_4O_{10}]$, which has a greasy feel and pearly luster. Talc is the chief constituent of soapstone (*steatite*), which is used for carvings by Inuit artists. The *serpentines*, $Mg_3(OH)_4[Si_2O_5]$, have sheet silicate structures in which the sheets tend to be curved. *Antigorite* has essentially flat sheets with gentle waves of radius of curvature 7.5 nm, whereas in chrysotile the curvature is all in one direction, creating tubular fibers with inner and outer diameters of about 10 and 20 nm, respectively, and a high aspect ratio (length/diameter) (Fig. 7.3).

7.1.1 Asbestos: Uses and Hazards

Chrysotile is a noncombustible fibrous solid that has been widely used as a fireproof thermal insulator, for brake linings, in construction materials, and for filters under the name of asbestos. It decomposes with loss of water at 600–800 °C, eventually forming forsterite and silica at 810–820 °C. Because it is more resistant to attack by alkalis than are the amphibole asbestoses, chrysotile has been used in chloralkali cell membranes and in admixture with Portland cement for making sewer pipes (Chapter 11).

Asbestos fibers have been linked to the high incidence of lung cancer and mesothelioma (an otherwise rare pleural cancer) in asbestos workers and are considered to pose a threat to the population at large. Some 5000 deaths per annum in the United States have been attributed to mineral fiber contacts. There is now evidence that the *true* asbestoses such as amosite or crocidolite have a much higher association with mesothelioma than does the chrysotile used in most commercial asbestos, but controversy continues regarding the health hazards of asbestoses. The problem is evidently a general one with dust particles and fibers of any kind that are small enough to enter the alveoli of the lungs and remain there for long periods without dissolution or biodegradation.[6,7] In any event, prudence dictates that anyone handling asbestos or other insoluble fibrous materials should wear appropriate breathing equipment and clothing, and asbestos filters should not be used in preparing foodstuffs such as wines.

7.2 Aluminosilicates

Silica occurs widely in Nature as *quartz*, often in large transparent crystals of characteristic shape but also in the translucent agglomerations of microscopic crystals known as *chalcedony*, which includes cherts and flint. Other natural crystalline varieties of SiO_2 include *tridymite* and *cristobalite* (opal is a semiprecious stone that consists of microcrystalline, hydrous cristobalite). All forms of silica involve three-dimensional networks of corner-linked SiO_4 tetrahedra.

If we were to replace some of the "Si^{4+}" ions with Al^{3+}, it would still be possible to have a three-dimensional Si—O—Al network, but cations would be needed to counterbalance the now anionic structural framework. Thus, for each substituent Al^{3+}, we must add, say, an Na^+, a K^+, or half a Ca^{2+} ion. The *feldspars*, which along with quartz and micas (see below) are typical constituents of granites, can be viewed in this way:

$$\begin{aligned} K[AlSi_3O_8] &\ \text{orthoclase} \\ Na[AlSi_3O_8] &\ \text{albite} \\ Ca[Al_2Si_2O_8] &\ \text{anorthite.} \end{aligned}$$

Plagioclase is a solid solution of albite in anorthite (or vice versa).

The *micas* have layer structures in which silicate sheets are combined with aluminate units; the aluminum ions can be octahedrally as well as tetrahedrally coordinated. For example, the mica *muscovite* contains both octahedral and tetrahedral Al^{3+}:

$$\underset{\substack{\uparrow \\ \text{octahedral}}}{KAl_2}(OH)_2[Si_3\underset{\substack{\uparrow \\ \text{tetrahedral}}}{Al}O_{10}].$$

The potassium ions are located between the flat aluminosilicate sheets (Fig. 7.4). Crystals of micas cleave easily parallel to the sheets, and the thin transparent flakes can be used for electrical insulation (e.g., in capacitors) or as furnace windows. *Phlogopite*, $KMg_3(OH)_2[Si_3AlO_{10}]$, has a similar structure but with Mg^{2+} in octahedral environments instead of Al^{3+}.

Many *clay minerals* have aluminosilicate layer structures. For example, in *kaolinite*, $Al_2(OH)_4[Si_2O_5]$ (Fig. 7.5), the Al^{3+} are all in octahedral locations. Clay minerals of the *smectite* or swelling type, such as *montmorillonite*, can absorb large amounts of water between the aluminosilicate

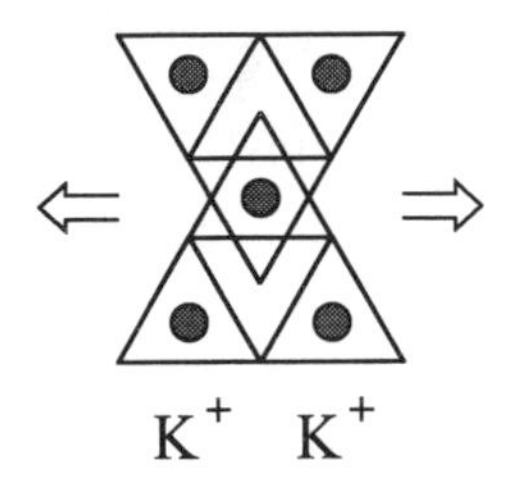

Figure 7.4 Structure of muscovite (sheet viewed edgewise).

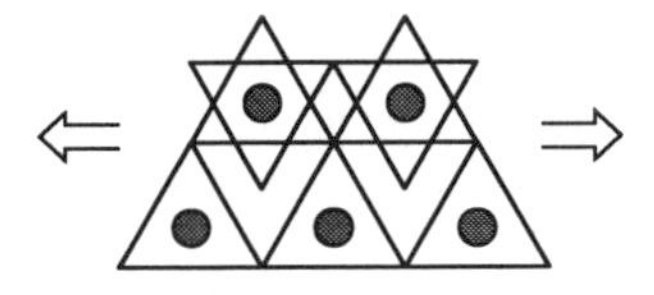

Figure 7.5 Layer structure of kaolinite, $Al_2(OH)_4[Si_2O_5]$ (sheet viewed edgewise): Al^{3+} are octahedrally coordinated, "Si^{4+}" tetrahedrally.

layers, with the result that the interlayer spacing Δd can expand greatly. In petroleum recovery operations, superheated water is sometimes injected into formations such as the Athabasca tar sands of northern Alberta with the object of reducing the viscosity of the oil content and sweeping it to the surface in a production well. If, however, smectite minerals form by hydrothermal reactions within the formation, the swollen products may block the flow path (perhaps to advantage, if regions stripped of oil are thereby sealed off). Typically, kaolinite and dolomite in a tar-sands matrix (which is mainly quartz) will react with high temperature water to form montmorillonite.

7.3 Zeolites

Zeolites merit special consideration in view of their growing importance in chemical engineering. They are natural or synthetic aluminosilicates in which the anionic Al—O—Si framework encloses cavities linked by channels (Fig. 7.6).[8] They are represented by

$$M_{a/n}[(AlO_2)_a(SiO_2)_b]\cdot xH_2O$$

where $1 \leq b/a \leq 5$. Again, when we have n Al^{3+} ions substituting for "Si^{4+}" in what would otherwise be SiO_2, we must have a counterion M^{n+}. Here, however, the M^{n+} occupy the cavity/channel system, along with the x water molecules, and are often easily replaceable with other cations that

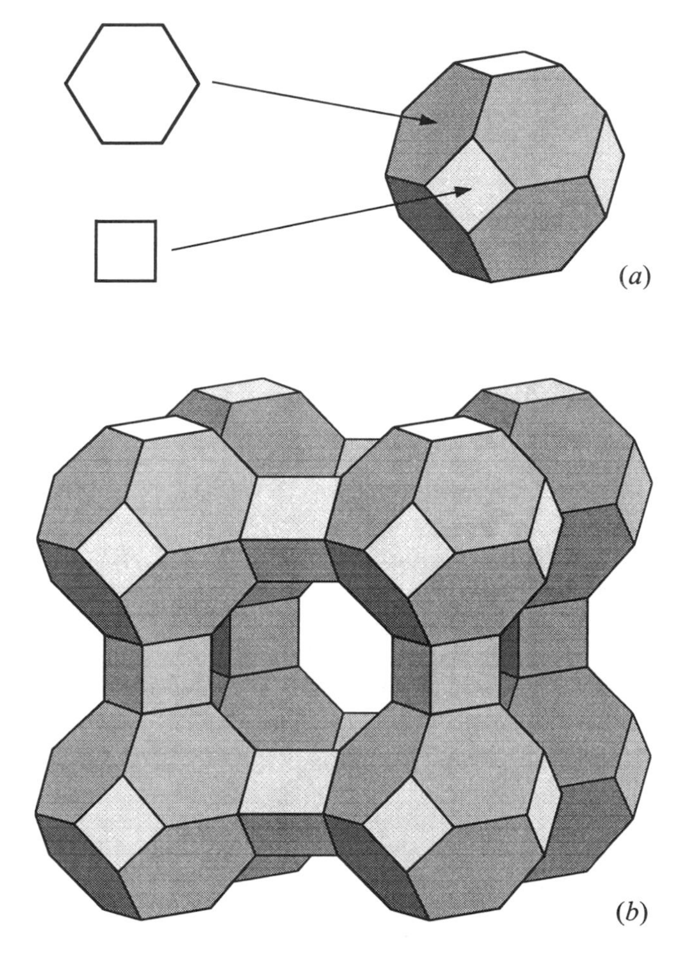

Figure 7.6 (*a*) Fusion of four-Si/Al and six-Si/Al rings to form the sodalite cage and (*b*) linking of eight sodalite cages to form the cavity-and-channel structure of zeolite-A, $M^{I}_{12}Al_{12}Si_{12}O_{48}{\cdot}27H_2O$. Each corner is an Si or Al atom; the edges represent bridging oxygens.

may diffuse into these *pores* from a solution in which the zeolite may be immersed. Similarly, the water in the pores is driven out on heating, and the name "zeolite" comes from the Greek, meaning "boiling stone."

Typical zeolites include the commonly encountered *analcite* (natural, also called *analcime*, $Na[AlSi_2O_6]{\cdot}H_2O$), *faujasite* {natural, (Na_2,Ca)-$[(AlO_2)_2(SiO_2)_4]{\cdot}8H_2O$, the structural analog of synthetic *zeolites X* and *Y*}, and *zeolite-A* {synthetic, $M^{I}_{12}[(AlO_2)_{12}(SiO_2)_{12}]{\cdot}27H_2O$}. The zeolite-like aluminosilicate framework mineral *sodalite*, $Na_8[(AlO_2)_6(SiO_2)_6]Cl_2$,

TABLE 7.1
Effective Channel Widths in Zeolite-A

Zeolite type	*Cation*	*Channel width (pm)*
3A	K^+	300
4A	Na^+ [a]	400
5A	$\frac{1}{2}Ca^{2+}$ [b]	500

[a] Na^+ is smaller than K^+, so pores are wider
[b] Ca^{2+} has a size similar to Na^+, but there only half as many

contains the truncated octahedral structural unit known as the *sodalite cage*, shown in Fig. 7.6a, which is found in several zeolites. The corners of the faces of the cage are defined by either four or six Al/Si atoms, which are joined together through oxygen atoms. The zeolite-A structure, as shown in (Fig. 7.6b), is generated by joining sodalite cages through the four-Si/Al rings, so enclosing a cavity or *supercage* bounded by a cube of eight sodalite cages and readily accessible through the faces of that cube (*channels* or *pores*). The structural frameworks of faujasite and zeolites X and Y are similarly generated by joining sodalite cages together through the six-Si/Al faces. In zeolite-A, the effective width of the pores is controlled by the nature of the cation M^+ or M^{2+} (see Table 7.1).

Natural zeolites form hydrothermally (e.g., by the action of hot water on volcanic ash or lava), and synthetic zeolites can be made by mixing solutions of aluminates and silicates and maintaining the resulting gel at temperatures of 100 °C or higher for appropriate periods. Zeolite-A can form at temperatures below 100 °C, but most zeolite syntheses require hydrothermal conditions (typically 150 °C at the appropriate pressure). It appears that the reaction mechanism involves redissolution of the gel and reprecipitation as the crystalline zeolite.[8d, 8e, 9] The identity of the zeolite produced depends on the composition of the solution. Aqueous alkali metal hydroxide solutions favor zeolites with relatively high Al contents, while the presence of organic molecules such as amines or alcohols favors highly siliceous zeolites such as silicalite or ZSM-5. Various tetraalkylammonium cations favor the formation of certain specific zeolitic structures and are known as *template* ions, although it should not be supposed that the channels and cages form simply by the wrapping of aluminosilicate fragments around suitably shaped cations.

Zeolites have many uses, most importantly as *cation exchangers* (e.g., in water softening), as *desiccants* (i.e., drying agents), and as solid *acid catalysts*.

7.3.1 Zeolites as Cation Exchangers

Water that contains significant amounts of Ca^{2+} or Mg^{2+} is said to be "hard." These ions cause soluble soaps such as sodium stearate, palmitate, and oleate to precipitate as insoluble scums, which are an unsightly nuisance as well as a waste of soap:

$$Ca^{2+}(aq) + 2C_{17}H_{35}CO_2{}^-(aq) \rightarrow (C_{17}H_{35}CO_2)_2Ca(s). \quad (7.1)$$

Synthetic detergents such as alkyl sulfonates may not be precipitated, but Ca^{2+} and Mg^{2+} ions tend to form anion–cation complexes with them in solution and so reduce their effectiveness as detergents. The detergent power derives from the ability to form *micelles* enveloping the greasy dirt particles with the anionic end of the detergent molecule outward, so forming a colloidal dispersion in water (Fig. 7.7, cf. Section 14.2). The Ca^{2+} and Mg^{2+} can be replaced by Na^+, which does not precipitate scum, by treatment with a zeolite in its sodium form, say, Na_2Z:

$$Na_2Z(s) + Ca^{2+}(aq) \rightleftharpoons CaZ(s) + 2Na^+(aq) \quad (7.2a)$$

$$Na_2Z(s) + Mg^{2+}(aq) \rightleftharpoons MgZ(s) + 2Na^+(aq). \quad (7.2b)$$

The equilibria of reaction 7.2 lie to the right, as the more highly charged cations have greater affinity for the anionic zeolite framework Z, and so a dilute solution of Ca^{2+} or Mg^{2+} (i.e., typical hard water) can be passed through a column of zeolite particles to replace essentially all of the divalent cations in solution with Na^+. All the Na_2Z will eventually be converted to CaZ or MgZ, but, since reaction 7.2 is reversible under sufficiently forcing conditions, the Na-form zeolite can be regenerated by back-flushing the column with concentrated brine. The resulting solution of $NaCl/MgCl_2/CaCl_2$ is then discarded.

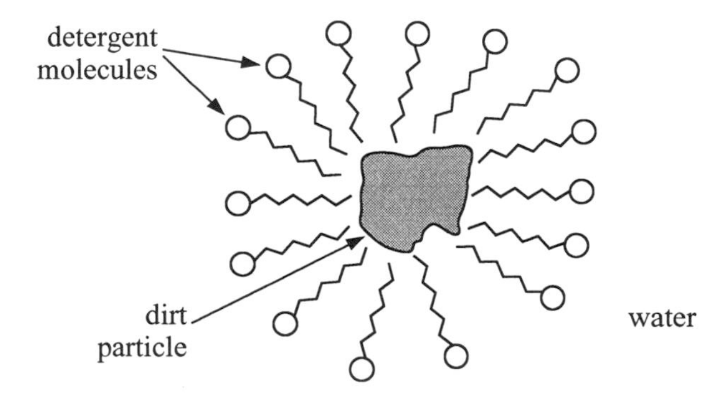

Figure 7.7 Detergent action.

Alternatively, finely powdered zeolites may be used in detergent powders as a *builder,** since zeolite particles smaller than 10 μm do not stick to clothing. Formerly, sodium polyphosphates (Section 7.7) were used extensively as detergent builders to "tie up" the Ca^{2+} and Mg^{2+} in hard water as soluble complexes (or as a precipitate that washes away). However, many communities ban phosphate detergents because of pollution problems (Sections 7.7 and 9.6), thus creating a major new market for zeolites.

7.3.2 Zeolites as Desiccants

If the water content is driven off (usually by heating to 350 °C in a vacuum), the dehydrated zeolite becomes an avid absorber of *small* molecules, especially water. The size of the molecules that can be absorbed is limited by the zeolite pore diameter, which is different for different zeolites (Table 7.1); a given zeolite (e.g., zeolite 3A) can be a highly selective absorber of, say, small amounts of water from dimethyl sulfoxide (DMSO) solvent. For this reason, dehydrated zeolites are often called *molecular sieves.*

To be retained by the charged zeolite framework, a molecule has to be polar, as well as small enough to penetrate the pores. Thus, the angular (104°) H_2O molecule, with its strongly polar bonds and permanent dipole moment (as well as an ability to form hydrogen bonds to the anionic framework), is strongly absorbed. However, monatomic helium, tetrahedral methane, and linear CO_2 have no permanent dipole moments and are not held in a typical zeolite, even though they can penetrate it easily. Consequently, zeolites are used to remove water from natural gas feedstock used in the cryogenic production of helium and in preparation of liquefied natural gas (LNG) for shipment. (Helium, from alpha particles produced in radioactive decay in rocks, tends to accumulate in natural gas.)

7.3.3 Zeolites as Solid Acid Catalysts

The effective area of the anionic aluminosilicate framework in the pores of a zeolite is at least 100 times the external surface area, and it can be as high as 1000 $m^2\ g^{-1}$. Consequently zeolites are unusually effective as catalysts for reactions that are favored by aluminosilicate surfaces. Substitution of "Si^{4+}" by Al^{3+} in a "silica" framework makes it *acidic* and, potentially, *coordinatively unsaturated.* Suppose, for example, that we heat the NH_4^+ form of a zeolite. Ammonia is driven off, and one H^+ remains to counterbalance each Al^{3+} that has substituted for a silicon. The protons are attached to oxygens of the aluminosilicate framework:

---

*A builder is an additive that improves the effectiveness of the detergent.

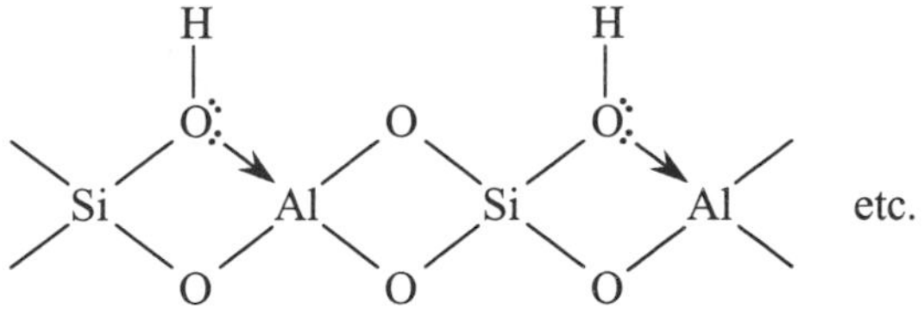

$$\begin{array}{c} \mathrm{H} \quad\quad\quad\quad \mathrm{H} \\ | \quad\quad\quad\quad\quad | \\ \diagdown \mathrm{Si} \diagup \ddot{\mathrm{O}}{:} \searrow \mathrm{Al} \diagup \mathrm{O} \diagdown \mathrm{Si} \diagup \ddot{\mathrm{O}}{:} \searrow \mathrm{Al} \diagup \\ \diagup \quad \diagdown \mathrm{O} \diagup \quad \diagdown \mathrm{O} \diagup \quad \diagdown \mathrm{O} \diagup \quad \diagdown \end{array} \quad \text{etc.}$$

However, further heating drives off the elements of water as water vapor:

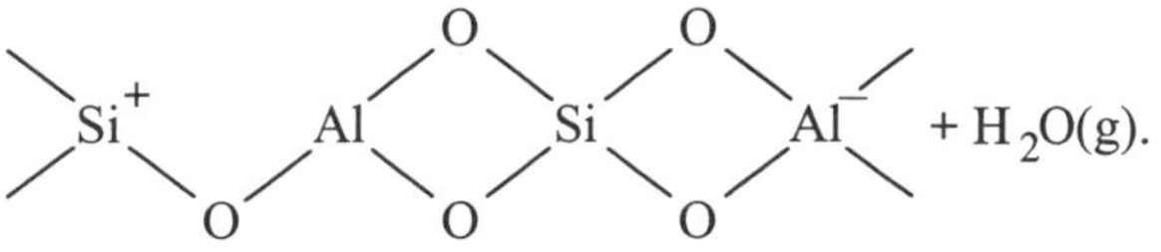

$$\diagdown \mathrm{Si}^{+} \quad \diagdown \mathrm{O} \diagup \mathrm{Al} \diagup \mathrm{O} \diagdown \mathrm{Si} \diagup \mathrm{O} \diagdown \mathrm{Al}^{-} \diagup \quad + \mathrm{H_2O(g)}.$$

Here, one Al has lost H^+ from the neighboring OH group, and one Si has lost OH^- and therefore becomes coordinatively unsaturated, as well as positively charged. This Si can therefore act as catalytic site for the numerous organic reactions that are catalyzed by *Lewis acids* (i.e., electron-pair acceptors).

Consider, for example, the zeolite-catalyzed dehydration of alcohols to give olefins:

$$\mathrm{RCH_2CH_2OH} + -\overset{|}{\underset{|}{\mathrm{Si}}}{}^{+} \longrightarrow \mathrm{RCH_2CH_2}^{+} + -\overset{|}{\underset{|}{\mathrm{Si}}}-\mathrm{OH} \tag{7.3}$$

$$\mathrm{RCH_2CH_2}^{+} \longrightarrow \mathrm{RCH{=}CH_2} + \mathrm{H}^{+} \tag{7.4}$$

$$\mathrm{H}^{+} + -\overset{|}{\underset{|}{\mathrm{Si}}}-\mathrm{OH} \longrightarrow \mathrm{H_2O} + -\overset{|}{\underset{|}{\mathrm{Si}}}{}^{+} \text{ (regenerated).} \tag{7.5}$$

At higher temperatures, C—H and C—C bonds may be similarly broken. Thus, zeolite catalysts may be used for (*i*) alkylation of aromatic hydrocarbons (cf. the Friedel–Crafts reactions with $AlCl_3$ as the Lewis acid catalyst), (*ii*) *cracking* of hydrocarbons (i.e., loss of H_2), and (*iii*) isomerization of alkenes, alkanes, and alkyl aromatics.

In isomerizations, zeolites have special merit in their ability to admit straight-chain but not branched-chain molecules into the pores. Thus, normal alkanes up to n-$C_{14}H_{30}$ can penetrate the pores of zeolite 5A to reach the cavities where the C—C or C—H bonds may be catalytically broken; the fragments, on reemerging from the pores, can recombine as isomerized molecules. The reverse process is not possible, since the isomers, having

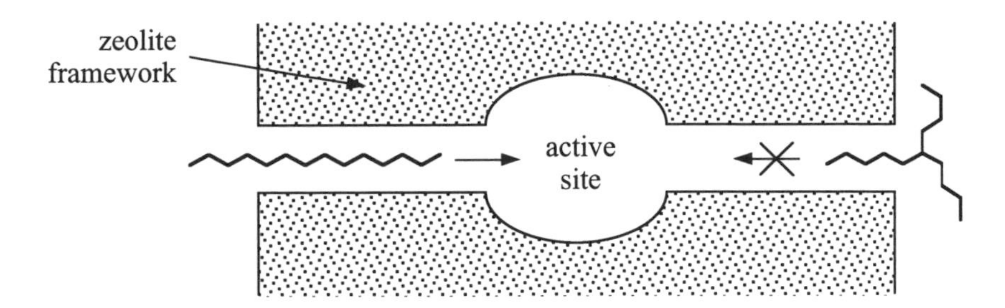

Figure 7.8 Shape-selective reforming. The straight-chain alkane (left) can enter the zeolite pore and penetrate to the catalytic site, whereas the branched-chain isomer cannot.

branched chains, cannot enter the pores (Fig. 7.8). This is known as *shape-selective reforming*[10, 11] and is important in the upgrading of the gasoline fraction in petroleum refining. The object is to minimize the amount of *n*-alkanes (especially *n*-heptane) in fuel for internal combustion engines, as they promote *engine knock* (violent detonation rather than smooth burning of the air–fuel mixture in the cylinder), and to maximize the branched-alkane content, especially isooctanes.

In practice, short-chain alkanes and alkenes are normally used as feedstock for shape-selective catalytic formation of isooctanes at relatively low temperatures. Until the 1980s, lead alkyls (Section 18.1) were added to most automotive fuels to help suppress engine knock, but they have been phased out in North America because of the chronic toxicity of lead and lead compounds. The most commonly used nonlead antiknock additive is now methyl *tert*-butyl ether [MTBE; $CH_3OC(CH_3)_3$], which is made by the reaction of methanol with 2-methylpropene, $(CH_3)_2C{=}CH_2$ (see Section 7.4). The latter is obtained by catalytic cracking of petroleum fractions to give 1-butene, which is then shape-selectively isomerized on zeolitic catalysts.

7.4 Clays

Clays are layer silicates (phyllosilicates) of particle size less than about 4 μm, produced by the weathering of aluminosilicate rocks. Clay minerals fall roughly into two structural classes: the kaolinite type, based on paired sheets of tetrahedral ($SiO_4{}^{4-}$) and octahedral [$AlO_n(OH)_{(6-n)}^{(3+n)-}$ or $MgO_n(OH)_{(6-n)}^{(4+n)-}$] units (Fig. 7.5), and the illite type in which the octahedral sheet is sandwiched between two layers of tetrahedra (cf. micas such as muscovite, Fig. 7.4). Many important clay minerals such as vermiculite, biotite, and smectites (notably montmorillonite and beidellite, the principal components of bentonite clay deposits) fall in the latter category. In the natural clay minerals, the space between the composite layers contains the charge-balancing cations—usually Na^+ or K^+—with variable amounts of absorbed water, and the total available surface area (mostly internal) can be as high as 800 m^2 g^{-1}. Consequently, like zeolites, clays can act as cation exchangers, absorbents (e.g., for water or to decolorize organic liquids), and (usually after some modification) as highly selective catalysts.

The cation-exchange capacity of montmorillonite is on the order of 1 mol monovalent cations per kilogram of clay, and replacement of alkali metal cations by H^+ (usually achieved by replacing M^+ with $NH_4{}^+$ and then driving off NH_3 at 200–300 °C, as noted earlier for zeolites) or by more highly charged cations such as Fe^{3+} can result in extremely acidic surface environments and hence potent catalytic surfaces. In the case of ion exchange for multiply charged cations, the acidity originates in an enhanced tendency of the hydrated cations to undergo "hydrolysis" (Section 13.6)—Cr^{3+} and Fe^{3+} are particularly acidic when absorbed into clays—and is further increased when the absorbed water content is low. If, however, the clay is strongly heated, loss of too much water results in reduced acidity and ultimately (above ~150 °C) collapse of the clay layers, and for this reason acidified natural clays are less satisfactory than zeolites for high temperature catalysis.

Collapse of the clay layers can be prevented by *pillaring*,[12] that is, by introducing molecular units that bind to the roof and floor of the interlayer space. For example, alumina pillars can be introduced in the form of the tridecaaluminum cation $(AlO_4)Al_{12}(OH)_{24}(H_2O)_{12}{}^{7+}$,* which is an important constituent of aqueous Al^{III} in the range pH 5–6. Zirconia pillars can be inserted using $ZrOCl_2$ and chromia pillars with hydrolytic polymers of Cr^{III}. The interlayer spacing Δd can be manipulated by choice of an appropriate pillar (e.g., chromia-pillared clays with gallery heights in excess of 2 nm have been reported[12b]), and the resulting porous solid can act as a shape-selective acidic catalyst in the manner of zeolites.

The acidity of a clay can be either of the Brønsted (H^+ donor) or Lewis (electron pair acceptor) type. Even at temperatures below 100 °C, tertiary carbocation intermediates can be generated on clays with high Brønsted acidity through protonation of the C=C double bond in secondary alkenes, as in the clay-catalyzed formation of MTBE from methanol and isobutene:

$$(CH_3)_2C{=}CH_2 \xrightarrow[\text{from clay}]{H^+} (CH_3)_3C^+ \xrightarrow[H^+ \text{ to clay}]{+\ CH_3OH} (CH_3)_3COCH_3. \qquad (7.6)$$

Primary and secondary carbocations can be formed similarly, but at higher temperatures.

*This ion has a very symmetrical structure in which the central ***tetrahedral*** AlO_4 unit is enclosed in four blocks of three ***octahedral*** edge-sharing AlO_6 units.

A wide range of proton-assisted organic reactions, including formation of amides, anhydrides, esters, heterocyclic compounds, lactones, and peptides, can be catalyzed by clays. Indeed, it has been argued that life may have originated from organic compounds formed catalytically from simple molecules such as ammonia, CO_2, and hydrocarbons in suspensions of clays in ancient seas. Lewis acid activity is evident in the catalysis by cation-exchanged clays of Friedel–Crafts reactions (e.g., the acylation of aromatic hydrocarbons with carboxylic acids), Diels–Alder cyclic additions, and aldol condensations, and clays have been used as catalysts in the cracking of petroleum.

If the cation introduced by ion exchange is capable of multiple valence, the clay may serve as a catalyst for oxidation or reduction reactions. For example, montmorillonite treated with iron(III) nitrate is so reactive that it has to be stored under an inert atmosphere; the clay catalyzes reactions of the nitrate ion, such as oxidation of secondary alcohols to ketones (via nitrite ester intermediates) and organic hydrazides to azides, and the nitration of phenols.

Finally, clays can be used as supports for other catalysts such as platinum metal or aluminum chloride, largely to facilitate recovery of the catalyst from a liquid after reaction (e.g., by filtration).

7.5 Silica and Silicate Glasses

The stablest form of silica[8f, 13] under ambient conditions is α-quartz. Although natural quartz is abundant, commonly occurring as large, transparent crystals and as sand, purer quartz is needed for the electronics industry as a frequency standard (e.g., in quartz watches) or piezoelectric* material. Large crystals of quartz can be grown hydrothermally in aqueous NaOH or Na_2CO_3 near its critical point, using a small temperature gradient between the "nutrient" quartz and the seed crystals on which the product is grown.

α-Quartz, which has a trigonal crystal structure, undergoes a rapid, reversible transition to hexagonal β-quartz at 573 °C and then slowly changes to hexagonal β-tridymite at about 870 °C; tridymite in turn goes over slowly to cubic β-cristobalite at 1470 °C, and this melts at 1713 °C. The reversion of cristobalite and tridymite to quartz is slow, so that these forms can exist at room temperature (as α-modifications). In addition, dense modifications with six-coordinate Si are found in shocked rocks associated with meteorite impact craters: coesite forms only above 450 °C and 3.8 GPa, and stishovite requires over 1200 °C and 13 GPa. Survival of those metastable polymorphs on the geological timescale is evidence of an extremely slow recrystallization rate.

---

*A piezoelectric solid generates a voltage in response to applied pressure.

Indeed, molten silica itself crystallizes very slowly, particularly once the temperature has fallen below about 1000 °C, so that, on cooling, it normally "vitrifies," that is, forms a *glass*.[14] The molecular structure of the glass, commonly called fused silica or fused quartz, consists of irregular —Si—O—Si— networks. Whereas the quartz structure contains rings consisting of six SiO_4 tetrahedra, linked regularly into other such rings in three dimensions with long-range order, the rings in silica glass are variable in size, the structure is random, and only very short-range order is apparent. Glasses such as fused silica are true solids (albeit noncrystalline), but from the standpoint of molecular structure they are often described as extremely viscous liquids, supercooled far below their normal freezing points. In fact, there is a definite temperature (or narrow range of temperatures), the *glass transition temperature*, at which the properties of a supercooling liquid change from those of a fluid to those of a rigid glass.

Fused silica glass is made by melting quartz and then cooling the liquid from about 1900 °C *under vacuum* to ensure that no air bubbles become entrapped in the viscous melt. It has a very low coefficient of thermal expansion and so (unlike ordinary soda lime glass, Section 7.5.1) does not shatter on shock heating or cooling. It can also be highly transparent to visible light if made in sufficiently pure form, for example, by reaction of redistilled $SiCl_4$ (a moisture-sensitive liquid, bp 57 °C; Section 17.8) in the vapor phase with oxygen above 1100 °C[14e]:

$$SiCl_4(g) + O_2(g) \xrightarrow{1100\,^\circ C} SiO_2(s) + 2Cl_2(g). \qquad (7.7)$$

If care is taken to exclude any hydrogen-containing material (which could produce light-absorbing OH groups in the glass) and any traces of transition metal ions (such as the ubiquitous iron), silica made in this way can be fused and drawn into optical fibers with 95% or better light transmission per kilometer. Optical fiber cables are now used to carry laser light, modulated with digitized telephone signals, over large distances; since the frequency of light is much higher than that of the radio waves or microwaves used in conventional telecommunications, the capacity of optical fibers for carrying information is correspondingly greater.

Silicon tetrachloride can be used to produce a smoke like powder known commercially as *silica fume* (*fumed silica, pyrogenic silica*) by reaction in an oxygen–hydrogen flame:

$$SiCl_4(g) + O_2(g) + 2H_2(g) \xrightarrow{\text{flame}} SiO_2(s) + 4HCl(g). \qquad (7.8)$$

The fluffy product is extremely finely divided (particle size typically 7–40 nm, powder density 0.03–0.06 g cm^{-3}) but nonporous (surface area ca. 200–400 m^2 g^{-1}). Like fused silica, silica fume is amorphous, and this, together with the small particle size, gives it relatively high solubility. It is used as a thixotropic thickening agent for epoxy resins, in special high strength cements (Section 11.2), and as a white, unreactive strengthening agent for rubbers (particularly silicone rubbers) or plastics in place of carbon black.

Other "synthetic" silicas can be made from sand by dissolving it in aqueous NaOH, reprecipitating it with an acid (H_2SO_4 or HCl), and heating the resulting gel to dehydrate it. Such products are used to the extent of about 1 million metric tons per year worldwide for thermal insulation, as thickening agents for paints and printing inks, as supports for catalysts, and as absorbents for gases and liquids. "Silica gel" is perhaps the most widely used commercial desiccant (drying agent).

7.5.1 Silicate Glasses

Ordinary bottle or window glass is a sodium calcium silicate (soda lime or soda lime–silica glass) that may contain variable amounts of other minor constituents. A typical composition is 72% SiO_2, 14% Na_2O, 10% CaO, and 3% MgO. The last two constituents are necessary to prevent the glass from being water-soluble; sodium silicates form viscous aqueous solutions called *waterglass*. Even so, ordinary glass is slightly soluble in hot water and is quite quickly attacked by aqueous alkali (e.g., in dishwashers). It is made by fusing a mixture of sand (SiO_2), sodium carbonate (soda ash), and lime (CaO, usually containing some MgO if dolomitized limestone was calcined). The most common impurity in glass is iron, which, as Fe^{2+}, imparts a green color.

Soda-lime glass has a large coefficient of thermal expansion and cracks if subjected to sudden changes of temperature. If, instead of lime, the glass is made by fusing SiO_2 (81%) with Na_2CO_3 (equivalent to 4.5% Na_2O), boric oxide (B_2O_3, 12.5%), and Al_2O_3 (2%), a glass is obtained that is higher melting than soda-lime glass and very resistant to thermal shock. This borosilicate glass is popularly known by the Corning Glass Company's tradename, *Pyrex*, and is widely used for kitchenware as well as for laboratory glassware.

Silicate glasses generally are brittle and have poor tensile strengths, but they are quite strong in compression. In "tempered" (thermally toughened) glasses for automobile windshields and similar uses, compressive stresses are created in the surface of the glass by heating the whole specimen to about 800 °C and then cooling it quickly with air jets. Alternatively, the glass object can be steeped in a concentrated solution of a potassium salt; ion exchange replacement of Na^+ by the larger K^+ ion results in compressive stresses at the surface. In either case, the compressive stresses strengthen the glass object, and, more importantly, if the stressed glass does fracture, it breaks into small, roughly cubic fragments rather than large, sharp shards that would cause serious injury to anyone nearby. Glass *fibers* have considerable strength, presumably because of compression by the surface layer chilled during spinning, but they are generally used only as reinforcement for cements and plastics.

Because glasses are metastable solids, they may eventually crystallize or *devitrify*, especially if held close to their melting points for extended periods (as novice glassblowers discover to their dismay). Devitrification renders glass brittle, opaque, and useless, but controlled crystallization in the presence of suitable nucleating agents such as TiO_2 can produce extremely tough microcrystalline glass ceramics, which usually include other oxides such as Al_2O_3.

7.6 Soluble Silicates and Aluminates

Silica and silicates are soluble in aqueous alkali.[9,15–17] The silicate anions produced in solution may be regarded as derived from silicic acid [H_4SiO_4 or $Si(OH)_4$], which has a first pK_a of 9.47* at 0.6 mol L^{-1} ionic strength:

$$H_4SiO_4 \rightleftharpoons H^+ + H_3SiO_4^- \tag{7.9}$$

and a second pK_a of about 12.65:

$$H_3SiO_4^- \rightleftharpoons H^+ + H_2SiO_4^{2-}. \tag{7.10}$$

This, however, is an oversimplification, since these monomeric anions are in rapid equilibrium with oligomers (small polymers) such as the dimer (cf. Fig. 7.2b), cyclic trimer (7.2c), and other small silicate oligomers, 22 of which have been identified by silicon-29 magnetic resonance methods and many of which have no counterparts in solid silicate minerals[9]:

$$2(HO)_3SiO^- \rightleftharpoons {}^-O(HO)_2SiOSi(OH)_2O^- + H_2O, \text{etc.} \tag{7.11}$$

Clearly, the higher the concentration of dissolved silica, the greater the fraction that will be present as oligomers. High excesses of alkali and high temperatures, however, favor the monomer ($H_3SiO_4^-$ and $H_2SiO_4^{2-}$) and the smaller oligomers. On acidification of silicate solutions, a gel of silicic acid, which is poorly soluble at ordinary temperatures, is produced.

The solubility of silicic acid in water, and hence of solid silica in its various forms, becomes quite important at 200 °C or more (under pressure). This must be considered when, for example, superheated water is injected into oil-bearing formations, since the dissolved silica will be reprecipitated on cooling. This effect is intensified if the water is alkaline, because of the much higher solubility of $H_3SiO_4^-$ (etc.) salts relative to H_4SiO_4; hydrothermal dissolution and reprecipitation of SiO_2 in aqueous alkali is used in growing large single crystals of quartz for the electronics industry, as noted earlier.

---

*$pK_a = -\log K_a$, where K_a is the equilibrium constant for acid ionization.

Soluble silicates are usually manufactured by fusing high-purity sand (quartz) with soda ash or caustic soda, with the ratio SiO_2:Na_2O ranging from 0.5 to 3.4, depending upon the end use. Aqueous solutions of the products have a soapy feel, and indeed soluble silicates improve the detergent power of soaps. They have long been incorporated as builders in household wash powders (10 to 15%) and more so in industrial detergents. Their mode of action is not entirely clear. Soluble silicates are also effective in suppressing the corrosion of iron, probably by precipitating insoluble sodium iron silicates such as *acmite* ($NaFeSi_2O_6$) on the affected surfaces. Sodium silicate solutions are also employed in the hydrothermal production of synthetic zeolites, by reaction with aluminate ions.

The aqueous chemistry of aluminum(III) above pH 6 differs from that of silicates in that the only important species, other than solid $Al(OH)_3$ at pH 5–8, is $Al(OH)_4^-$, which, although isoelectronic with $Si(OH)_4$, shows no tendency to catenate. On the other hand, below pH 5 Al^{III}, unlike the poorly soluble $Si(OH)_4$, is freely soluble as Al^{3+}(aq) [actually $Al(OH_2)_6{}^{3+}$, Section 13.2], while at intermediate pH hydrolytic Al species, including the ion $Al_{13}O_4(OH)_{24}(OH_2)_{12}{}^{7+}$ referred to above, predominate in solution. However, $Al(OH)_4^-$ units can readily insert themselves into silicate anion species in solution. The result is usually the prompt precipitation of an aluminosilicate gel (a typical zeolite precursor), although over some limited Al, Si, and OH^- concentration ranges quite high concentrations of dissolved aluminosilicates can be maintained over many months.[9]

An interesting aspect of aluminosilicate solution chemistry is that dissolved silica appears to provide protection for living things against aluminum poisoning.[18] The Al^{3+} ion is a potent toxin if it penetrates living cells, apparently because it interferes with the normal physiological functions of phosphate ions (see below). Fortunately, at physiological pH, almost all dissolved Al is in the form of $Al(OH)_4^-$, which, as a negatively charged species, is much less likely to penetrate biological membranes or bind phosphate ions than is Al^{3+}(aq). For these and other reasons, Al^{III} normally does not penetrate the cells of key organs of humans such as the brain, but if the normal protective mechanisms of the digestive and excretory systems are subverted, as in the case of patients with kidney failure on dialysis, Al ions may accumulate in the body, causing deterioration of the bones (osteomalacia) and the brain (leading to dementia and possible death). Similarities between Al-induced dialysis dementia and Alzheimer's disease have reinforced suspicion that Al is involved in some way in the latter. Be that as it may, it has been shown that silica is an essential nutrient, probably (at least in part) because soluble aluminosilicates form at physiological pH and assist excretion of Al. Grasses such as barley accumulate silica, and Birchall and co-workers made the welcome point that one of the best dietary sources of bioavailable silica is therefore *beer*.[19]

7.7 Phosphates and Aluminophosphates

Following the periodic table across from Al and Si, we come to phosphorus, which differs from the other two in exhibiting *two* main valence states, III and V. Thus, both PCl_3 and PCl_5 are well known; the latter in the solid state consists of tetrahedral cations and octahedral anions, $[PCl_4^+][PCl_6^-]$, but the oxidation state in both is still five. Both compounds hydrolyze in water to H_3PO_3 (phosphorous acid—note the spelling!) and H_3PO_4 (phosphoric acid), respectively. The pK_a values for the first, second, and third ionizations of H_3PO_4 at 25 °C are 2.16, 7.21, and 12.32, respectively, so the relevant species at physiological pH are $H_2PO_4^-$ and HPO_4^{2-} while PO_4^{3-} is predominant only in strongly alkaline solutions.

The tetrahedral phosphate ion PO_4^{3-}, like the silicate ion with which it is isoelectronic, can form chains and networks, and indeed its parent oxide, a white, extremely hygroscopic solid commonly called phosphorus pentoxide (P_2O_5), consists of discrete P_4O_{10} molecular units in which four PO_4 units each share three corner oxygens such that the four P centers themselves form a tetrahedron. We customarily distinguish the following classes of phosphates:

(*a*) *orthophosphates*, containing the discrete ion PO_4^{3-};

(*b*) *pyrophosphates* or *diphosphates*, containing the ion $P_2O_7^{2-}$ (cf. the corresponding silicate structure, Fig. 7b;

(*c*) *polyphosphates*, anions consisting of open chains of PO_4 tetrahedra, $O(PO_3)_n^{(n+2)-}$, as in Fig. 7e [e.g., the triphosphate ion, $P_3O_{10}^{5-}$ or $^-O{-}PO(O^-){-}O{-}PO(O^-){-}O{-}PO(O^-){-}O^-$; and

(*d*) *cyclic polyphosphates (metaphosphates)*, $(PO_3)_n^{n-}$, notably the cyclic trimer $(PO_3)_3^{3-}$ (structure as in Fig. 7c) and tetramer $(PO_3)_4^{4-}$.

All living things are totally dependent on phosphates. The genetic material of life, DNA (deoxyribonucleic acid), consists of sugar units (deoxyribose) linked with phosphate units to form a helical chain; basic side groups (adenine, cytosine, guanine and thymine), in a definite order corresponding to coded genetic information, link up with the side groups in a second intertwined chain, giving a double helix. Westheimer[20] points out that phosphate ion is the only readily available molecule that can serve in this regard, as what is required is a unit that can be divalent and yet still have a remaining negative charge to stabilize the DNA against hydrolysis and facilitate its retention within biological membranes.

In DNA and the related single-stranded RNA (ribonucleic acid), the phosphate units are monomeric and play a strictly structural role. Phosphates, however, are also essential to the metabolism of living cells, in which formation and subsequent hydrolysis of phosphate–phosphate bonds serves as a storage and release mechanism for biochemical energy. Thus, two adenosine diphosphate (ADP) units can be converted reversibly to one adenosine triphosphate (ATP) and one adenosine monophosphate (AMP) unit, with the reactions being greatly accelerated by enzymes (biological catalysts). Again, phosphate is the only feasible ion for this purpose.[20] A further biological function of phosphates is in the formation of bones and teeth, the essential structural material of which is *hydroxyapatite*, $Ca_5(PO_4)_3OH$.

Phosphate, then, is a key nutrient for all living things, but it is generally less freely available than the other main essential elements C and N; in other words, phosphorus is usually the growth-limiting nutrient. About half of the world's phosphates occur as igneous deposits (e.g., in the Kola Peninsula, Russia) and the rest as sedimentary deposits containing fish scales and bones from former seas (e.g., in Texas, Montana, and Idaho). Such phosphates [mostly *apatite*, $Ca_5(PO_4)_3(OH,F,Cl)$*] are highly insoluble and hence biologically unavailable. We return to this point in connection with agricultural fertilizers (Section 9.6). By the same token, releases of *soluble* phosphate wastes (notably sodium triphosphate, an excellent builder for dishwasher and other detergents) into rivers and especially lakes can result in explosive growth of algae and eventual death of the aquatic ecosystem through deoxygenation by decomposition of dead algae. Consequently, phosphates have acquired a negative image in connection with public concern over environmental degradation.

7.7.1 Phosphate Fibers

Griffith[7] chronicled in lively fashion the growth and sudden demise of an extraordinary project of the Monsanto Co. to make phosphate fibers as substitutes for the much maligned asbestoses. The object was to make materials with most of the desirable properties of asbestos, but that would hydrolyze *slowly* in, say, the alveoli of the lungs to form a soluble and biologically beneficial product (phosphate ions), so avoiding the tendency of insoluble aluminosilicate fibers to remain indefinitely in biological material with the attendant risk of cancer.

Most of the research centered on long-chain phosphates $[Ca(PO_3)_2]_n$ and $[NaCa(PO_3)_3]_n$, but $[Na(PO_3)]_n$ and $[Mg(PO_3)_2]_n$ also appeared promising. These formulas imply "infinite" chain lengths, but, as noted earlier, a finite phosphate chain of n units will have the formula $O(PO_3)_n^{(n+2)-}$. Thus, one can limit the chain length in, say, $[Na(PO_3)]_n$ to a particular value n by preparing a melt in which the ratio $[Na_2O]/[P_2O_5]$ is *precisely* set to $(n+2)/n$; for long chains, this will be very close to unity. The technique

*Geologists often take the name "apatite" to mean fluoroapatite, which is the most common form, especially in igneous deposits.

involves allowing a sodium polyphosphate melt of the right composition to crystallize, remelting it (at 700 °C), and quenching it quickly to ambient temperature. This procedure can be repeated several times to give closer control of chain length (up to $n \approx 400$ in this case), and cross-linking of the chains can be achieved by adding very small amounts of P_2O_5. Careful milling produces fibers with diameters of 0.1 μm $\{[NaCa(PO_3)_3]_n\}$ to 10 μm $\{[Ca(PO_3)_2]_n\}$ with aspect ratio ~100 that can be spun into threads, for example, for fireproof cloth or as reinforcement for plastics. Other proposed uses range from brake linings to surgical implants. Phosphate fibers may tend to hydrolyze slowly in moist (particularly acidic) conditions, but this is not necessarily a disadvantage.

The Monsanto project is said to have been canceled, despite the effectiveness and designed "environmental friendliness" of the products, because of the sinister reputations (not wholly deserved) of both asbestos and phosphates in the popular view. A similar problem with public perception of silicone surgical implants was noted in Section 3.5. The tenor of our times is such that much beneficial technology may be lost through litigation (or the mere fear of it) regardless of any favorable scientific evidence that is put forward.

7.7.2 Aluminophosphates

As noted earlier, the intracellular toxicity of Al^{3+} is a manifestation of strong affinity for PO_4^{3-}. Aluminum orthophosphate, $AlPO_4$, is isoelectronic with silica and shares many of its properties. Just as zeolite like structures can be made with compositions approximating that of silica, so there are $AlPO_4$ phases with channel-and-cavity architecture, and these show promise as catalysts.[21] In general, however, $AlPO_4$ phases are less acidic than the corresponding silicas unless some of the Al^{3+} ions are replaced with Si^{IV}. In one study, $AlPO_4$ needles were successfully grown hydrothermally in various orientations on a gold surface and shown to have zeolite like microporosity; the device shows promise as a selective sensor or as a catalytic membrane with molecule-specific activity.[22]

Exercises

7.1 Show that the empirical formula of the silicate framework of a pyroxene will be $(SiO_3)^{2-}$; that of an amphibole, $(Si_4O_{11})^{6-}$; and that of a sheet silicate, $(Si_2O_5)^{2-}$.

7.2 How much zeolite-A in its Na^+ form would be needed to soften completely 1.000 tonne of water, the hardness of which was equivalent to 150.0 ppm (i.e., 150.0 mg/kg) dissolved $CaCO_3$? Assume 100%

efficiency of ion exchange. (The calculated mass will therefore be a minimum, in practice.)
[*Answer:* 547 g.]

7.3 Why must recrystallization of quartz for use in the electronics industry be carried out hydrothermally?

7.4 Sketch the structure of the phosphorus pentoxide molecule, on the basis of the description given in Section 7.7.

7.5 At what pH would $H_2PO_4^-$ and HPO_4^{2-} be present in equal concentrations in aqueous phosphate solutions at 25 °C?
[*Answer:* pH 7.21.]

References

1. V. M. Goldschmidt, "Geochemistry." Oxford Univ. Press, London, 1954.

2. W. A. Deer, R. A. Howie and J. Zussman, "Rock-Forming Minerals," Vols. 1–5. Longmans, London, 1978. A condensed version is presented in W. A. Deer, R. A. Howie, and J. Zussman, "An Introduction to the Rock-Forming Minerals" Longmans, London, 1966.

3. J. E. Fergusson, "Inorganic Chemistry and the Earth," pp. 169–174, and 305 ff. Pergamon, Oxford, 1982.

4. L. Smart and E. Moore, "Solid State Chemistry." Chapman & Hall: London, 1992.

5. A. F. Wells, "Structural Inorganic Chemistry," 5th Ed. Oxford Univ. Press, London, 1984.

6. R. C. Brown, J. A. Hoskins, and J. Young, Not allowing the dust to settle. *Chem. Br.* **28**, 910–915 (1992); B. T. Mossman, J. Bignon, M. Corn, A. Seaton, and J. B. L. Gee, Asbestos: Scientific developments and implications for public policy. *Science* **247**, 294–301 (1990).

7. E. J. Griffith, "Phosphate Fibers." Plenum, New York, 1995.

8. (*a*) R. P. Townsend (ed.), "Properties and Applications of Zeolites," Special Publication 33. Chemical Society, London, 1980; (*b*) D. W. Breck, "Zeolite Molecular Sieves." Robert E. Krieger Publ., Malabar, Florida, 1984; (*c*) P. A. Jacobs and R. A. van Santen (eds.), "Zeolite Facts, Figures and Future." Elsevier, Amsterdam, 1989; (*d*) R. M. Barrer, "Hydrothermal Chemistry of Zeolites." Academic Press,

London, 1982; (*e*) M. L. Occelli and H. E. Robson (eds.), "Zeolite Synthesis," ACS Symp. Ser. 398. American Chemical Society, Washington, D.C., 1989; (*f*) P. Kleinschmit, Silicas and zeolites. *In* "Industrial Inorganic Chemicals: Production and Uses" (R. Thompson, ed.), pp. 327–349. Royal Society of Chemistry, Cambridge, 1995.

9. T. W. Swaddle, J. Salerno, and P. A. Tregloan, Aqueous aluminates, silicates and aluminosilicates. *Chem. Soc. Rev.* **23**, 319–325 (1994).

10. S. M. Csicsery, Shape selective catalysis in zeolites *Chem. Br.* **21**, 473–477 (1985).

11. J. Haggin, Shape selectivity key to designed catalysts. *Chem. Eng. News*, December 13, 9–15 (1982).

12. (*a*) R. W. McCabe, Clay chemistry. *In* "Inorganic Materials" (D. W. Bruce and D. O'Hare, eds.), pp. 295–351. Wiley, Chichester, U. K., 1992; (*b*) R. Burch (ed.), Pillared clays. Special one-topic issue of *Catal. Today* **2**, No. 2–3 (1988); (*c*) P. Laszlo, Chemical reactions on clays. *Science* **235**, 1473–1477 (1987).

13. R. K. Iler, "The Chemistry of Silica." Wiley, New York, 1979.

14. (*a*) D. Kolb and K. E. Kolb, The chemistry of glass. *J. Chem. Educ.* **56**, 604–608 (1979); (*b*) C. M. Melliar-Smith, Optical fibers and solid state chemistry. *J. Chem. Educ.* **57**, 574–579 (1980); (*c*) R. L. Tiede, Glass fibers—are they the solution? *J. Chem. Educ.* **59**, 198–200 (1982); (*d*) P. G. Harrison and C. C. Harrison, Glasses, ceramics, and hard metals. *In* "Insights into Speciality Inorganic Chemicals" (D. Thompson, ed.), pp. 325–368. Royal Society of Chemistry, Cambridge, 1995; (*e*) M. Jerzembeck and A. Kornick, Chemical reactions in the manufacture of waveguides for long distance optical data transmission. *Angew. Chem. Int. Ed. Engl.*, 1991, **30**, 745–753 (1991).

15. R. Thompson (ed.), "The Modern Inorganic Chemicals Industry." Special Publication 31, pp. 320–352. Chemical Society, London, 1977.

16. J. S. Falcone, Jr. (ed.), "Soluble Silicates," ACS Symp. Ser. 194. American Chemical Society, Washington, D.C., 1982.

17. L. S. Dent Glasser, Sodium silicates. *Chem. Br.* **18**, 33–39 (1982).

18. J. D. Birchall, The essentiality of silicon in biology. *Chem. Soc. Rev.* **24**, 351–358 (1995).

19. J. P. Bellia, J. D. Birchall, and N. B. Roberts, *Lancet* **343**, 235 (1994).

20. F. H. Westheimer, Why Nature chose phosphates. *Science 235*, 1173–1178 (1987).

21. S. T. Wilson, B. M. Lok, C. A. Messina, T. R. Cannan, and E. M. Flanigen, Aluminophosphate molecular sieves: a new class of microporous crystalline inorganic solids. *J. Am. Chem. Soc.* **104**, 1146 (1982).

22. S. Feng and T. Bein, Vertical aluminophosphate molecular sieve crystals grown at inorganic–organic interfaces. *Science* **265**, 1839–1841 (1994).

Chapter 8

The Atmosphere and Atmospheric Pollution

THE IMPORTANCE of main group inorganic chemicals may be gauged in terms of production tonnage, as they make up most of the top dozen or so industrial chemicals (Table 1.2).[1–3] Indeed, sulfuric acid consumption was at one time used as an informal measure of a nation's economic activity, since H_2SO_4 is used in a host of chemical, metallurgical, and general manufacturing operations. Prominent among these "heavy" inorganic chemicals are products obtained from the atmosphere: elemental oxygen and nitrogen (isolated from air mostly by fractional distillation of liquid air, although gas-phase membrane diffusion processes and sorption/desorption cycles using zeolites are increasingly used where high purity of the product is unimportant[2,4]) and the nitrogen-derived chemicals ammonia, nitric acid, ammonium nitrate, and urea.[1–3] Conversely, many of the more serious atmospheric pollutants (CO, CO_2, O_3, NO_x, SO_2) are also main group inorganic substances.[5–11] We consider this latter aspect first.

8.1 Carbon Dioxide and Greenhouse Gases

8.1.1 Carbon Dioxide and the Greenhouse Effect

Ordinary air contains a variable amount of water vapor, but its composition on a dry basis by *volume* is 78.08% N_2, 20.95% O_2, 0.934% Ar, and, presently, only 0.036% (360 parts per million, ppm) CO_2, plus minor amounts of Ne, He, CH_4, Kr, H_2, N_2O, and Xe. Nevertheless, carbon dioxide is believed to exert a major influence on climate through the "*greenhouse effect*": sunlight penetrates the atmosphere, is mostly absorbed by the Earth, and is reemitted as infrared radiation, but an important fraction of this is trapped by water vapor, CO_2, and certain other "greenhouse gases" such as methane in the atmosphere, leading to a significant warming of the environment.

An extreme example of an apparent CO_2 greenhouse effect is found on the planet Venus, which has an atmosphere of CO_2 and N_2 at 100 bars total pressure and a ground temperature of approximately 460 °C. Venus, however, is much closer than the Earth to the Sun. It is believed that the atmospheres of Mars, Venus, and Earth were all originally composed mainly of CO_2, N_2, and H_2O. Venus' water is now gone, however, and Earth's CO_2 was mostly either locked up as carbonates in sedimentary rocks formed in ancient oceans or partly converted to O_2 some 2×10^9 years ago through photosynthesis by primitive plants and algae. Biological cycling of O_2 continues today, and O_2 has a residence time of about 5000 years in the atmosphere. On the Earth, some 400–500 million years ago (in the early Paleozoic era), CO_2 levels were 12–20 times those of the present, but they subsequently dropped sharply during the Carboniferous and Permian periods when the rapidly expanding vascular plant population flourished and became buried on a vast scale, forming the massive coal deposits associated with rocks of that age. The consequent diminution in atmospheric CO_2 concentrations and hence lessening of the greenhouse effect may well account for the uniquely long period of glaciation in Permian–Carboniferous time.[12]

Normally, the mean atmospheric CO_2 concentration is buffered by the oceans, which hold CO_2 in solution at the surface and also precipitate it in the depths as limestone ($CaCO_3$) or dolomite [$CaMg(CO_3)_2$], and by green plants (biomass), which convert it to carbohydrates. With the spread of industrialization and increases in human population, however, atmospheric CO_2 levels have increased markedly through the burning of fossil fuels (oil, natural gas, and especially coal) and loss of the CO_2-consuming forests—either through felling for timber and fuel or through burning to clear land for agriculture. Figure 8.1 shows that there is an annual cycle of about 6 ppm in the atmospheric CO_2 concentration, due to seasonal variation in biological uptake and release, but that the cycle overlies a long-term upward trend of 1–2 ppm per year.

At present, an average of 1.1 metric tons of CO_2 is released annually for each person on Earth; for North Americans, this figure is 5 metric tons. This is expected to lead to a doubling of the CO_2 concentration and a warming of all climates over the twenty-first century. There has been much controversy[13] as to whether the global warming of 0.3–0.6 °C that has already occurred since the mid-nineteenth century is causally connected to the well-documented increase in atmospheric CO_2 levels since the industrial revolution. Recent years have been particularly warm—1995 was the warmest year (mean global surface temperature 14.8 °C) on record to that date—and further rises in the mean global temperature of 1.0 to

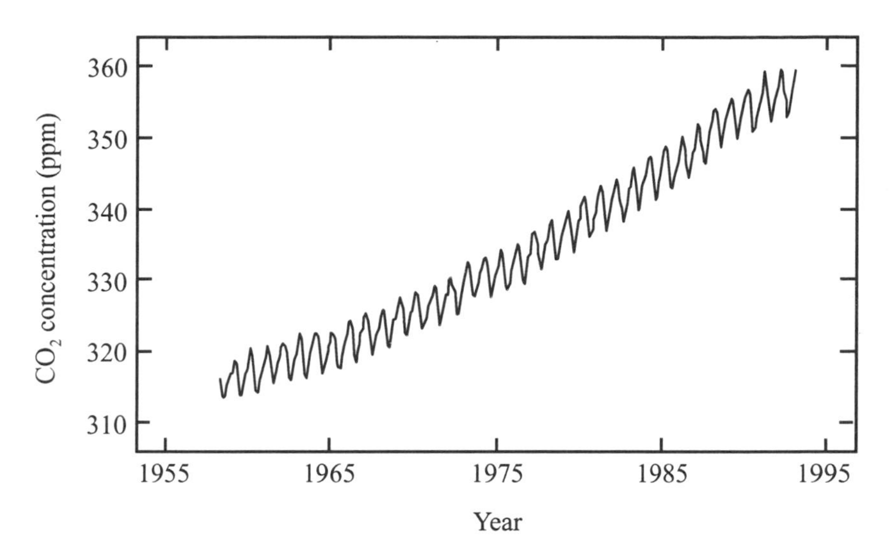

Figure 8.1 Annual variation in atmospheric carbon dioxide concentration as recorded at the Mauna Loa Observatory, Hawaii.

3.5 °C by the year 2100 have been predicted on the basis of revisions of various models (earlier versions predicted a 1.5 to 4.5 °C increase by the mid-twentyfirst century). At the time of writing (1996), an international consensus is emerging to the effect that global warming is indeed occurring, and that it is probably correlated with the atmospheric CO_2 concentration.

It is, however, extremely difficult to separate the anticipated effects of CO_2 greenhouse warming from numerous other influences, notably variations in solar input, effects of heat retention by clouds, and cooling effects of the haze of particulate matter (notably sulfate aerosols) produced by the burning of coal and other fossil fuels.[14] There is also uncertainty over the fate of CO_2 released to the atmosphere; an unexpectedly strong terrestrial biospheric CO_2 sink in temperate northern latitudes that can account for about half the global output of CO_2 from fossil fuel burning has been discovered (though not clearly identified—increasing biomass in temperate and boreal forests is a possibility).[15] Thus, we are still a long way from understanding the biogeochemical CO_2 cycle and its possible effects on climate, yet there is widespread public concern, at least in Western countries, over the threat of climatic change. The political dilemma is a thorny one: should industrial activity, particularly in those developing countries whose economies are dependent on coal-burning,* be sharply curtailed in an attempt to forestall anticipated climatic changes that may yet turn out

*About 40% of the world's energy demand is met with coal.

to have little dependence on CO_2 greenhouse warming? Indeed, as often discussed among Canadians in winter, would the consequences of global warming necessarily be bad?

The consequences of global warming are hard to predict, but a rise in mean sea level (partly through thermal expansion of seawater, partly through melting of glaciers and polar ice caps) can be expected. Certainly, mountaineers have long noted a worldwide recession of glaciers, and a rate of rise of 4 mm per year in mean sea level has been measured.[16] The potential rise in sea level through the twentyfirst century is thus not expected to exceed 0.5 m; but, in the longer term, low-lying land (e.g., in Bangladesh) could become permanently submerged, as could some oceanic islands such as Tongatapu (the main island of the Kingdom of Tonga) and the Tuamotu Archipelago, and coastal cities everywhere would be at risk. On the continents, higher temperatures may cause spreading of deserts, although warmer seas may mean more precipitation in other areas. Life-threatening heat waves may plague major cities, as happened in Chicago in the summer of 1995. The incidence of tropical or subtropical diseases, particularly malaria, will probably increase and spread into areas such as southern continental Europe and the eastern United States.

On the other hand, there is clear evidence that plant growth is favored by higher CO_2 levels, and the spreading of forests into higher latitudes has been forecast.[17] Extension of boreal forests into present-day tundra should be especially marked in Alaska, northern Canada, and Siberia. Such developments would presumably mitigate the greenhouse effect by removing some of the anthropogenic CO_2, acting in effect as a negative feedback mechanism to limit greenhouse warming. Some of the striking improvement in crop yields during the twentieth century may be attributable to the increased concentration of atmospheric CO_2, rather than solely to improved agricultural practices; in other words, the CO_2 pollutant may function as a fertilizer.

The United Nations Panel on Climate Change made several recommendations to limit CO_2 emissions, including the following:

(*a*) use natural gas as fuel rather than coal or oil;

(*b*) use solar or nuclear energy instead of fossil fuels for electricity generation;

(*c*) reduce rates of deforestation; and

(*d*) limit use of automobiles.

The reality is, however, that it will be politically difficult or impossible to stabilize CO_2 emissions at 1990 levels, let alone bring about a significant reduction. Rubin *et al.*[18] discuss some realistic mitigation options.

8.1.2 Other Greenhouse Gases

In addition to water vapor and CO_2, chlorofluorocarbons (see Sections 8.3 and 12.3), nitrous oxide (Section 8.4), and methane contribute to the greenhouse effect, and their concentrations in the atmosphere are also increasing markedly. Methane is released to the atmosphere by natural mechanisms such as decay of vegetation in wetlands and digestion of vegetable matter by termites (which constitute a remarkably large fraction of the biomass, especially in the tropics).[19] Human activities, however, contribute important amounts of CH_4 from agriculture (notably rice cultivation, which accounts for about 25% of all methane emissions to the atmosphere, and animal husbandry—the digestive processes of ruminants generate much methane), sewage disposal, and natural gas leakages. The atmospheric methane concentration is currently only about 1.7–1.8 ppm, but methane is 30 times more efficient at trapping radiation than is CO_2, so that it adds another 15% or so to the expected greenhouse effect.

Methane is removed continually from the atmosphere by reaction with ·OH radicals (Section 8.3). In contrast, chlorofluorocarbons and related volatile compounds are inert under the conditions of the lower atmosphere (troposphere), so atmospheric concentrations of these refrigerants and solvents will tend to increase as long as releases continue. The chief concern over chlorofluorocarbons is that they are a major factor in destruction of the stratospheric ozone layer (Section 8.3). They have been banned under the Montréal Protocol of 1988, but it is important that whatever substitutes (inevitably greenhouse active) are introduced to replace them degrade relatively quickly in the troposphere to minimize any contribution they may be capable of making to greenhouse warming.

The concentration of nitrous oxide (N_2O) in the atmosphere is only about 0.30 ppm but it increases by about 0.2% per year and so is at least partly of anthropogenic origin. Nitrous oxide is a product of the degradation of nitrate fertilizers, but some 7×10^5 tonnes of N_2O are released annually to the atmosphere during production of nylon.[20] The residence time of N_2O in the atmosphere is about 150 years, and it could in the future contribute up to 10% of the anticipated greenhouse warming.

8.1.3 Supercritical Carbon Dioxide

Carbon dioxide has a conveniently low critical point (31 °C, 7.39 MPa), and supercritical CO_2 has become the most widely used fluid where supercritical solvent properties are required, as it is also inexpensive and nontoxic. The solvent powers of supercritical fluids generally increase with increasing density, which can be regulated at will by varying the pressure. The absence of a gas–liquid interface and associated surface tension in a supercritical fluid enables the fluid to penetrate porous solids freely, and also to evaporate from fragile porous solids without disrupting their structure. As seen in Section 19.1, this means that supercritical fluids such as CO_2 can be used to remove water from aqueous gels without significant damage to the gel structure.

Supercritical CO_2 at temperatures of 31–50 °C and pressures of about 10 MPa is now widely used as a nontoxic extractant of excess fats from foodstuffs and in decaffeinating coffee, but its largest scale future use is likely to be the enhancement of recovery of oil that cannot be extracted from wells by conventional techniques. Supercritical CO_2 is finding increasing favor as a solvent for chemical syntheses, for example, in the radical-promoted polymerization of fluoroacrylic monomers in homogeneous solution, for which ozone layer-destroying chlorofluorocarbons had been the only effective solvents previously.[21a]

Although supercritical CO_2 is an effective solvent for oils, fats, and similar substances, it is a poor one for nonvolatile hydrophilic (water-loving) substances such as proteins or metallic salts. Adding water as such to the supercritical CO_2 is of little help, as the solubility of water in it is limited. Johnson and co-workers[21b] overcame the latter limitation by forming water-in-CO_2 emulsions with the aid of an added nontoxic perfluoropolyether surfactant that forms reverse micelles around the water microdroplets, in effect combining the special properties of supercritical CO_2 with the solvent power of water. These emulsions can dissolve a variety of biomolecules at near-ambient temperatures, without loss of their biological activity.

8.1.4 Other Properties of Carbon Dioxide

The greenhouse effect apart, CO_2 is an innocuous substance—it is present in substantial concentrations in the air we exhale, and it is widely used in the food and beverage industry (e.g., to make effervescent drinks). When evolved in caves or other enclosures, however, CO_2 can displace air by virtue of its higher density and thus present a risk of asphyxiation. An extreme example of mass asphyxiation by CO_2 occurred in August, 1986, in Cameroon (West Africa) when Lake Nyos "turned over," so decompressing a large volume of CO_2-laden water from its depths, whereon about 1×10^8 m^3 of CO_2 gas spread over the surrounding land, killing 1746 people and several thousand head of livestock. A similar occurrence in nearby Lake Monoun in August, 1984, killed 45 people.

Uses of CO_2 besides those noted above include

(*a*) manufacture of urea fertilizer (Section 9.3);

(*b*) preparation of dry ice (i.e., solid CO_2), which has a vapor pressure of 100 kPa at −78 °C and so volatilizes at that temperature and atmospheric pressure without melting; and (*c*) provision of a blanketing inert atmosphere (e.g., in welding).

Carbon dioxide is usually marketed as the liquid under pressure.

8.2 Carbon Monoxide

The bonding in carbon monoxide is best explained by pointing out that it is isoelectronic with (has the same number of electrons as) N_2, and the bond is effectively triple. The difference in nuclei, however, results in much higher reactivity for CO than for N_2 and in the propensity of CO for forming coordinate bonds to many transition metal centers, as in $Fe(CO)_5$. Molecular orbital theory explains this in terms of the ability of a low lying empty antibonding orbital concentrated on the C end to accept electrons from a filled *d* orbital of the metal, while the resulting buildup of negative charge on C allows its unused electron pair to be donated back to the metal:

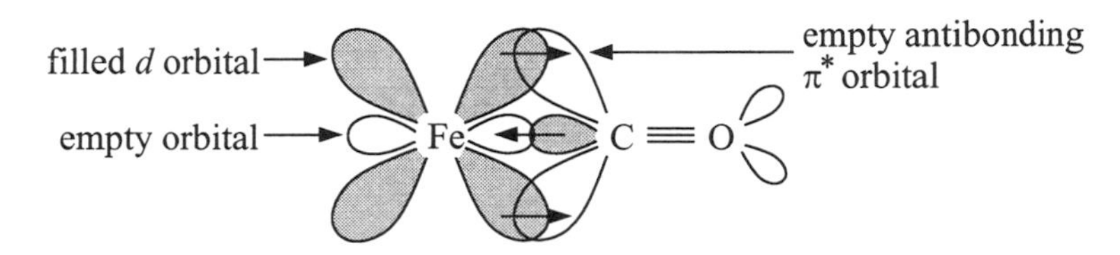

The toxicity of CO (and, in a sense, of the isoelectronic cyanide ion, CN^-) is a direct consequence of this, as CO will compete effectively with O_2 for the iron centers of the hemoglobin (Hb) in blood:

$$HbO_2 + CO \xrightleftharpoons{K \gg 1} HbCO + O_2 \tag{8.1}$$

HbO_2 =

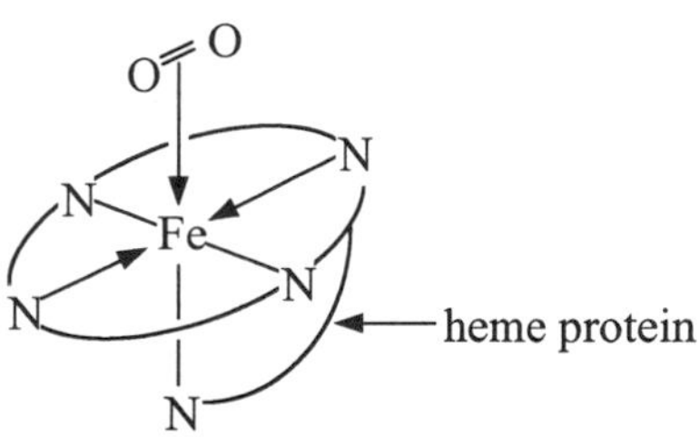

heme protein

Quite small concentrations of the odorless, colorless CO in air can therefore lead to headache and extreme drowsiness, and eventually to death by preventing O_2 from being delivered to tissues where it is required for metabolism. Fortunately, reaction 8.1 is readily reversible, so first aid for CO poisoning consists in moving the victim to fresh air and making the person breathe vigorously. The victim will recover completely if this is done soon enough.

Nevertheless, CO poisoning is a serious hazard associated with incomplete combustion of carbon or carbon compounds (particularly at high temperatures, where it is the favored oxide of C if the O_2 supply is limited—Section 17.7), for example, in automobile exhausts, tobacco smoke, domestic wood or coal stoves, and industrial operations. Natural processes, such as forest fires, photolysis of CO_2 over lakes, and decay of vegetation, account for the production of about six times as much CO as is produced through human activity. However, the *concentrations* are rarely high enough to be toxic—the average natural CO concentration in the troposphere is about 1 ppm (cf. 360 ppm for CO_2)—whereas accumulations of CO in city streets, parking structures, and buildings can produce toxic symptoms. Nature also *removes* CO from the air by the action of soil bacteria, so *dilution* of CO is sufficient to protect health and the environment. The high concentrations of CO in automobile exhausts may be reduced by use of a *catalytic converter* (see Section 8.4), which accelerates the air oxidation of CO to the relatively benign CO_2.

8.3 Ozone

Ozone (O_3), an *allotrope* (or modification) of elemental oxygen, is a very endothermically formed colorless gas with a characteristic sharp smell. It is formed by the action of ultraviolet light or electrical discharges on ordinary oxygen gas:

$$\tfrac{3}{2}O_2 \xrightarrow[\text{discharge}]{\text{UV or electrical}} O_3 \quad \Delta H_f^\circ = +142.7 \text{ kJ mol}^{-1}. \tag{8.2}$$

Consequently, substantial concentrations of ozone can accumulate around electrical equipment and photocopying machines. Rooms containing these must be well ventilated, since ozone is toxic and can cause headaches and irritation of the mucous membranes, even in low concentrations. It has been suggested that ozone may be the cause of the excess death rate from leukemia (blood cancer) in electrical workers. Ozone is also extremely damaging to plant life, much more so than acid rain (Sections 8.4 and 8.5). It also attacks rubber and may cause rubber insulation on electrical equipment to perish rapidly.

Nevertheless, ozone is beneficial in at least three ways. It is an excellent disinfectant and is used increasingly to sterilize municipal water supplies (Section 14.7). It can also be used as a powerful oxidant in the chemical process industries. Finally, O_3 is an important constituent of the *ozone layer* in the stratosphere.* The ozone layer absorbs short-wavelength ultra-

*The stratosphere is that part of the atmosphere between about 12 and 47 km in altitude—the lower boundary varies from 17 km at the equator to 7.5 km near the poles. The region below that level is called the troposphere.

violet (UV) light that would otherwise be very damaging to living things on the Earth's surface.

8.3.1 The Stratospheric Ozone Layer

It is customary to distinguish three kinds of UV light:

(*a*) UV-A (320–400 nm), which causes tanning and concomitant premature aging of human skin but otherwise probably does not present a serious health hazard;

(*b*) UV-B (290–320 nm), which produces sunburn and can cause skin cancer and eye damage (cataracts); and

(*c*) UV-C (200–290 nm), which causes cancer and severe skin damage.

Until the 1980s, the stratospheric ozone layer has blocked the UV-C and much of the UV-B in incoming sunlight. UV-A and some UV-B are normally present in solar radiation at ground level—more so at high altitudes—and light-skinned people in particular, when outdoors, have always required protection against them. Protection may be provided by shady hats, body covering, and any of the excellent sunscreen lotions now available.

Anthropogenic pollutants, however, have caused marked thinning of the ozone layer. Since 1987, it has been recognized that a vast "hole" in the stratospheric ozone layer develops over Antarctica during local spring, and breakaway sections of this hole have swept as far north as Australia, New Zealand, and South Africa. A similar phenomenon occurs in the Arctic, although the extent of stratospheric ozone depletion there is less dramatic than in the much stronger vortex that forms in southern winter and spring over Antarctica.[22] With partial loss of the protection of the stratospheric ozone layer, exposure to UV-B and possibly UV-C at ground level is increasing dangerously (for example, the summer UV-B levels in Toronto, Canada, rose by 35% between 1990 and 1994), and the incidence of cataracts, skin cancer, and various dermatological problems can be expected to rise sharply.* The protective measures outlined above are therefore becoming increasingly necessary.

These human medical problems do not necessarily spell disaster—most skin cancers can be cured, eye damage can often be corrected, and we may come to regard the dry, wrinkled skin of a former tanning enthusiast as beautiful. However, the great unknown factor is the effect of increased UV-B and UV-C exposure on ecosystems at large. Already, it is a matter of common experience worldwide that frogs and other amphibians have suddenly become scarce; although no causal connection with ozone layer

*In seafront communities in northern New South Wales, Australia, the incidence of skin cancer has reportedly reached 30%.

depletion can be made, it has been shown that the DNA in eggs of these amphibians is very susceptible to UV radiation damage, often leading to death. The effect of UV-B irradiation on whole ecosystems is hard to anticipate. For example, growth of both freshwater algae and the insect (chironomid) larvae that feed on them is reduced by exposure to UV-B, but the latter are the more severely affected, with the net result that the algae actually thrive because of the reduction in grazing.[23] Thus, we cannot predict what the consequences of ozone layer depletion might be, but the possibility exists that the entire ecosystem could be disrupted and the sources of our food lost.

Following the demonstration in 1970 by Paul Crutzen that nitrogen oxides destroy ozone catalytically, there has been much concern that the ozone layer could be depleted by introduction of excessive amounts of nitrogen oxides from supersonic commercial aircraft operating in the stratosphere:

$$NO + O_3 \rightarrow NO_2 + O_2. \tag{8.3}$$

In 1974, F. Sherwood Rowland and Mario Molina, who shared the 1995 Nobel Prize in Chemistry with Crutzen, showed that chlorine from photolyzed chlorofluorocarbons (CFCs) such as CF_2Cl_2 and $CFCl_3$, which were used as supposedly *inert* refrigerants, solvents for cleaning electronic components, plastic foam blowing agents, and aerosol spray propellants, can also catalyze ozone loss. Subsequently, the chlorine monoxide molecule ClO, which is involved in the chlorine-catalyzed ozone destruction cycle, has been shown to be present in the holes in the ozone layer and to correlate inversely with the level of surviving O_3, as expected from the following simplified reaction scheme[5, 6, 24, 25]:

$$CF_2Cl_2 \text{ or } CFCl_3 \xrightarrow[\text{stratosphere}]{\text{photolysis in}} \cdot Cl \tag{8.4}$$

$$\cdot Cl + O_3 \longrightarrow \cdot ClO + O_2 \tag{8.5}$$

$$O_3 \xrightarrow{\text{UV}} O(^1D) + O_2(^1\Delta) \tag{8.6}$$

$$O + \cdot ClO \longrightarrow \cdot Cl + O_2 \tag{8.7}$$

$$O(^1D) + CF_2Cl_2 \text{ or } CFCl_3 \longrightarrow \cdot ClO, \text{ etc.} \tag{8.8}$$

Note that since the atomic chlorine consumed in reaction 8.5 is regenerated in 8.7, it acts as a *catalyst* for ozone decomposition. Reaction 8.6 represents the decomposition of ozone by ultraviolet light, so that reactions 8.6 and 8.2 form a dynamic system that accounts for the absorption of ultraviolet light by the ozone layer.

Inevitably, the chemistry of the stratospheric ozone layer is more complicated than reactions 8.2–8.8 suggest.[24] First, it has been found that HCl and $ClONO_2$ (chlorine nitrate) serve as chlorine reservoirs in stratospheric polar clouds:

$$HCl + ClONO_2 \xrightarrow{\text{surface}} Cl_2 + HNO_3. \tag{8.9}$$

Second, reaction 8.9 and other relevant reactions appear to occur preferentially on available solid surfaces, which are often ice crystals but may also be particles of sulfate hazes from volcanic eruptions or human activity. Third, volatile bromine compounds are even more effective (via Br atoms) than chlorine sources at destroying ozone; methyl bromide is released into the atmosphere naturally by forest fires and the oceans, but anthropogenic sources include the use of organic bromides as soil fumigants (methyl bromide, ethylene dibromide) and bromofluorocarbons as fire extinguishers (*halons* such as CF_3Br, CF_2BrCl, and $C_2F_4Br_2$).

The Montréal Protocol of 1988 prescribed a phaseout of production of specified CFCs in industrialized countries by the end of 1995. Although record ozone losses were reported for the northern winter of 1995, atmospheric chlorine levels began to decline later that year, so that the ban seems to be having the desired effect. The production and use of some ozone-threatening chlorinated solvents such as CH_3CCl_3 have also been curtailed. Unfortunately, use of CFCs for refrigeration in many developing countries will have to continue for some time for economic reasons, and no satisfactory substitute for halons in firefighting has been found to date.

8.3.2 Ozone as a Pollutant

Reaction 8.6 produces atomic oxygen in an excited (^1D) electronic state, *singlet oxygen.* This is important in the troposphere too, as $O(^1D)$ reacts with water vapor to produce chemically reactive hydroxyl radicals, ·OH:

$$O(^1D) + H_2O \rightarrow 2{\cdot}OH. \tag{8.10}$$

The main oxidant produced in the decomposition of ozone in liquid water (as, e.g., in the use of ozone for disinfection of municipal water supplies or for destruction of viruses in donated blood) is again the hydroxyl radical. An excited state of *molecular* oxygen, $O_2(^1\Delta)$, is also produced in reaction 8.6 and is also called *singlet oxygen. Singlet* means that all electron spins are paired; both atomic and molecular oxygen have two unpaired electrons in their ground states, which are called *triplet* states. Triplet atomic oxygen, $O(^3P)$, is produced in the troposphere when NO_2 is photolyzed.

The hydroxyl radical is normally present only in low concentrations in the troposphere, as it reacts with further ozone to form the hydroperoxy radical HOO· which in turn gives hydrogen peroxide H_2O_2). Ozone, the hydroxyl radical, and hydrogen peroxide are the main oxidizing species in the troposphere, from the standpoint of environmental chemistry. The hydroxyl radical in particular performs an important function as a natural cleansing agent for the atmosphere.[26] In elevated concentrations, however, it can be harmful to living things, and ·OH radicals and singlet oxygen are probably responsible for much of the biological havoc wreaked by ozone, which is finally receiving recognition as one of the most objectionable of the common air pollutants. Tropospheric ozone is at least partially responsible for the widespread damage to forests by polluted air that is occurring in northern Europe and northeastern North America.

Ozone is produced in substantial concentrations by industrial activity and, indirectly, from automobile exhausts. The most important sequence of reactions producing tropospheric ozone begins with hydrocarbon vapors, nitric oxide, and sunlight:

(*a*) photolysis of hydrocarbons RH in the presence of O_2 gives alkylperoxy radicals ROO· (reaction 8.17);

(*b*) ROO· reacts with NO to give NO_2 (reaction 8.18);

(*c*) sunlight breaks up the NO_2 to form $O(^3P)$ atoms, regenerating NO; and

(*d*) $O(^3P)$ is scavenged by the ordinary O_2 of the air, forming O_3.

Thus, rather surprisingly, the key to reducing tropospheric ozone pollution is to minimize the release of hydrocarbon vapors from such sources as unburnt fuel in automobile exhausts, vaporization of fuel at service station pumps (modern pumps recover fuel vapors from automobile fuel tanks while refilling them), kitchen exhausts from fast-food restaurants and, as in Mexico City, leakages of liquefied petroleum gas (mainly butane) used for domestic heating and cooking.[26]

8.4 Nitrogen Oxides

8.4.1 Nitrous Oxide

Nitrous oxide (N_2O, see Section 2.11) is a colorless, odorless gas with mildly anaesthetic properties (laughing gas). It is formed in Nature by bacterial reduction of nitrates. The electronic structure of this linear molecule is best understood by noting that it is isoelectronic with CO_2, which is also linear. It is rather easily decomposed into N_2 and O_2, and so can support combustion.

Nitrous oxide is nontoxic—it used as the propellant in whipped-cream spray cans—and so might seem to be an unlikely pollutant. However, as noted earlier, it may contribute significantly to greenhouse warming. Furthermore, on diffusing to the stratosphere, N_2O becomes involved in the ozone cycle (reactions 8.2, 8.3, and 8.6) following its conversion to nitric oxide (NO):

$$N_2O + O(^1D) \rightarrow 2NO. \tag{8.11}$$

There is therefore concern that the ever-increasing use of synthetic nitrate fertilizers may result in further depletion of the ozone layer. Eventually, stratospheric NO is returned to the Earth as nitric acid (see Section 8.4.2), but the overall dynamics of the complex atmospheric chemistry are still not fully understood.

8.4.2 Nitric Oxide and Nitrogen Dioxide

The oxides NO (nitric oxide), NO_2 (nitrogen dioxide), and N_2O_4 (dinitrogen tetroxide), commonly lumped together as "NO_x," pose pressing pollution problems. The colorless gas NO can form by direct combination of atmospheric N_2 and O_2; however, the reaction is very endothermic and therefore can not give a significant yield of NO except at high temperatures. Both the equilibrium and the rate constants for NO formation at 298 K are extremely unfavorable:

$$N_2 + O_2 \rightleftharpoons 2NO \quad \Delta H^0 = +180.6 \text{ kJ mol}^{-1}. \tag{8.12}$$

Even at 3000 K, the yield of NO from air is still only a few percent, but the equilibrium is *rapidly* established (see Exercises 8.2 and 9.1). Consequently, whenever air is subjected to very high temperatures and then rapidly quenched to 1000 K or below, the small but significant high temperature NO yield is "frozen in," as happens in internal combustion engines and in high-temperature furnaces when vented. Pollution by NO_x is therefore usually blamed on automobile traffic, but heavy industry is a major culprit too.

Nitric oxide reacts quite rapidly with oxygen in air to form the red-brown gas NO_2:

$$2NO + O_2 \rightarrow 2NO_2 \tag{8.13a}$$

$$NO + O_3 \rightarrow NO_2 + O_2. \tag{8.13b}$$

This in turn dimerizes to yellow N_2O_4, which is normally encountered as a gas (the liquid boils at 21 °C at 1 bar):

$$2NO_2 \rightleftharpoons N_2O_4. \tag{8.14}$$

This dimerization is not unexpected, since NO_2 has one unused valence electron left on the N. Reaction 8.14 is exothermic, so that the NO_2/N_2O_4 mixture is browner at high temperature and at high dilutions. In high concentrations near room temperature, it is more yellow. The brown smog layers often seen over cities are due primarily to dilute NO_2, with some contribution from particulate matter.

In high concentrations, the pungent-smelling NO_2 is toxic and can cause septic pneumonia. Obviously, the same would apply to high concentrations of inhaled NO; in any case, NO would be toxic for the same reason that CO is poisonous—it binds strongly to the hemoglobin of the blood (see Eq. 8.1). On the other hand, NO generated transiently at *low concentrations* in the bodies of mammals through enzyme-catalyzed oxidation by O_2 of L-arginine performs essential messenger functions in the transmission of nerve impulses, despite its cytotoxicity. This is why amyl nitrite, nitroglycerin, sodium nitroprusside $\{Na_2[Fe(CN)_5NO]\}$, and other potential sources of NO are effective as vasodilators (i.e., act to widen constricted blood vessels and so relieve angina pectoris, high blood pressure, and other circulatory problems); they release additional NO that activates guanylate cyclase and so induces muscular relaxation in the blood vessel walls. The oxidized (NO^+, nitrosonium ion) and reduced (NO^-, nitroxyl ion) forms of NO are also biologically important.[27]

Environmental concern centers on two sequences of reactions that NO_x undergoes in the atmosphere.

(*a*) NO_2 in the troposphere reacts with hydroxyl radicals (from reaction 8.10, via reactions 8.2, 8.6, and 8.10) to form nitric acid:

$$NO_2 + \cdot OH \rightarrow HNO_3. \tag{8.15}$$

This is eventually carried into the soil, some distance from the source of NO_x, by rain or snow. Thus, NO_x, along with oxides of sulfur (Section 8.5), is connected with the phenomenon of *acid precipitation*,* which is at least partly responsible for the current destruction of temperate forests (Section 8.3). It also contributes to acidification of lakes, leading to fish kills through disruption of the food chain or, more commonly, dissolution of aluminum from normally insoluble soil minerals [nominally $Al(OH)_3$]—$Al^{3+}(aq)$ ions kill fish by displacing the Ca^{2+} that regulates gill permeability, causing loss of Na^+ and clogging of the gills with mucus:

$$Al(OH)_3 + 3H^+ \rightleftharpoons Al^{3+}(aq) + 3H_2O. \tag{8.16}$$

That Al^{3+} is the main toxic agent in many acidified lakes is supported by observations of improved fish survival rates when the silica content of the water is increased, as dissolved silica can form either soluble or insoluble aluminosilicates (see Section 7.6). Mobilized aluminum has also been linked to forest damage, since, in sufficient concentration, it is directly toxic to roots of spruce trees and many other plants.

On the other hand, nitrate ion (NO_3^-) is a valuable fertilizer, and some HNO_3 precipitation may be beneficial to vegetation so long as

*The term *acid precipitation* includes acid-bearing snow, fog, etc., as well as the widely discussed *acid rain*.

the excess acidity is neutralized (e.g., by a natural $CaCO_3$ content of the soil or by deliberate liming). Besides, it seems that SO_2 (Section 8.5) and/or ozone (Section 8.3) are the chief causes of airborne environmental damage, although there is evidence that NO_x and SO_x are more damaging when acting together than when acting separately.

(*b*) Automotive emissions of NO_x and unburned hydrocarbons can react photochemically under strong sunlight to form *photochemical smog*, a white aerosol that is intensely irritating to the eyes and mucous membranes. This problem is serious in urban areas like Los Angeles and Mexico City, where a combination of local weather and topography can keep automobile-contaminated air trapped under intense sunlight for extended periods. The chemistry of smog formation is complex, involving over 50 radical gas reactions (see, e.g., Fergusson[8]), but can be crudely summarized thus:

$$\underset{\text{unburnt fuel}}{\mathrm{R{-}H}} \xrightarrow{\text{light}} \cdot\mathrm{H} + \cdot\mathrm{R} \xrightarrow{\mathrm{O_2}} \mathrm{ROO\cdot} \tag{8.17}$$

$$\left.\begin{array}{l}\mathrm{NO + O_3}\ \text{or} \\ \mathrm{NO + ROO\cdot}\end{array}\right\} \longrightarrow \mathrm{NO_2} \tag{8.18}$$

$$\mathrm{ROO\cdot + NO_2} \longrightarrow \underset{\text{peroxynitrates}}{\mathrm{ROONO_2}} \tag{8.19}$$

$$\mathrm{ROO\cdot + NO \rightarrow RO\cdot} \xrightarrow{\mathrm{O_2}} \underset{\text{aldehydes}}{\mathrm{R'CHO}} \tag{8.20}$$

$$\mathrm{R'CHO + \cdot OH} \xrightarrow{\mathrm{O_2}} \mathrm{R'CO{-}OO\cdot} \xrightarrow{\mathrm{NO_2}} \underset{\text{PAN}}{\mathrm{R'CO{-}OONO_2}}. \tag{8.21}$$

The worst irritants in photochemical smogs are peroxyacyl nitrates, PAN. They produce singlet oxygen when they hydrolyze, and this may account for their biological action. Other objectionable components of photochemical smogs include aldehydes, organic hydroperoxides (ROOH), and peroxynitrates.

Given that dependence on the internal combustion engine for transportation is likely to continue for many years (General Motors introduced their first commercially viable electric automobile at the end of 1995), the best single approach to these problems is to reduce NO_x levels in automobile emissions.[5–7] Three basic strategies are (*i*) reducing the combustion temperature (since reaction 8.12 is endothermic), for example, by lowering the compression ratio of the engine to ~8.5:1 from the more usual 10.5:1, with consequent loss of engine efficiency; (*ii*) making the concentration of O_2 after combustion as low as possible by using fuel-rich mixtures (e.g.,

11:1 fuel to air by weight, rather than the efficient 15:1 which is near the theoretical value)—a wasteful strategy that leaves much unburnt fuel and CO; (*iii*) removing NO and unburnt fuel from the exhaust gases with a catalytic converter.[28]

The last option has been implemented in North America since 1970. Originally, catalytic converters had two stages, first *reducing* NO to N_2 (using unburnt fuel or CO as the reductant) and then *oxidizing* the remaining CO and hydrocarbons to CO_2 with injected air at about 400 °C (the exhaust gas recirculation or EGR system). To maintain reducing conditions in the first stage, the engine is tuned for an air–fuel mixture slightly richer than the stoichiometric 14.7:1. The catalysts usually employed are platinum or platinum–rhodium alloys, which also catalyze oxidation of sulfur in the fuel (~0.04%) to SO_3 (Section 10.2); some sulfuric acid (H_2SO_4) is produced, but this appears not to be a serious problem.

Since 1981, *three-way* catalytic systems have been standard in new cars sold in North America.[6,28a] These systems consist of platinum, palladium, and rhodium catalysts dispersed on an activated alumina layer ("washcoat") on a ceramic honeycomb monolith; the Pt and Pd serve primarily to catalyze oxidation of the CO and hydrocarbons, and the Rh to catalyze reduction of the NO. These converters operate with a near-stoichiometric air–fuel mix at 400–600 °C; higher temperatures may cause the Rh to react with the washcoat. In some designs, the catalyst bed is electrically heated at start-up to avoid the problem of temporarily excessive CO emissions from a cold catalyst. Zeolite-type catalysts containing bound metal atoms or ions (e.g., Cu/ZSM-5) have been proposed as alternatives to systems based on precious metals.

8.5 Sulfur Dioxide and Trioxide

Sulfur dioxide is a pungent, easily liquefiable (bp −10 °C) gas that is produced in large amounts through combustion of fossil fuels (notably coal), roasting of sulfide ores, pulp and paper mill discharges, and natural volcanic activity. The direct toxicity of SO_2 is moderate, but prolonged exposure can be fatal, especially to victims who already have bronchial problems. Some 4000 premature deaths were attributed to the SO_2 content of the disastrous London smog of December 5–9, 1952. The toxic effects of SO_2 are exacerbated by its adsorption on smoke particles that can lodge in the pulmonary tract, but it is toxic enough on its own; the World Health Organization recommends annual mean limits of 40–60 μg SO_2 m^{-3}. Nowadays, coal burning in British cities is strictly regulated, and coal-smoke smogs have virtually disappeared there.

The United Kingdom, however, still produces 3×10^6 metric tons of sulfur emissions a year, and atmospheric sulfur dioxide causes further problems in the form of acid precipitation, much of which falls downwind in Scandinavia and causes acidification of lakes and destruction of fish. By 1975, brown trout had disappeared from 50% of Scandinavian lakes. A similar situation arises in the northeastern United States and eastern Canada, downwind of the industrial area around the Great Lakes. In these cases, the precipitation falls in areas of primarily igneous rocks that, unlike, say, limestone ($CaCO_3$), cannot neutralize the acidity. Similar situations exist in many other parts of the world; the anthropogenic global emissions of SO_2 total some 1×10^8 metric tons per year, to which about half as much again is added by natural sources (primarily volcanoes).

Sulfur dioxide dissolves readily in water to form sulfurous acid, H_2SO_3, which is a fairly strong acid itself, with $pK_a(1) = 1.91$ and $pK_a(2) = 7.18$, at 25 °C and infinite dilution.*

$$H_2SO_3 \underset{}{\overset{K_a(1)}{\rightleftharpoons}} H^+ + \underset{\text{bisulfite}}{HSO_3^-} \tag{8.22a}$$

$$2HSO_3^- \rightleftharpoons H_2O + \underset{\text{metabisulfite}}{S_2O_5^{2-}} \tag{8.22b}$$

$$HSO_3^- \overset{K_a(2)}{\rightleftharpoons} H^+ + \underset{\text{sulfite}}{SO_3^{2-}} \tag{8.22c}$$

The sulfur-derived acidity in acid rain, however, is due mainly to H_2SO_4 rather than H_2SO_3. The direct reaction of gaseous SO_2 with oxygen to give SO_3 (a colorless substance, mp 17 °C, bp 45 °C), and hence H_2SO_4 by hydrolysis, is extremely slow.[29]

Brandt and van Eldik[30] review of the chemistry of sulfur(IV) oxidation, with emphasis on the catalytic role of metal ions such as Fe^{2+}. We consider here only a simplified summary of the principal atmospheric oxidation processes. It is likely that oxidation is effected primarily through the action of hydrogen peroxide or ozone in water droplets in clouds, through the photochemical effect of ultraviolet light, or by heterogeneous catalysis of the SO_2–O_2 reaction by dust particles:[9,30,31]

$$H_2O + SO_2 + \tfrac{1}{2}O_2 \text{ (or } \tfrac{1}{3}O_3\text{, or } H_2O_2\text{)} \rightarrow H_2SO_4(\text{l}) \text{ (aerosol)}. \tag{8.23}$$

Oxidation by H_2O_2 is probably the dominant process; as noted earlier, H_2O_2 in clouds is formed from hydroxyl radicals, so the importance of $\cdot OH$ as the main cleansing agent in the troposphere is again evident. In dry air (i.e., in the absence of clouds), the chief oxidant is probably $\cdot OH$ itself (cf. reaction 8.15):

$$SO_2 + \cdot OH \rightarrow \cdot SO_3H \xrightarrow{\cdot OH} H_2SO_4(\text{aerosol}). \tag{8.24}$$

---

* $pK_a = -\log K_a$

The aerosols of sulfuric acid so formed increase the reflectivity (albedo) of the Earth's atmosphere, cutting down the solar radiation that reaches the Earth's surface and so counteracting to some extent the greenhouse warming due to CO_2 emissions that accompany the SO_2, as mentioned earlier. Airborne sulfuric acid may be neutralized by traces of ammonia in the air, giving particulate NH_4HSO_4 and $(NH_4)_2SO_4$ hazes, but in the absence of such neutralization the aqueous sulfuric acid droplets in tropospheric clouds may reach pH 1.5 or lower.

The residence times of SO_2 and H_2SO_4 in the troposphere are typically only a few days, but sulfuric acid aerosols reaching the stratosphere can be very persistent; together with nitric acid, they provide the solid surfaces in polar stratospheric clouds on which reaction 8.9 and related processes occur heterogeneously. Indeed, studies suggest that NO_x emissions of commercial supersonic aircraft in the lower stratosphere may pose less of a threat to the ozone layer than previously supposed; however, the accompanying formation of sulfuric and nitric acid aerosols may exacerbate ozone loss by increasing the available catalytic surface area.

Sulfuric acid is a stronger acid than sulfurous [$pK_a(1) < 0$, $pK_a(2) = 1.99$ at 25 °C and infinite dilution]; rain as acidic as pH 2.1 has been recorded at Hubbard Brook, New Hampshire, and the pH of water droplets in clouds can be as low as 1.5 (for comparison, the pH of rainwater saturated with atmospheric CO_2 is about 5.6 at 15 °C). Acid rain destroys building materials (especially marble), kills fish and vegetation, accelerates metallic corrosion (Sections 16.5 and 16.7), and can be directly harmful to humans (e.g., it causes the "alligator skin" condition reported in Cubatão, Brazil). Sulfate rain is not completely without redeeming features, as many soils (e.g., in southern Alberta, Canada) are sulfur-deficient. On balance, however, its acidity is unacceptable, and sulfur oxide emissions must be controlled at the source. Several control measures are possible:

(*a*) *Desulfurization of fuels.* Desulfurization of fuels is difficult and generally too expensive.[32] In coals, the organic sulfur content (thiophene, etc.) can be extracted from the crushed coal with a suitable solvent. The inorganic sulfur component, which is primarily pyrite (FeS_2), can be oxidized with aqueous O_2, iron(III) sulfate (at 100 °C, the Meyers process), hydrogen peroxide/sulfuric acid, or nitric acid. Sulfur dioxide itself can be used in aqueous HCl to oxidize pyrite to elemental sulfur and aqueous $FeCl_2$ (which is then air oxidized to solid Fe_2O_3); the sulfur is then burned to give more SO_2. "Caustic leaching" consumes aqueous NaOH and gives a mixture of aqueous sulfide and thiosulfate ions:

$$8FeS_2 + 30NaOH \rightarrow 4Fe_2O_3 + 14Na_2S + Na_2S_2O_3 + 15H_2O. \quad (8.25)$$

An interesting variant on the latter, the Battelle hydrothermal coal process,[33] involves hydrothermal conversion of both inorganic and organic forms of sulfur to sulfide ion (with efficiencies of 99% and 70%, respectively) and thence to hydrogen sulfide gas, followed by recovery of the easily handled element from H_2S by the Claus process (Section 10.1).

$$\begin{array}{rcl} \text{S in coal} + \text{NaOH(aq)} & \xrightarrow{220\text{–}350\,^\circ\text{C}} & \text{S}^{2-}\text{(aq)} \\ & & \downarrow \text{CO}_2\text{(g)} \\ \text{S}_8 & \xleftarrow{\text{Claus}} & \text{H}_2\text{S(g)} + \text{Na}_2\text{CO}_3\text{(aq)} \end{array} \tag{8.26}$$

(*b*) *Scrubbing of stack gases with limestone.* Sulfur-containing fuels are burnt as received, and the SO_2 is removed by passing the flue gases up a column of wet limestone:

$$\text{CaCO}_3\text{(s)} + \text{SO}_2\text{(g)} \xrightarrow{\text{H}_2\text{O}} \text{CaSO}_3\text{(s)} + \text{CO}_2\text{(g)}. \tag{8.27}$$

This is easier than pretreatment of the fuel, but the waste is calcium *sulfite* which, unlike the more familiar sulfate (gypsum, Plaster of Paris), is quite toxic to plants and is therefore difficult to dispose of safely.

(*c*) *Recovery of SO_2 as H_2SO_4.* Direct conversion of SO_2 in flue gases to useful sulfuric acid (as in Section 10.2) is economically attractive, but usually the SO_2 will contain fly ash and other impurities that may poison the catalyst necessary to oxidize SO_2 to SO_3. A better procedure involves scrubbing the stack gases with a cool spray of an SO_2 absorbent—typically a solution of an organic amine—that releases the SO_2 on heating in a separate column; the SO_2 can then be converted to elemental sulfur by the Claus process or to sulfuric acid (Chapter 10). Union Carbide's CanSolv process is of this type; it has been implemented successfully at the Suncor, Inc., oil sands plant at Fort McMurray, Alberta, Canada, where petroleum coke of high S content (7%) is disposed of by burning to make steam.

(*d*) *Fluidized bed combustion.* In the temperature range 820 to 870°C, limestone reacts with SO_2 and air to give calcium sulfate rather than sulfite, and this product can be safely disposed of or utilized, for example, as roadbed cement. Fennelly[34] describes a two-stage fluidized bed furnace (Fig. 8.2) in which coal fluidized in an airstream is burned at 1100 to 1300°C (the temperature at which combustion is most efficient) around the boiler tubes, giving highly efficient heat transfer without formation of hot spots. The gases then pass to the

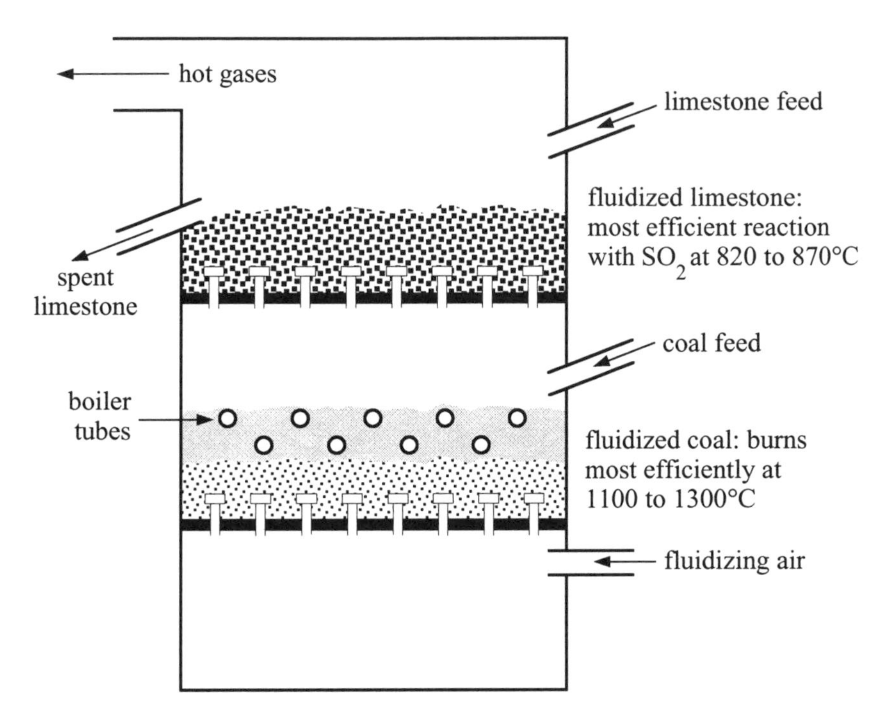

Figure 8.2 Two-stage fluidized bed coal-fired boiler that minimizes sulfur dioxide emissions. Adapted from P. F. Fennelly, *Am. Sci.* **72**, 254 (1984).

fluidized limestone bed where over 90% of the SO_2 is removed:

$$CaCO_3 \rightarrow CaO + CO_2 \tag{8.28}$$

$$CaO + SO_2 + \frac{1}{2}O_2 \rightarrow CaSO_4. \tag{8.29}$$

Reaction 8.28 is endothermic and 8.29 exothermic, and the combination is almost thermoneutral ($\Delta H \approx 0$). Since the combustion temperature is lower than in traditional boiler designs, less NO_x is emitted, and the ash does not melt and so cannot foul the heat transfer surfaces.

(*e*) *Use of low-sulfur fuels.* Sulfur dioxide emissions from coal-burning utilities could be substantially reduced by the simple expedient of using low-sulfur coals (e.g., from western Canada). Transportation costs and political considerations must, however, be taken into account.

Exercises

8.1 Wet limestone scrubbing of stack gases to remove SO_2 gives hydrated calcium sulfite

$$CaCO_3 + SO_2 + \tfrac{1}{2}H_2O \rightarrow CaSO_3{\cdot}\tfrac{1}{2}H_2O + CO_2 \qquad (1)$$

rather than the comparable hydrated sulfate

$$CaCO_3 + SO_2 + \tfrac{1}{2}H_2O + \tfrac{1}{2}O_2 \rightarrow CaSO_4{\cdot}\tfrac{1}{2}H_2O + CO_2. \qquad (2)$$

Assume the gases are at 298 K and that other standard conditions prevail, and determine whether the prevalence of reaction 1 at near-ambient conditions is a result of thermodynamic or kinetic constraints.

	$CaSO_3{\cdot}\frac{1}{2}H_2O$	$CaSO_4{\cdot}\frac{1}{2}H_2O$	O_2	
ΔH_f°	−1311.7	−1576.74	0	kJ mol^{-1}
S°	121.3	130.5	205.138	J K^{-1} mol^{-1}

8.2 Given that the standard heat of formation of nitric oxide is 90.25 kJ mol^{-1} and that the standard entropies of N_2, O_2 and NO are 191.61, 205.138, and 210.761 J K^{-1} mol^{-1}, respectively, derive the equilibrium constant K for the formation of 1 mol of NO at 298 K and 1 bar, assuming ideal gas behavior. With the composition of air given in Section 8.1, calculate the equilibrium pressure P_{NO} of NO that would form in air at 298 K and 100 kPa if the reaction were kinetically feasible.
[*Answers:* $K = 6.8 \times 10^{-16}$; $P_{NO} = 2.8 \times 10^{-11}$ Pa.]

8.3 From the standpoint of abatement of the contribution of CO_2 to the greenhouse effect, natural gas (methane) is preferred as a fuel over gasoline and, particularly, coal. Explain why this is so.

8.4 There has been much discussion of the possibility of converting CO_2 emissions to useful organic compounds, but most proposals involve endothermic reactions that would require input of heat, probably generated at the expense of yet more CO_2 emissions. The proposed formation of liquid acetic acid ($\Delta H_f^\circ = -485.76$ kJ mol^{-1}, $S^\circ = 178.7$ J K^{-1} mol^{-1}) from CO_2 and methane has been claimed to be exothermic. Is this reaction feasible in the industrial context, assuming an appropriate catalyst can be found? (Use the data of Appendix C.)

References

1. W. Büchner, R. Schliebs, G. Winter, and K. H. Büchel, "Industrial Inorganic Chemistry." VCH, New York, 1989 (translated by D. R. Terrell).

2. R. Thompson (ed.), "The Modern Inorganic Chemicals Industry," Special Publication 31. Chemical Society, London, 1977; R. Thompson (ed.), "Industrial Inorganic Chemicals: Production and Uses." Royal Society of Chemistry, Cambridge, 1995.

3. "Chemistry in the Economy." American Chemical Society, Washington, D.C., 1973.

4. W. J. Thomas (ed.), "Separation of Gases: Fifth BOC Priestley Conference," Special Publication 80. Royal Society of Chemistry, Cambridge, 1990.

5. I. Bodek, W. J. Lyman, W. F. Reehl, and D. H. Rosenblatt, "Environmental Inorganic Chemistry." Pergamon, New York, 1988; N. J. Bunce, "Environmental Chemistry." Wuerz, Winnipeg, Manitoba, 1991; S. E. Manahan, "Environmental Chemistry." Lewis Publishers, Chelsea, Michigan, 1991; S. E. Manahan, "Fundamentals of Environmental Chemistry." Lewis Publishers, Boca Raton, Florida, 1993.

6. R. M. Harrison (ed.), "Pollution: Causes, Effects and Control," 2nd Ed. Royal Society of Chemistry, Cambridge, 1990.

7. R. W. Boubel, D. L. Fox, D. B. Turner, and A. C. Stern, "Fundamentals of Air Pollution," 3rd Ed. Academic Press, New York, 1994.

8. J. E. Fergusson, "Inorganic Chemistry and the Earth." Pergamon, Oxford, 1982.

9. A. Cocks and T. Kalland, The chemistry of atmospheric pollution. *Chem. Br.* **24**, 884–888 (1988).

10. M. B. Hocking, "Modern Chemical Technology and Emission Control." Springer-Verlag, New York, 1984.

11. J. W. Birks, J. G. Calvert, and R. W. Sievers (eds.), "The Chemistry of the Atmosphere: Its Impact on Global Change." American Chemical Society, Washington, D.C., 1993.

12. D. M. Hunten, Atmospheric evolution of the terrestrial planets. *Science* **259**, 915–920 (1993); J. F. Kasting, Earth's early atmosphere. *Science* **259**, 920–926 (1993); R. A. Berner, Atmospheric carbon dioxide levels over phanerozoic time. *Science* **249**, 1382–1386 (1990); R. A. Berner, Paleozoic atmospheric CO_2: importance of solar radiation and plant evolution. *Science* **261**, 68–70 (1993).

13. B. Hileman, Web of interactions makes it difficult to untangle global warming data. *Chem. Eng. News* April 27, 7–19 (1992); R. Pocklington, K. Drinkwater, and R. Morgan, Reasons for scepticism about

greenhouse warming. *Can. Chem. News* October, 19–22 (1993); P. C. Novelli, T. J. Conway, E. J. Dlugokencky, and P. P. Tans, Recent changes in carbon dioxide, carbon monoxide and methane and the implications for global climate change. *W. M. O. Bull.* **44**, 32–38 (1995); B. Hileman, Climate observations substantiate global warming models. *Chem. Eng. News*, November 27, 18–22 (1995); N. E. Graham, Simulation of recent global temperature trends. *Science* **267**, 666–671 (1995).

14. E. Friis-Christensen and K. Lassen, Length of the solar cycle: An indicator of solar activity closely associated with climate. *Science* **254**, 698–700 (1991); R. F. Service, Sun's role in warming is discounted. *Science* **268**, 28–29 (1995); R. A. Kerr, A fickle sun could be altering the Earth's climate after all. *Science* **269**, 633 (1995); D. J. Thomson, The seasons, global temperature, and precession. *Science* **268**, 59–68 (1995); V. Ramanathan, B. Subasilar, G. J. Zhang, W. Conant, R. D. Cess, J. T. Kiehl, H. Grassl, and L. Shi, Warm pool heat budget and shortwave cloud forcing: a missing physics? *Science* **267**, 499–502 (1995); J. Lelieveld and J. Heintzenberg, Sulfate cooling effect on climate through in-cloud oxidation of anthropogenic SO_2. *Science* **258**, 117–120 (1992); J. T. Kiehl and B. P. Briegleb, The relative roles of sulfate aerosols and greenhouse gases in climate forcing. *Science* **260**, 311–314 (1993).

15. P. Ciais, P. P. Tans, M. Trolier, J. W. C. White, and R. J. Francey, A large northern hemispheric terrestrial CO_2 sink indicated by the $^{13}C/^{12}C$ ratio of atmospheric CO_2. *Science* **269**, 1098–1102 (1995). P. E. Kauppi, K. Mielikäinen, and K. Kuusela, Biomass and carbon budget of European forests, 1971 to 1990. *Science* **256**, 70–74 (1992).

16. R. S. Nerem, Global mean sea level variations from TOPEX/POSEIDON altimeter data. *Science* **268**, 708–710 (1995).

17. E. Culotta, Will plants benefit from high CO_2? *Science* **268**, 654–656 (1995).

18. E. S. Rubin, R. N. Cooper, R. A. Frosch, T. H. Lee, G. Marland, A. H. Rosenfeld, and D. D. Stine, Realistic mitigation options for global warming. *Science* **257**, 148–149, 261–266 (1992).

19. A. Brauman, M. D. Kane, M. Labat, and J. A. Breznak, Genesis of acetate and methane by gut bacteria of nutritionally diverse termites. *Science* **257**, 1384–1387 (1992).

20. M. H. Threinen and W. C. Trogler, Nylon production—an unknown source of N_2O. *Science* **251**, 932 (1991).

21. J. M. DeSimone, Z. Guan, and C. S. Elsbernd, Synthesis of fluoropolymers in supercritical carbon dioxide. *Science* **257**, 945–947 (1992); K. P. Johnson, K. L. Harrison, M. J. Clarke, S. M. Howdle, M. P. Heitz, F. V. Bright, C. Cartier, and T. W. Randolph, Water-in-carbon dioxide microemulsions: An environment for hydrophiles including proteins. *Science* **271**, 624–626 (1996); see also commentary by E. J. Beckman, p. 613.

22. M. R. Schoeberl and D. L. Hartmann, The dynamics of the stratospheric polar vortex and its relation to springtime ozone depletions. *Science* **251**, 46–52 (1991); W. H. Brune, J. G. Anderson, D. W. Toohey, D. W. Fahey, S. R. Kawa, R. L. Jones, D. S. McKenna, and L. R. Poole, The potential for ozone depletion in the Arctic polar stratosphere. *Science* **252**, 1260–1266 (1991); R. D. Bojkov, The international ozone assessment—1994. *W. M. O. Bull.* **44**, 42–50 (1995); F. S. Rowland and M. Molina, Ozone depletion: 20 years after the alarm. *Chem. Eng. News*, August 15, 8–13 (1994).

23. B. Hileman, Amphibian population loss tied to ozone thinning. *Chem. Eng. News* March 7, 5 (1994); M. L. Bothwell, D. M. J. Sherbot, and C. M. Pollock, Ecosystem response to solar ultraviolet-B radiation: influence of trophic-level interactions. *Science* **265**, 97–100 (1994).

24. F. S. Rowland, Chlorofluorocarbons and the depletion of stratospheric ozone *Am. Sci.* **77**, 36–45 (1989); T.-L. Tso, L. T. Molina, and F. C.-Y. Wang, Antarctic stratospheric chemistry of chlorine nitrate, hydrogen chloride and ice: release of active chlorine. *Science* **238**, 1253–1260 (1987); J. G. Anderson, D. W. Toohey, and W. H. Brune, Free radicals within the Antarctic vortex: the role of CFCs in Antarctic ozone loss. *Science 251*, 39–46 (1991); P. S. Zurer, Complexities of ozone loss continue to challenge scientists. *Chem. Eng. News* June 12, 20–23 (1995).

25. H. F. Hemond and E. J. Fechner, "Chemical Fate and Transport in the Environment." Academic Press, New York, 1994.

26. J. O. Nriagu and M. S. Simmons (eds.), "Environmental Oxidants." Wiley, New York, 1994; A. M. Thompson, The oxidizing capacity of the Earth's atmosphere: probable past and future changes. *Science* **256**, 1157–1165 (1992).

27. E. W. Ainscough and A. M. Brodie, Nitric oxide—some old and new perspectives. J. Chem. Educ. **72**, 686–692 (1995); P. L. Feldman, O. W. Griffith, and D. J. Stuehr, The surprising life of nitric oxide. *Chem. Eng. News* December 20, 26–38 (1993); A. R. Butler, NO—its role in the control of blood pressure. *Chem. Br.* **26**, 419–421 (1990);

J. S. Stamler, D. J. Singel, and J. Loscalzo, Biochemistry of nitric oxide and its redox-activated forms. *Science* **258**, 1898–1902 (1992).

28. (*a*) R. J. Farrauto, R. M. Heck, and B. K. Speronello, Environmetal catalysts. *Chem. Eng. News* September 7, 34–44 (1992); (*b*) M. Shelef, Selective catalytic reduction of NO_x with N-free reductants. *Chem. Rev.* **95**, 209–225 (1995); (*c*) M. Bowker and R. W. Joyner, Automobile catalysts. *In* "Insights into Speciality Inorganic Chemicals" (D. Thompson, ed.), pp. 145–169. Royal Society of Chemistry, Cambridge, 1995.

29. R. Steudel, Sulfuric acid from sulfur trioxide and water—a surprisingly complex reaction. *Angew. Chem. Int. Ed. Engl.* **34**, 1313–1315 (1995).

30. C. Brandt and R. van Eldik, Transition metal-catalyzed oxidation of sulfur(IV) oxides. Atmospheric-relevant processes and mechanisms. *Chem. Rev.* **95**, 119–190 (1995).

31. J. H. Seinfeld, Urban air pollution: state of the science. *Science* **243**, 745–752 (1989); S. E. Schwarz, Acid deposition: unraveling a regional phenomenon. *Science*, *243*, 753–763 (1989).

32. R. A. Meyers, "Coal Desulfurization." Dekker, New York, 1977; L. D. Smoot (ed.), "Fundamentals of Coal Combustion." Elsevier, Amsterdam, 1993; R. C. Eliot, "Coal Desulfurization Prior to Combustion." Noyes Data, Park Ridge, New Jersey, 1978.

33. E. P. Stambaugh, J. F. Miller, S. S. Tam, S. P. Chauhan, H. F. Feldmann, H. E. Carlton, J. F. Foster, H. Nack, and J. H. Oxley, Hydrothermal process produces clean fuel. *Hydrocarbon Process.* **54**, 115-116 (1975); A. L. Hammond, Hydrothermal process produces clean fuel. *Science*, **189**, 128–130 (1975).

34. P. F. Fennelly, Fluidized bed combustion. *Am. Sci.* **72**, 254 (1984).

Chapter 9

Nitrogen, Phosphorus, and Potash in Agriculture

THE FOOD we eat is derived ultimately from plants, algae, and microorganisms that convert inorganic nutrients such as carbon dioxide, water, nitrogen, sulfates, and phosphates into organic compounds such as carbohydrates, proteins, fats, and vitamins. The complex biochemical processes involved are usually driven by sunlight (through photosynthesis). We can identify groups of plant nutrients as follows:

(*a*) *Macronutrients*—C, H, and O (from CO_2 of the air, and water);

(*b*) *Primary nutrients*—N, P, and K (the topics of this chapter);

(*c*) *Secondary nutrients*—Ca, Mg, and S (often deficient in tropical soils); and

(*d*) *Micronutrients*—B, Cl, Co, Cu, Fe, Mn, Mo, Si, and Zn.

Agriculture therefore depends on there being a sufficient supply of inorganic nutrients to plants. Cereals, vegetables, fruit-bearing trees or plants, and animal fodder require *bioavailable* nutrients, that is, nutrients in forms that they can use. Since intensive agriculture depletes many natural nutrients, synthetic nutrients (fertilizers) must be supplied.[1–7] In particular, we need to "fix" the inert N_2 of the atmosphere as soluble, reactive compounds such as nitrates, ammonia, and ammonium salts. Other major fertilizer components are sulfate, potassium, and phosphate ions. It may also be necessary to provide trace nutrients, such as cobalt compounds, or to remove excess soil acidity by treatment with lime (CaO). World fertilizer demand in the year 2001 is expected to be about 1.5×10^8 metric tons N, 7.6×10^7 metric tons P_2O_5, and 6.7×10^7 metric tons K_2O; these projections represent an increase of about a factor of 10 in 40 years. Clearly, inorganic fertilizer production is an extremely important part of the global economy.

9.1 Natural Sources of Fixed Nitrogen

Apart from minor sources such as thunderstorms (Section 9.2), atmospheric nitrogen is fixed in Nature by certain soil bacteria, blue-green algae, and microorganisms in the root nodules of legumes. This is accomplished either by oxidation to nitrate

$$\tfrac{1}{2}N_2(g) + \tfrac{1}{2}H_2O(l) + \tfrac{5}{4}O_2(g) \rightarrow HNO_3(aq) \qquad \Delta H^\circ = -65 \text{ kJ mol}^{-1} \quad (9.1)$$

or reduction to ammonia or ammonium salts

$$\tfrac{1}{2}N_2(g) + \tfrac{3}{2}H_2O(l) \rightarrow NH_3(aq) + \tfrac{3}{4}O_2(g) \qquad \Delta H^\circ = +348 \text{ kJ mol}^{-1}. \quad (9.2)$$

Reaction 9.1 might seem to be thermodynamically favored, but in fact no kinetically easy route from triply bonded N_2 to NO_3^- exists, since the endothermic intermediate NO (Section 8.4.2) is likely to be involved. As written, reaction 9.2 has prohibitive energetics, but in practice the process is more complex than this. For example, the fact that free O_2 is not formed, but is in effect consumed in other biochemical reactions, makes for a favorable energy balance. The limiting factor is again kinetics, as plausible intermediates such as hydrazine ($H_2N—NH_2$) are endergonic compounds.

Natural nitrogen-fixing systems overcome these kinetic barriers with certain *enzymes* known as nitrogenases which contain iron and molybdenum atoms. Enzymes are biological catalysts, usually having complicated protein-based structures, often with complexed metal ions as the active sites. The exact structure and mode of action of the various nitrogenases are the subjects of intensive current research. The essential features of molybdenum nitrogenase are (i) twin molybdenum–iron units $MoFe_7S_8$ held in a protein (total molar mass ~230,000) and (ii) an Fe_4S_4 cluster tied by sulfur linkages into another protein (total molar mass ~65,000).[8] The Fe_4S_4 unit is of roughly cubic geometry, with Fe and S atoms at alternate corners, and is typical of the active centers of a group of biological redox (i.e., oxidation–reduction or electron transfer) reagents known as *ferredoxins.* The iron protein supplies electrons, together with energy obtained from the hydrolysis of ATP to ADP and inorganic phosphate (Section 7.7), to the iron–molybdenum protein where N_2 is sequestered and then reduced to ammonia, presumably in a stepwise manner. Vanadium–iron and indeed all-iron nitrogenases have been found, but most N_2-fixing enzymes are of the Mo–Fe class, of which over a dozen different examples are known.

The prospect of mimicking the ability of nitrogenases to fix N_2 in mild aqueous conditions is attractive in view of the high energy and capital requirements of current NH_3 and HNO_3 manufacture, as described in Sections 9.2–9.5.

9.2 Direct Combination of Nitrogen and Oxygen

As noted in Section 8.4, atmospheric N_2 and O_2 combine endothermically in small but significant yield and at a sufficiently rapid rate above about 2000 K, but the gases must be quenched rapidly if the high–temperature yield of NO is to be recovered for subsequent conversion to nitric acid:

$$\tfrac{1}{2}N_2 + \tfrac{1}{2}O_2 \rightleftharpoons NO \quad \Delta H^\circ = 90.3 \text{ kJ mol}^{-1}. \tag{9.3}$$

Lightning strikes generate transient temperatures of several thousand kelvins, and may contribute significant amounts of fixed N to the biosphere by reaction 9.3 in regions where thunderstorms are frequent, for example, in tropical rain forests. Electric arcs have been used in this way in industry, notably in Norway, where hydroelectric power is inexpensive. In the United States in the 1950s, the *Wisconsin process*[9] was developed for NO production. In this process, air is blasted over hot refractory pebbles (MgO, preheated to ~2200 °C by burning fuel oil or natural gas) and is then quenched rapidly. Increases in energy costs since the 1950s, however, have made the Wisconsin and arc processes increasingly uneconomical when compared to the catalytic oxidation of ammonia in nitric acid production (Section 9.4).

9.3 Ammonia Synthesis

The direct gas-phase synthesis of ammonia from nitrogen and hydrogen (the *Haber process*, 1908) is presently the cornerstone of the fertilizer industry:

$$\underset{1\text{ vol}}{N_2} + \underset{3\text{ vol}}{3H_2} \rightleftharpoons \underset{2\text{ vol}}{2NH_3} \quad \Delta H^\circ = -92.2 \text{ kJ mol}^{-1}. \tag{9.4}$$

It is implicit in reaction 9.4 that the equilibrium yield of ammonia is favored by high pressures and low temperatures (Table 9.1). However, compromises must be made, as the capital cost of high pressure equipment is high and the rate of reaction at low temperatures is slow, even when a catalyst is used. In practice, Haber plants are usually operated at 80 to 350 bars and at 400 to 540 °C, and several passes are made through the converter. The catalyst (Section 6.2) is typically finely divided iron (supplied as magnetite, Fe_3O_4 which is reduced by the H_2) with a KOH promoter on a support of refractory metallic oxide. The upper temperature limit is set by the tendency of the catalyst to sinter above 540 °C. To increase the yield, the gases may be cooled as they approach equilibrium.

TABLE 9.1
Equilibrium Yields of Ammonia According to Equation 9.4

Temperature (°C)	*Yield (%)*	
	at 25 bars	*at 400 bars*
100	91.7	99.4
400	8.7	55.4
500	2.9	31.9

The nitrogen required is obtained by fractional distillation of liquid air.[1b] The hydrogen used to be obtained by electrolysis of liquid water; if inexpensive surplus electrical capacity becomes available in the future, this method may well be reintroduced. Catalytic photolysis of water using sunlight is another possible future source of H_2. The *Haber–Bosch process* of 1916 used *water–gas*, which is a mixture of H_2, CO, and CO_2 made by alternating blasts of steam and air over coke at red heat:

$$H_2O(g) + C(s) \xrightarrow{\sim 800\,°C} H_2(g) + CO(g) \quad \Delta H = +175 \text{ kJ mol}^{-1} \quad (9.5)$$

$$2H_2O(g) + C(s) \rightarrow 2H_2(g) + CO_2(g) \quad \Delta H = +178 \text{ kJ mol}^{-1} \quad (9.6)$$

The periodic air blasts to reheat the coke are necessary because of the marked endothermicity of reactions 9.5 and 9.6. The CO content is used to generate further H_2 in the *water gas shift reaction*:

$$H_2O(g) + CO(g) \rightleftharpoons CO_2(g) + H_2(g) \quad \Delta H = -41 \text{ kJ mol}^{-1}. \quad (9.7)$$

The exothermicity of reaction 9.7 dictates a low temperature (450 °C or less) and use of a catalyst (Fe_2O_3/Al_2O_3, usually) if good yields are to be obtained in a reasonable time. Usually, the reaction is carried out in two steps. First temperatures of 450 to 500 °C are used and a substantial degree of conversion quickly results. Then the temperature is dropped to about 200 °C to optimize the yield. The CO_2 is removed by scrubbing the gas with water, in which CO_2 is much more soluble than H_2.

At present, H_2 is usually made from natural gas (methane, CH_4) or naphtha by *steam reforming*, first introduced in the 1930s:

$$CH_4(g) + H_2O(g) \rightleftharpoons CO(g) + 3H_2(g) \quad \Delta H = +206 \text{ kJ mol}^{-1}. \quad (9.8)$$

A temperature of 700 to 800 °C is optimal, with a nickel catalyst. Alternatively, methane can be partially oxidized with pure oxygen:

$$CH_4 + \tfrac{1}{2}O_2 \rightleftharpoons CO + 2H_2 \quad \Delta H = -36 \text{ kJ mol}^{-1}. \quad (9.9)$$

At first sight, steam reforming seems to give more H_2 per CH_4 than reaction 9.9, but it is endothermic and requires that additional methane be burned to provide the heat input. In practice, reaction 9.8 is still preferred, inasmuch as it is more economical to provide the needed heat than to recover the waste heat of reaction 9.9. In either case, the CO may be used to make more H_2 by reaction 9.7.

Ammonia is readily liquefied under pressure or on cooling (normal bp −33 °C, critical point 132.5 °C and 11.4 MPa). The liquid (*anhydrous ammonia*) was formerly used as a refrigerant, and today ammonia is transported in liquid form by truck, barge, and pipeline. It is often actually injected directly as fertilizer into soils to a depth of 15–30 cm, where it is retained by virtue of its high solubility (456 g per kg water) in the water content of the soil. However, this technology is hazardous; tank pressures may reach 1.5 MPa on a warm day, both liquid and gaseous ammonia are corrosive to the flesh, and the gas forms an explosive mixture with air over the composition range 16–25% NH_3. Ammonia is therefore commonly converted in aqueous solution to either ammonium sulfate or ammonium nitrate, which are then recovered by evaporation as crystalline solids for safe shipment, storage, and use:

$$2NH_3(g) + H_2SO_4(aq) \rightarrow (NH_4)_2SO_4(aq). \qquad (9.10)$$

The use of urea has gained favor as an agricultural source of nitrogen, largely because it can be produced in forms that release nitrogen slowly to the soil. The "SCU" fertilizers familiar to gardeners are urea pellets coated with ~2% paraffin wax containing elemental sulfur that oxidizes away slowly in the soil; typically, they contain 32% N and 30% S. Alternatively, urea can be made to react with formaldehyde to form urea–formaldehyde polymers that decompose slowly in the ground. Urea is made by reaction of ammonia with the CO_2 by-product of the water–gas shift reaction (9.7):

$$CO_2 + 2NH_3 \rightarrow \underset{\text{ammonium carbamate}}{NH_4[CO_2NH_2]} \xrightarrow[200\,^\circ C]{150 \text{ bars}} \underset{\text{urea}}{(H_2N)_2CO} + H_2O. \qquad (9.11)$$

9.4 Nitric Acid and Ammonium Nitrate

The *catalytic* oxidation of ammonia by air over platinum gauze at ~900 °C gives nitric oxide (reaction 9.12), which is then oxidized to nitric acid by air and *liquid* water in a "nitrous gas absorber" (reactions 9.13 and 9.14):

$$NH_3 + \tfrac{5}{4}O_2 \rightarrow NO + \tfrac{3}{2}H_2O(g) \qquad \Delta H^\circ = -226.4 \text{ kJ mol}^{-1} \qquad (9.12)$$

$$NO + \tfrac{1}{2}O_2 \rightarrow NO_2 \qquad \Delta H^\circ = -57.1 \text{ kJ mol}^{-1} \qquad (9.13)$$

$$3NO_2 + H_2O(l) \rightarrow 2HNO_3(l) + NO \qquad \Delta H^\circ = 71.7 \text{ kJ mol}^{-1}, \qquad (9.14)$$

where ΔH° refers to 25 °C rather than the reaction temperature. The net reaction is thus

$$NH_3 + 2O_2 \rightarrow HNO_3 + H_2O(l) \qquad (9.15)$$

with $\Delta H^\circ = -447.1$ kJ mol^{-1} if the product is aqueous nitric acid (hypothetical 1.0 molal) or -413.8 kJ mol^{-1} for 100% liquid HNO_3. *Specific* catalysis of reaction 9.12 is *essential*, otherwise the thermodynamically favored oxidation of ammonia to N_2 will occur instead (recall that the formation of NO from its elements is *endergonic*; Sections 8.4 and 9.2):

$$NH_3 + \tfrac{3}{4}O_2 \rightarrow \tfrac{1}{2}N_2 + \tfrac{3}{2}H_2O(g) \quad \Delta H^\circ = -316.6 \text{ kJ mol}^{-1}. \qquad (9.16)$$

For the same reason, the gas leaving the catalyst should be substantially free of NH_3; otherwise, reaction 9.17 will occur while the gas is still hot:

$$6NO + 4NH_3 \rightarrow 5N_2 + 6H_2O. \qquad (9.17)$$

Note that the most important function of the catalyst, here and in many other instances, is not so much its overall catalytic activity as its *selectivity* in promoting the one reaction (9.12) over its competitors. In biological systems, enzymes have evolved to extraordinary degrees of selectivity; often, they will catalyze one specific biochemical reaction and no others.

The *nitrous gas absorption* step (reactions 9.13 and 9.14) is slow, especially if concentrated HNO_3 is required, since cooling to 2 °C is then necessary. Consequently, large countercurrent towers of stainless steel are needed, with associated high capital cost. The recovery of the heat of reaction of this step is inefficient because of the low temperature of the source gases that must be maintained. It has been suggested that the energy of reaction 9.12 could be more effectively recovered if it is run in a fuel cell (see Exercise 15.8).

Some nitric acid is used for the manufacture of explosives and chemicals, but much is converted on-site to the potentially explosive high nitrogen fertilizer ammonium nitrate (Section 2.11). Ammonia gas from the Haber plant is absorbed in aqueous HNO_3, and the NH_4NO_3 solution is evaporated to a liquid melt ($< 8\%$ H_2O) for crystallization, but care must be taken to keep the pH of the solution above about 4.5 and to exclude any material (chlorides, organic compounds, metals) that might catalyze the explosive decomposition of NH_4NO_3. It is also wise to keep the melt mass low and to vent it to avoid pressure buildup. The solid product should be stored well away from the main plant.

Solid NH_4NO_3 is very *hygroscopic* (i.e., it picks up water from the air). Nonoxidizable drying agents such as clays are usually added to suppress this effect and the consequent caking. Calcium carbonate (chalk, crushed limestone) may be added to form a nonexplosive product with 26% N. Alternatively, ammonium nitrate may be marketed as an aqueous solution, also containing ammonia and urea.

9.5 Sulfates

The manufacture of sulfuric acid is discussed in Chapter 10. We note here simply that 90% of the sulfur produced industrially is made into H_2SO_4. Of this, two-thirds is consumed in fertilizer manufacture, either directly in making ammonium sulfate fertilizer, or indirectly in producing superphosphate (Section 9.6), potassium sulfate (Section 9.7), and other products. Sulfur is an important plant nutrient, and some soils are sulfur-deficient and accordingly require additions of sulfate or elemental sulfur. However, sulfur is less frequently the limiting factor in plant growth than is fixed N or soluble phosphates. The chief role of $SO_4{}^{2-}$ in $(NH_4)_2SO_4$ fertilizer is as a benign vehicle for the ammonium ion (cf. K_2SO_4, Section 9.7).

9.6 Phosphates

Phosphorus is essential for plant growth and is often the limiting nutrient in aquatic ecosystems. Consequently, if abundant supplies of phosphate (e.g., from detergents in sewage plant discharges) are introduced into lakes, explosive growth of algae usually results (eutrophication of the lake). As the algae consume most of the oxygen in the lake, other aquatic life is virtually wiped out. Excessive aquatic weed growth can also result from phosphate discharges. For example, aquatic weeds in the Bow River downstream of the city of Calgary, Alberta, thrive on phosphates remaining in discharged treated sewage, resulting in this case in a world-class trout fishery, but weed growth interferes with the water intakes of downstream water users. The simplest means of removing phosphates (present mainly as monohydrogen phosphate ion, $HPO_4{}^{2-}$) from waste water is to add lime, since *basic* (i.e., OH^--containing) calcium phosphates are insoluble in water:

$$5Ca(OH)_2 + 3HPO_4{}^{2-}(aq) \rightarrow Ca_5(PO_4)_3OH(s)_\downarrow + 6OH^- + 3H_2O. \quad (9.18)$$

Conversely, the usual mineral source of phosphorus is insoluble rock phosphate [apatite, $Ca_5(PO_4)_3(F,OH)$—often abbreviated to $Ca_3(PO_4)_2$]. Rock phosphate must be converted to *acidic* calcium phosphates (i.e., to Ca^{2+} salts of $HPO_4{}^{2-}$ and $H_2PO_4{}^-$) in order to make a fertilizer soluble enough to be readily utilized by plants. This is done by treating rock phosphates with concentrated H_2SO_4; apatite of sedimentary origin is preferred over igneous phosphates in this respect, as the larger crystals of the latter are less easily digested by the sulfuric acid. The F^- content is either driven off as gaseous HF and SiF_4* on heating and scrubbed from the stack gases with water, or retained in the liquid phase as H_2SiF_6. The *superphosphate* product that results is typically 32% $CaHPO_4$ plus $Ca(H_2PO_4)_2{\cdot}H_2O$, 3%

*Some silica or silicates will inevitably be present in the rock phosphate.

absorbed H_3PO_4, and 50% $CaSO_4$. Traditionally, the P content is expressed as a percentage of phosphorus pentoxide (here, 20% P_2O_5).

Alternatively, we may first make concentrated phosphoric acid (H_3PO_4) itself (a syrupy liquid much like concentrated H_2SO_4) and use this to convert further rock phosphate to *concentrated superphosphate* [83% $CaHPO_4$ or $Ca(H_2PO_4)_2$, 3% H_3PO_4, and only 5% $CaSO_4$; in effect, 48% P_2O_5]. The key principle in either case is that H_2SO_4 is a much stronger acid than H_3PO_4 (e.g., *in water*, the first pK_a values are <0 and 2.12, respectively, at 25 °C), and so phosphate ions are protonated at the expense of H_2SO_4:

$$Ca_3(PO_4)_2(s) + 3H_2SO_4(l) \rightarrow 2H_3PO_4(l) + 3CaSO_4(s). \tag{9.19}$$

For the same reason, the fluoride ion is converted to HF (pK_a 3.20), which is volatile and is driven off by heat as HF gas (under vacuum if necessary).

The $CaSO_4$ may form as gypsum ($CaSO_4{\cdot}2H_2O$) or anhydrite ($CaSO_4$), depending on reaction conditions. The important engineering point is to obtain an easily filterable solid. Another possibility is to neutralize the H_3PO_4 with ammonia to obtain $(NH_4)_3PO_4$, a readily soluble fertilizer with very high available N and P contents.

Agricultural phosphates can also be obtained from steelmaking slags (Section 17.7) and from guano, the droppings of seabirds, accumulated over millennia on certain South Pacific islands such as Ocean Island and Nauru. The latter source has been heavily exploited and is now almost exhausted, bringing a catastrophic decline in the fortunes of the islands' inhabitants.

Most of the world production of phosphates goes into fertilizer, but some is used as detergent builders (Section 7.7). In toothpastes, calcium pyrophosphate has proved effective as a mild abrasive in eliminating tartar, while $Na_2[FPO_3]$, made by reaction of NaF with cyclic sodium metaphosphates ($NaPO_3$), is widely used as a fluoridating agent to suppress dental caries (Section 12.3). A minor amount of rock phosphate is used to make elemental phosphorus by reduction with coke in the presence of silica in the electric furnace (see Section 17.7):

$$2Ca_3(PO_4)_2(s) + 6SiO_2(s) + 10C \xrightarrow{1450\,^\circ C} P_4(g) + 6CaSiO_3(l) + 10CO(g). \tag{9.20}$$

The phosphorus vapor condenses as the white solid allotrope, which is toxic and, above 31 °C, spontaneously flammable. When white phosphorus is heated to about 400 °C in a sealed vessel, it transforms into a red allotrope that is nontoxic and rather inert. Red phosphorus is therefore the preferred form of the element for commercial use.

9.7 Potash

Fertilizers are usually designated with three numbers that indicate, respectively, the nitrogen, phosphate, and potash* contents. Traditionally, the P content is expressed as $\%P_2O_5$ ($= 2.29 \times \%P$) and the K content as $\%K_2O$ ($= 1.20 \times \%K$). Thus, a 20–3–4 fertilizer contains 20% N, 3% P as P_2O_5, and 4% potassium as K_2O. About 95% of all potash produced goes into fertilizers.

In Saskatchewan, where the world's largest reserves of potash are located, it is encountered mainly as deposits of the chloride, KCl (*sylvite*—or *sylvinite*, if mixed with NaCl), surrounded by a mineralization zone of carnallite, $KCl{\cdot}MgCl_2{\cdot}6H_2O$. The potash is recovered either by mining the solid or, at depths greater than about 1 km, by solution mining; the latter involves more expenditure of energy. The solid KCl (density 1.98 g cm^{-3}) can be freed of the denser NaCl (2.16 g cm^{-3}) by flotation in a fluid of intermediate density (typically, a homogeneous mixture of brine and magnetite of density 2.08 g cm^{-3}) or by crystallization from solution. The solubility of KCl in 100 mL water rises from 34.7 g at 20 °C to 56.7 g at 100 °C, whereas that of NaCl is little affected by temperature. Thus, a solution saturated with KCl and NaCl at 20 °C will dissolve substantially more KCl from sylvinite on heating to 100 °C, and relatively pure KCl will crystallize out on cooling. Alternatively, since KCl crystals have a preferential affinity for primary alkyl amine salts, crushed KCl coated with such compounds will seek the air–water interface in a froth blown in a saturated aqueous KCl solution; the crystals can be recovered from the skimmed froth (*froth flotation*).

When chlorides are undesirable (e.g., where excess Cl^- tends to accumulate in the soil as in the Netherlands or in arid, irrigated regions, or where a Cl^--sensitive crop such as grapes is grown), KCl is converted to K_2SO_4 by various methods.

(*a*) In the *Mannheim process* the solid is heated at relatively low temperatures with concentrated H_2SO_4 to get solid $KHSO_4$, which is then heated strongly with more KCl to yield K_2SO_4:

$$KCl + H_2SO_4 \rightarrow KHSO_4 + HCl \quad \text{(exothermic)} \qquad (9.21)$$

$$KHSO_4 + KCl \rightarrow K_2SO_4 + HCl \quad \text{(endothermic).} \qquad (9.22)$$

This illustrates the general principle that a nonvolatile powerful protonator (H_2SO_4, b.p. 290–317 °C) will displace a volatile weaker acid (hydrogen chloride, b.p. −85 °C) on heating.

*The term potash in its strict sense refers to potassium carbonate (K_2CO_3), which was formerly made by leaching wood ***ash*** with water in iron ***pots***. In common usage, potash refers to any source of bioavailable potassium.

(*b*) The *Hargreaves process* follows a similar principle:

$$4KCl(s) + 2SO_2 + O_2 + 2H_2O(g) \rightarrow 2K_2SO_4 + 4HCl. \qquad (9.23)$$

(*c*) A third method uses complex precipitation processes involving aqueous solutions of (usually) magnesium sulfate minerals, for example, kainite, $KCl{\cdot}MgSO_4{\cdot}3H_2O$:

$$KCl + KCl{\cdot}MgSO_4{\cdot}3H_2O \rightarrow K_2SO_4 + MgCl_2 + 3H_2O. \qquad (9.24)$$

Alternatively, KCl can be converted to KNO_3, which obviously has high fertilizer value:

$$3KCl + 4HNO_3(65\%) \xrightarrow{75\,^\circ C} 3KNO_3 + Cl_2 + NOCl + 2H_2O. \qquad (9.25)$$

The nitrosyl chloride is oxidized separately with air to NO_2 and thence to HNO_3, so that the net reaction becomes

$$2KCl + 2HNO_3 + \tfrac{1}{2}O_2 \rightarrow 2KNO_3 + Cl_2 + H_2O. \qquad (9.26)$$

Exercises

9.1 The rate of reaction of air over a hot MgO pebble bed to form nitric oxide is given by

$$\frac{dP_{NO}}{dt} = k_f P_{N_2} P_{O_2} - k_r P_{NO}^2$$

where the rate constants in $bar^{-1}\ s^{-1}$ are as follows:

Temperature	k_f	k_r
2000 °C	0.6	370
2200 °C	7	2000

Assume the partial pressure of NO to be negligible relative to the partial pressures of N_2 and O_2, which can be taken to be 0.78 and 0.21 bar, respectively. On this basis, what is the equilibrium partial pressure of NO (*a*) at 2000 °C and (*b*) at 2200 °C? (*c*) What is the heat of formation of (one mole of) NO in this temperature regime? What will be the initial rate of loss of NO if air at equilibrium is shock cooled (*d*) from 2200 to 2000 °C, (*e*) from 2000 to 1800 °C, and (*f*) from 2000 to 1500 °C? What do your results imply for the operation of a Wisconsin process plant?
[*Answers:* (*a*) 0.016 bar; (*b*) 0.024 bar; (*c*) +90 kJ mol^{-1}; (*d*) −0.11 bar s^{-1}; (*e*) -8×10^{-3} bar s^{-1}; (*f*) -3×10^{-4} bar s^{-1}.]

9.2 Use the enthalpy, entropy, and heat capacity data of Appendix C together with Eqs. 2 and 5 from Exercise 2.1 to calculate the equilibrium constant K for the formation of ammonia from hydrogen and nitrogen (*a*) at 25 °C and (*b*) at 500 °C. Use these results to calculate the equilibrium percentage yield of ammonia (*c*) at 25 °C and (*d*) at 500 °C, if the total pressure is 100 bars at all times and the hydrogen and nitrogen partial pressures are in the ratio 3:1. Assume ideal gas behavior and that ΔC_P° is independent of temperature, and note that, if the partial pressure of nitrogen is x, those of hydrogen and ammonia will be $3x$ and $(100 - 4x)$ bars, respectively. Problem (*c*) can be solved using an approximation, but (*d*) is best solved iteratively.
[*Answers:* (*a*) 5.9×10^5 bar^{-2}; (*b*) 1.098×10^{-5} bar^{-2}; (*c*) 99.4%; (*d*) 8.9%.]

9.3 From the thermodynamic data of Appendix C, show that the product of the reaction of ammonia gas with oxygen would be nitrogen, rather than nitric oxide, under standard conditions and in the absence of kinetic control by, for example, specific catalysis of NO formation by platinum. (Assume the other product to be water *vapor*.)
[*Answer:* The free energy changes per mole of NH_3 are −326.4 and −239.8 kJ, respectively, so that the first reaction is strongly favored over the second, but *both* are feasible.]

9.4 What are the N–P–K ratings of the following anhydrous fertilizers?

(*a*) ammonium nitrate with 3.0% clay added as anticaking agent;

(*b*) pure diammonium hydrogen phosphate;

(*c*) pure potassium nitrate.

[*Answers:* (*a*) 34–0–0; (*b*) 21–54–0; (*c*) 14–0–47.]

9.5 Consider the formation of ammonia from stoichiometric amounts of hydrogen and nitrogen in the gas phase (Eq. 9.4). If, after establishment of equilibrium, the *total* pressure is increased, does the partial pressure of nitrogen increase, decrease, or stay the same?

References

1. (*a*) R. Thompson (ed.), "The Modern Inorganic Chemicals Industry," Special Publication 31. Chemical Society, London, 1977; (*b*) R. Thompson (ed.), "Industrial Inorganic Chemicals: Production and Uses." Royal Society of Chemistry, Cambridge, 1995.

2. J. E. Fergusson, "Inorganic Chemistry and the Earth." Pergamon, Oxford, 1982.

3. J. A. Kent (ed.), "Riegel's Handbook of Industrial Chemistry," 9th Ed. Van Nostrand–Reinhold, New York, 1992.

4. N. Basta, "Shreve's Chemical Process Industries Handbook," 6th Ed. McGraw-Hill, New York, 1994.

5. "Fertilizer Manual." International Fertilizer Development Center, United Nations Industrial Development Organization, New York, 1980.

6. W. Büchner, R. Schliebs, G. Winter, and K. H. Büchel, "Industrial Inorganic Chemistry." VCH, New York, 1989 (translated by D. R. Terrell).

7. F. T. Nielsson (ed.), "Manual of Fertilizer Processing." Dekker, New York, 1987.

8. S. J. Lippard and J. M. Berg, "Principles of Bioinorganic Chemistry." University Science Books: Mill Valley, California, 1994; E. I. Stiefel and G. N. George, Ferredoxins, hydrogenases and nitrogenases: metal-sulfide proteins. *In* "Bioinorganic Chemistry," I. Bertini, H. B. Gray, S. J. Lippard, and J. S. Valentine, (eds.). pp. 365–453. University Science Books, Mill Valley, California, 1994; J. J. R. Fraústo da Silva and R. J. P. Williams, "The Biological Chemistry of the Elements." Oxford Univ. Press, Oxford, 1993.

9. E. D. Ermenc, Wisconsin process pebble furnace fixes atmospheric nitrogen. *Chem. Eng. Prog.* **52***(4)*, 149–153 (1956).

Chapter 10

Sulfur and Sulfur Compounds

10.1 Elemental Sulfur

ELEMENTAL SULFUR[1–4] occurs naturally in association with volcanic vents and, in Texas and Louisiana, as underground deposits. The latter are mined by injecting air and superheated water, which melts the sulfur and carries it to the surface in the return flow (the *Frasch process*). Most of the sulfur used in industry, however, comes as a by-product of the desulfurization of fossil fuels. For example, Albertan "sour" natural gas, which often contains over 30% (90%, in some cases) hydrogen sulfide (H_2S), as well as hydrocarbons (mainly methane) and small amounts of CO_2, carbonyl sulfide (COS), and water, is "sweetened" by scrubbing out the H_2S and then converting it to elemental S in the *Claus process*.[5] The Claus process is applicable in any industrial operation that produces H_2S (see Section 8.5); it converts this highly toxic gas to nontoxic, relatively unreactive, and easily transportable solid sulfur.

Hydrogen sulfide gas has a familiar rotten-eggs smell, and there is a tendency to underestimate it as a mere malodorous nuisance with comic associations. In fact, H_2S is *extremely toxic*—about three times as toxic, weight for weight, as hydrogen cyanide. Hydrogen sulfide kills gas field workers who walk unwittingly into high concentrations of the gas. Breathing is instantly arrested (reportedly, the warning sense of smell is paralyzed by overexposure to the gas); in the short term, however, the heartbeat continues, and the victim can be revived without permanent injury if removed promptly from the area of high H_2S concentration and given *immediate* artificial respiration.

Hydrogen sulfide must be removed from natural gas that is to be burned as fuel because SO_2 emissions must be minimized (Section 8.5)—a limit of 4 ppm H_2S is set for pipeline natural gas. The organic sulfur content of fuel oil must also be minimized; this is typically done by converting it to H_2S, for example, by hydrodesulfurization (Section 6.3), followed by conversion to sulfur by the Claus process.

Hydrogen sulfide is weakly acidic and may be removed from natural gas by absorption in a countercurrent of an appropriate basic solution, usually aqueous monoethanolamine (MEA, $HO—CH_2—CH_2—NH_2$) or diethanolamine (DEA, $HO—CH_2—CH_2—NH—CH_2—CH_2—OH$):

$$HOCH_2CH_2NH_2(aq) + H_2S(g) \rightleftharpoons HOCH_2CH_2NH_3^+(aq) + HS^-(aq). \tag{10.1}$$

The H_2S can be regenerated by stripping the MEA solution with steam, since reaction 10.1 is readily reversible. (This would not be feasible if the absorbant were a strong base such as NaOH.) There is a problem in that CO_2 and COS react slowly with MEA to give oxazolidone and thiooxazolidone, respectively:

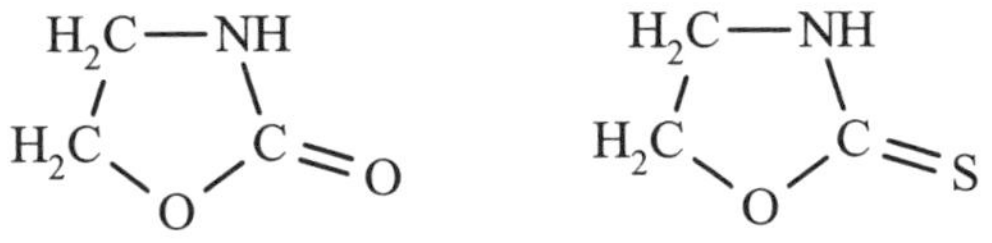

```
H2C—NH          H2C—NH
 /    \          /    \
H2C    C=O      H2C    C=S
  \   /           \   /
    O               O
```

The oxazolidones must be reconverted to MEA periodically with NaOH. Diethanolamine has the advantage of giving an easily regenerable compound with COS.

The recovered H_2S is burned partially to SO_2 in a limited air supply in the *front-end furnace*,

$$2H_2S + 3O_2 \rightarrow 2SO_2 + 2H_2O, \tag{10.2}$$

and the SO_2 and unchanged H_2S are then made to react:

$$2H_2S + SO_2 \rightarrow 3S + 2H_2O. \tag{10.3}$$

Because reaction 10.3 is exothermic, low temperatures give more complete conversion, but it is then unacceptably slow. A compromise is to work at a somewhat elevated temperature (450 °C) with an appropriate catalyst (Fe_2O_3 or γ-Al_2O_3). The net reaction is

$$2H_2S + O_2 \rightarrow 2H_2O + 2S. \tag{10.4}$$

In practice, conversions of up to 98% are achieved; residual H_2S is then burned to SO_2 and (environmental considerations permitting) is usually disposed of up the stack.

Elemental sulfur exhibits complicated allotropy, that is, it exists in many modifications.[4] The stable, prismatic crystal form at room temperature, α-S or orthorhombic sulfur, is built up of stacks of S_8 rings (Section 3.4). If heated quickly, it melts at 112.8 °C. If it is heated *slowly*, however, it changes to needlelike crystals of β-S or monoclinic sulfur, which is the stable form above 95.5 °C and which melts at 119 °C. Both β-S and the yellow mobile melt (below 160 °C) are composed exclusively of S_8 rings. Solids containing S_7, S_9, S_{10}, S_{12}, and other rings are known, but all slowly revert to S_8 below 160 °C.

Above 160 °C, however, molten S becomes increasingly brown and extremely viscous, with maximum viscosity at about 200 °C (Fig. 10.1). This is caused by opening of the S_8 rings and formation of chains of up to 100,000 S atoms. Beyond 200 °C, the average chain length decreases, and the sulfur becomes somewhat more fluid again. The deep red color above 250 °C is due largely to S_3 and S_4 fragments, while above the normal boiling point of 444.60 °C* the vapor contains mainly S_3, S_4, S_5, and S_7 units. If molten sulfur is poured into ice water, much of the long-chain structure is "frozen in"; the white, plastic product is known as catena- or λ-sulfur, but it eventually reverts to α-sulfur. Obviously, these facts have important consequences for the industrial handling of elemental sulfur. It must also be remembered that sulfur can be set on fire in air.

Most sulfur (90%) is converted to sulfuric acid. When the price of sulfur is low, as was the case through much of the 1970s, it is often uneconomic to transport it to distant markets. Many large-scale uses for surplus sulfur have been suggested, including "Thermopave" (S-containing asphalt), "Sulfurcrete" (a concrete made from pebbles and molten sulfur), and foamed sulfur for roadbed insulation in very cold climates.[6]

10.2 Sulfuric Acid

Most sulfuric acid[1–3, 7, 8] is currently produced by burning elemental sulfur from the Claus or Frasch processes in air to obtain sulfur dioxide, catalytically oxidizing the SO_2 with air to sulfur trioxide, and then hydrolyzing the SO_3 to H_2SO_4. Alternative sources of SO_2 include roasting of sulfide ores (e.g., pyrite, FeS_2) and burning of high-sulfur fossil fuels.

As noted in Section 8.5, there are two important complications in the conversion of SO_2 to liquid H_2SO_4: the oxidation of SO_2 is generally slow and must be catalyzed, and the direct reaction of SO_3 with water tends to produce intractable aerosols (mists) of H_2SO_4. The *lead chamber process*, which dates back to 1746, employs nitrogen oxides as the catalyst; the intermediate HO—SO_2—O—NO, or "nitrosylsulfuric acid," is easily

*This temperature is cited to 0.01 °C because it is one of the calibration points on the international temperature scale.

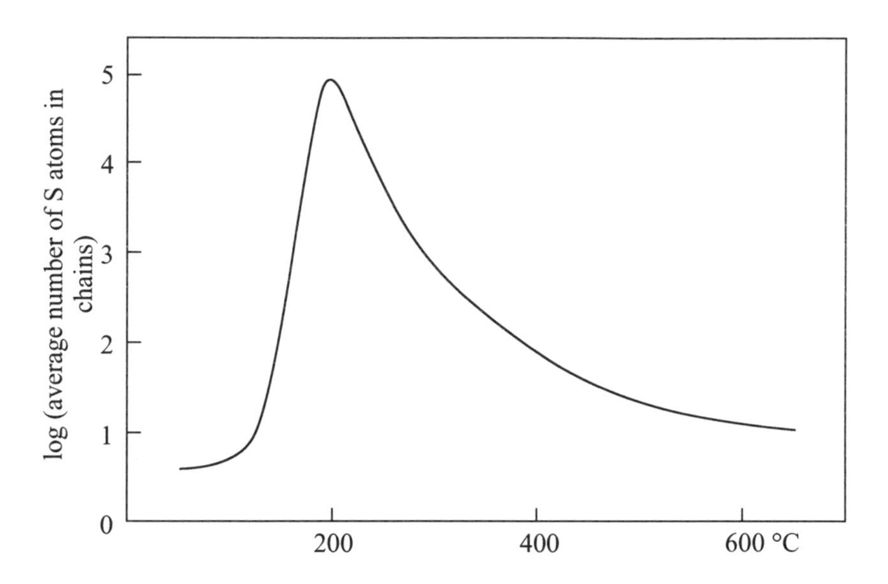

Figure 10.1 Dependence of molecular chain length, and hence viscosity, of molten sulfur on temperature.

hydrolyzed to liquid (or aqueous) H_2SO_4 and nitrogen oxides, which are recycled. Details of the chemistry remain obscure even today, and plant operation is said to be more of an art than a science.

At present, almost all sulfuric acid is made by the *contact process*, which has been in use since 1831. The first step is exothermic air oxidation of SO_2 catalyzed by vanadium pentoxide (V_2O_5) or platinum (reaction 10.5). The yield of SO_3 is limited on the first pass to some 60% because the temperature rises to 600 °C or more; usually, three more passes over the catalyst are made, and the yield can be increased to 98%. The SO_3 vapor is then absorbed into 100% H_2SO_4 (reaction 10.6), and water is added to the resulting mixture of disulfuric ($H_2S_2O_7$) and sulfuric acids (known as *oleum*) until the $H_2S_2O_7$ is all hydrolyzed to H_2SO_4 (reaction 10.7). This obviates the aerosol problem.

$$2SO_2 + O_2 \xrightleftharpoons[V_2O_5]{\text{Pt or}} 2SO_3 \qquad (10.5)$$

$$SO_3 + O{=}\overset{\overset{\displaystyle OH}{|}}{\underset{\underset{\displaystyle O}{\|}}{S}}{-}OH \rightarrow O{=}\overset{\overset{\displaystyle OH}{|}}{\underset{\underset{\displaystyle O}{\|}}{S}}{-}O{-}\overset{\overset{\displaystyle OH}{|}}{\underset{\underset{\displaystyle O}{\|}}{S}}{=}O \qquad (10.6)$$

$$H_2S_2O_7 + H_2O \rightarrow 2H_2SO_4 \qquad (10.7)$$

Concentrated sulfuric acid is a powerful protonating agent, and oleum is even more so. Oleum is used to catalyze important petrochemical reactions that proceed by formation of a carbocation intermediate (Section 7.3). The protonating power of concentrated H_2SO_4 results in strong desiccating ability; it absorbs water avidly from the air and chars paper or sugar by extracting the *elements* of water from it [paper and sugar are carbohydrates, i.e., have the *empirical* formula $C_x(H_2O)_y$].

10.3 Other Products from Elemental Sulfur

Sulfur dioxide, which is mostly made by burning sulfur and is usually marketed as the liquid (normal bp $-10\,°C$), has several uses:

(*a*) in extractive mineralogy, SO_2 serves as a mild reductant, for example, in extracting iodine from Chilean sodium nitrate deposits (Section 12.4.3) and in processing uranium;

(*b*) in winemaking, sulfur dioxide is a widely used preservative, although it has the disadvantage that some people suffer severe allergic reactions to sulfur dioxide or sulfites (wines and dried fruits that contain sulfites must therefore be clearly labeled to that effect);

(*c*) SO_2 is important in pulping and paper making (Section 10.4); and

(*d*) in water treatment, SO_2 can reduce, for example, excess chlorine used in water sterilization, or chromium(VI) in wastewater from electroplating plants or tanneries.

Not all sulfur trioxide is converted to sulfuric acid. Although the normal liquid range of SO_3 is narrow (17–45 °C), it is produced as the liquid for use in the sulfonation of some aromatic organic compounds. The vapor in an airstream is used to produce alkyl sulfonates for use as detergents.

10.4 Sulfur Chemicals in the Pulp and Paper Industry

Although the finest quality paper usually has a substantial rag content, most paper is made from wood.[1,2,9–11] Wood consists of three main components: (*a*) fibers of length 3–5 mm (North American softwoods*) or 1–2 mm (hardwoods) consisting of *cellulose* (i.e., polysaccharides, polymerized sugars that hydrolyze in aqueous acid to simple sugars like glucose); (*b*) *lignins*, which bind the fibers together and are of variable composition but contain many linkages of the following type:

*Softwoods include cedar, fir, larch, pine, and spruce; typical hardwoods are ash, birch, maple, oak, poplar, and willow.

TABLE 10.1
Principal Chemical Wood Pulping Processes

Kraft process	*Sulfite process*
Alkaline: carbon-steel vessels can be used, though stainless steel is better.	Acidic: must use corrosion-resistant (stainless steel) vessels.
Gives long fibers, therefore strong paper (*kraft* means strength in Swedish and German).	Shorter fibers because of acid hydrolysis; paper is weaker, and residual acid may cause long-term embrittlement.
Any kind of wood may be used.	Spruce or fir chips only.
Product is brown unless bleached.	Product is white.
Typical uses: packaging; high quality white paper (when bleached).	Typical use: general purpose white paper.

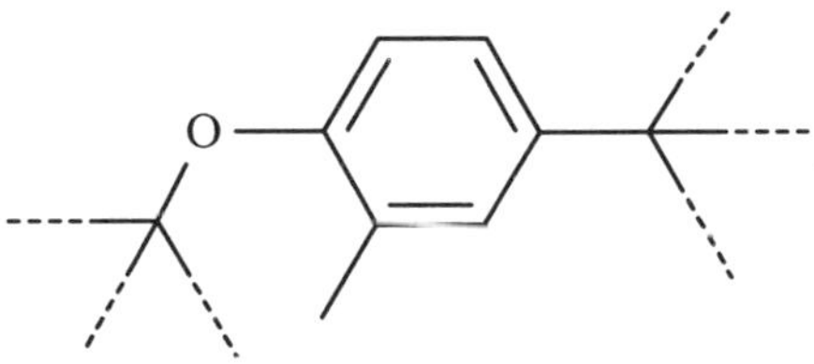

O

and (*c*) oils, fats, waxes, and proteins that occur inside the wood cells and break down in processing to give terpenes, fatty acids, tannins, etc.

The objective of wood pulping is to break down the lignin structure to give soluble anions such as

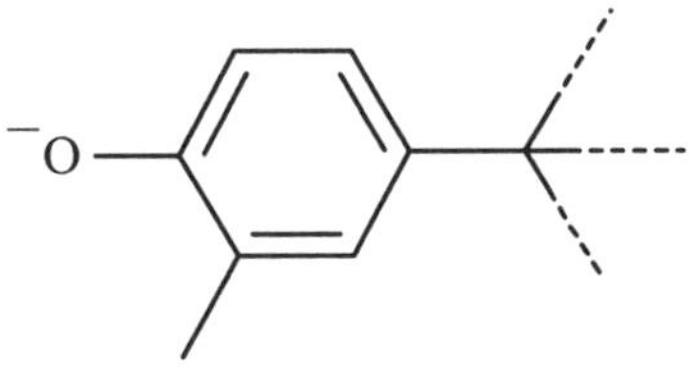

$^{-}$O

without excessive hydrolysis of cellulose. Although low-grade paper such as newsprint can be made by pulping wood mechanically (*groundwood pulp*), better quality material is produced by chemical pulping, since lignin remaining in paper hastens embrowning and decay. The two chief methods are the *kraft* (or *sulfate*) *process* and the *sulfite process*, summarized in Table 10.1. The manner in which the chemicals are recycled and the by-products recovered (as much for environmental protection as for economic reasons) provides an instructive example of chemical engineering practice.

10.4.1 The Kraft Process

The kraft pulping process depends on aqueous alkali sulfides. It was discovered serendipitously by Dahl in Germany in 1879 but attracted little interest there and was not implemented until 1890, in Sweden. Wood chips are digested at about 180 °C (800 kPa vapor pressure) in aqueous $NaOH/Na_2S/Na_2CO_3$ (pH >12) for 1 to 3 hours—the choice of "cooking time" (the technical term) depends on the temperature and the composition of the particular wood. A typical cooking liquor might contain "total titratable alkali" of 120 g L^{-1} ([NaOH] + $[Na_2S]$ + $[Na_2CO_3]$) of which 100 g L^{-1} might be "active alkali" ([NaOH] + $[Na_2S]$), the "sulfidity" would be around 24–28% {100 × $[Na_2S]$/([NaOH] + $[Na_2S]$)}, and the percentage "causticizing efficiency" about 78–80% {100 × [NaOH]/([NaOH] + $[Na_2CO_3]$)}, all components being expressed throughout as Na_2O equivalent. Thus, the actual concentration of NaOH as such would be around 2.5 mol L^{-1}.

The aryl–ether linkages and methoxy groups of the lignin are attacked by OH^- and S^{2-} (or HS^-) to give soluble alcohols, acid anions, and some mercaptans (RSH) and organic sulfides (R_2S), which often cause an exceedingly unpleasant stench in the neighborhood of the kraft mill. However, some 1.5 to 3.2 kg dimethyl sulfide per metric ton cellulose can usually be recovered from the exhaust gases and may be oxidized in the gas phase with air, using nitric oxide as catalyst, to give the versatile solvent dimethyl sulfoxide [DMSO, $(CH_3)_2S{=}O$]:

$$(CH_3)_2S + NO_2 \rightarrow (CH_3)_2SO + NO. \qquad (10.8)$$

$$\text{(NO} \xrightarrow{+\frac{1}{2}O_2} NO_2 \text{, recycled)}$$

Turpentine may also be recovered from the exhaust gas condensate.

The spent aqueous solution, known as *black liquor*, is centrifuged to give *tall oil*, a source of oleic and linoleic acids which are used in making soaps and greases. The aqueous phase is then evaporated to dryness, and the residue is ignited so that the organic content is charred to elemental carbon. To this molten residue, which contains NaOH, *salt cake* (Na_2SO_4) is added. Salt cake is obtainable from natural brines in Texas or Searle's Lake, California, or can be made[2,8] by heating ordinary salt with concentrated sulfuric acid (cf. K_2SO_4 manufacture, Section 9.7):

$$2NaCl(s) + H_2SO_4(l) \rightarrow 2HCl(g) + Na_2SO_4. \qquad (10.9)$$

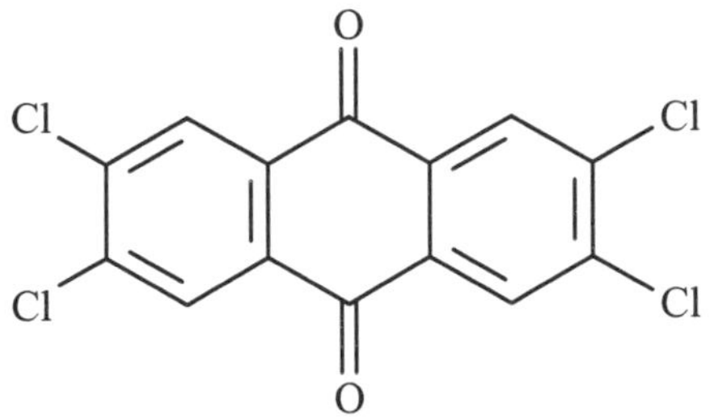

Figure 10.2 Structure of 2,3,7,8-tetrachlorodibenzo-p-dioxin (TCDD).

The role of the carbon is to *reduce* the sulfate ion of the salt cake to sulfide (cf. pyrometallurgy, Chapter 17). The carbon forms carbonate rather than CO_2 because of the alkaline medium:

$$Na_2SO_4 + 2C + 4NaOH(l) \rightarrow 2Na_2CO_3 + Na_2S + 2H_2O. \qquad (10.10)$$

The product is cooled and dissolved in water to give *green liquor*—the green color being due to complexing of traces of iron by sulfide. (The same green material develops around the yolks of overcooked hard-boiled eggs, where slow decomposition of the egg-white proteins gives H_2S.) Addition of hydrated lime [$Ca(OH)_2$, Chapter 11] regenerates NaOH

$$Ca(OH)_2 + Na_2CO_3(aq) \rightarrow CaCO_3(s)_{\downarrow} + 2NaOH(aq) \qquad (10.11)$$

by virtue of the poor solubility of $CaCO_3$, which precipitates and is returned to the lime kiln. The solution of NaOH and Na_2S plus some remaining Na_2CO_3 is called *white liquor*, as the iron sulfides are removed by the liming. White liquor is sent back to the digester for the next batch operation.

The brownish pulp is acceptable for making packaging material but requires bleaching for use as writing or printing paper. Chlorine has traditionally been used as a pulp bleach and is readily available on-site in pulp mills that make their own NaOH by electrolysis of brine (Section 11.4). However, chlorine tends to chlorinate as well as to oxidize organic compounds in the pulp, producing small amounts of various organochlorine compounds including *dioxins*. Laboratory tests have demonstrated that 2,3,7,8-tetrachlorodibenzo-*p*-dioxin (or TCDD, Fig. 10.2), in particular, is an extremely potent toxin and carcinogen toward test animals.[12] Therefore much concern has arisen over the direct use of chlorine for pulp bleaching.* The modern use of chlorine dioxide gas (ClO_2, Section 12.2) followed by

*It should be recognized, however, that dioxins also occur naturally, from forest fires (trees contain chloride ions) and the action of peroxidase enzymes on chlorophenols.

further treatment with aqueous "chloride of lime" [crude $Ca(ClO)_2$, made by the reaction of hydrated lime with chlorine] gives lower dioxin levels as well as a better, acid-free paper.[13]

A completely Cl-free alternative bleaching agent, however, is alkaline aqueous hydrogen peroxide (H_2O_2). Pure liquid H_2O_2 is too unstable for industrial use, and even in aqueous alkali it is susceptible to metal ion-catalyzed decomposition. Hydrogen peroxide can be stabilized by adding a complexing agent to sequester trace metal ion contaminants (Section 13.5); sodium silicate is commonly used but tends to form solid silicate deposits on equipment.* Using H_2O_2 has obvious advantages since no TCDD can be formed and since the H_2O_2 can itself be disposed of (as necessary) simply by catalyzing its decomposition to water and O_2 (Section 13.5). Furthermore, the product will be acid free (see next section) because the reaction conditions are alkaline. Hydrogen peroxide is readily made by electrolyzing aqueous H_2SO_4 or, more commonly, by the anthraquinone process.[3,7b] No doubt, the hydrogen peroxide method will soon largely replace the others—at the time of writing, the annual consumption of H_2O_2 in North America is about 500,000 metric tons and growing by approximately 10% a year.

The paper is then "filled" with a finely divided white solid to give a durable, opaque white paper. Usually either titanium dioxide (Section 17.8) or precipitated calcium carbonate (Section 11.1) is used as the filler. The latter, in addition to being less expensive, has the advantage of being somewhat alkaline, so providing continuing protection against degradation of the cellulose by acids in the air (see next section).

10.4.2 The Sulfite Process

Sulfite pulping was introduced commercially in Sweden in 1874 and by the end of the nineteenth century had become the main chemical pulping process. Where supplies of soft white conifer woods are abundant (e.g., in Canada), the sulfite process is still used to some extent, but in all countries it has been largely superseded by the kraft process, which produces acid free paper (see later) and is less environmentally objectionable. Sulfite process chemistry involves attack of sulfurous acid (H_2SO_3) on C=C double bonds and ketone groups in lignin to form soluble sulfonates. Since the acidic solutions cause severe corrosion of ordinary steels, stainless steel fittings are required, but, as explained in Chapter 16, even these will be attacked if the conditions become too reducing. For the same reason, digesters lined with acid-resistant brick are needed. The solutions also cause some undesirable hydrolysis of the cellulose fibers. It is therefore usual to use a buffer solution of HSO_3^- and H_2SO_3 in 3:1 ratio rather than straight aqueous H_2SO_3. This

---

*Proprietary complexing agents that circumvent this problem are now available; see *Chemical and Engineering News* October 9, 7 (1995).

keeps the hydrogen ion concentration low (see Exercise 10.3) and minimizes the volume of solution by increasing the solubility of SO_2.

The sequence is as follows. (*a*) Sulfur dioxide, made by burning sulfur or roasting pyrite (FeS_2), is passed over wet limestone ($CaCO_3$) until a total of 7% SO_2 has dissolved:

$$SO_2(g) + H_2O(l) \rightarrow H_2SO_3(aq) \xrightarrow{CaCO_3} Ca(HSO_3)_2(aq) + CO_2(g). \tag{10.12}$$

(*b*) The resulting liquor is used to digest the wood chips. Live steam may be injected at 7 bars to give a temperature of 150 °C over a 6- to 12-hour period. Alternatively, the solution may be heated indirectly to around 130 °C with steam coils for about 24 hours, giving a somewhat stronger pulp. (*c*) Recycling the chemicals involves recovery of SO_2 from the relief gases and evaporation of the liquor to a syrupy consistency before addition of slaked lime to precipitate $CaSO_3$, which is returned to the digester. The dissolved calcium lignin sulfonates can be worked up to give useful products such as vanillin, but usually the chief concern is environmental protection. Dimethylsulfide can also be recovered from the spent liquor and converted to DMSO as in reaction 10.8.

Sulfite paper has a relatively short life span, since residual acid will continue to hydrolyze the cellulose and cause embrittlement. Further sources of acid include aluminum sulfate (which is added together with resin to suppress bleeding or feathering of ink into the paper) and SO_2 and NO_x from the atmosphere. Much of the world's library collections and archives will soon be lost as the paper crumbles. Various deacidification treatments (e.g., with ammonia, morpholine, cyclohexylamine carbamate, or diethylzinc) have been proposed and tried, but at best they can only halt the process of embrittlement and cannot reverse it.[14] With the move to kraft pulping, alkaline peroxide bleaching, and increasing use of precipitated calcium carbonate as a filler, the high quality papers produced today are intrinsically "acid free" and should also resist subsequent acidification by SO_2-polluted air fairly well.

10.4.3 Recycling Waste Paper

In the latter half of the twentieth century, paper wastes, which make up a major fraction of domestic garbage (mostly in the form of corrugated cardboard, newspapers and office papers), have placed severe strains on the capacity of municipalities to dispose of them (typically in sanitary landfills). This, in addition to concern over excessive harvesting of forests to feed pulp mills, has generated a requirement for the recycling of used paper.

In processing scrap paper, there is no need for delignification with sulfur chemicals, since lignin was removed in manufacture of the new paper. If the scrap paper is to be used for packaging materials or chipboard, the color is unimportant and deinking is then unnecessary. A minor fraction (about 20%) of recycled paper, however, is used for printing paper, and deinking is usually achieved by either ***ink washing*** or ***ink flotation***. In ink washing, the waste paper is repulped with aqueous NaOH solution containing hydrogen peroxide, sodium silicate (primarily as a stabilizer for the H_2O_2), and a detergent like solute such as stearate ion that helps emulsify the ink components in micelles (Fig. 7.7). The emulsion is drawn off, the ink particles precipitated by flocculation (Section 14.2), and the liquid phase recycled. Ink flotation uses alkaline peroxide solutions similar to those in ink washing but involves forming a froth with a suitable detergent, whereupon the ink particles seek the water-air interface in the foam which can then be skimmed off (this is similar to the froth flotation technology used in concentrating metallic ores, Section 17.1). The flotation method works with relatively large ink particles ($> 5\mu m$), whereas micelle formation requires small particles. Other aspects of paper recycling (removal of metal staples, gums, rubber, sizing, and filler binders) lie outside the scope of this book.[10]

Exercises

10.1 On the basis of the following information, estimate the equilibrium constant for formation of sulfur trioxide from the dioxide and oxygen in the gas phase (*a*) at 25 °C and (*b*) at 800 °C. (*c*) What assumption(s) did you have to make in answering parts (*a*) and (*b*)?

	ΔH_f (kJ mol^{-1})	$S°$ (J K^{-1} mol^{-1})
$SO_2(g)$	−296.830	248.22
$SO_3(g)$	−395.72	256.76
$O_2(g)$	0	205.138

[*Answers:* (*a*) 2.59×10^{12} bar$^{-1/2}$; (*b*) 0.80 bar$^{-1/2}$.]

10.2 The dense clouds in the atmosphere of Venus, which obscure the surface of the planet totally from our view, are evidently composed of a sulfuric acid aerosol. Why are such clouds possible on Venus but not on Earth? (See Section 8.1.)

10.3 In the sulfite pulp process, the digestion fluid is typically made by dissolving sulfur dioxide in water in the presence of limestone until the solution contains 7.0% by weight SO_2, three-quarters of which is in the form of bisulfite ion. Why is this done? In support of your answer to this question, calculate the pH of (*a*) the digestion solution prepared as above, (*b*) a hypothetical 7% solution of SO_2 alone, and (*c*) an actual saturated solution of SO_2 alone (2.9% SO_2), if the first acid dissociation constant of sulfurous acid is 1.7×10^{-2}, at ambient temperature and pressure.

[*Answers:* (*a*) 2.25; (*b*) 0.88; (*c*) 1.09.]

10.4 H. P. Stephens and J. W. Cobble [*Inorg. Chem.* **10**, 619–625 (1971)] give the following expression for the pK_a of the aqueous hydrosulfide ion HS^- over the temperature range 0–100 °C at infinite dilution:

$$pK_a = (4500/T) + 12.6\ \log(T/298.15) - 1.29,$$

where T is in kelvins. Assume that this expression can be extrapolated to typical kraft pulping conditions of 175 °C and [NaOH] ≈ 2.5 mol L^{-1}, and so predict the predominant form of sulfide in the pulping liquor (i.e., obtain an estimate of the ratio $[S^{2-}]/[HS^-]$). To do this, you will need a value for the self-ionization constant of water; see Exercise 2.1b. Apart from the intended temperature limits of the Stephens–Cobble equation, what other factor is being disregarded in this semiquantitative exercise?
[*Answer:* $[S^{2-}]/[HS^-] \approx$ 10:1.]

10.5 On the basis of the information given in the first paragraph of Section 10.4.1, calculate the approximate concentrations of NaOH, Na_2S, and Na_2CO_3 in a typical kraft pulping medium.

References

1. N. Basta, "Shreve's Chemical Process Industries Handbook," 6th Ed. McGraw-Hill, New York, 1994.

2. J. A. Kent (ed.), "Riegel's Handbook of Industrial Chemistry," 9th Ed. Van Nostrand-Reinhold, New York, 1992.

3. W. Büchner, R. Schliebs, G. Winter, and K. H. Büchel, "Industrial Inorganic Chemistry." VCH, New York, 1989 (translated by D. R. Terrell).

4. F. A. Cotton and G. Wilkinson, "Advanced Inorganic Chemistry," 5th Ed. Wiley (Interscience), New York, 1988.

5. E. H. Baughman, Sulfur recovery to reduce SO_2 pollution. *In* "Pollution Prevention in Industrial Processes" (J. J. Breen and M. J. Dellarco, eds.), ACS Symp. Ser. 508, pp. 79–85. American Chemical Society, Washington, D.C., 1992.

6. T. W. Swaddle, Sulfur utilization—a challenge to Canadians. *Chem. Can.* **26**, 22–24 (1974).

7. (*a*) R. Thompson (ed.), "The Modern Inorganic Chemicals Industry," Special Publication 31. Chemical Society, London, 1977; (*b*) R. Thompson (ed.), "Industrial Inorganic Chemicals: Production and Uses." Royal Society of Chemistry, Cambridge, 1995.

8. J. A. Fergusson, "Inorganic Chemistry and the Earth," p. 180. Pergamon, Oxford, 1982.

9. F. A. Lowenheim and M. K. Moran, "Faith, Keyes and Clark's Industrial Chemicals," 4th Ed. Wiley (Interscience), New York, 1975.

10. C. J. Biermann, "Essentials of Pulp and Papermaking." Academic Press, New York, 1993.

11. A. M. Thayer, Paper chemicals. *Chem. Eng. News* November 1, 28–41 (1993).

12. D. J. Hanson, Dioxin toxicity: new studies prompt debate, regulatory action. *Chem. Eng. News* August 12, 7–14 (1991).

13. D. A. Perham, Changes in chemical requirements of the bleached kraft pulp industry *In* "Modern Chloralkali Technology" (T. C. Wellington, ed.), Vol. 5, pp 13–22. SCI/Elsevier Applied Sciences, New York, 1992.

14. C. J. Shanhani and W. K. Wilson, Preservation of libraries and archives. *Am. Sci.* **75**, 240–251 (1987); M. Sun, The big problem of brittle books. *Science*, **240**, 598–600 (1988); R. Wedinger, "Preserving our written heritage." *Chem. Br.* **28**, 898–900 (1992).

Chapter 11

Alkalis and Related Products

THE MOST IMPORTANT industrial alkalis are the weak alkali *ammonia* (Section 9.3), *caustic soda* (sodium hydroxide), and *lime* (calcium oxide).[1–6] For many industrial and agricultural purposes, the most economical source of alkali is lime, which is used in steelmaking and other metallurgical operations (~45% of U.S. production of lime), in control of air pollution from smokestack gases (Chapter 8), in water and sewage treatment (Sections 9.6 and 14.5), in pulp and paper production (Section 10.4), in reduction of soil acidity, in cement and concrete manufacture (indirectly, as discussed later), and in many chemical processes such as paper making (Section 10.4). In short, lime is one of the most important of all chemical commodities.

11.1 Lime Burning

Like chalk, limestone is largely calcium carbonate and dissociates reversibly to lime and carbon dioxide at sufficiently high temperatures (*calcination*):

$$CaCO_3(s) \rightleftharpoons CaO(s) + CO_2(g). \tag{11.1}$$

The reaction is endothermic, but the temperature is not usually allowed to exceed 1100 K, despite incompletion of reaction, since a more highly crystalline and hence less reactive modification of lime would be formed. (The formation of less reactive modifications is common when inorganic solids are strongly heated.) Fortunately, removal of the gaseous product helps drive the reaction to the right. The carbon dioxide may be recovered along with the lime for on-site use, for example, in the *Solvay process* for Na_2CO_3 (Section 11.3), but usually it is sent up the stack. In any event, the stack gases should be freed of particulate matter by electrostatic precipitation or scrubbing with water. A pure limestone source is important; ubiquitous silica (SiO_2) or silicate impurities form molten calcium silicate ($CaSiO_3$) in the kiln, which may prevent free flow of the lime.

Magnesium ion tends to substitute for Ca^{2+} in minerals, and magnesium calcium carbonate containing equal numbers of Mg^{2+} and Ca^{2+} ions is a distinct mineral [*dolomite*, $CaMg(CO_3)_2$], calcination of which gives a mixed Ca–Mg oxide. The value of CaO, however, lies largely in the moderate solubility of $Ca(OH)_2$ (giving ~0.05 mol OH^- per liter). Calcium hydroxide is known as *hydrated lime* in the United States and *slaked lime* in the United Kingdom:

$$CaO(s) + H_2O(l) \rightarrow Ca(OH)_2(s) \tag{11.2}$$

$$Ca(OH)_2(s) \rightleftharpoons Ca^{2+}(aq) + 2OH^-(aq). \tag{11.3}$$

Unlike $Ca(OH)_2$, however, $Mg(OH)_2$ is poorly soluble (1.6×10^{-4} mol $Mg(OH)_2\ L^{-1}$). This can be turned to advantage if $Mg(OH)_2$ is required for use as such or for reconversion by heating to magnesium oxide (*magnesia*), which is used as a refractory (i.e., high temperature resistant) structural material (mp 2800 °C). In particular, magnesium may be recovered from seawater (0.13% Mg^{2+}) if it is treated with calcined dolomite, (Ca,Mg)O. The $Mg(OH)_2$ that forms from the (Ca,Mg)O remains largely undissolved, but the $Ca(OH)_2$ so produced dissolves and causes precipitation of the $Mg^{2+}(aq)$ as further $Mg(OH)_2$. Not all commercial magnesia is made by this route, however; if the mineral *magnesite* ($MgCO_3$) is available in sufficient purity, it can readily be calcined to give MgO and CO_2.

Demand for magnesium hydroxide is growing because of environmental concerns. It is finding new applications as a substitute for NaOH, CaO and Na_2CO_3 in treating waste aqueous acid streams, since, because of its poor solubility, it neutralizes acids without the risk of raising the pH of the solution above 9, the limit set by newer pollution restrictions in the United States. In anticipation of moves to phase out organochlorine compounds, manufacturers are also introducing magnesium hydroxide in place of chlorinated organic flame retardants in plastics; endothermic decomposition of $Mg(OH)_2$ to nontoxic MgO and water vapor stops the spread of combustion.

Reaction 11.2 is strongly exothermic; water poured onto fresh lime (*quicklime*) may boil. Lime is therefore often used as a water-removing agent (desiccant) for gases or organic solvents. Solutions or suspensions of hydrated lime in water ("milk of lime") absorb CO_2, forming solid $CaCO_3$; indeed, "precipitated calcium carbonate," made in this way, is in increasing demand as an inexpensive acid-neutralizing white filler for paper (Section 10.4).

11.2 Cement and Concrete

Lime kilns are frequently associated with cement-making.[7] The lime–sand cements in use since Roman times gain mechanical strength from the slow reaction of $Ca(OH)_2$ with CO_2 of the air to form interlocking crystals of $CaCO_3$. The sand acts primarily as a matrix around which this process occurs.

11.2.1 Portland Cement

Portland cement, which was discovered in 1824, sets from within. It is made by firing limestone (~75%) with clays at about 1450 °C, causing partial fusion and the formation of a clinker, which is then powdered. This material typically contains the equivalents of about 67% CaO, 22% SiO_2, 5% Al_2O_3, 3% Fe_2O_3, and the balance other materials, and the four chief components are the following anhydrous phases:

(*a*) dicalcium silicate, β-Ca_2SiO_4, known to cement chemists as *alite* (50–70% of clinker mass);

(*b*) tricalcium silicate Ca_3SiO_5, or *belite* (15–30%);

(*c*) an *aluminate* phase (5–10%), nominally tricalcium aluminate ($Ca_3Al_2O_6$) but with extensive substitution of Ca^{2+} and Al^{3+} by Na^+, K^+, Fe^{3+}, and "Si^{4+}";

(*d*) a *ferrite* phase (5–15%), nominally Ca_2AlFeO_5 but again with substitution by adventitious ions.

High ferrite contents give the cement a dark color.

The anhydrous phases of Portland cement react exothermically with water at different rates by complicated mechanisms[7b–d] to form ultimately a hard mass of hydrated compounds. In particular, the aluminate phase hydrates rapidly, so it is usually necessary to add $CaSO_4$ (anhydrite or gypsum) to the powder to prevent it from setting too quickly when moistened. For the first 4 weeks, hydration of alite is the most important process giving strength to the cement; reaction of belite is slow but adds substantially to the strength after the first month. One can therefore design a rapid-hardening cement by maximizing the alite content and by grinding the clinker as finely as possible—this leads to a relatively high temperature rise in the early stages of hydration and hence accelerated chemical reaction rates. (It is also possible to accelerate both the setting and hardening of Portland cement by adding calcium chloride, but chloride ion must not be used in reinforced concrete; see Section 11.2.2.)

Conversely, coarsely ground clinkers with lower alite content are preferred for situations where cement emplacement is slow or temperatures can be high, as in the lining of oil wells. In such circumstances, organic compounds such as sucrose (ordinary sugar), calcium lignosulfonate (a by-product of the pulp and paper industry, Section 10.4), or melamine sulfonate/formaldehyde polycondensate may be added to retard setting; they appear to act by adsorption on the active components of the cement through oxygen atoms on the organic molecules (an example of a hard–hard interaction, Section 2.9). The use of such organic "superplasticizers" is especially important to ensure workability of high-strength concretes in the field.[7e]

The alite:belite ratio can be controlled by the composition of the kiln feed. The raw material consists initially of high purity limestone (essentially calcite, $CaCO_3$) and clay minerals (silicates and aluminosilicates such as kaolinite and pyrophyllite) plus some quartz (SiO_2) and iron oxides such as goethite [α-FeO(OH)]. At around 800 °C the calcite decomposes, releasing CO_2 and reacting with the clay minerals to give belite and some excess lime. The aluminate and ferrite phases also form at this stage. It is only when the temperature passes about 1300 °C that the excess lime reacts with the belite to give alite until the lime is consumed, and simultaneously the aluminate and ferrite phases melt, causing the solid calcium silicate particles to stick together as nodules. The solid aluminates and ferrites reappear as the clinker is cooled.

Rapid cooling of the clinker is preferred for many reasons, notably to prevent the reversion of alite to belite and lime in the 1100–1250 °C regime and also the crystallization of periclase (MgO) at temperatures just below 1450 °C. The magnesium content of the cement should not exceed about 5% MgO equivalent because most of the Mg will be in the form of periclase, which has the NaCl structure, and this hydrates slowly to $Mg(OH)_2$ (brucite), which has the CdI_2 layer structure (Section 4.6). Incorporation of further water between the OH^- layers in the $Mg(OH)_2$ causes an expansion that can break up the cement. Accordingly, only limestone of low Mg content can be used in cement making; dolomite, for example, cannot be used. Excessive amounts of alkali metal ions, sulfates (whether from components of the cement or from percolating solutions), and indeed of free lime itself should also be avoided for similar reasons.

"Pastes" of Portland cement—meaning water–cement mixtures that can set (stiffen) and then harden (develop compressive strength)—typically have water:solid ratios of 0.3–0.6 by weight. Setting occurs in a few hours, while hardening is slower and may still be incomplete after a year or more. X-Ray diffraction methods (Section 4.1) show that $CaSO_4$ phases (such as gypsum added to moderate the rate of setting) disappear within the first 24 hours, and alite over about 30 days, with the belite content hydrating more slowly. The hydrated product shows surprisingly few sharp X-ray diffraction features other than those of $Ca(OH)_2$, and the hydrated silicate mass is better described as an extremely stiff *gel* rather than a crystalline material. Some CO_2 from the air may diffuse through the water in the pores of the cement and react with the $Ca(OH)_2$ to form crystalline $CaCO_3$ as described at the beginning of this section; this can add significantly to mechanical strength of Portland cement, but the essential feature of Portland cement is the strength of the silicate gel. In fact, the carbonation process may occur at the expense of some of the calcium silicate hydrate, leading to shrinkage and consequent crazing of the surface of the cement or concrete.

Many cements used today are composites of Portland cement and industrial waste materials that can enter into the hydration reactions and contribute to the strength of the hardened product. These substances include pulverized fuel ash (PFA) from burning of pulverized coal in thermal power stations, crushed blast-furnace slag (Section 17.7), and natural or artificial *pozzolanas*—that is, volcanic ash and similar finely particulate siliceous or aluminosilicate materials that can react with the $Ca(OH)_2$ in Portland cement to form hydrated calcium silicates and aluminates. As noted earlier, the solubility of $Ca(OH)_2$ is such that the pH of pore water in Portland cements will be about 12.7, at which the Si–O–Si or Si–O–Al links in the solid pozzolanas will be attacked slowly by OH^- to form discrete silicate and aluminate ions and thence hydrated calcium silicate or aluminate gels.

Condensed *silica fume* or *microsilica*, a finely divided SiO_2 by-product of silicon production (see Sections 7.5 and 17.8), is an example of a waste material with very high pozzolanic activity and more reproducible composition than PFA or slag; the use of silica fume is essential if concretes with strengths greater than 100 MPa are required. Another effective pozzolana is ash from burning of rice husks, since cereals, like all grasses, accumulate substantial amounts of silica. Such composite cements offer environmentally and economically sound solutions to some problems of waste disposal, and also open possibilities for design of cements with modified properties.

An alternative to silicate-based Portland cement is the calcium aluminate cement, *ciment fondu*, which originated with the Lafarge company in France in 1908. Ciment fondu is typically made by heating limestone with *bauxite*, which is mainly AlO(OH) but contains much iron oxide (see Section 17.2). As noted above, calcium aluminate hydrates and hardens much more rapidly than alite, and so ciment fondu, either as such or mixed with Portland cement, can be used whenever a rapidly setting cement is required, for example, for construction at low temperatures. Concretes made from aluminate cements remain serviceable at higher temperatures than Portland cements and so are used to make cast refractories for pyrometallurgical applications.

The tensile strength of Portland cements is limited by the presence of relatively large pores (0.1–1.0 mm). Modern research has produced cements that are substantially free of large pores (*macro-defect-free* or *MDF* cements) and have very promising mechanical properties; for example, MDF cements can be formed into cement springs.[7c,d]

11.2.2 Concrete

Concrete is commonly thought of simply as a mixture of a cement with some coarsely particulate material or *aggregate*, but it is more than that: components of the paste bind chemically to the surface of the aggregate, giving added strength to the whole. The nature of this bonding seems to be similar to that occurring within the paste itself.

An important additional feature of most structural concrete is the presence of steel reinforcement rods or cables, and a common cause of concrete failure is corrosion of the reinforcing steel. Apart from the loss of reinforcement after the steel has corroded, the expansion caused by formation of rust can crack the surrounding concrete. The mechanism of corrosion of iron is considered in detail in Chapter 16; we note here simply that the alkaline environment of the cement will normally suppress corrosion rates to acceptable levels but ingress of CO_2 may reduce the pH and so favor corrosion. More importantly, the corrosion rate is greatly increased by chloride ions in particular. This means that $CaCl_2$ should not be used as an accelerator for setting and hardening when steel reinforcement is present; also, severe deterioration must be anticipated in concrete structures such as bridge decks, parking structures, and marine piers that are exposed to deicing salts or seawater. Furthermore, if there is an oxygen concentration gradient along the reinforcing steel (as when prestressing cables are anchored outside the specimen), loss of Fe^{2+} from the steel in the O_2-deficient regions may lead to sudden failure of the steel. Such problems can be minimized if the steel is properly covered to isolate it electrically.

Concretes are also susceptible to damage by sulfate ions, although in this case the reaction is with the Portland cement rather than with the reinforcing steel; gypsum is formed (among other things), causing an expansion that can destroy the concrete. The sulfate ions may come from percolating water, but more usually they originate in the raw materials used to make the cement or in the aggregate. Percolation of water per se can damage concrete by slowly dissolving away (leaching) the more soluble components of the cement, such as $Ca(OH)_2$. A particularly insidious cause of concrete deterioration is the so-called *alkali–silica reaction* (ASR),[7f] which can cause expansion and cracking of concrete in moist environments. Although the detailed mechanism of ASR is unclear, it is primarily a reaction between aqueous alkali metal hydroxides and silica similar to the pozzolanic reaction described above, except that the silicate gel forms in an environment poor in Ca^{2+}, such as may arise if the cement precursors had high alkali metal ion contents. Under such circumstances, the OH^- concentration in the water in the pores is not limited by the solubility of $Ca(OH)_2$, and the pH can be as high as 13.7. The alkali metal silicate or aluminosilicate product of the ASR is capable of imbibing large amounts of water, with consequent expansion and damage to the concrete.

11.3 Soda Ash

Sodium carbonate is a widely used source of mild alkali, either in hydrated form of big, glassy $Na_2CO_3 \cdot 10H_2O$ crystals (*washing soda*) or as the powdery anhydrous solid Na_2CO_3 (*soda ash*). In North America, most soda ash is now derived from natural alkaline brines (California) or from underground mineral deposits of *trona*, $Na_2CO_3 \cdot NaHCO_3 \cdot 2H_2O$ (Wyoming):

$$2(Na_2CO_3 \cdot NaHCO_3 \cdot 2H_2O) \xrightarrow{\text{heat}} 3Na_2CO_3 + CO_2 + 3H_2O. \quad (11.4)$$

Formerly, most sodium carbonate was made by the famous *Solvay process*, which has been used since 1869 but is no longer competitive with trona. Nevertheless, the Solvay process merits study as a classic example of chemical engineering practice. The *net* Solvay reaction

$$2NaCl + CaCO_3 \xrightarrow{NH_3(aq)} Na_2CO_3 + CaCl_2 \quad (11.5)$$

cannot proceed directly but involves several steps in which ammonia (which was an expensive by-product of coal distillation before the advent of the Haber process, Section 9.3) is used and recycled.

(*a*) *Ammonia absorption.* NH_3 in the recycled gases is dissolved in brine until a 6 mol L^{-1} NH_3 solution is obtained. Since this is an exothermic process, cooling is necessary.

(*b*) *Carbonation.* Carbon dioxide from an on-site lime kiln (Section 11.1) is absorbed in the ammoniated brine:

$$NaCl + NH_3 + CO_2 + H_2O \xrightarrow{aq} NaHCO_{3\downarrow} + NH_4Cl. \quad (11.6)$$

Reaction 11.6 is exothermic as well, so cooling to 20 °C is required. The precipitated $NaHCO_3$ is filtered. The reaction is stopped at about 75% completion, otherwise NH_4HCO_3 is also precipitated.

(*c*) *Calcination.* The $NaHCO_3$ product is heated at approximately 150 °C to form soda ash:

$$2NaHCO_3(s) \xrightarrow{150\,^\circ C} Na_2CO_3(s) + H_2O(g) + CO_2(g). \quad (11.7)$$

(*d*) *Ammonia recovery.* The solution phase ("mother liquor") is heated to expel excess CO_2, and then lime from the kiln is added to regenerate the ammonia for recycling:

$$2NH_4Cl(aq) + Ca(OH)_2 \rightarrow CaCl_2(aq) + 2NH_3(g) + 2H_2O(l). \quad (11.8)$$

The Solvay process consumes only brine (aqueous NaCl) and limestone, which are inexpensive, and energy. The only waste product is calcium chloride, which can be used for deicing roads. Inevitably, there are complications; in particular, the plant design must provide for the elimination of solid contaminants such as clay minerals (from the limestone) and $CaSO_4$ (from the sulfate ion usually present in brines).

11.4 Caustic Soda: The Chloralkali Industry

Annual production of caustic soda (sodium hydroxide) in the United States is about 10 million metric tons, of which 50% is consumed by the chemical industry and a further 20% by pulp and paper plants. Sodium hydroxide is made by *electrolyzing* strong brine, that is, decomposing it by passing an electric current.[1–6] Saturated brine contains about 360 g or 6.2 mol NaCl per kilogram water.

The principles of electrolysis are considered in detail in Chapter 15. In brief, aqueous solutions conduct electricity by virtue of movement of positively charged ions (cations) to the negative electrode (cathode), where they or another component of the solution—often the solvent itself—may pick up electrons and so be *reduced*, while the corresponding negative ions (anions) are attracted to the positive electrode (anode) where they or some other species may be *oxidized*. Thus, the electric current within the solution is carried by the movement of anions and cations in opposite directions under the influence of an applied electrical potential, but the current can only be sustained if something is oxidized at the anode and something reduced at the cathode. In the case of aqueous NaCl, the chloride ion (Cl^-) migrates to the anode where it loses an electron to form, first, a chlorine atom and then half of a chlorine molecule (Cl_2), whereas the sodium ion migrates to the cathode but the net electrochemical reaction there is ultimately reduction of solvent water to hydrogen gas and hydroxide ion, although elemental sodium may be a transient intermediate in that multistep process:

$$Cl^-(aq) \longrightarrow \tfrac{1}{2}Cl_2(g) + e^- \tag{11.9}$$

$$Na^+(aq) + e^-[\rightarrow Na] \xrightarrow{H_2O} Na^+(aq) + OH^-(aq) + \tfrac{1}{2}H_2(g). \tag{11.10}$$

Passage of 1.0 mol of electrons (one *faraday*, 96,485 A s) will produce 1.0 mol of oxidation or reduction—in this case, 1.0 mol of Cl^- converted to 0.5 mol of Cl_2, and 1.0 mol of water reduced to 1.0 mol of OH^- plus 0.5 mol of H_2. Thermodynamically, the electrical potential required to do this is given by the difference in standard electrode potentials (Chapter 15 and Appendix D) for the anode and cathode processes, but there is also an additional voltage or *overpotential* that originates in *kinetic* barriers within these multistep gas-evolving electrode processes. The overpotential can be minimized by catalyzing the electrode reactions; in the case of chlorine evolution, this can be done by coating the anode with ruthenium dioxide.

Since the products of the electrolysis of aqueous NaCl will react if they come in contact with each other, an essential feature of any *chloralkali cell* is separation of the anode reaction (where chloride ion is oxidized to chlorine) from the cathode reaction (in which OH^- and H_2 are the end products). The principal types of chloralkali cells currently in use are the *diaphragm* (or *membrane*) *cell* and the *mercury cell.*

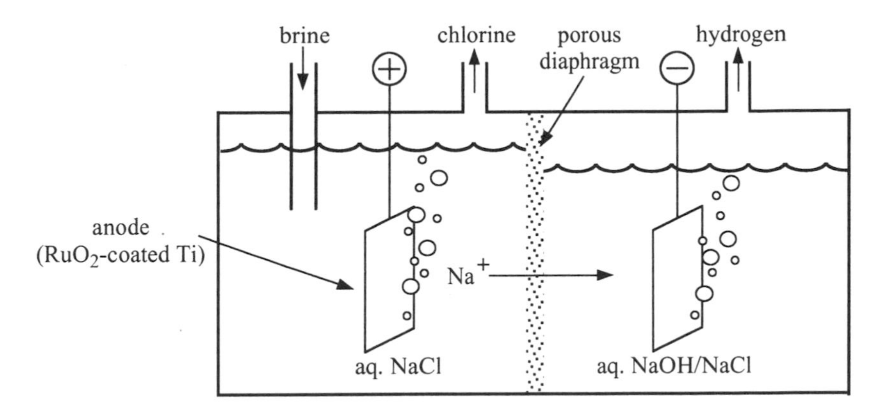

Figure 11.1 Principle of diaphragm chloralkali cell.

11.4.1 Diaphragm Cell

In a diaphragm cell, the anode and cathode compartments are separated by a porous diaphragm (Figs. 11.1 and 11.2). Formerly, diaphragms were made of asbestos, but now special polymers created for chloralkali electrolysis have been introduced.

In the obsolescent asbestos diaphragm cell, the product on the cathode side is typically 11% in NaOH and 16% NaCl (i.e., about 2.7 mol of each per kilogram of solution). Evaporation to about 50% NaOH causes most of the NaCl to crystallize, leaving about 1% NaCl in solution; this caustic soda is sufficiently pure for many industrial uses. The Si–O links in the asbestos are attacked by the alkali (Section 3.5 and Chapter 7), and the diaphragm soon deteriorates.

Most newer diaphragm materials are membranes made of highly alkali-resistant fluorocarbon polymers that incorporate cation-exchanging functional groups such as sulfonate.[8] One such material, Nafion (Du Pont tradename), is

$$
\begin{array}{l}
\cdots\!-\!CF\!-\!CF_2\!-\!\cdots \quad \} \text{ inert fluorocarbon ``backbone''} \\
\quad\;\, | \\
\quad\;\, O \\
\quad\;\, | \\
\quad\;\, CF_2 \\
\quad\;\, | \\
\quad\;\, CF\!-\!CF_3 \\
\quad\;\, | \\
\quad\;\, O \\
\quad\;\, | \\
\quad\;\, CF_2\!-\!CF_2\!-\!SO_3^- \quad Na^+ \text{ (exchangeable).}
\end{array}
$$

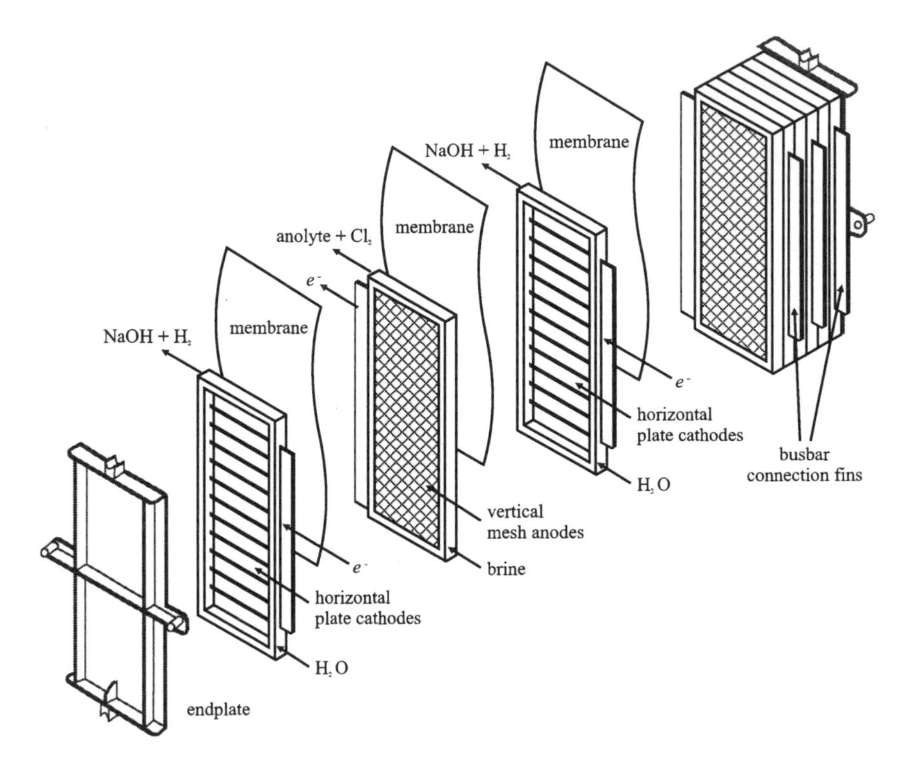

Figure 11.2 Lurgi membrane chloroalkali electrolyzer. Adapted from R. Dworak, K. Lohrberg, and R. Müller, Ref. 8*b*, p. 258, with permission.

Because the cations are readily exchangeable, the membranes allow rather free passage of Na^+ from anode to cathode compartments to match current flow in the external circuit. Since OH^- or Cl^- penetration is negligible, substantially pure NaOH solution can be made in a membrane cell.

Other fluorocarbon cation-exchanging membranes have been made in which carboxylate groups ($-CO-O^-$) substitute for the sulfonate functions (e.g., Flemion membranes developed by Asahi Glass Co. in Japan). Carboxylate membranes excel at preventing anion penetration; on the other hand, the sulfonated types retain larger amounts of water and allow higher rates of passage of cations. Thus, in practice, Du Pont industrial chloralkali membranes are constructed with a relatively thick fluorocarbon sulfonate sheet facing the anode compartment to provide high ion conductivity, a thin fluorocarbon carboxylate sheet facing the cathode compartment to maximize anion rejection, and a reinforcing mesh of fluorocarbon polymer on the sulfonate side to provide mechanical support. The ionic capacity of the membrane is equivalent to about 100 g per liter brine. These membranes can tolerate current densities of up to 4 kA m^{-2} at 80–95 °C in 30–35% NaOH solution. They are, however, susceptible to damage by Ca^{2+} or Mg^{2+}, the hydroxides of which have limited solubility and so tend to precipitate within the membrane and elsewhere. Figure 11.2 shows the structure of a typical modern membrane chloralkali cell.

11.4.2 Mercury Cell

Mercury chloralkali cells (Fig. 11.3) depend on the ability of mercury metal (which is liquid at ambient temperatures) to dissolve sodium metal to form an *amalgam* (sodium amalgam, Na/Hg), which is much less reactive toward water than is metallic sodium. Thus, the elemental sodium, which is arguably an intermediate in reaction 11.10, can be trapped as Na/Hg, transported out of the electrolysis cell (the dilute amalgam is also free flowing), and made to react with pure, chloride-free water in a separate vessel to give pure aqueous NaOH and hydrogen gas. A graphite *denuder* is used to accelerate the last process.

At the anodes:

$$\mathrm{Cl^-} \rightarrow \tfrac{1}{2}\mathrm{Cl_2} + e^-. \tag{11.11}$$

(some O_2 is also liberated)

At the cathode:

$$\mathrm{Na^+(aq) + Hg(l)} + e^- \rightarrow \mathrm{Na/Hg(l)}. \tag{11.12}$$

At the denuder:

$$\mathrm{Na/Hg(l) + H_2O(l) \rightarrow NaOH(aq) + Hg(l)} + \tfrac{1}{2}\mathrm{H_2(g)}. \tag{11.13}$$

The mercury cell thus gives very pure NaOH and, in terms of energy consumption, is also more economical than the diaphragm cell. However, the inevitable leakage of some mercury into local rivers or lakes can have (and has had) serious consequences because of the bacterial conversion of Hg to methylmercury ion, CH_3Hg^+, which then becomes concentrated in successive steps of the food chain:

$$\mathrm{Hg} \xrightarrow[\text{sediments}]{\text{aquatic}} \mathrm{CH_3Hg^+} \rightarrow \text{fish} \rightarrow \text{humans}.$$

Some oceanic fish, notably swordfish and tuna, tend to accumulate significant mercury levels from *natural* sources in this way, but there is no evidence that moderate consumption of these fish poses any toxicological hazard to humans. However, mercury contamination from *industrial* sources can reach levels high enough to pose a risk to health. Between 1956 and 1975, 143 people died in Minamata, Japan, and many more suffered permanent disability, a result of consuming fish contaminated with methylmercury of industrial origin. Methylmercury poisoning is now commonly called *Minamata disease.* In Canada, the White Dog and Grassy

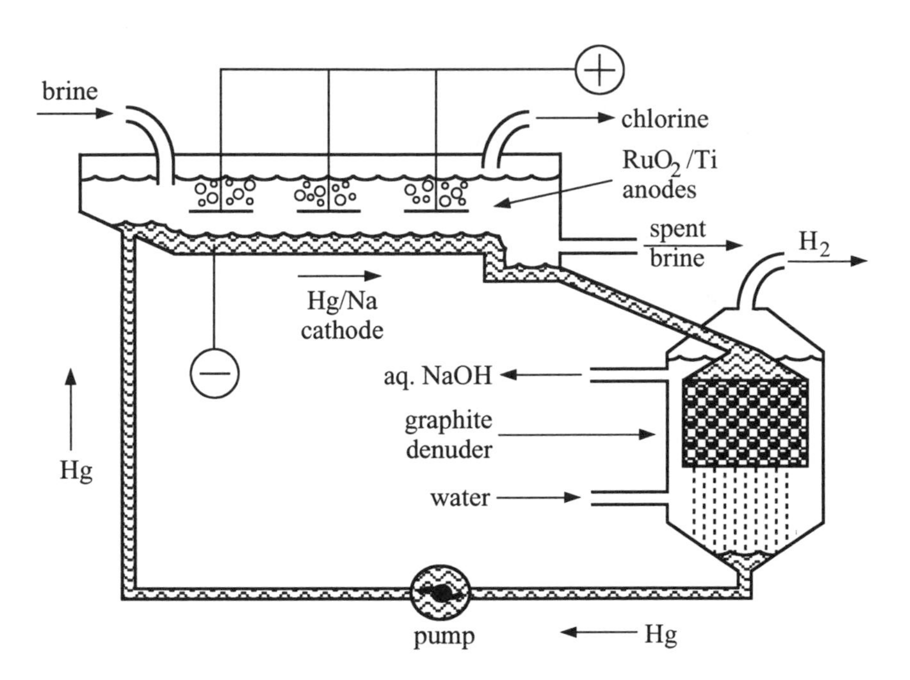

Figure 11.3 Mercury chloroalkali cell.

Narrows Indian bands have been deprived of traditional fish supplies because of mercury contamination of the English and Wabigon Rivers system by pulp mills. Recall that pulp mills consume NaOH and often use chlorine-derived products for bleaching (Section 10.4).

Even without bacterial conversion to soluble CH_3Hg^+, elemental mercury itself is very toxic. Since it is a cumulative poison, repeated exposure to the vapor of metallic mercury (even though the vapor pressure is low) can have serious, potentially fatal, long-term consequences. Mental derangement, extreme diffidence, immune system deficiencies, tremors, loose teeth, and walking problems are among the symptoms of mercury poisoning.* Metallic mercury poisoning and Minamata disease are extremely serious problems, and industrial use of mercury must be tightly controlled or, where alternative technology exists, eliminated.

*The expression "mad as a hatter" may have originated in the use of mercury(II) nitrate to finish once-fashionable beaver hats, although mercury-induced mental problems are characterized by timidity and loss of self-confidence rather than raving. Perhaps the "mad scientist" stereotype has similar origins, since mercury was indispensable in alchemy and in early chemistry and physics. Even Sir Isaac Newton, in the middle of his career in 1692–1693, suffered a bout of psychosis that has been attributed to mercury poisoning from his alchemical work (he had a predilection for tasting the products of experiments).[9]

11.4.3 Future of the Alkali Industry

The primary objective of the chloralkali industry is production of NaOH, which is in increasing demand for environmental protection purposes such as neutralization of acid wastes as well as industrial production of commodities such as paper. Chlorine and hydrogen are essentially by-products that have often been marketed at cost to dispose of them. As recently as 1990, consumption of chlorine was in balance with the use caustic soda (about 40 million and 45 million metric tons, respectively, worldwide), but, as discussed in Chapter 12, the global demand for chlorine is likely to fall behind that for NaOH by about 1% per year because of environmental concerns. For example, although the pulp and paper industry is consuming more and more NaOH, its demand for chlorine is expected to fall by a factor of about 4 from 1990 to 2000 as non-chlorine bleaching methods are introduced (Section 10.4).[8b] Since chlorine is not easy to dispose of in an economically and environmentally acceptable way (unlike the other by-product, hydrogen, which can be burned as fuel), shortages of caustic soda are inevitable unless soda ash or lime is used in place of caustic soda wherever possible, or unless soda ash is converted to caustic soda with lime (reaction 10.11) on a large scale.

Exercises

11.1 Here are some standard thermochemical data relevant to lime burning (limestone can be taken to be pure, solid $CaCO_3$ in its calcite form):

	Calcite	$CO_2(g)$	$CaO(s)$	
ΔH_f°	−1206.92	−393.509	−635.09	$kJ\ mol^{-1}$
S°	92.9	213.74	39.75	$J\ K^{-1}\ mol^{-1}$
C_P°	81.88	37.11	42.80	$J\ K^{-1}\ mol^{-1}$

(*a*) Write down the algebraic expression for the equilibrium constant K for formation of solid lime from solid limestone.

(*b*) From the above data, evaluate K for standard conditions (1 bar, pure solid phases).

(*c*) Assume the heat capacity of reaction (ΔC_P°) to be negligible and calculate the approximate temperature at which the CO_2 pressure will reach 1 bar (i.e., the standard pressure) at equilibrium.

(*d*) Evaluate ΔC_P°. In view of the temperature range you estimated, was it reasonable to neglect it, for the purposes of part (*c*)?

[*Answers:* (*b*) 1.43×10^{-23} bar; (*c*) 1110 K, or 837 °C.]

11.2 (*a*) From the data of Appendix C, show that dolomite is thermodynamically more stable than an equimolar mixture of calcite ($CaCO_3$) and magnesite ($MgCO_3$) at ambient temperature.

(*b*) Write balanced equations to show how magnesia may be made from dolomite and seawater.

11.3 Calculate the overall Ca and Si contents (expressed as CaO and SiO_2) of a Portland cement clinker that has 55% alite, 30% belite, 5% aluminate phase, and 10% ferrite (assume ideal compositions for the latter two). Would you expect this to behave as a fast or a slow setting cement?
[*Answers:* 65.6% CaO, 27.1% SiO_2; slow setting.]

11.4 Could the waste from a fluidized-bed limestone/coal furnace (Section 8.5) be used in Portland cement composites? What possible problem(s) might arise?

11.5 The solubility of lime in water at 25 °C is 1.13 g L^{-1}, expressed as CaO; the corresponding enthalpy of dissolution is −13.8 kJ mol^{-1}. (*a*) What is the solubility product (K_{sp}) of $Ca(OH)_2$ at 25 °C? (*b*) What is the solubility of CaO at 0 °C? (*c*) What is the pH of the saturated lime solution at 25 °C? (Ignore ionic strength effects.)
[*Answers:* (*a*) 3.29×10^{-5}; (*b*) 1.34 g L^{-1}; (*c*) 12.60.]

11.6 A mercury chloralkali cell was found to require a current of 37.2 kA in order to produce 1 metric ton of caustic soda per day.

(*a*) What is the current efficiency of the cell?

(*b*) Why is the efficiency less than 100%?

[*Answer:* (*a*) 95.4%.]

11.7 How much chlorine is produced per metric ton of caustic soda, in chloralkali operations?
[*Answer:* 0.89 metric ton.]

References

1. J. J. Leddy, The chlor-alkali industry. *J. Chem. Educ.* **57**, 640–641 (1980).

2. R. Thompson (ed.), "The Modern Inorganic Chemicals Industry," Special Publication 31. Chemical Society, London, 1977; R. Thompson (ed.), "Industrial Inorganic Chemicals: Production and Uses." Royal Society of Chemistry, Cambridge, 1995.

3. J. A. Kent (ed.), "Riegel's Handbook of Industrial Chemistry," 9th Ed. Van Nostrand-Reinhold, New York, 1992.

4. J. E. Fergusson, "Inorganic Chemistry and the Earth." Pergamon, Oxford, 1982.

5. N. Basta, "Shreve's Chemical Process Industries Handbook," 6th Ed. McGraw-Hill, New York, 1994.

6. W. Büchner, R. Schliebs, G. Winter, and K. H. Büchel, "Industrial Inorganic Chemistry." VCH, New York, 1989 (translated by D. R. Terrell).

7. (*a*) C. Hall, On the history of Portland cement after 150 years. *J. Chem. Educ.* *53*, 222–223 (1976); (*b*) H. F. W. Taylor, "Cement Chemistry." Academic Press, New York, 1990; (*c*) J. D. Birchall, A. J. Howard, and K. Kendall, New cements—inorganic plastics of the future. *Chem. Br.* **18**, 860–863 (1982); (*d*) D. M. Roy, New strong cement materials: Chemically bonded ceramics. *Science* **235**, 651–658 (1987); (*e*) S. P. Shah and S. H. Ahmad (Eds.), "High Performance Concrete: Properties and Applications." McGraw-Hill, New York, 1994; (*f*) D. W. Hobbs, "Alkali–Silica Reaction in Concrete." Thomas Telford, London, 1988.

8. (*a*) A. Eisenberg and H. L. Yeager (eds.), "Perfluorinated Ionomer Membranes," ACS Symp. Ser. 180. American Chemical Society, Washington, D.C., 1982; (*b*) T. C. Wellington (ed.), "Modern Chloralkali Technology," Vol. 5. SCI/Elsevier Applied Science, New York, 1992; (*c*) R. W. Curry (ed.), "Modern Chloralkali Technology," Vol. 6, Special Publication 164. Royal Society of Chemistry, Cambridge, 1995; (*d*) W. N. Brooks, The chloralkali cell: From mercury to membrane. *Chem. Br.* **22**, 1095–1098 (1986).

9. W. J. Broad, Sir Isaac Newton: Mad as a hatter. *Science* **213**, 1341–1344 (1980). See, however, letters to the editor: L. J. Goldwater, *Science* **214**, 742 (1981); M. R. Laker, *Science* **215**, 1185 (1982).

Chapter 12

The Halogens

12.1 The Chlorine Controversy

AS NOTED IN the preceding chapter, almost all elemental chlorine is made as a by-product of caustic soda production, although the obsolete *Deacon process* of 1868 has been revived (with improvements) for recycling Cl_2 on-site in plants where chlorination of hydrocarbons forms gaseous HCl, an objectionable waste product:

$$2HCl(g) + \tfrac{1}{2}O_2(g) \rightleftharpoons Cl_2(g) + H_2O(g). \tag{12.1}$$

The reaction is exothermic (see Exercise 12.1), but, since it is very slow, a catalyst is necessary. Nitric oxide, once again, can serve as an oxygen carrier, as in the lead chamber process (Section 10.2) and in reaction 10.8, where $(CH_3)_2S$ generated in the kraft process is converted to DMSO. Even so, at the elevated temperatures required, reaction 12.1 needs to be forced to completion by absorption of the steam in concentrated sulfuric acid or some other desiccant. In variants of the Deacon process, copper chloride acts as the catalyst or as an intermediate for chlorine regeneration.

Two-thirds of the chlorine produced in North America is consumed by the organic chemicals industry (25% goes to ethylene dichloride production alone). Pulp and paper mills account for another 15%, while 5% of the total is used in water treatment. All of these applications, however, have environmental implications that led to demands from activist groups that the production of chlorine derivatives be reduced (even, in the extreme view, phased out entirely). While these concerns are being taken seriously in all quarters, and some chlorinated products have already been banned, the fact is that chlorine-based technologies make, and will continue to make, important positive contributions to human health and prosperity. Indeed, at the time of writing in 1996, chlorine consumption continues to rise, although, as noted in Section 11.4.3, there is a growing mismatch between growth in NaOH production and slower growth in demand for the chlorine by-product.[1,2] The perceived problems may be summarized briefly as follows.

(*a*) Chlorofluorocarbons (CFCs), once widely used as refrigeration fluids, intermediates and solvents for fluorocarbon production, cleaning fluids (particularly in the manufacture of electronic components), aerosol propellants, and plastic foam blowing agents, have been shown unambiguously[3] to be the major factors in causing destruction of the stratospheric ozone layer, as discussed in Section 8.3. Production of several hitherto widely used CFCs has been terminated as of January 1, 1996, under the Montréal Protocol of 1988, but these very inert compounds will be with us for some time. The urgent questions of how to dispose of stocks and what to replace them with are considered in Section 12.3. Efficient, nontoxic refrigeration fluids are of great importance in providing for minimization of food losses through spoilage, especially in tropical regions where many of the world's malnourished people live. New, affordable, "ozone-friendly" refrigeration technology must be made available soon, but the technical requirements of changing the technology are neither trivial nor inexpensive. In the meantime, the use of condemned CFCs must continue in some developing countries. It should be noted that the ban on importation of CFCs into the United States has quickly led to criminal activities involving smuggling of CF_2Cl_2.

(*b*) Chlorinated solvents such as 1,1,1-trichloroethane (CH_3CCl_3) also threaten the stratospheric ozone pool, and several such compounds (e.g., chloroform, $CHCl_3$) are carcinogenic to some degree. Production of many such substances is already being phased out.

(*c*) Many chlorine-containing synthetic pesticides, herbicides, and fungicides (e.g., DDT, alachlor, hexachlorobenzene), as well as TCDD (Section 10.4) and some chlorinated hydrocarbons, are known to disrupt the endocrine systems of animals, including humans. The use of many such compounds, such as polychlorobiphenyls (PCBs, formerly used as electrical transformer oils) and DDT, has been prohibited, but they linger in the environment and interfere with the reproductive cycles of animals, particularly those higher up in the food chain, notably birds of prey and salmonid fish. It seems that these molecules can mimic natural hormones.

(*d*) Many extremely useful plastics, such as polyvinyl chloride (PVC), are made from chlorinated monomers. Some 3.5 millon metric tons of chlorine goes into PVC products in North America annually. Materials such as PVC are themselves benign; however, small amounts of production by-products, such as TCDD, are toxic to various degrees, and there is the additional problem of how to dispose of waste chlorinated plastics safely.

(*e*) We noted the problem of chlorinated organic emissions from chlorine-based pulp and paper bleaching in Section 10.4. A similar problem arises from using chlorine to sterilize water supplies (Section 14.7). Aqueous chlorine (actually its disproportionation product, HOCl) is an extremely effective and economical disinfectant, and it has the advantage over alternatives such as ozone of persisting long enough to give lasting protection in municipal water distribution systems. Trace organic compounds in the water, however, will tend to be chlorinated; one of the likely products is the known carcinogen, chloroform. One has to weigh the slightly increased risk of cancer that *may* arise from decades of exposure to traces of chlorocarbons in drinking water against the very real benefits to public health that disinfection with chlorine has brought—barely 100 years ago, epidemics of cholera and other water-borne diseases were common in Europe and North America, and even today they break out from time to time in countries where water treatment procedures are inadequate.

In addition, chlorine derivatives are important as intermediates in the chemical industry, and there are numerous chlorine-containing pharmaceuticals for which no substitutes are presently available. Furthermore, organochlorine compounds, some very toxic, do occur naturally on a large scale. Clearly, however, we must endeavor to avoid adding unnecessarily to the natural load of toxins; as the old adage goes, it is the dose that makes the poison.[4] Certainly, problems exist that require an intelligent and chemically informed resolution, but the total ban advocated by some on the use of chlorine and chlorinated compounds is neither necessary nor acceptable.

12.2 Oxides and Oxoacids of Chlorine

Chlorine, when dissolved in water, undergoes rapid *disproportionation* (i.e., it oxidizes and reduces itself simultaneously) to hypochlorous acid and hydrochloric acid, with an equilibrium constant of 5×10^{-4} mol^2 L^{-2}:

$$\underset{\mathrm{Cl(0)}}{\mathrm{Cl_2(aq)}} + \mathrm{H_2O(l)} \rightleftharpoons \underset{\mathrm{Cl(I)}}{\mathrm{HOCl(aq)}} + \mathrm{H^+} + \underset{\mathrm{Cl(-I)}}{\mathrm{Cl^-}}. \qquad (12.2)$$

Hypochlorous acid is a weak acid ($\mathrm{p}K_\mathrm{a} = -\log K_\mathrm{a} = 7.4$) and decomposes slowly to give oxygen, even in dilute solution. It cannot be isolated in the free state.

$$\mathrm{HOCl} \rightarrow \tfrac{1}{2}\mathrm{O_2} + \mathrm{H^+} + \mathrm{Cl^-} \qquad (12.3)$$

Since hypochlorous acid is a powerful disinfectant, chlorine is widely used to render municipal water safe for drinking.

If chlorine is dissolved in *cold, dilute* aqueous NaOH, *kinetically* stable solutions of the hypochlorite ion are obtained:

$$Cl_2 + 2OH^- \rightarrow OCl^- + Cl^- + H_2O. \quad (12.4)$$

Such solutions are marketed as household bleaches and disinfectants (Javex and Clorox, for example) and are used industrially as oxidants. Given the chemical and biological activity of hypochlorite solutions, however, it is necessary to destroy any significant quantities of hypochlorites in wastewater before it is discharged into rivers or lakes. This can be done by reduction with sulfite ion or hydrogen peroxide:

$$NaOCl + Na_2SO_3 \rightarrow NaCl + Na_2SO_4 \quad (12.5)$$

$$NaOCl + H_2O_2 \rightarrow NaCl + O_2 + H_2O. \quad (12.6)$$

However, since ClO^-(aq) is *thermodynamically* unstable with respect to decomposition to chloride ion and oxygen gas (see Exercise 12.2), catalytic decomposition is an attractive alternative to chemical reduction. ICI's HYDECAT system achieves this with a proprietary supported nickel catalyst.[5]

If chlorine is passed into *hot, concentrated* alkali, the *chlorate* ion ClO_3^- is formed:

$$\underset{6Cl(0)}{3Cl_2} + 6OH^- \rightarrow \underset{5Cl(-I)}{5Cl^-} + \underset{Cl(V)}{ClO_3^-} + 3H_2O. \quad (12.7)$$

Most chlorate salts are made in industry by direct electrolysis of acidified, concentrated brines at 40 to 45 °C in cells designed to give *good mixing* of the anode and cathode reaction products (contrast normal chloralkali practice, Section 11.4).[5–8] Chlorate ion is the conjugate base of chloric acid, a moderately strong acid that cannot be isolated from aqueous solution; attempts to make it directly (e.g., by treating solid $KClO_3$ with concentrated H_2SO_4) produce gaseous chlorine oxides and may result in a violent explosion.

Chlorates are strong oxidizing agents. They are used in fireworks[9] (Section 2.11) and as unselective herbicides, but the main (and rapidly expanding) use of $NaClO_3$ is as a source of chlorine dioxide for bleaching paper pulp (Section 10.4) and for sterilizing water (Section 14.7). The chlorate ion must be reduced to ClO_2 with a mild reductant, as powerful reductants may reduce the ClO_2 in turn to chlorine or chloride ion. In the *Mathieson process*, the mild reductant is SO_2:

$$2NaClO_3 + SO_2 + H_2SO_4(aq) \rightarrow 2ClO_2(g) + 2NaHSO_4(aq). \quad (12.8)$$

Methanol in dilute sulfuric acid is the reductant in the *Solvay process* (not to be confused with the Solvay soda ash process described in Chapter 11)

$$6NaClO_3 + CH_3OH + 6H_2SO_4(aq) \rightarrow 6ClO_2 + CO_2 + 6NaHSO_4 + 5H_2O, \quad (12.9)$$

while in the *Rapson process* hydrochloric acid reduces chlorate ion, being itself oxidized to chlorine. Chemetics' integrated chlorine dioxide process[5] uses hydrogen gas from the electrolytic production of $NaClO_3$ to reduce the chlorine from the Rapson reaction to HCl, which is then recycled:

$$NaCl + 3H_2O \xrightarrow{\text{electrolyze}} NaClO_3 + 3H_2 \quad (12.10)$$

$$2NaClO_3 + 4HCl \longrightarrow 2ClO_2 + 2NaCl + Cl_2 + 2H_2O \quad (12.11)$$

$$H_2 + Cl_2 \longrightarrow 2HCl. \quad (12.12)$$

The orange-yellow gas ClO_2 is explosive in high concentrations. Chlorine dioxide should therefore be made only where and when required, and it should be diluted with N_2 or CO_2 for safe handling. To circumvent the shipping problem, ClO_2 may be converted to solid sodium *chlorite*, $NaClO_2$ (containing chlorine in the III oxidation state), by reduction with, for example, hydrogen peroxide in aqueous NaOH. Like ClO_2, chlorites can be used in place of chlorine for pulp bleaching, water disinfection, and so forth, and the market for $NaClO_2$ is growing.* Chlorous acid ($HClO_2$) is unstable, so that, although ClO_2^- is stable in neutral or alkaline solution, it disproportionates to ClO_2 and Cl^- (O_2 and ClO_3^- are possible by-products) in aqueous acid:

$$5HClO_2(aq) \longrightarrow 4ClO_2(g) + HCl(aq) + 2H_2O. \quad (12.13)$$

Further electrolysis of chlorate solutions gives the perchlorate anion ClO_4^-, which is the anion of perchloric acid, the strongest common acid. The anion also has a negligible tendency to form complexes with metal ions in solution and so is a great favorite of physical-inorganic chemists who may want to adjust the acidity or ionic strength (Sections 2.2 and 13.2) of a solution of metal ions without interference by the added anion in the reaction being studied. Solid perchlorates such as NH_4ClO_4 are used as oxidizers in solid rocket propellants. (The fuel that is oxidized is typically powdered aluminum.) However, perchlorates can present a serious explosion hazard (Section 2.11). All perchlorates, solid or liquid, are potentially dangerous oxidants. This is especially true of *concentrated* (72%) and, worse still, of 100% $HClO_4$. It is essential to keep perchlorates and organic or other oxidizable matter apart, unless one knows exactly what one is doing. For example, a drop of concentrated $HClO_4$ produces a resounding explosion with the commonly used solvent DMSO (Section 10.4).

*Users of sodium chlorite should be aware of a possible frictional explosion hazard; see R. H. Simoyi, *Chemical and Engineering News* March 22, 4 (1993).

12.3 Fluorine and Fluorine Compounds

Fluorine is less familiar than chlorine, yet it has the greater crustal abundance (Table 1.1), occurring mainly as fluoroapatite [$Ca_5(PO_4)_3(F,OH)$], fluorite (fluorspar, CaF_2), and fluoridated micas, and is of major industrial importance.[7,10–12] The name of the element reflects the fluorescent properties of fluorite and fluoroapatite, and indeed the light-emitting coating in fluorescent lighting tubes is usually fluoroapatite that has been doped with transition metal or lanthanide ions to produce light with a color balance that approximates daylight.

12.3.1 Fluoride Ion and Dental Health

Fluoride ion tends to replace the isoelectronic ion OH^- in solids. In particular, hydroxyapatite, $Ca_5(PO_4)_3OH$, the chief constituent of tooth enamel, reacts slowly with aqueous fluoride to form fluoroapatite, which is harder and more resistant to tooth decay. The dental profession therefore advocates fluoride treatments at the time of regular dental checkups, and also (primarily to provide for the dental health of children) the presence of 0.7–1.0 ppm natural or added F^- in drinking water. Too much (>4 ppm) F^- in water, however, is acknowledged to cause mottling of teeth and bone sclerosis. Thus, the range of concentrations over which fluoride in public drinking water is considered to be beneficial is rather narrow.

Many municipalities now introduce fluoride artificially to public water supplies, often as Na_2SiF_6 or H_2SiF_6, by-products of phosphate fertilizer production (Section 9.6) that hydrolyze to F^- and silica. Controversy over the benefits and alleged hazards of artificial fluoridation of municipal water supplies has been extraordinarily emotional and shows no sign of abating.[13] Certainly, fluoride concentrations in the parts-per-thousand range can cause acute toxic symptoms such as nausea—clear proof that one can indeed have too much of a good thing—but it is difficult to assess claims of adverse effects of long-term exposure to fluoride in drinking water at the 1 ppm level, not least because even people living in municipalities with unfluoridated water now receive substantial doses of F^- from food and drink prepared in locales where fluoridation is practiced. In the United States, the National Research Council concluded in 1993 that fluoride ion concentrations in drinking water below the limit of 4 ppm recommended by the Environmental Protection Agency do not cause cancer, bone disease, or kidney failure (as some opponents of fluoridation have claimed). In 1994, the World Health Organization confirmed that water fluoridation is the most effective preventive measure against tooth decay.

This much is certain: there has been a dramatic decline in the incidence of dental caries since fluoridation of water supplies became widespread. This is doubly fortunate inasmuch as the safety of mercury amalgams, the most commonly used, hardest wearing, and least expensive of the materials currently available for filling dental cavities, is now being questioned. These amalgams are typically made from equal weights of mercury and an alloy of silver and other metals (e.g., one proprietary alloy contains 41% Ag, 31% Sn, and 28% Cu; another consists of 68% Ag, 18% Sn, 13% Cu, and 1% Zn). Although the mercury in the amalgam is in effect chemically combined with the other metals and quite inert, mercury vapor can be detected in the mouths of subjects with amalgam fillings after the mildly abrasive actions of chewing food or brushing the teeth, so inevitably they are ingesting minute amounts of mercury daily (the toxicity of mercury was noted in Section 11.4.2). In some individuals who are hypersensitive to mercury, removal of such amalgam fillings has led to a rapid improvement in health, but for the general public a balance has to be struck between the effectiveness of amalgam fillings in restoring the grinding surfaces of teeth and the possibility of mild toxic effects.* A macabre related issue is the recent concern in the United Kingdom that mercury vapor from incinerated dental fillings may threaten the health of people living in the vicinities of crematoria.

12.3.2 Fluorocarbons

The chemistry of fluorine is dominated by its electronegativity, which is the highest of all elements. The colorless gas F_2 has an estimated standard electrode potential $E°$(Chapter 15) of +2.85 V for reduction to F^- (cf. +1.36 V for Cl_2 to Cl^-), and thus F_2 immediately oxidizes water to oxygen ($E° = +1.23$ V), and 2% aqueous NaOH to the gas F_2O. Obviously, F_2 cannot be made by electrolysis of aqueous NaF. The usual preparation involves electrolysis of HF–KF melts in a Monel (Cu–Ni alloy) or copper apparatus.

Hydrocarbons inflame spontaneously in F_2, giving fluorocarbons and HF. Controlled fluorination of organic compounds is possible by electrolyzing them in liquid HF (bp 20 °C) at a nickel anode, at voltages just below that needed to evolve F_2:

$$CH_3COOH \xrightarrow[\text{electrolyze}]{\text{HF(l)}} CF_3COOF \qquad (12.14)$$

$$C_8H_{18} \longrightarrow C_8F_{18}. \qquad (12.15)$$

Perfluoroalkanes (i.e., alkanes in which all H atoms are replaced by F) have high thermal stability and resistance to oxidation and hydrolysis. This is

*Release of relatively large amounts of mercury vapor when amalgam fillings are drilled out may present an acute hazard worse than long-term exposure to trace levels of Hg. This paradox is reminiscent of the heightened hazard from airborne fibers when asbestos is removed from buildings.

because F_2 is itself a better oxidant than O_2, while the high strength of the C—F bond makes it hard to hydrolyze, despite its polarity (see Exercise 12.5). Furthermore, the presence of C—F bonds in perfluoroalkanes strengthens C—C bonds by some 10% relative to alkanes, and the F atoms, which are about twice as large as H atoms, are more effective in blocking access of attacking reagents to the carbon chain.[14]

It should not be thought, however, that perfluorocarbons are completely inert toward combustion. Even the very inert perfluorocarbon polymer *polytetrafluoroethylene* [PTFE, Du Pont's Teflon; $F(CF_2CF_2)_nF$] is thermodynamically unstable in oxygen with respect to CO_2 and CF_4 (Exercise 12.6) and can burn in a 95% O_2/5% N_2 mixture at 0.1 MPa, although combustion is hard to initiate because of the nonvolatility of PTFE and the resistance of the thermal degradation products to oxidation. Conflagrations involving more reactive, volatile fluorocarbons such as perfluorotoluene have been reported.[15]

Fluorocarbon oils and solid polymers are expensive but are invaluable where conditions demand them.[14] In addition to the excellent thermal stability that the fluorine atoms confer, the very high electronegativity and hence low polarizability of F atoms results in unusually low intermolecular forces in fluorocarbons, so that the smaller molecules are more volatile than might be expected from the molar masses while the polymers have very low surface energies. The familiar waxy solid PTFE (average molar mass $\sim 1 \times 10^6$) has an excellent thermal range (mp 327 °C, with excellent mechanical and electrical stability up to 250 °C at least), is inert to almost all chemical reagents other than molten alkali metals, and has little tendency to be wetted by or to adhere to other materials. It is also insoluble in all common solvents, although perfluorocarbon oils will cause it to swell at about 300 °C. It may be synthesized by the following sequence, in which the catalyst is typically antimony trifluoride, SbF_3:

$$HF(g) + CHCl_3(g) \xrightarrow{\text{catalyst}} CHClF_2 \xrightarrow[1000\,^\circ C]{500 \text{ to}} C_2F_4 \tag{12.16}$$

$$C_2F_4 \xrightarrow[\text{or ROO}\cdot]{\text{thermal,}} \cdots - CF_2 - CF_2 - \cdots . \tag{12.17}$$

Because PTFE melts at such a high temperature and the melt is very viscous, it is difficult to work by conventional plastics techniques such as injection molding or extrusion. It is usually formed into useful shapes by sintering at about 380 °C; sometimes liquid alkanes are used as a carrier, but the product then tends to be porous (this emerges as a serious problem when thin-walled objects are machined out of PTFE stock). Accordingly, several fluorocarbon thermoplastics have been developed that have lower melt viscosities, at the expense of somewhat poorer thermal and mechanical properties than PTFE. For example, the thermoplastic *FEP* (*fluorinated ethylene propylene*), made by copolymerization of $F_2C{=}CF_2$ and

$CF_3CF{=}CF_2$, has a lower melt viscosity than PTFE and can be readily formed by extrusion; it is mechanically stable to 200 °C. Brady[14] has provided an excellent concise account of these materials.

12.3.3 Chlorofluorocarbons

The necessity of phasing out production and use of CFCs is explained in Sections 8.3 and 12.1. The lesson to be learned is that *inert* does not necessarily mean *harmless;* no product or by-product of industry can be released into the environment without careful consideration of the consequences. The questions that remain are how can existing stocks of these very inert fluids be destroyed,[16] and what can take their place as refrigerants and solvents?[17]

Suggested methods of destroying CFCs include incineration (which gives noxious corrosive products) and dehalogenation with molten sodium (which is a hazardous substance in itself). Burdeniuc and Crabtree[16] found that CFC vapors are completely "mineralized" (converted to inorganic salts) in a single pass through a bed of powdered sodium oxalate at 270 °C:

$$CF_2Cl_2(g)+2Na_2C_2O_4(s) \rightarrow 2NaF(s)+2NaCl(s)+4CO_2(g)+C(s). \quad (12.18)$$

This reaction is easily controlled, uses an inexpensive reagent, and gives benign products that are easily disposable. The method also shows promise as a means of converting inert saturated fluorocarbons into more reactive fluoroalkenes or fluoroarenes for use in syntheses.

Commodore CFC Technologies of Columbus, Ohio, developed an instructive method of converting CFCs to ammonium halides and formate by reaction with a solution of sodium metal in liquid ammonia. Sodium dissolves in $NH_3(l)$ to give a blue solution that owes its color to *solvated electrons*; in time, or if catalyzed with iron oxide, for example, this decomposes to give hydrogen gas and sodium amide, $NaNH_2$, much as one would expect by analogy with the much more vigorous reaction of sodium with water (Eq. 11.10). The metastable blue solution of ammoniated electrons, however, is a very potent basic reductant and can achieve 99.99% destruction of CFCs. The solid ammonium salts recovered are converted with aqueous NaOH to NaCl, NaF, and NaOOCH for disposal and NH_3 for recycling.

Fluorocarbons containing hydrogen but no chlorine (HFCs)* pose no

---

*HFC stands for hydrofluorocarbon (no Cl content). CFCs that contain hydrogen are often termed HCFCs. The arcane number code for CFCs and HCFCs works as follows: the *first digit* gives the number of C atoms *minus one* (blank if zero), the *second digit* gives the number of H atoms *plus one*, and the *third digit* is the number of F atoms. (The number of Cl atoms is not explicitly stated, but is taken to equal the number of remaining valences.) Thus, $C_2H_2Cl_2F_2$ is CFC-132 (or Freon 132), and all single-carbon CFCs or HFCs are represented by two-digit numbers. Isomers are designated a, b, etc.

threat to the ozone layer (although they can be expected to act as persistent greenhouse gases in the troposphere) and are favored as replacements for the $CFCl_3$ (CFC-11) and CF_2Cl_2 (CFC-12) that have been widely used as refrigerants. The U.S. domestic appliance and automotive industries have settled on CF_3—CH_2F (HFC-134a) as the preferred substitute for CFC-12. Unfortunately, whereas CFC-12 (formerly the most commonly used refrigerant) is easily made from inexpensive CCl_4 and HF, HFC-134a requires a costly multistep synthesis; for example, starting from tetrachloroethylene, one of the several possible synthetic routes goes via CF_3CHCl_2 and then CF_3CHFCl. Furthermore, the change to use of HFC-134a requires some modification of existing refrigeration units to operate at higher pressures and with different lubricants.

Chlorofluorocarbons containing some hydrogen (HCFCs) break down much more rapidly in the troposphere than do H-free CFCs and so deliver less chlorine to the stratosphere. Examples are $CHCl_2CF_3$ (HCFC-123) and CH_3CCl_2F (HCFC-141b). These fluids may serve as interim CFC substitutes, notably for blowing polyurethane foams for insulation in refrigerator walls, but in the longer term they, too, will have to be phased out as *some* ozone layer damage would result from their use, and they would also contribute to the greenhouse effect.

Nonfluorine CFC substitutes have been considered, but few are fully satisfactory. For example, we could go back 50 years to the use of anhydrous ammonia as a refrigerant, but NH_3 is as toxic now as it ever was. Cyclopentane could be used as a foam-blowing agent, but it is less effective than HCFC-141b and besides would contribute to the volatile organic compound load in the troposphere, which is the root cause of ozone pollution (Section 8.3.2). On the other hand, supercritical CO_2 is emerging as an alternative to CFCs in various steps in the preparation of fluorocarbon polymers (Section 8.1.3).

12.3.4 Fluorine and Nuclear Energy

Fluorine is used in the nuclear industries of many countries to make *uranium hexafluoride* for enrichment of uranium in the fissile ^{235}U isotope:

$$UO_2(s) + 4HF(aq) \rightarrow UF_4(s) + 2H_2O \qquad (12.19)$$

$$UF_4(s) + F_2(g) \rightarrow UF_6. \qquad (12.20)$$

UF_6 is a solid (mp 64 °C) but has a high vapor pressure (15.3 kPa at 25 °C). Since ^{19}F is the only stable isotope of fluorine, the only molecular species by mass in UF_6 are $^{235}UF_6$ and $^{238}UF_6$. Repeated diffusion of $UF_6(g)$ through porous plugs (or centrifugation of the vapor) concentrates $^{235}UF_6$ relative to $^{238}UF_6$, since the speed of diffusion varies inversely as the square root of the molecular mass. This enrichment is not needed for fueling power reactors with good neutron economy, such as the Canadian CANDU system. Gaseous diffusion plants for UF_6 are expensive, not least because UF_6 is itself an aggressive fluorinating agent and thus must be contained in special fluorine-resistant materials.

12.4 Bromine and Iodine

12.4.1 Extraction of Bromine

Bromine is a dense, red, volatile, corrosive liquid (bp 59 °C) that is best made by oxidizing the small amount of Br^- in seawater with chlorine (higher bromide concentrations occur in the Dead Sea and in certain natural brines, e.g., in Arkansas and Michigan). The vapor of the resulting Br_2 is then carried off in an air stream:

$$Cl_2(g) + 2Br^-(\text{seawater}) \xrightarrow{\text{pH } 3.5} 2Cl^-(aq) + Br_2(l). \tag{12.21}$$

Reaction 12.21 illustrates the trend toward lower electronegativity (less oxidizing power) as one descends a periodic group. The bromine vapor is trapped in aqueous Na_2CO_3 (in effect, a mild source of alkali) as bromide and bromate ions

$$3Br_2 + 6OH^-(aq) \rightarrow 5Br^- + BrO_3{}^- + 3H_2O \tag{12.22}$$

and is recovered on acidification:

$$BrO_3{}^- + 5Br^- + 6H^+ \rightarrow 3Br_2 + 3H_2O. \tag{12.23}$$

A qualitative similarity to the aqueous chemistry of chlorine will be evident. For each oxoanion of chlorine, there is a corresponding bromine species, although perbromate salts form only under certain strongly oxidizing conditions (e.g., oxidation of bromate ion in alkaline solution with F_2 or XeF_2) and in fact were unknown until 1968.

12.4.2 Uses and Hazards of Bromine

Elemental bromine finds use as a volatile, moderate oxidant, and as such it is corrosive to flesh. The most important uses of bromine compounds until the 1990s were as agricultural soil fumigants (ethylene dibromide or methyl bromide), flame retardants, and fire extinguishing fluids, with lesser amounts going into drilling fluids, pharmaceuticals, photographic materials, and water treatment. Formerly, the largest scale use for a bromine product was addition of ethylene dibromide as a lead scavenger to gasolines containing tetraethyl lead as an antiknock agent, but such fuels have now been banned in many countries because of the toxicity of lead.

Atom for atom, bromine is even more efficient at destroying ozone than is chlorine. There has therefore been much concern that releases of volatile bromine compounds such as methyl bromide may contribute disproportionately to thinning of the stratospheric ozone layer. Whereas there is no longer any doubt over the role of human activity in stratospheric pollution by CFCs, which are exclusively anthropogenic, attempts to assess the importance of human activity in pollution by methyl bromide have been confused by large natural releases of CH_3Br from oceans and forest fires. Besides, unlike the case of CFCs released into the environment, a major fraction of the methyl bromide injected into soils to kill pests is destroyed in the ground.

Concern for the ozone layer has led to banning of the manufacture (though not yet the *use*) of *halons*, which are bromofluorocarbons, analogous to CFCs and, like them, unmistakably anthropogenic. Halons 1301 (CF_3Br) and 1211 (CF_2BrCl) have been of great value in extinguishing fires, particularly in aircraft and other special environments. They are fast acting, nontoxic, noncorrosive, and leave no residue. They suppress fires by scavenging the radicals ·H and ·OH, which are the main reactive species in combustion:

$$CF_3Br \rightarrow \cdot CF_3 + \cdot Br, \tag{12.24}$$

$$\cdot Br + \cdot H \rightarrow HBr, \tag{12.25}$$

$$HBr + \cdot H \rightarrow H_2 + \cdot Br, \tag{12.26}$$

$$HBr + \cdot OH \rightarrow H_2O + \cdot Br. \tag{12.27}$$

Up to the time of writing in 1996, no satisfactory alternative to halon fire-fighting agents has been put on the market, although perfluoroalkylamines, which contain neither Br nor Cl but break down thermally to give the ·H and ·OH scavenger $\cdot CF_3$, appear to be effective flame suppressants.[18]

12.4.3 Iodine

Solid iodine is purple, as is its vapor (I_2), but iodine is often brown in solution, for example, in oxygen-containing solvents such as ethanol (*tincture of iodine* antiseptic) or in water, in which its solubility is increased by formation of a complex ion I_3^- with iodide ion. Iodine is obtained by oxidizing the ash of dried seaweeds (kelp); alternatively, sodium iodate, which is present in the $NaNO_3$ deposits of the Atacama desert in Chile, may be reduced to iodide with aqueous HSO_3^- ion, followed by the iodide/iodate analog of reaction 12.23 if elemental iodine is wanted. Most of the U.S. production, however, comes from chlorination of natural I^--bearing brines in Michigan (cf. reaction 12.21). Since iodine is a fairly volatile solid, it can be conveniently purified by sublimation.

Iodine is concentrated in humans by the thyroid gland to form the iodo-amino acid *thyroxine*, which is essential to normal health and development. Iodine is a rather rare element (crustal abundance 0.00003 weight %, cf. Table 1.1), so the thyroid gland has become very efficient at scavenging iodide ion. As iodine is deficient in the diet in some locations, a small amount of iodide ion is routinely added to commercial table salt ("iodized salt").

The efficiency of the thyroid gland in concentrating iodide becomes a liability in the event of fission product releases following a nuclear power plant accident (or explosion of a nuclear weapon). Thermal-neutron-induced fission of a uranium-235 nucleus produces two unequal fragments of various masses averaging respectively about 95 and 137 (plus 2 or 3 more neutrons that can keep the nuclear chain reaction going, and of course much energy). The fragments are neutron rich and so undergo β^- radioactive decay (emission of an electron from within the nucleus, in effect converting a neutron into a proton) at various rates to improve the neutron–proton balance. Thus, radioactive iodine isotopes such as ^{129}I (half-life 1.6×10^7 years)* and ^{131}I (half-life 8 days), owing to the volatility of iodine, can spread rapidly over the surrounding country. Concentration of iodine radioisotopes by the thyroid (typically via milk from cattle grazing on contaminated grass) can lead to cancer of the gland. Fortunately, there is a fairly simple countermeasure: the affected population can be given sodium or potassium iodide tablets to swallow, thereby diluting the radioiodine intake and causing most of it to be excreted with the excess iodide. Other fallout fission products, however, such as strontium-90 and cesium-137, are less easily dealt with.

Exercises

12.1 From the data in Appendix C, calculate (*a*) the enthalpy of reaction and (*b*) the equilibrium constant for reaction 12.1 (the Deacon process) at standard conditions. (*c*) Above what temperature does the reaction cease to be thermodynamically favored (i.e., ΔG° becomes positive), if ΔC_P° can be ignored?
[*Answers:* (*a*) -57.20 kJ mol^{-1}; (*b*) 4.49×10^6 bar$^{-1/2}$; (*c*) 614 °C.]

12.2 Use the data of Appendix C to show that the aqueous hypochlorite ion is thermodynamically unstable with respect to exothermic decomposition ($\Delta H^\circ = -60.1$ kJ mol^{-1}, $\Delta G^\circ = -95.0$ kJ mol^{-1}) to aqueous chloride and oxygen gas under standard conditions.

12.3 The active ingredient in a well-known brand of toothpaste that re-

*The half-life of a radioactive isotope is the time it takes for half of the amount of the isotope to decay away.

duces sensitivity in teeth is strontium chloride, $SrCl_2$. How do you think it works? (Appendix F may help.)

12.4 A proprietary foam used by dentists in treating teeth topically with fluoride contains 2.15% NaF and 0.23% HF, with a stated pH of 3.5. (*a*) What pH would you expect on the basis of the literature pK_a value of 3.20 at 25 °C and infinite dilution? Explain any difference. [*Hint*: see Section 2.2.] (*b*) Why is the foam designed to be acidic? (*c*) Why do the implanted fluoride ions not wash off the teeth quickly in drinks or when the teeth are brushed?
[*Answer:* (*a*) pH 3.85.]

12.5 What do the following mean single bond energies tell us about the susceptibility of fluorocarbons to hydrolysis?

C—F	490 kJ mol^{-1}
O—H	459
H—F	565
C—O	358

[*Note:* HF(aq) is a "weak" acid.]

12.6 Show that combustion of PTFE in pure O_2 at 0.1 MPa is exothermic, with $\Delta H°(298\text{ K}) = -498$ kJ per mole of each of the products CO_2 and CF_4.

References

1. B. Hileman, Concerns broaden over chlorine and chlorinated hydrocarbons. *Chem. Eng. News* April 19, 11–20 (1993); B. Hileman, J. R. Long, and E. M. Kirschner, Chlorine industry running flat out despite persistent health fears. *Chem. Eng. News* November 21, 12–26 (1994); I. Amato, The crusade against chlorine. *Science* **261**, 152–154 (1993).

2. L. S. McCarty, Chlorine and organochlorines in the environment: A perspective. *Can. Chem. News* **46***(3)*, 22–25 (1994); R. Pocklington and D. Wimberly, Some concerns about chlorination. *Can. Chem. News* **46***(3)*, 26–27 (1994); B. Child, Chlorine: Major benefits for society, major challenges for industry. *Can. Chem. News* **46***(3)*, 28–30 (1994).

3. J. M. Russell III, M. Luo, R. J. Cicerone, and L. E. Deaver, Satellite confirmation of the dominance of chlorofluorocarbons in the global stratospheric chlorine budget. *Nature (London)* **379**, 526–529 (1996).

4. See correspondence by G. Gribble, T. G. Spiro, and V. M. Thomas, Organochlorine compounds. *Chem. Eng. News* February 13, 4–5 (1993), and references cited therein.

5. T. C. Wellington (ed.), "Modern Chloralkali Technology," Vol. 5, SCI/Elsevier Applied Sciences, New York, 1992.

6. N. Basta, "Shreve's Chemical Process Industries," 6th Ed. McGraw-Hill, New York, 1994.

7. W. Büchner, R. Schliebs, G. Winter, and K. H. Büchel, "Industrial Inorganic Chemistry." VCH Publishers, New York, 1989 (translated by D. R. Terrell).

8. C. J. Biermann, "Essentials of Pulping and Papermaking," Chapters 5 and 17. Academic Press, New York, 1993.

9. J. A. Conkling, Chemistry of fireworks. *Chem. Eng. News* June 29, 24 (1981).

10. R. Thompson (ed.), "The Modern Inorganic Chemicals Industry," Special Publication 31. Chemical Society, London, 1977; R. Thompson (ed.), "Industrial Inorganic Chemicals: Production and Uses." Royal Society of Chemistry, Cambridge, 1995.

11. J. A. Kent (ed.), "Riegel's Handbook of Industrial Chemistry," 9th Ed. Van Nostrand-Reinhold, New York, 1992.

12. J. E. Fergusson, "Inorganic Chemistry and the Earth." Pergamon, Oxford, 1982.

13. B. Hileman, Fluoridation of water. *Chem. Eng. News* August 1, 26–42 (1988); see also Fluoridation of water [correspondence]. *Chem. Eng. News* October 10, 2–4 (1988).

14. R. F. Brady, Jr., Fluoropolymers. *Chem. Br.* **26**, 427–430 (1990).

15. E. R. Larsen, Perfluorocarbon flammability. *Chem. Eng. News* February 17, 2–3 (1992), and references cited therein.

16. J. Burdeniuc and R. H. Crabtree, Mineralization of chlorofluorocarbons and aromatization of saturated fluorocarbons by a convenient thermal process. *Science*, **271**, 340–341 (1996).

17. R. Pool, The elusive replacements for CFCs. *Science* **242**, 666–668 (1988); P. S. Zurer, Looming ban on production of CFCs, halons spurs switch to substitutes. *Chem. Eng. News* November 15, 12–18 (1993).

18. M. Freemantle, Search for halon replacements stymied by complexities of fires. *Chem. Eng. News* January 30, 25–31 (1995); H. Fukaya, T. Ono, and T. Abe, New fire suppression mechanism of perfluoroalkylamines. *J. Chem. Soc. Chem. Commun.* 1207–1208 (1995).

Chapter 13

Ions in Solution

13.1 Energetics of Solvation

WE SAW IN Section 4.7 that the dominant factor in the energy balance governing the existence of ionic solids is the *lattice energy*, U. For ionic compounds in solution,[1–4] the corresponding quantity is the *solvation energy*, which results from interaction of the ions with the solvent, rather than with each other. There must, of course, be equal sums of anionic and cationic charges in the solution to preserve electrical neutrality, and the ions will interact with one another significantly except at high dilutions (Section 2.2). However, it is the *free energy of solvation* $\Delta G^\circ_{\mathrm{solv}}$ at infinite dilution that replaces U when the Born–Haber cycle is adapted for solutions.

13.1.1 Born Theory of Solvation

Suppose we have a spherical conductor of radius R in a vacuum and we bring up a total charge q from infinite distance, in infinitesimal increments dq. The work W_0 done in charging the sphere against the charge itself, as it builds up, will be

$$W_0 = \int_0^q \frac{q}{4\pi\varepsilon_0 R}\, dq = \frac{q^2}{2R(4\pi\varepsilon_0)} \tag{13.1}$$

where ε_0 is the permittivity of a vacuum. For the same process in a solvent of *relative permittivity* (*dielectric constant*) D, the corresponding work W_{D} is given by

$$W_{\mathrm{D}} = \frac{q^2}{2RD(4\pi\varepsilon_0)}. \tag{13.2}$$

TABLE 13.1
Relative Permittivities (Dielectric Constants) D and Normal Liquid Ranges of Common Solvents

Solvent	*mp (°C)*	*bp (°C, 1 bar)*	*D (at 25°C)*
Hydrocyanic acid, HCN	−13	25	123
Sulfuric acid, 100% H_2SO_4	10	290–317	100.0
Water, H_2O	0	100	78.3
Propylene carbonate	−49	242	64.4
DMSO, $(CH_3)_2SO$	18	189	46.6
Acetonitrile, CH_3CN	−45	82	36.2
Methanol, CH_3OH	−98	65	32.6
Acetone, $(CH_3)_2CO$	−94	56	20.7
Sulfur dioxide, SO_2(l)	−96	−10	15.4
Ammonia, NH_3(l)	−78	−33	16.9
Acetic acid, CH_3COOH	17	118	6.2
n-Hexane, C_6H_{14}	−95	69	1.9

Consequently, if we transfer the charged sphere from a vacuum to the solvent, its electrostatic self-energy is *lowered* by an amount ΔW:

$$\Delta W = \frac{q^2}{2R(4\pi\varepsilon_0)}\left(1 - \frac{1}{D}\right). \tag{13.3}$$

Born pointed out that a mole of gaseous ions of radius r and charge ze would be similarly stabilized on transfer to the solvent, the work difference being $-\Delta G^\circ_{\text{solv}}$. Strictly speaking, the foregoing argument is not really applicable to ions, especially if $z = \pm 1$, since one cannot charge up an ion with increments less than $\pm e$, but the result is still valid:

$$-\Delta G^\circ_{\text{solv}} = \frac{N(ze)^2(1 - D^{-1})}{8\pi\varepsilon_0 r}. \tag{13.4}$$

For polar solvents like water, DMSO, or 100% sulfuric acid, D^{-1} is quite small compared to unity (Table 13.1) so the electrostatic self-energy of a gaseous ion is almost entirely eliminated on transferring the ion to a polar solvent. For an ionic compound to be freely soluble in a given solvent, the solvation energies of its anions and cations must outweigh the lattice energy sufficiently, otherwise an ionic solid results instead. Ionic solids are therefore not usually very soluble in solvents of low D.

Solubility equilibria can be treated in terms of the free energy of solution, as outlined in Section 2.3, and the temperature dependences can be related to the enthalpy and entropy of solution. The Born–Haber cycle for solutions, in terms of enthalpies, can be written as follows:

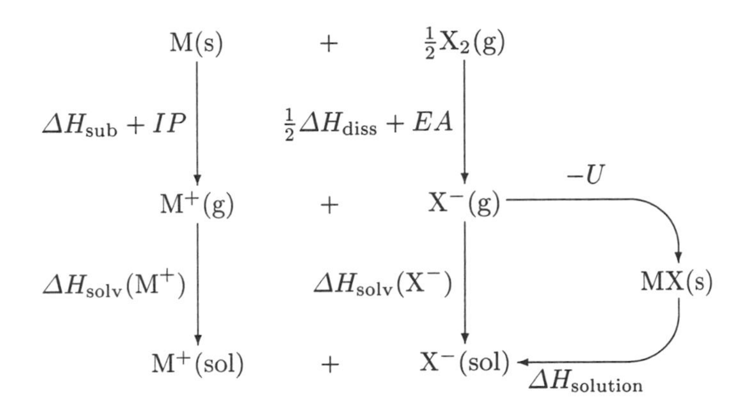

For water, typical values of the heats of hydration (i.e., solvation by water) are

$$\begin{aligned} M^+ &= -400 \text{ kJ mol}^{-1} \\ M^{2+} &= -1700 \\ M^{3+} &= -4500 \\ X^- &= -400\,. \end{aligned}$$

The sums of these enthalpies for typical aqueous MX or MX_2 are seen to be comparable with the corresponding $-U$ values given in Section 4.7. The balance, however, is delicate, and solubility equilibria are not accurately predictable from the simple Born model. The data of Table 13.1 may serve as a rough guide to the effectiveness of various liquids as solvents for electrolytes: HCN should be excellent but is too volatile (and extremely toxic); H_2SO_4 should be better than water, if its extreme acidity can be tolerated; propylene carbonate is good and offers an unusually wide liquid range, and so has been extensively investigated as a solvent for electrolytes in advanced fuel cells (Section 15.5); liquid ammonia is generally poor, acetic acid is worse, and hydrocarbons such as hexane are useless.

13.1.2 Limitations of the Born Theory

Table 13.1 does not explain, however, why the ionic solid silver chloride, which is well known to be poorly soluble in water, dissolves readily in liquid ammonia, despite a much less favorable relative permittivity. The reason is that the silver ion interacts strongly with specific ammonia molecules to give a *complex ion*, usually formulated as $Ag(NH_3)_2^+$. Indeed, Ag^+ interacts with water in aqueous solution to give a complex $Ag(OH_2)_4^+$, but the energy of the interaction is less than with NH_3.

The upshot is that the Born theory of solvation fails because it regards the solvent as a continuous dielectric, whereas in fact solute ions (especially metal cations with $z > 1$) often interact in a specific manner with solvent molecules. In any event the molecular dielectric is obviously very "lumpy" on the scale of the ions themselves. The Born theory and other "continuous dielectric" models work reasonably well when metal ion solute species are treated as solvent complexes such as $Cr(OH_2)_6^{3+}$ rather than naked ions such as Cr^{3+}, but the emerging approach to solvation phenomena is to simulate solvation dynamically at the molecular level using computer methods.

As we see in the next section, metal–solvent complexes are real, characterizable species, formed by donation of unused valence-shell electron pairs (*lone pairs*) to the metal ion by "donor" molecules to form chemical bonds, but it is less clear how solvation of *anions* should be treated, since solvent lone pairs will tend to be oriented *away* from negatively charged species. For protonic solvents (i.e., those with ionizable H atoms) like water, it is expected that solvent protons, with their partial positive charge, will seek anions and form *hydrogen bonds* with them. In a hydrogen bond, the proton is attached to both an atom A in its original molecule by a two-electron bond and to an atom B that donates a pair of electrons to form a second bond to the proton. This is similar to metal complex formation, but with H^+ in place of the metal ion. The hydrogen bond is usually asymmetric ($A—H\cdots B$), but in the bifluoride ion HF_2^- the hydrogen bond is unusually strong and is symmetric: $[F\cdots H\cdots F]^-$.

Molecules of protonic solvents such as water are typically hydrogen bonded to one another to some degree, giving some short-range order, but these hydrogen bonds form and break up continually. The presence of hydrogen bonding in liquids is manifested in boiling points that are abnormally high; for example, the normal boiling points of the isoelectronic liquids water (two lone pairs available for hydrogen bonding), ammonia (one lone pair), and methane (none) are 100, −33, and −182 °C, respectively. The Born theory does not take into account the effects of hydrogen bonding. There is evidence from X-ray diffraction studies on solutions that chloride ion is associated specifically with six water molecules, presumably through hydrogen bonding, and certainly thermodynamic data show that protonic solvents are much better at solvating anions than are nonprotic donor solvents such as DMSO or acetonitrile, however effective the latter may be in solvating cations. Solvents such as hexane that have low relative permittivities and neither lone pairs for donor action nor ionizable protons for hydrogen bonding are ineffective in dissolving ionic compounds.

13.2 Metal Complexes

Cations, particularly those with charge number $z > 1$, are considerably smaller than the parent neutral atoms, and so the electric potential ($\propto z/r$) at cation surfaces is high. Consequently, the electron cloud of a neighboring molecule or anion becomes displaced toward a cation, and, if there is an unused pair (lone pair) of electrons in the valence shell of the molecule or anion, it may actually become involved in a sort of covalent bond with the cation, using empty valence-shell orbitals on the cation.[1,2,5,6]

Such a bond, in which the donor molecule (or anion) provides both bonding electrons and the *acceptor* cation provides the empty orbital, is called a *coordinate* or *dative* bond. The resulting aggregation is called a *complex*. Actually, any molecule with an empty orbital in its valence shell, such as the gas boron trifluoride, can in principle act as an electron pair acceptor, and indeed BF_3 reacts with ammonia (which has a lone pair, $:NH_3$) to form a complex $H_3N{:}\!\rightarrow\!BF_3$. Our concern here, however, is with metal cations, and these usually form complexes with from 2 to 12 donor molecules at once, depending on the sizes and electronic structures of the cation and donor molecules. The bound donor molecules are called *ligands* (from the Latin *ligare*, to bind), and the acceptor and donor species may be regarded as *Lewis acids* and *Lewis bases*, respectively.

The nature of complex compounds of metals was first made clear by Alfred Werner of the University of Zürich around 1900, largely from investigations of cobalt(III) complexes, in which ligand substitution is slow [a multitude of mixed-ligand complexes of cobalt(III) can therefore be isolated and characterized]. For example, Werner recognized that the various cobalt(III) chloride *ammines* (ammonia complexes; see Appendix E for explanation of the nomenclature of complexes) that were known all contained a six-coordinate cobalt(III) complex with, say, x chloride ions bound as ligands, $(6 - x)$ ammonia ligands, and $(3 - x)$ chloride ions as such in the crystal lattice. On dissolving one of these ammines in water, the $(3 - x)$ free Cl^- ions could be precipitated at once as solid silver chloride on adding

TABLE 13.2
Cobalt(III) Chloride Ammines

Solid compound	*Color*	*Ionized* Cl^-	*Formulation as Complex*
$CoCl_3{\cdot}6NH_3$	yellow	3	$[Co(NH_3)_6]Cl_3$
$CoCl_3{\cdot}5NH_3$	purple	2	$[Co(NH_3)_5Cl]Cl_2$
$CoCl_3{\cdot}4NH_3$	green	1	*trans*-$[Co(NH_3)_4Cl_2]Cl$
$CoCl_3{\cdot}4NH_3$	violet	1	*cis*-$[Co(NH_3)_4Cl_2]Cl$

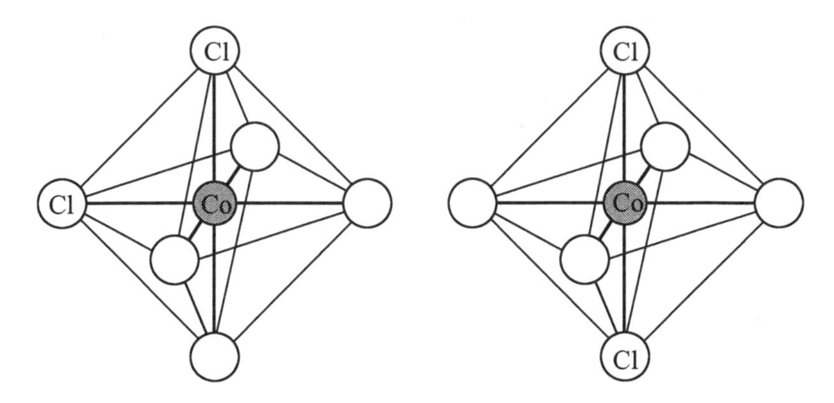

Figure 13.1 Cis (left) and trans isomers of dichlorotetraamminecobalt(III) ion. Unmarked spheres represent NH_3 ligands.

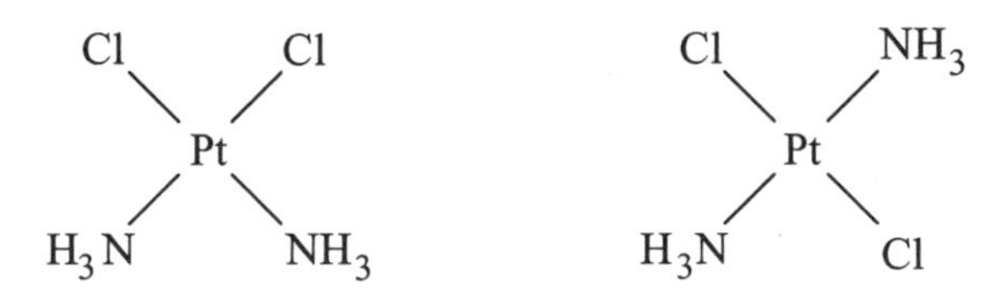

Figure 13.2 Square planar cis- (left) and trans-dichlorodiammineplatinum(II).

excess silver nitrate solution, whereas the remaining x chlorides would be precipitated only on boiling (Table 13.2).

The green and violet tetraammines have the same chemical composition, that is, they are isomers and are the *only* two isomers with this composition. Werner realized that this was possible only if the six ligands were deployed about the cobalt(III) center in an *octahedral* arrangement (cf. octahedral coordination in solids, Sections 4.3 and 4.4); for example, a flat hexagonal complex $Co(NH_3)_4Cl_2^+$ would have *three* isomers, like ortho-, meta-, and para-disubstituted benzenes. Werner correctly identified the green compound as the trans isomer (chloro ligands on opposite sides of the octahedron) and the violet as cis (same side), as in Fig. 13.1.

In the same way, the existence of two isomers, yellow and orange, of $Pt(NH_3)_2Cl_2$ showed that these complexes are square planar, rather than tetrahedral (Fig. 13.2); the yellow one has no electric dipole moment and therefore is the trans isomer, whereas the orange does and is cis. A tetrahedral $Pt(NH_3)_2Cl_2$ would have just one isomer (with a dipole moment). Four-coordinate complexes of platinum(II), palladium(II), and gold(III) are virtually always square planar, but tetrahedral complexes such as the purple tetrathiocyanatocobaltate(II) ion in $Hg[Co(NCS)_4]$ are often encountered.

Most commonly, metal ions M^{2+} and M^{3+} (M = a first transition series metal), Li^+, Na^+, Mg^{2+}, Al^{3+}, Ga^{3+}, In^{3+}, Tl^{3+}, and Sn^{2+} form octahedral six-coordinate complexes. Linear two coordination is associated with univalent ions of the coinage metal (Cu, Ag, Au), as in $Ag(NH_3)_2{}^+$ or $AuCl_2{}^-$. Three and five coordination are not frequently encountered, since close-packing considerations tell us that tetrahedral or octahedral complex formation will normally be favored over five coordination, while three coordination requires an extraordinarily small radius ratio (Section 4.5). Coordination numbers higher than six are found among the larger transition metal ions [i.e., those at the left of the second and third transition series, as exemplified by $TaF_7{}^{2-}$ and $Mo(CN)_8{}^{4-}$] and in the lanthanides and actinides [e.g., $Nd(H_2O)_9{}^{3+}$ as well as $UO_2F_5{}^{3-}$ which contains the linear *uranyl* unit $O{=}U{=}O^{2+}$ and five fluoride ligands coordinated around the uranium(VI) in an "equatorial" plane]. For most of the metal complexes discussed in this book a coordination number of six may be assumed.

As noted earlier, metal ions in polar solvents will form complexes with the solvent molecules. X-Ray diffraction, EXAFS, and visible absorption spectroscopy show that nickel(II) ion in dilute aqueous solution is present as the green hexaaqua complex $Ni(H_2O)_6{}^{2+}$, just as in solids such as $NiSO_4{\cdot}7H_2O$, which is actually $[Ni(H_2O)_6]SO_4{\cdot}H_2O$. In the crystal, the extra water molecule is loosely associated with the sulfate ion independently of the nickel–aqua complex; it is sometimes referred to as *lattice* water, as distinct from *complexed* water.

Similarly, the grayish-purple $Cr^{3+}(aq)$ is $Cr(H_2O)_6{}^{3+}$ just as in solid $[Cr(H_2O)_6](NO_3)_3{\cdot}3H_2O$ or $[Cr(H_2O)_6]Cl_3$ (no lattice water). There is, however, an isomeric solid $CrCl_3{\cdot}6H_2O$ which is green; this is actually *trans*-$[Cr(H_2O)_4Cl_2]Cl{\cdot}2H_2O$, and in solutions of $Cr^{3+}(aq)$ containing fairly high chloride concentrations we find not only $Cr(H_2O)_6{}^{3+}$ but also dull-green $Cr(H_2O)_5Cl^{2+}$, leaf-green *cis*- and *trans*-$Cr(H_2O)_4Cl^{2+}$, and possibly complexes containing three or more chloro ligands in place of water. As it happens, ligand substitution at chromium(III) centers is quite slow, occurring on a time scale of hours at room temperature, just as in the cobalt(III) and platinum(II) cases discussed earlier, and the various aquachlorochromium(III) species can be separated from one another, for example, by ion-exchange chromatography (Section 17.3.2). Not all metal ions undergo ligand substitution quite so slowly, however; in solutions of Ti^{3+}, V^{2+}, V^{3+}, Cr^{2+}, Mn^{2+}, Mn^{3+}, Fe^{2+}, Fe^{3+}, Co^{2+}, Ni^{2+}, Cu^{2+}, Zn^{2+}, Cd^{2+}, Hg^{2+}, and all 1+, 2+, and 3+ ions (including the lanthanides and actinides) outside the transition series, simple ligand substitution normally reaches equilibrium in less than about a second, sometimes in less than a microsecond. The relative rates of ligand substitution in metal complexes can be rationalized in terms of ligand field theory.[1,2,6]

In general, then, metal ions in solution form complexes (frequently six coordinate) with the solvent molecules, their counterions, and other donor molecules that happen to be in the solution. For example, in ammoniacal aqueous solution, Ag^+ forms $Ag(NH_3)_2{}^+$ (as noted above), Cu^{2+} forms a series of aquaammines but most notably the royal blue *trans*-$Cu(NH_3)_4(OH_2)_2{}^{2+}$, and cobalt(II) forms $Co(NH_3)_x(H_2O)_{6-x}{}^{2+}$ complexes which react quite rapidly with oxygen in air to give the strawberry-red cobalt(III) complex $Co(NH_3)_5OH_2{}^{3+}$ or (if much chloride ion is present) the $Co(NH_3)_5Cl^{2+}$ ion mentioned above.

The HSAB classification introduced in Section 2.9 is useful in making educated guesses as to what kinds of ligands will form strong complexes with a particular metal ion. Metal ions that are classified as soft acids form more stable complexes with soft than with hard bases. For example, soft mercury(II) forms a particularly stable complex $[HgI_4]^{2-}$ with soft aqueous iodide ion (to the extent that the normally very insoluble red solid HgO dissolves easily in aqueous KI, liberating hydroxide ion quantitatively), but no analogous complex forms with hard aqueous fluoride ion. Conversely, hard Al^{3+} ion is strongly complexed by aqueous F^-, but not by I^-. Thus, as a first step in designing a drug to treat chronic mercury(II) poisoning, one would look to ligands with sulfur donor atoms; indeed, it is bonding with sulfur proteins that causes Hg to accumulate in the body.

13.3 Chelation

Many complexing agents exist that have more than one potential donor atom in the molecule. If the molecular geometry is appropriate, then these can act as ligands that attach themselves to a metal ion through two or more separate points. Such ligands are called *chelating agents* (from the Greek *khele* meaning a claw), and the resulting complexes are known as *metal chelates*. If there are two points of attachment, we speak of *bidentate* chelation; if three, *terdentate;* if four, *quadridentate*, and so on.* Suppose, for example, we join two ammonia ligands together with a short hydrocarbon chain; we now have a bidentate chelating agent, for example, $H_2N{-}CH_2{-}CH_2{-}NH_2$, which is commonly called *ethylenediamine* (abbreviated en), although it contains no C=C double bond. Figure 13.3 shows the structure of $Co(en)_3{}^{3+}$. Note that there are two nonsuperposable ways in which $Co(en)_3{}^{3+}$ can be drawn, namely, as left-handed and right-handed mirror images. These are known as *optical isomers* of $Co(en)_3{}^{3+}$ because a solution of one isomer will rotate the plane of polarization of light to the left as it passes through, and a solution of the other will rotate it to the right. The two isomers will be present in equal amounts as

*The preferred prefixes are Latin (to match the Latin stem *dens*, tooth), rather than the Greek di-, tri-, tetra-, etc., but either the Latin or the Greek system is acceptable.

$Co(en)_3{}^{3+}$ salts are usually prepared, but they can be resolved (i.e., separated from one another) by fractional crystallization of a salt of $Co(en)_3{}^{3+}$ from solution with an anion that is itself an optical isomer, for example, the *d*-tartrate anion. The fact that optical isomers of $Co(en)_3{}^{3+}$ and other chelates do exist was used by Werner as incontrovertible proof of his theory of octahedral coordination in cobalt(III) and other complexes.

Bidentate ligands. Besides en, the following bidentate ligands are common:

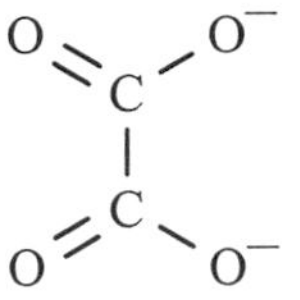

oxalate (ox) 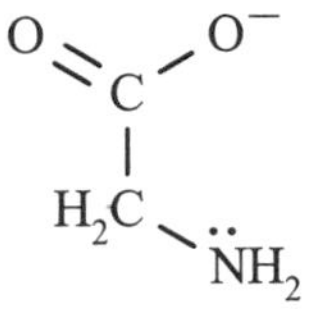 glycinate 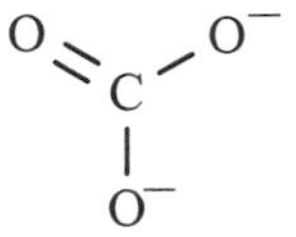 carbonate

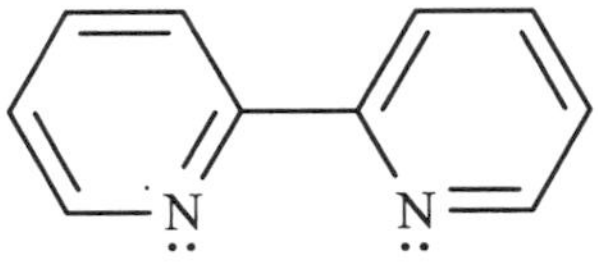

2,2'-bipyridine (bpy) 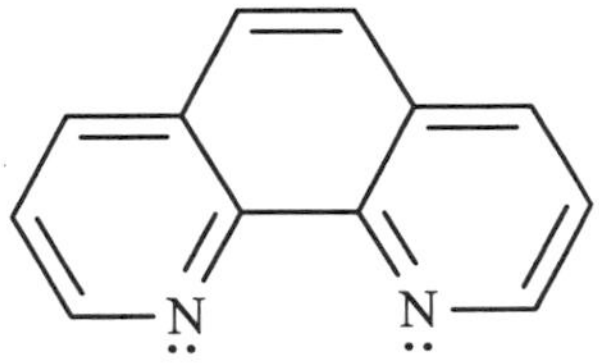 1,10-phenanthroline (phen)

Note that most of these ligands form a five-membered ring when coordinated to a metal ion. A familiar rule of thumb in organic chemistry is that six-membered rings are preferred, but this applies to molecules in which the bond angles in the ring are 109.5° (as in the puckered cyclohexane ring) or 120° (as in benzene). Here, we have at least one 90° bond angle, (donor atom)—(metal)—(donor atom), if the coordination geometry is octahedral, so five-membered rings are generally more stable. The carbonato ligand forms a somewhat strained four-membered ring when bidentate and, significantly, is often encountered as a unidentate ligand. Similarly, carboxylates $R—CO_2{}^-$ can be bidentate, with both oxygens coordinated, as in the complex $Th(O_2CCH_3)_6{}^{2-}$, where the coordination number of thorium(IV) is actually 12. In general, though, individual carboxylate groups tend to be unidentate. The glycinate ion, a sort of hybrid of oxalate and ethylenediamine, is the anion of glycine, the simplest of the α-amino acids. These compounds are extremely important in that they are the fundamental building blocks of proteins and other biological molecules. The fact that such compounds are potential chelating agents means that metal ions often interact strongly (and not necessarily beneficially) with biological systems.

Terdentate ligands. By analogy with ethylenediamine and glycine, we expect *diethylenetriamine* (dien) and *iminodiacetate* ion (IDA^{2-}) to be

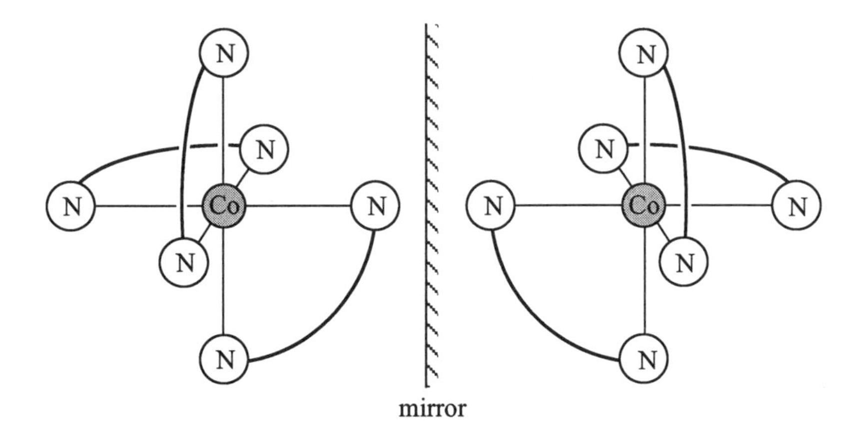

Figure 13.3 Optical isomers of $Co(en)_3{}^{3+}$. N—N represents $H_2NCH_2CH_2NH_2$. The two molecules shown are not superposable (cf. left and right hands).

effective terdentate chelating agents:

$$H_2\ddot{N}-CH_2-CH_2-\ddot{N}H-CH_2-CH_2-\ddot{N}H_2$$

dien

$$O{=}C(O^-)-CH_2-\ddot{N}H-CH_2-C(O^-){=}O$$

IDA^{2-}

Quadridentate ligands. One quadridentate derivative of en and dien is *tren*. It has a carboxylato analogue in the *nitrilotriacetate* ion (NTA^{3-}):

$$:N(CH_2CH_2\ddot{N}H_2)_3$$

tren

$$:N(CH_2COO^-)_3$$

NTA^{3-}

These molecules are flexible because rotation about single bonds is free, so they can wrap themselves around a metal ion to obtain four comfortable donor-atom-to-metal links within five-membered rings. Nitrilotriacetic acid is easily synthesized industrially from ammonia, formaldehyde, and hydrocyanic acid (the *Strecker synthesis*) and therefore is potentially an inexpensive but effective chelating agent.

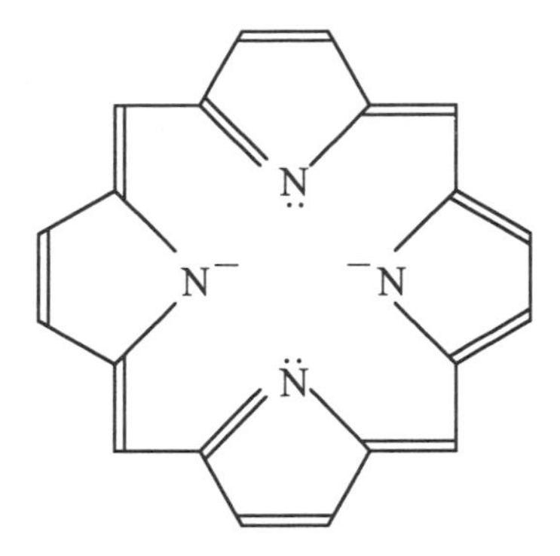

Figure 13.4 The anion of porphine. Porphyrins are porphine rings with peripheral substituents. The porphine ring is flat, and metal ions can be bound within the square planar arrangement of nitrogen atoms.

There are numerous biologically important quadridentate chelating agents in which four nitrogen donor atoms are locked into a square-planar arrangement in a flat *porphyrin* ring; the basic structure is outlined in Fig. 13.4. These units are found, for example, in the O_2-carrier hemoglobin, where the complexed metal ion is iron (see Section 8.2), and in chlorophyll, the green substance that mediates photosynthesis in plants and in which the coordinated metal ion is Mg^{2+}.

Sexidentate chelation. The outstanding example of a powerful and versatile chelating agent is the *ethylenedinitrilotetraacetate* anion, $EDTA^{4-}$, sometimes called *ethylenediaminetetraacetate*, or, in medical circles, *Versene* (Dow trade name):

```
                             CH2COO⁻
                            /
⁻OCOCH2     H2C — N
        \       |     \
         N — CH2      CH2COO⁻
        /
⁻OCOCH2
```

Successive pK_a values of the parent acid H_4(EDTA) are 2.0, 2.7, 6.2, and 10.3 at room temperature. The first two represent removal of protons from two of the four —COOH groups. It turns out that the other two ionizable protons reside on the nitrogens, so that in Na_2H_2(EDTA) (the form usually supplied by manufacturers) all the carboxyl groups are ionized to —COO^- and the two nitrogens are protonated (giving a *zwitterion* structure, common among amino acids). When $H_2(EDTA)^{2-}$ complexes a metal ion, however, all the protons are displaced, and the flexible molecule wraps

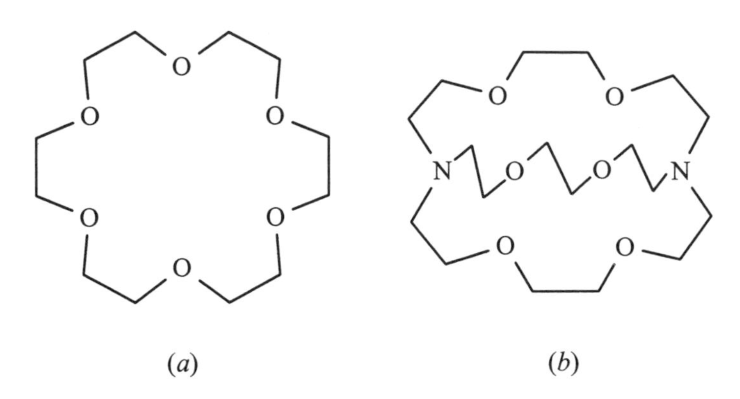

Figure 13.5 (*a*) A crown ether (18-crown-6) and (*b*) a cryptand (2,2,2-crypt).

itself around the ion in such a way that, usually, the —COO$^-$ and ≡N: donors occupy six octahedral sites. In a few examples, one carboxylate function is attached, not to the metal, but to a proton, so that the EDTA ligand is only *quinquedentate* (five coordinate). On the other hand, in lanthanide complexes such as $La(OH_2)_3(EDTA)^-$, the EDTA is sexidentate, but the overall coordination number is more than six.

A sexidentate complexing agent that is typical of a rapidly expanding class of new compounds called *crown ethers* is the flexible 18-crown-6, sketched in Fig. 13.5a. Crown ethers form strong complexes even with alkali metal ions, which (being large, as cations go, and of low charge) interact only weakly with most complexing agents. Crown ethers are useful in solubilizing alkali metal salts sufficiently to dissolve in solvents of low relative permittivity. Crown-ether-like molecules consisting of several rings fused to form a cage that can enclose a metal ion are called *cryptands* (Fig. 13.5b), and such complexes are called *cryptates.*

Chelate complexes are much more stable than comparable complexes of unidentate ligands. This *chelate effect* is difficult to express quantitatively because of problems with units (Section 13.4), but we can see qualitatively that, once one donor atom of (say) a sexidentate ligand is attached to a metal ion, the local concentration of five more donors is suddenly very high, with the result that they readily coordinate with the metal ion. Had the six donors been floating around independently in the solution, no such cooperative action would have been possible. Some authors have pointed out that one sexidentate ligand can displace six unidentate ligands, such as aqua or other coordinated solvent molecules, resulting in a positive entropy change (one molecule pinned down for six set loose) and hence a more negative ΔG than for analogous unidentate–unidentate substitutions, but again these arguments turn out to be dependent on the choice of standard state. The essential points are that multidentate ligands are extraordinarily effective complexing agents, and that the advantage of high *denticity* is especially marked at low concentrations (Exercise 13.5).

13.4 Stability Constants

The equilibrium constant for the *formation* of a metal complex is known as its *stability constant.* (Some authors, however, present the datum as its reciprocal, the *instability constant* of the complex, by analogy with the dissociation of a weak acid.) There are two kinds of stability constants: *stepwise* (K_1, K_2, K_3,..., K_n) and *overall* (β_n). We will assume that there are six aqua ligands to be replaced by some other unidentate ligand X^{x-}, in an aqueous solution of M^{m+}:

$$M(H_2O)_6{}^{m+} + X^{x-} \overset{K_1}{\rightleftharpoons} M(H_2O)_5X^{(m-x)+} + H_2O. \tag{13.5}$$

For clarity, we can ignore the solvent water and charges:

$$\frac{[MX]}{[M][X]} = K_1 \ (\mathrm{L\ mol^{-1}}) \tag{13.6}$$

$$MX + X \overset{K_2}{\rightleftharpoons} MX_2 \tag{13.7}$$

$$\frac{[MX_2]}{[MX][X]} = K_2 \ (\mathrm{L\ mol^{-1}}) \tag{13.8}$$

$$MX_2 + X \overset{K_3}{\rightleftharpoons} MX_3. \tag{13.9}$$

In principle, this continues until we reach MX_6. Alternatively, we can consider the formation of MX_6 as a single step

$$M + 6X \overset{\beta_6}{\rightleftharpoons} MX_6 \tag{13.10}$$

$$\frac{[MX_6]}{[M][X]^6} = \beta_6 \ (\mathrm{L^6\ mol^{-6}}), \tag{13.11}$$

and the connection between the stability constants is therefore

$$\beta_n = K_1{\cdot}K_2{\cdot}K_3...K_n. \tag{13.12}$$

If a sexidentate ligand were to coordinate to M^{m+}, β_n would just be $\beta_1 = K_1$, with units L mol^{-1}. Consequently, we cannot compare β_1 for formation of a sexidentate ligand with the overall stability constant β_6 for six comparable unidentate ligands because the units are incompatible.

The stability constants are defined here in terms of concentrations and hence have dimensions. True thermodynamic stability constants K_n° and β_n° would be expressed in terms of activities (Section 2.2), and these constants can be obtained experimentally by extrapolation of the (real) measurements to (hypothetical) infinite dilution. Such data are of limited value, however, as we cannot restrict our work to extremely dilute solutions. At practical concentrations, the activities and concentrations of ions in solution differ significantly, that is, the activity coefficients are not close to unity; worse still, there is no thermodynamically rigorous means of separating anion and cation properties for solutions of electrolytes. Thus, single-ion activity coefficients are not experimentally accessible, and hence, strictly speaking, one cannot convert equations such as 13.6 or 13.8 to thermodynamically exact versions.

Modern theories of electrolytes based on the *Debye–Hückel* approach (Section 2.2) show that ionic activity coefficients are governed mainly by the *ionic strength*, I, given by $0.5 \sum c_i z_i^2$ where c_i is the concentration and z_i the charge on ions of the ith kind in the solution. These i kinds include all *ions* in the solution, as well as M^{m+} and X^{x-}. Accordingly, we can vary $[M^{m+}]$ and $[X^{x-}]$ with constant activity coefficients to derive K_n or β_n by maintaining an effectively constant ionic strength I with a swamping concentration of some inert electrolyte. Usually a perchlorate such as $LiClO_4$ or $NaClO_4$ is used because, of all the commoner anions, ClO_4^- has the least tendency to become involved as a ligand itself (Section 12.2).

Tabulations of stability constants[7,8] usually list $\log K$ values for various M^{m+} and X^{x-} (x may be zero, of course) *at specified ionic strengths*, often including extrapolated values for $I = 0$. The temperature is also noted, since ΔH° for complex formation, though rarely large, cannot normally be disregarded.

13.5 Uses of Complexing Agents

Most applications of complexing agents depend upon the fact that they will "tie up" most of the metal ion present in a solution as a complex, leaving only a low concentration of free metal ion.

13.5.1 Chelating Agents

Ligands of the NTA^{3-} and $EDTA^{4-}$ type (*complexones* or *sequestering agents*, as they are sometimes called) are often able to dissolve deposits of metal oxides, hydroxides, sulfides, and carbonates because they displace solubility equilibria such as reaction 13.13 to the right by reducing the free metal ion concentration to very low levels[9]:

$$M(OH)_m(s) \overset{K_{sp}}{\rightleftharpoons} M^{m+}(aq) + mOH^-(aq) \tag{13.13}$$

$$[M^{m+}][OH^-]^m = K_{sp} \tag{13.14}$$

$$M^{m+}(aq) + EDTA^{4-}(aq) \overset{\beta_1}{\rightleftharpoons} [M(EDTA)]^{(4-m)-} \tag{13.15}$$

$$\beta_1 = \frac{[M(EDTA)^{(4-m)-}]}{[M^{m+}][EDTA^{4-}]}. \tag{13.16}$$

Thus, if β_1 is large enough, for a given solubility product K_{sp}, and the pH and free EDTA concentration are appropriate, the solid $M(OH)_m$ of reaction 13.13 will pass entirely into solution (see Exercise 13.6). Since the donor atoms in these ligands are "hard" oxygen, they are particularly effective against hard M^{m+} ions such as Ca^{2+} or Fe^{3+}.

Equations 13.13 to 13.16 could be applied to the dissolution of brown iron stains or corrosion deposits [nominally $Fe(OH)_3$], and a parallel set of equations could be written for the dissolution of $CaCO_3$ (boiler scale) or of the sulfide tarnish on copper or brass. Many domestic cleaning agents contain chelating agents; bathroom cleaners typically contain 1–5% $Na_4(EDTA)$, and $Na_3(NTA)$ is an excellent detergent *builder* (Section 7.3.1), acting to keep Ca^{2+} and Mg^{2+} tied up as complexes so that they cannot interfere with detergent action. In this regard, $Na_3(NTA)$ offers a biodegradable substitute for $Na_5P_3O_{10}$, an effective traditional builder but one that presents ecological problems (Section 7.7). Claims (probably unjustified) that NTA^{3-} is sufficiently carcinogenic to present a significant public health hazard have led to a ban on use of $Na_3(NTA)$ as a detergent builder in some jurisdictions. It might also be argued that, by its very nature, NTA in wastewater may mobilize normally insoluble toxic heavy metal compounds as NTA complexes in rivers, lakes, etc. However, an important attribute of $H_3(NTA)$ and $H_4(EDTA)$ is that they are amino acids, and hence readily biodegradable; most NTA in wastewater would be destroyed by bacteria in secondary sewage treatment.

Boiler tubes are often cleaned with EDTA or NTA solutions to remove both $CaCO_3$ scale and corrosion products. In pressurized heavy water nuclear power reactors, radioactive corrosion deposits (in effect, magnetite in which some of the Fe^{2+} has been replaced by radioactive $^{60}Co^{2+}$) can be removed from the coolant water circuits with an aqueous mixture of oxalic and citric acids (both good chelators for Fe^{3+}) and EDTA. In home laundry operations, bloodstains on clothing can be removed by treatment with oxalic acid, which takes up the iron from the hemoglobin (Section 8.2) as $Fe(ox)_3{}^{3-}$. By the same token, oxalates are toxic when taken internally, as are many other complexing agents. For example, EDTA is used as a means of removing lead, a cumulative poison, from the body of a person with suspected lead poisoning, but it will strip out many other essential metal ions at the same time; thus, such Versene treatments are very risky and should not be undertaken unless the diagnosis is definite. (Unfortunately, the maximum allowable level of lead in blood or urine is only about twice the normal level, so diagnosis requires careful chemical analysis for Pb.)

On the other hand, the insolubility of the various forms of Fe_2O_3, $Fe(OH)_3$, and $FeO(OH)$ makes it difficult for plants or animals to get the necessary amount of iron *into* their systems. This is especially true for plants in soils that are alkaline, since OH^- suppresses $Fe(OH)_3$ dissolution (reaction 13.13). Gardeners therefore feed iron to evergreens as soluble chelate (usually EDTA) complexes. Certain bacteria and fungi have evolved to produce iron-chelating substances, in which the donor atoms are usually oxygen, just as in oxalate.[10] In humans, iron reserves are brought into the body by a complicated protein chelating agent (*transferrin*) and are stored in the spleen as $FeO(OH)$ plus iron phosphates inside a protein envelope (*ferritin*).

Transition metal ions, even in trace amounts, catalyze many chemical reactions, such as the decomposition of hydrogen peroxide

$$H_2O_2(aq) \xrightarrow[\text{(etc.)}]{Fe^{3+}} H_2O + \tfrac{1}{2}O_2 \qquad (13.17)$$

or the deterioration of foodstuffs (e.g., salad dressing). Catalysis by metal ions can be suppressed by addition of a small amount of a chelating agent, usually $Na_2H_2(EDTA)$, to tie up any traces of free metal ions in an anionic form.

Many of the less obvious uses of EDTA and NTA depend on their ability to sequester Ca^{2+} and Mg^{2+} ions. For example, ophthalmic solutions for wetting contact lenses or irrigating the eyes contain EDTA ions (among other things) to suppress gram-negative bacteria, notably *Pseudomonas* species, that are particularly harmful to the eyes. These bacteria may develop resistance to most disinfectants because of a protective cell wall consisting of lipopolysaccharides held firmly held together with Ca^{2+} and Mg^{2+} ions. The main function of the EDTA ions is to remove those cations.

13.5.2 Unidentate Complexing Agents

Although chelating ligands are especially effective, there are many unidentate complexing agents of commercial importance. As noted in Section 13.1, silver ion interacts strongly with ammonia to form $[Ag(NH_3)_2]^+$, so that substantial amounts of AgCl or AgBr, which are poorly soluble in water, can be dissolved in aqueous ammonia. Thus, at 25 °C and low ionic strength, we have (Exercise 13.4)

$$\frac{[Ag(NH_3)_2{}^+]}{[Ag^+(\text{free})][NH_3(\text{free})]^2} = \beta_2 = 1.07 \times 10^7 \text{ L}^2 \text{ mol}^{-2}, \qquad (13.18)$$

and the solubility product K_{sp} for AgBr gives us

$$[Ag^+(\text{free})][Br^-] = K_{sp} = 5.3 \times 10^{-13} \text{ mol}^2 \text{ L}^{-2} \qquad (13.19)$$

where $[Br^-]$ must equal the total Ag^+ concentration in solution, that is, $[Ag^+(\text{free})]+[Ag(NH_3)_2{}^+]$, and the common factor $[Ag^+(\text{free})]$ can be eliminated. The tarnish that forms on silverware is mainly Ag_2S, and it can similarly be dissolved in aqueous ammonia, which is present in many commercial silver cleaning fluids. Silver ion also complexes strongly with *thiosulfate* ion (Section 3.4), formerly called *hyposulfate*, and photographers' "hypo" fixing solution makes use of the resulting ability of aqueous $Na_2S_2O_3$ to dissolve undeveloped solid AgCl or AgBr from photographic film, as described in Section 13.5.3.

Cyanide ion is a very powerful unidentate complexing agent, especially for transition metal ions such as Fe^{2+} and Fe^{3+}, which form hexacoordinate complexes commonly called *ferrocyanide* $[Fe(CN)_6{}^{4-}]$ and *ferricyanide* $[Fe(CN)_6{}^{3-}]$, respectively. These complexes (especially ferrocyanide) have low toxicity despite the cyanide content, because the cyanide is so strongly bound. Indeed, a traditional antidote for swallowed cyanide is to drink freshly mixed aqueous ammonia and iron(II) sulfate, which will tie up the CN^- as $Fe(CN)_6{}^{4-}$. The ammonia serves to increase the pH of the stomach contents, thus minimizing formation of undissociated HCN which would be absorbed rapidly by the body. The toxicity of cyanide is itself due to its complexing with the iron(III) centers in porphyrin units in cytochrome *c*, which is responsible for crucial steps in the transfer of electrons to molecular oxygen in mitochondria (respiratory organelles) of cells that make up the bodies of the higher animals. Complexation by CN^- takes place rapidly and completely prevents reduction of the iron(III) center to iron(II). Since this reduction is essential to the respiratory cycle, death follows quickly.[11]

Cyanides are widely used in industry, for example, in gold refining (Section 17.2) and in electroplating (Section 15.7), and they must be handled with appropriate care. In particular, contact with acids must be avoided, otherwise HCN (*prussic acid*, pK_a 9.3), which is highly volatile and can be absorbed through the skin as well as through the lungs, will form. Hydrogen cyanide is also produced when nitrile-based synthetic fibers (Orlon, Acrilan, etc.), adhesives, and rubbers burn. Many fire victims who die of "smoke inhalation" are actually killed by HCN. Finally, HCN is found in peach pits and bitter almonds, the characteristic odor of whichis actually HCN; hence, such foods could be fatally toxic if eaten in large amounts.

13.5.3 Photographic Chemistry

Complexing agents play critical roles in silver halide-based photography, a brief outline of which follows.[12] Black-and-white photographic film consists of finely divided silver halide grains ($\leq$1 μm diameter) dispersed in gelatin (the *emulsion*), on a polyester film base.* The silver halide is formed within the gelatin by reaction of aqueous silver nitrate with a potassium halide. Bromide is the usual choice of halide, as AgCl is sensitive to light only of wavelengths shorter than 420 nm; conversely, though, AgCl is well suited for use in slow photographic printing papers, which can be processed under a red safelight in the darkroom. Even so, AgBr is sensitive mainly to blue and ultraviolet light; addition of 3% AgI extends sensitivity to about 530 nm, but green or cyan dyes must be added to panchromatic film emulsions to ensure balanced coverage of the whole visible spectrum.

Grains that are exposed to light (i.e., to the bright parts of the image projected onto the film in the camera) decompose very slightly to free silver and halogen, forming a *latent image* of grains with minute spots of silver, usually on the crystal surfaces, that serve as nuclei for reduction of the whole grains to silver metal by subsequent action of a "developer." In the language of Chapter 5, the detailed sequence of events is probably as follows:

(*a*) an incident photon excites electrons (effectively from bromide ions) in an AgBr crystal into the conduction band,

(*b*) the electrons move through the crystal until trapped by a defect, typically on the surface, so forming a negatively charged sensitivity speck,

(*c*) interstitial Ag^+ ions move to the sensitivity specks, where they become silver atoms, and

(*d*) the bromine atoms formed in this process are scavenged by the gelatin.

At least four Ag atoms seem to be necessary to form a nucleus for subsequent reduction of an AgBr grain. Clearly, the larger the grain size, the "faster" the film will be, since a greater mass of AgBr will be sensitized per speck, but the image will then be coarser; thus, fine-grained films offering exquisitely detailed images are typically rather slow.

The exposed film is treated with a developing solution in a dark tank for a measured time at a controlled temperature. Development can amplify the latent silver image up to 10^9-fold, but excessively long or rapid development will result in unsensitized AgBr being reduced too. Developing agents (i.e., mild reductants for AgBr) include *p*-aminophenol, *N*-methyl-*p*-aminophenol sulfate (Metol), hydroquinone, pyrogallol, phenidone, *p*-phenylenediamine, and derivatives of these compounds. Metol–hydroquinone

*The traditional cellulose acetate is no longer used as a film base.

mixtures are especially popular. In all cases, sodium metabisulfite (Section 8.5) is added to prevent oxidation of the developer by the oxygen of the air, and, since developing agents work much faster at high pH, sodium carbonate, borate, or sulfite is included to keep the solution mildly alkaline. Often, EDTA salts or related complexing agents are added to developer solutions to suppress any adverse effects of hard water or adventitious metal ions. When development has progressed far enough, the alkaline developing solution is replaced with an acidic wash solution or "stop bath," buffered at pH 3–5.

The unreduced silver halide must then be completely removed by complexation with $Na_2S_2O_3$ or $(NH_4)_2S_2O_3$ solution:

$$AgBr(s) + 2S_2O_3{}^{2-}(aq) \rightarrow Ag(S_2O_3)_2{}^{3-}(aq) + Br^-(aq); \qquad (13.20)$$

otherwise, it will slowly photolyze to silver metal when the film is brought out of the darkroom. The fixing solution usually contains some sodium sulfite or bisulfite to suppress sulfur deposition from decomposition of thiosulfate:

$$HSO_3{}^-(aq) + S(s) \rightleftharpoons S_2O_3{}^{2-}(aq) + H^+(aq). \qquad (13.21)$$

The final result is a black deposit of Ag of the appropriate density in the emulsion wherever light had struck the film inside the camera. The image is therefore a *negative* one and must be projected onto silver halide-coated paper, and the development/fixing process repeated, to secure a permanent positive image. Thorough washing of the print is necessary to remove thiosulfate ions and silver salts completely, otherwise the image or the paper will deteriorate in time.

Color photography is considerably more complex, but color films also use silver halides for capturing the latent image, which is initially developed in much the same way as in black-and-white processing. Subsequently, dyes are formed to give a color-positive image for transparencies or a color-negative image for prints. In modern color films, there are usually three layers within the emulsion (an *integral tripack*), sensitive to red, green, and blue light, respectively, since these correspond to the color sensitivity of retinal cones in the human eye. A color-negative image consists of the *complementary colors* of these, namely, cyan, magenta, and yellow, respectively. The blue-sensitive layer is usually uppermost and is separated from the green-sensitive and basal red-sensitive layers by a yellow filter. The primary development stage is followed by one of several different processes, all of them quite complicated, and only a simplified overview of a hypothetical color-negative process is given here (see, e.g., Haist[12b] for details of commercial processes).

Within (say) the red-sensitive layer of a color-negative film, silver halide grains become sensitized by red light to form a latent image and are reduced to silver metal by a developer just as in black-and-white photography. In this case, however, the developer oxidized in this step is made to react immediately with a *coupler*, present either in the emulsion or in the developer solution, to produce an insoluble cyan dye (since a *negative* color image is required) in the immediate vicinity of the Ag halide grain. Not all developers can be made to react with couplers; developers of the *p*-phenylenediamine class are generally suitable. The silver is then removed by "bleaching" [oxidation with, usually, a solution of potassium ferricyanide, $K_3Fe(CN)_6$]. The color-negative image in complementary colors is then projected on to color-negative photographic paper, and the process is repeated to make a color-positive print.

13.6 Hydrolysis of Aqueous Cations

A water molecule is strongly polarized by coordination to a metal cation, toward which its electrons are attracted:

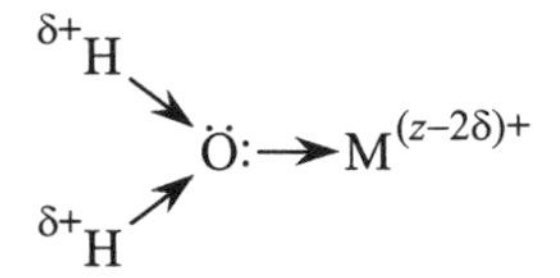

Consequently, aqua ligands are quite acidic:

$$M(H_2O)_6{}^{z+}(aq) \overset{K_a}{\rightleftharpoons} M(H_2O)_5OH^{(z-1)+}(aq) + H^+(aq). \tag{13.22}$$

Thus, a typical aqueous metal ion $M^{z+}(aq)$ can act as a *Brønsted* (i.e., proton-donating) *acid*, of which $MOH^{(z-1)+}(aq)$ is the conjugate base.[1,2,13] The polarizing power of M^{z+} will be greater, the larger its charge z and the smaller its ionic radius r, so that the greatest acid dissociation or "*hydrolysis*" *constants* K_a, or smallest pK_a values ($= -\log K_a$), will be expected for small, highly charged cations:

$$K_a = \frac{[MOH^{(z-1)+}][H^+]}{[M^{z+}]} \tag{13.23}$$

$$pK_a = pH + \log \frac{[M^{z+}]}{[MOH^{(z-1)+}]}. \tag{13.24}$$

From Eq. 13.24, it is seen that pK_a corresponds to the pH value at which hydrolysis of $M^{z+}(aq)$ is just half complete—assuming no other reactions intervene (but see later). Typical values of pK_a are about 14 for $z = 1$, 9 ± 3 for $z = 2$, and 3 ± 2 for $z = 3$, at least for the lighter elements. The pK_a values for the lanthanide(III) ions, for example, are on the order of 7 to 9 because of their larger radii:

	Sc^{3+}	Y^{3+}	La^{3+}	
r	75	90	103	pm
pK_a	4.5	7	9	

However, simplistic generalizations based on z and r are only partly successful in understanding metal ion hydrolysis. For instance, it is not obvious why K_a for $Fe^{3+}(aq)$ ($r = 64.5$ pm) is about 100-fold greater than K_a for $Cr^{3+}(aq)$ ($r = 61.5$ pm).

Like stability constants and other thermodynamic properties of metal ions in solution, hydrolysis constants are affected by ionic strength and temperature, and these should be specified when quoting precise pK_a values. For the ballpark figures cited here, 25 °C and high dilution are assumed.

Equation 13.22, however, does not tell the whole story about metal ion hydrolysis. Not only can further proton dissociations occur

$$\begin{aligned} M^{z+}(aq) &\rightleftharpoons MOH^{(z-1)+}(aq) + H^+ \\ &\rightleftharpoons M(OH)_2{}^{(z-2)+}(aq) + 2H^+ \\ &\rightleftharpoons \text{etc.} \end{aligned} \tag{13.25}$$

until insoluble $M(OH)_z(s)$ comes out of solution, but the conjugate base species $MOH^{(z-1)+}(aq)$, $M(OH)_2{}^{(z-2)+}(aq)$, etc., can polymerize by sharing OH^- ligands as bridging groups:

$$2M(H_2O)_{n-1}OH^{(z-1)+} \rightleftharpoons (H_2O)_{n-2}M\begin{matrix}\overset{H}{O} \\ \diagup\quad\diagdown \\ \diagdown\quad\diagup \\ \underset{H}{O}\end{matrix}M(OH_2)_{n-2}{}^{(2z-2)+} + 2H_2O. \tag{13.26}$$

Condensation reactions (i.e., elimination of bound water or its elements) like reaction 13.26 can build up large, polymeric, highly charged cationic molecules that might be expected to come out of solution as solids containing the available anion. In fact, however, they tend to remain dispersed as *colloids* because mutual electrostatic repulsions (actually, electrical double-layer effects*) prevent their *coagulation*. Colloidal solutions or *sols* can be kinetically stable more or less indefinitely and may be mistaken for true solutions; however, the colloidal particles may be large enough to scatter light, so that a beam passing through the sol shows up clearly (the *Tyndall effect*). Coagulation can often be induced by adding highly charged ions of

*A charged object immersed in an electrolyte solution attracts ions of opposite charge and repels ions of like charge, thereby creating an electrical double layer. Thus, the resistance of two colloidal ions to coagulation is due primarily to repulsion of the interpenetrating electrical double layers.

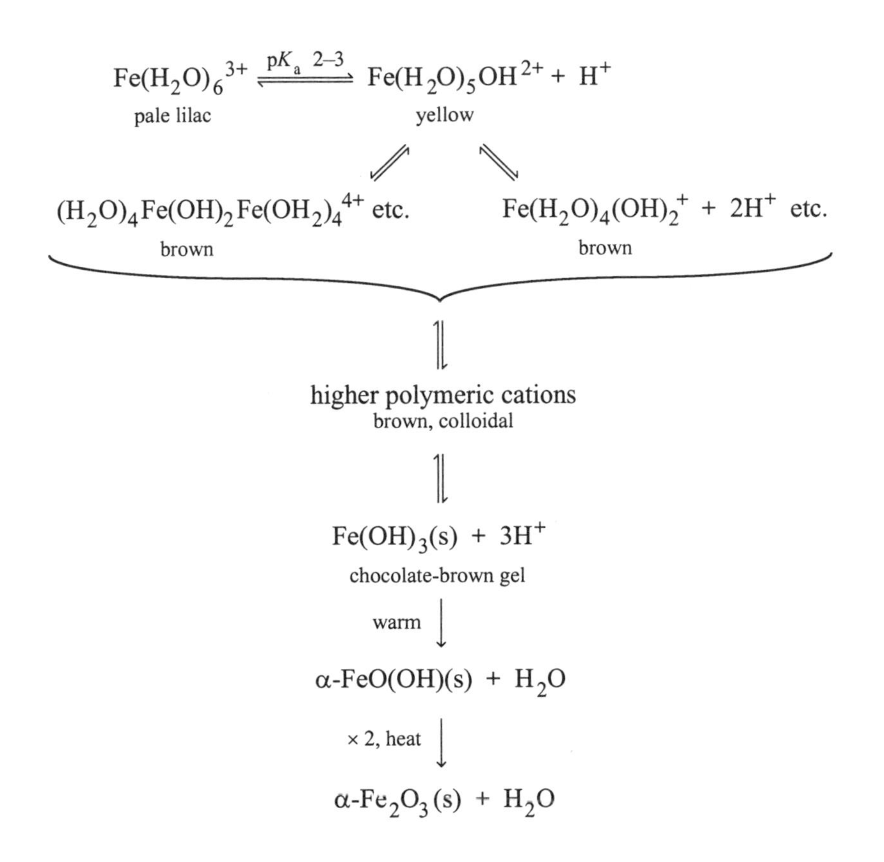

Figure 13.6 Hydrolysis of iron(III).

the opposite charge (in the case of cationic colloids, anions such as sulfate, polyphosphate, or silicate polymers can be used).

These phenomena are exemplified in the aqueous chemistry of iron(III). The ion $Fe(H_2O)_6{}^{3+}$ actually has a beautiful pale lilac color, as seen in solid "ferric alum" [$KFe(SO_4)_2 \cdot 12H_2O$] or iron(III) nitrate [$Fe(NO_3)_2 \cdot 9H_2O$]. In solution, however, iron(III) is typically yellow (at high acidities) or brown, because of hydrolysis and hydrolytic polymerization. The higher polymers form relatively slowly, so that the properties of the solution change on aging. At pH 5 or above, a dark brown, gelatinous precipitate of $Fe(OH)_3$ forms. This tends to dehydrate to form the yellow-brown solid *goethite* (α-FeO(OH)), and this in turn can be further dehydrated on heating to give red-brown *hematite* (α-Fe_2O_3). These hydrolytic solids are important constituents of many soils.

The hydrolytic scheme shown in Fig. 13.6 does not take into account complex formation between Fe^{3+} and other solutes, which tends to prevent the iron(III) from precipitating. In natural waters, for example, organic substances known as *humic acids*[14] may be present as a result of the decay of vegetation, especially in streams that drain marshes, muskeg, and peaty moorland.* These humic acids have complicated structures but typically include carboxylate and phenolic oxygen atoms (cf. lignin, Section 10.4) and they can act as effective chelating agents for metal ions, particularly iron(III). As a result, surface waters rich in humic acids are usually brown because of the complexed iron(III) they contain.

Hydrolysis of metal ions of charge 4+ or more can be expected to be extensive, involving the loss of several protons even in very acidic solution, so that species such as "$Ti(H_2O)_6{}^{4+}$" are simply not encountered. The hydrolyzed ions, however, tend to be *oxo* rather than *hydroxo* species, that is, $(H_2O)_5TiO^{2+}$ rather than $(H_2O)_4Ti(OH)_2{}^{2+}$. Such oxo cations are usually designated with the *-yl* suffix, as in zirconyl (ZrO^{2+}), vanadyl (VO^{2+}), uranyl ($UO_2{}^{2+}$), and neptunyl ($NpO_2{}^{+}$) ions. The concept of oxo complexes as hydrolysis products can be stretched further to include such anionic metal species as chromate ($CrO_4{}^{2-}$), permanganate ($MnO_4{}^{-}$) and vanadate ($VO_3{}^{-}$), which may be regarded as fully hydrolyzed "Cr^{6+}(aq)," "Mn^{7+}(aq)," and "V^{5+}(aq)," respectively. Ions such as $HCrO_4{}^{-}$ can be considered either as incompletely hydrolyzed chromium(VI) ($HOCrO_3{}^{-}$) or, more traditionally, as manifestations of weakness of the corresponding acid (chromic acid, H_2CrO_4; cf. H_2SO_4, Section 8.5).

Exercises

13.1 Estimate the standard free energy of hydration of one mole of gaseous ionized NaCl. The necessary information is given in Sections 4.7 and 13.1 and in Appendix F. Compare the answer with the corresponding lattice energy (Section 4.7).
[*Answer:* -1051 kJ mol^{-1} .]

13.2 (*a*) Sketch the $Co(en)_3{}^{3+}$ ion (Fig. 13.3) as viewed down a three-fold axis (i.e., looking down onto a triangular face, as in Fig. 7.1) of the octahedron of nitrogens, and show that it can be drawn as a screwlike structure with either a left- or a right-handed thread.

(*b*) Show that *cis*-$Co(en)_2Cl_2{}^{+}$, but not the trans isomer, can exhibit optical isomerism.

(*c*) How many isomers of $Cr(H_2O)_3Cl_3$ are there?

---

*In soil chemistry, a distinction is often made between *humic* compounds, which are effectively gels with three-dimensional polymeric structure, and *fulvic* compounds, which are dissolved polyfunctional complexants. Thus, the topic of this section is mainly the fulvic compounds.

13.3 Silver(I) forms a dithiosulfato complex for which $\beta_2 = 2 \times 10^{13}$ in dilute aqueous solution at 25 °C. Given that the solubility product of AgBr is 5.3×10^{-13} (and assuming that ionic strength effects are negligible), show that the minimal total concentration of $Na_2S_2O_3$ needed to dissolve 1.00 g silver bromide in 1.00 L of water at 25 °C is 0.0123 mol L^{-1}. Note that Na^+ is introduced simply as the counter-ion of $S_2O_3{}^{2-}$ and does not enter into the calculations.

13.4 (*a*) Calculate the minimum total concentration of ammonia that must be present in 1.000 L of water at 25 °C in order to dissolve 1.000 g silver bromide ($K_{sp} = 5.3 \times 10^{-13}$ mol^2 L^{-2}), assuming that the diamminesilver(I) ion ($\beta_2 = 1.07 \times 10^7$ L mol^{-1}) forms a freely soluble bromide salt.

(*b*) By what factor does the presence of the ammonia increase the solubility of the silver bromide?

[*Answers:* (*a*) 2.247 mol L^{-1}; (*b*) 7316-fold.]

13.5 The sexidentate chelating agent $EDTA^{4-}$ [ethylenedinitrilotetraacetate, $(^-OCOCH_2)_2NCH_2CH_2N(CH_2COO^-)_2$] can be regarded as being formed by joining together two *N*-methyliminodiacetate units [$MIDA^{2-}$, $(^-OCOCH_2)_2NCH_3$] through the methyl groups; MIDA itself is an effective terdentate chelating agent. Suppose we have a solution containing 2 mol L^{-1} $MIDA^{2-}$ and 1 mol L^{-1} $EDTA^{4-}$, in other words, the number of potential ligand atoms (O or N) associated with each of these chelating agents is the same.

(*a*) How would a small amount of Ni^{II} distribute itself among the forms Ni^{2+}(free), $Ni(MIDA)_2{}^{2-}$, and $Ni(EDTA)^{2-}$? (Assume, for the purposes of this exercise, that the MIDA and EDTA ions remain unprotonated.)

(*b*) How would this distribution change on pouring the solution into 1 $\times 10^9$ times its volume of pure water? What does this tell you about the effectiveness of chelating agents in very dilute solutions?

	Stability constants	
	log K_1	*log* K_2
$Ni^{2+} + MIDA^{2-}$	8.73	7.22
$Ni^{2+} + EDTA^{4-}$	18.56	

[*Answer:* (*a*) $1 : 3.6 \times 10^{16} : 3.6 \times 10^{18}$.]

13.6 Suppose we have two solutions of Ni^{2+} and of Fe^{3+} in water, each 0.001 *M* in metal ion and 0.1 *M* in EDTA. If the pH of the solutions is adjusted to 12, will either or both of the metal ions be precipitated as their hydroxides or stay in solution as the EDTA complexes? [At pH 12, $H_4(EDTA)$ can be regarded as completely dissociated to $EDTA^{4-}$.] The relevant data are as follows:

For $M^{n+} + EDTA^{4-} = M(EDTA)^{(4-n)-}$, $\log K = 18.6$ for Ni^{2+}
$= 25.1$ for Fe^{3+}.

The solubility products of $Ni(OH)_2$ and $Fe(OH)_3$ are 6.5×10^{-18} and 1×10^{-36}, respectively.

13.7 The pK_a of aqueous iron(III) ion in very dilute solution at 25 °C is 2.17. On the basis of this information alone, estimate (*a*) the minimal H^+ concentration required to ensure that 99% of the iron(III) is present as the hexaaqua complex, and (*b*) the $[H^+]$ at which you would expect $Fe(H_2O)_6{}^{3+}$ and $Fe(H_2O)_5OH^{2+}$ to be present in equal concentrations. (*c*) What factors (polymerization would be one) could interfere with the accuracy of the predictions, and how would you attempt to minimize their effects?
[*Answers:* (*a*) 0.67 mol L^{-1}; (*b*) 6.76×10^{-3} mol L^{-1}.]

13.8 Use the thermodynamic data of Appendix C to derive stability constants β_6 for the ferrocyanide and ferricyanide ions at 25 °C and infinite dilution.
[*Answers:* 3.5×10^{45}, 2.8×10^{52}.]

References

1. J. Burgess, "Metal Ions in Solution." Ellis Horwood, Chichester, 1978.
2. J. P. Hunt, "Metal Ions in Aqueous Solution." Benjamin, New York, 1963.
3. J. O'M. Bockris and A. K. N. Reddy, "Modern Electrochemistry," Vol. 1. Plenum, New York, 1973.
4. R. A. Robinson and R. H. Stokes, "Electrolyte Solutions," 2nd Ed. (revised). Butterworth, London, 1965.
5. F. A. Cotton and G. Wilkinson, "Advanced Inorganic Chemistry," 5th Ed., Chapter 2. Wiley (Interscience), New York, 1988.
6. F. Basolo and R. G. Pearson, "Mechanisms of Inorganic Reactions," 2nd Ed. Wiley, New York, 1967.
7. L. G. Sillén and A. E. Martell, "Stability Constants," Special Publication 17. Chemical Society, London, 1964, and Supplement (Special Publication 25), 1971. [An updated electronic version of the compilations, "IUPAC Stability Constants Database," is available from G. and L. D. Pettit, Academic Software, Sourby Old Farm, Otley, Yorkshire LS21 2PW, U. K. (E-mail: acadsoft@bcs.org.uk).]

8. A. E. Martell and R. M. Smith, "Critical Stability Constants," Vols. 1–4, Suppl. 1 and 2. Plenum, New York, 1974–1988.

9. J. R. Hart, EDTA-type chelating agents in everyday consumer products. *J. Chem. Educ.* **61**, 1060–1061 (1984); J. R. Hart, *J. Chem. Educ.* **62**, 75–76 (1985).

10. T. Emery, Iron metabolism in humans and plants. *Am. Sci.* **70**, 626–632 (1982).

11. D. A. Labianca, On the nature of cyanide poisoning. *J. Chem. Educ.* **56**, 788–791 (1979).

12. (*a*) B. H. Carroll, G. C. Higgins, and T. H. James, "Introduction to Photographic Theory: The Silver Halide Process." Wiley, New York, 1980; (*b*) G. Haist, "Modern Photographic Processing," Vols. 1 and 2. Wiley, New York, 1979; (*c*) L. Stroebel, J. Compton, I. Current, and R. Zakia, "Photographic Materials and Processes." Focal Press, Boston, Massachusetts, 1986; (*d*) G. T. Austin, "Shreve's Chemical Process Industries," 5th Ed., Chapter 23. McGraw-Hill, New York, 1985; (*e*) G. T. Eaton, "Photographic Chemistry," 2nd Ed. Morgan and Morgan, Dobbs Ferry, New York, 1965 (written for lay readers); (*f*) S. F. Ray, "Photographic Chemistry and Processing." Butterworth-Heinemann (Focal Press), Oxford, 1994 (a dictionary-style handbook).

13. C. F. Baes and R. E. Mesmer, "The Hydrolysis of Cations." Wiley (Interscience), New York, 1976.

14. C. Steelink, Humates and other natural organic substances in the aquatic environment. *J. Chem. Educ.* **54**, 599–603 (1977).

Chapter 14

Water Conditioning

14.1 Importance of Water Treatment

WATER IS a vital resource for industry and agriculture, as well as for domestic use. The ever-increasing demands for water are outstripping the supply in many places where abundant water was once taken for granted. It is therefore is becoming increasingly important to recycle water whenever possible, to ensure that water returned to rivers or lakes is fit for reuse by someone else (as well as safe for aquatic life), and to learn how to make use of impure but abundant resources such as brackish water or even seawater.[1–11] The value of the water treatment market has been estimated at $30 billion annually, worldwide.

Industry has huge demands for water, but its net consumption is actually relatively small. Some 70% of the water requirement is for cooling. Since such water is simply discharged after use, it may cause thermal pollution but usually it does not cause chemical pollution of natural waterways. Furthermore, much industrial water is now recycled within the plant, so that the gross intake-to-use ratio in North American pulp and paper mills, for example, is about 5:1 and rising, while the petrochemical industry should reach a ratio of 40:1 by the beginning of the twenty-first century. Agricultural irrigation, on the other hand, leads to large water losses through evaporation, and the water that does return to waterways will contain high concentrations of whatever salts the water originally contained or were leached from the soil, together with fertilizer, herbicide, and pesticide residues. Similarly, although soils that have become too saline can be improved by backflushing with water, the discharged salt solutions may cause unacceptable pollution of rivers or lakes.

Therefore, the objectives of water treatment are (*a*) to prepare available water for use in boilers, chemical processes, and cooling systems, for laundry, and for domestic consumption,[1–4] and (*b*) to clean up wastewater for discharge into natural waterways.[5–10] The appropriate treatment of wastewater prior to discharge is highly dependent on how the water was used, that is, on what specific contaminants were introduced. Since the number of possible contaminants is vast, a comprehensive discussion of wastewater treatment is beyond the scope of this book. Instead, we focus on treatment of water *intake* for typical contaminants: suspended and colloidal matter, dissolved solids, dissolved gases, bacteria and algae.

14.2 Suspended and Colloidal Matter

Suspended matter down to bacterial size can best be removed by subsidence in settling ponds and by skimming off floating material such as grease, which may form a paste that coats heat-exchange surfaces. Filtration would, no doubt, do a better job (see later), but on a large scale it is not always cost-effective (besides, filters tend to clog).

Colloidal suspensions (typically, of fine particles of clay minerals smaller than 10 μm) and emulsions (i.e., colloidal solutions of liquid in liquid, such as oil in water) cannot be broken by filtration; however, because colloidal particles are electrically charged, they can be brought out of solution by coagulation with a highly charged counterion (Section 13.6). Each charged colloidal particle is surrounded by a *diffuse electrical double layer* of low-charged counterions; these double layers resist interpenetration and so prevent coagulation of the colloid. The thickness of the diffuse electrical double layer will be greatly reduced if the counterions are highly charged, and coagulation is then facilitated.

The force keeping the colloidal particles apart is called the ζ (zeta) potential, and for colloidal particles in natural waters ζ ranges from approximately -14 to -30 mV at pH 5–8. Coagulants reduce this potential, and coagulation occurs as the ζ potential approaches zero; too much coagulant, however, will result in a significantly *positive* ζ potential and consequent redispersal of the colloid. The correct amount of coagulant to be added should therefore be determined by measuring the ζ potential, typically by studying the motion of colloidal particles under an applied electric field, using a microscope.

Coagulation alone may not produce a precipitate that will separate by settling. It is usual to employ a *flocculant*, a substance that can form bridges between the coagulated particles and create an open-structured, settleable, hydrolytic precipitate or *floc*. Thus, clay mineral particles generally are negatively charged in water, through dissociation of some alkali metal cations from the anionic aluminosilicate framework (Section 7.4), and can be coagulated with highly charged cations such as Al^{3+} or Fe^{3+} and precipitated in an $Al(OH)_3$ or $Fe(OH)_3$ floc. Usually, alum [$KAl(SO_4)_2 \cdot 12H_2O$] is added to the water to be treated, whereupon it hydrolyzes and forms a cationic floc that coagulates the anionic clay colloid. Alternatively, lime and $FeSO_4{\cdot}7H_2O$ (*copperas*) may be added to the aerated water to form a mixed $Fe(OH)_2/Fe(OH)_3$ floc. Sodium aluminate $\{Na[Al(OH)_4]\}$ may be used where an anionic floc is required. Oil-in-water emulsions can usually be broken by addition of alum, iron(III) salts, or mineral acids. Water-in-oil emulsions are broken with mineral acids or organic polymers (polyelectrolytes or polyalcohols).

Filtration can remove fine suspended solids and microorganisms, and "microfiltration" membranes of cellulose acetate or polyamides are available that have pores 0.1–20 μm in diameter. Clogging of such fine filters is an ever-present problem, and it is usual to pass the water through a coarser conventional filter first. "Ultrafiltration" with membranes having pores smaller than 0.1 μm requires application of pressures of a few bars; to keep the membrane surface free of deposits, water flows *parallel* to the membrane surfaces, with only a small fraction passing through the membrane. The membranes typically consist of bundles of hollow cellulose acetate or polyamide fibers set in a plastic matrix. Ultrafiltration bears some resemblance to reverse osmosis technology, described in Section 14.4, with the major difference that reverse osmosis can remove *dissolved* matter, whereas ultrafiltration cannot.

14.3 Origin and Effects of Dissolved Solids

The most significant dissolved ionic solid in most natural freshwater is *calcium carbonate* (actually present mainly as the bicarbonate), and so, for the purpose of water treatment, it is convenient to express the concentration of any ionic solute as a $CaCO_3$ equivalent. The formula weight of the solute is divided by the valence and by the number of its cations—for $CaCO_3$ itself, this comes to 50.0—and the concentration (usually in mg kg^{-1}, ppm) is converted to an equivalent concentration of $CaCO_3$ on this basis.[2]

The impact of water hardness due to calcium or magnesium ions on detergents was explained in Section 7.3.1 The source of most Ca^{2+} and Mg^{2+} in hard water is the dissolution of limestone ($CaCO_3$) or dolomite [$CaMg(CO_3)_2$]. Magnesium carbonate is fairly soluble (1.26 mmol L^{-1} at ambient temperature), but $CaCO_3$ is much less so (0.153 mmol L^{-1}). However, if the water contains dissolved CO_2 (as indeed it will if it is exposed to the air; see Exercise 14.9), the relatively freely soluble $Ca(HCO_3)_2$ forms, and the limestone slowly dissolves away:

$$CaCO_3(s) + H_2O(l) + CO_2(aq) \rightleftharpoons Ca(HCO_3)_2(aq). \tag{14.1}$$

Reaction 14.1 is reversible, however, and, if the solution is boiled, the CO_2 is swept out in the steam, and $CaCO_3$ is reprecipitated until its own solubility is reached. Thus, part of the Ca^{2+}-derived hardness is removed by boiling;

this is called *temporary hardness.* The deposit of $CaCO_3$ may be seen as "*fur*" in domestic kettles or as "*scale*" (not to be confused with the oxide scales of Section 5.6) or sludge in boilers. The deposit is highly undesirable in that it reduces heat transfer efficiency and may even cause blockages of tubes or valves. Boiler sludge has an unfortunate tendency to form intractable pastes with any oil in the feedwater. Frequent blowdown of boilers, as well as descaling, is therefore necessary to minimize sludge accumulation, unless the feedwater is appropriately conditioned.

Boiling does *not* remove magnesium salts, *nor* $CaCO_3$ at its own solubility level, *nor* "noncarbonate" calcium (i.e., Ca^{2+} that is counterbalanced by Cl^- or another anion of which the Ca^{2+} salt is freely soluble). The hardness that remains after boiling is called *permanent hardness.* Permanent hardness plus temporary hardness is called the *total hardness.*

It is incorrect to regard hard water as containing specific salts such as $Ca(HCO_3)_2$, $MgCO_3$, $MgSO_4$, and $CaCl_2$ because these are present in solution as mixtures of ions. The principal ions present in typical surface waters are

Cations: Na^+, (K^+), Mg^{2+}, Ca^{2+}, Fe^{2+}/Fe^{3+}

Anions: $CO_3{}^{2-}/HCO_3{}^-/CO_2$, Cl^-, $SO_4{}^{2-}$.

Others, such as Al^{3+} (which hydrolyzes), F^- (which precipitates as poorly soluble CaF_2), $HPO_4{}^{2-}$, $NO_3{}^-$, and $NH_4{}^+$, are minor components, usually not exceeding 1 ppm (i.e., 1 mg per kilogram of water). Aluminum ion is not usually significant in surface waters, except as a consequence of acid rain or the use of alum or sodium aluminate in the treatment of wastewater. Iron is fairly soluble as Fe^{2+} salts, which often occur in substantial concentrations in well waters. At pH values typical of surface waters, iron(II) is oxidized by the air to hydrolyzed iron(III) and hydrolytic polymers of iron(III). Iron in water imparts a bad taste, causes discoloration of appliances, and leaves iron oxide/hydroxide deposits in pipes or heat exchanges. Iron(III) can be retained in significant concentrations in surface waters through complexation by humic acids (Section 13.6).

Nitrate ion from fertilizers and treated sewage has reached disquietingly high concentrations in the water supplies of some countries of the European Community, which now sets an allowable upper limit of 50 mg L^{-1} for $NO_3{}^-$ in potable waters; higher nitrate levels can cause methemoglobinemia ("blue baby syndrome") in bottle-fed infants, and they can also lead to eutrophication of lakes and streams if phosphorus is not the limiting nutrient (Section 7.7). The problem of controlling nitrate pollution is proving to be refractory.[12]

Ionic solids apart, the most important dissolved solid in natural waters is *silica*, SiO_2. The solubility of silica depends strongly on the solid-state form;

for example, it is 120 ppm as amorphous SiO_2, but only 10 ppm as quartz, at 20 °C and pH 6. The solubility is greater at higher pH (Section 7.6) and also rises markedly with increasing temperature, so that quartz dissolution can cause problems in steam-injection oil recovery (Section 7.2). Silica is also appreciably soluble in high-pressure steam, with the solubility increasing essentially in direct proportion to the density of the steam. Consequently, in steam turbines, silica can be deposited on the delicately balanced blades as the steam expands. It is therefore critically important that feedwater for turbine boilers be substantially free of dissolved silica.

The total dissolved solids (TDS) content of rivers varies widely, often even within the same stream.[2] For example, Colorado River water typically has a TDS level of 92 ppm at Hot Sulfur Springs, Colorado, but when delivered to the city of Los Angeles, California, the TDS is 661 ppm, with 523 ppm ($CaCO_3$ equivalent) dissolved electrolytes. Surface waters drained from igneous or metamorphic rocks typically have low TDS. For example, the Saguenay river in Québec, coming off the granitic Canadian Shield, has 35 ppm TDS, whereas the Missouri River at Great Falls, Montana, coming down from the sedimentary eastern Rocky Mountains where there is much limestone and dolomite, has 234 ppm TDS, including $CaCO_3$ equivalents of 100 ppm Ca, 49 ppm Mg, and 47 ppm Na (total electrolyte 196 ppm $CaCO_3$ equivalent). All such concentration data, however, vary substantially with the seasons; for example, rapid runoff from melting snow or a downpour of rain will tend to dilute the minerals contributed by springs. Clearly, the anion and cation concentrations must balance: for example, in the domestic water of London, England, 270 ppm total (Ca^{2+} + Mg^{2+}) hardness and 30 ppm Na^+ are matched by 250 ppm "alkalinity" ($CO_3{}^{2-}$ + $HCO_3{}^-$), 20 ppm Cl^-, and 30 ppm sulfate/nitrate (all these concentrations are $CaCO_3$ equivalents).

The TDS of lake waters can be high because of evaporation, as in the Great Salt Lake of Utah or the Dead Sea (Israel/Jordan). Well waters often contain high concentrations of electrolytes leached from the rocks—notably iron salts, which are a major nuisance—and the highly saline waters associated with oil- or gas-bearing formations are frequently better described as brines.

Conveniently, the TDS content of public water supplies parallels the total electrolyte concentration, so that both the TDS and total electrolyte concentrations can be gauged, at least approximately, by measuring the *conductivity* of the water; 1.00 $\mu S\ cm^{-1}$* corresponds to 0.65 ppm TDS.[2] Calcium and magnesium contents were traditionally determined by titration with $EDTA^{4-}$ at pH 10, at which both Ca^{2+} and Mg^{2+} are complexed, and then in a fresh sample at pH 12–13, at which $Mg(OH)_2$ precipitates

*The conductivity of pure water at 25 °C is 0.0551 $\mu S\ cm^{-1}$, attributable to self-ionization.

and only the Ca^{2+} is measured. Instrumental methods such as atomic absorption spectrometry are now routinely used for water analysis.

14.4 Treatment for Dissolved Solids

The traditional way to free water of dissolved solids is to distil it, either at atmospheric pressure or by multistage flash evaporation at reduced pressure. Distillation removes virtually all solutes but is wasteful of energy unless the low grade heat can be economically recovered from the condensers. Flash evaporation is attractive in countries such as Saudi Arabia where energy is inexpensive and the only plentiful source of water is the sea, but problems usually arise with deposition of $CaCO_3$, $Mg(OH)_2$, and $CaSO_4$ scales.

Alternatively, potable water can be extracted from seawater by freezing; salts, which depress the freezing point of water, remain in the liquid phase. Generally, though, it is more practical to remove the relatively small amount of solutes (typically, 0.02% for river water) from the great excess of water, rather than vice versa. Seawater is an exceptional case, with about 3.5% dissolved solids. Water *softening* is concerned primarily with removal of Ca^{2+} and Mg^{2+}, but for some purposes removal of all dissolved solids (*deionization* or *demineralization*) is necessary.

14.4.1 Removal of Inorganic Solutes

The most widely used techniques for removing dissolved inorganic solids are *boiling*, addition of *washing soda*, *lime–soda softening*, *complexation*, *sodium ion exchange*, *demineralization*, *reverse osmosis*, *electrodialysis*, *adsorption* onto suspended solids, and *aeration*.

Boiling. Boiling will remove temporary hardness, but the resulting precipitation of $CaCO_3$ may be precisely what one is trying to avoid.

Addition of washing soda. Washing soda ($Na_2CO_3{\cdot}10H_2O$) precipitates most of the $MgCO_3$ and $CaCO_3$ by the common ion effect, that is, by forcing up the concentration of ${CO_3}^{2-}$(aq):

$$Ca^{2+}(aq) + {CO_3}^{2-}(aq) \rightleftharpoons CaCO_3(s) \tag{14.2}$$

$$[Ca^{2+}][{CO_3}^{2-}] = K_{sp}. \tag{14.3}$$

This is fairly costly, as substantial excesses of Na_2CO_3 are needed, and furthermore the solution is left quite alkaline (and hence likely to transport silica):

$${CO_3}^{2-} + H_2O \rightleftharpoons {HCO_3}^{-} + OH^{-}. \tag{14.4}$$

Lime–soda softening. Lime–soda softening involves removal of the temporary hardness by adding the calculated amount of hydrated lime (Section 11.1):

$$Ca(HCO_3)_2(aq) + Ca(OH)_2(aq) \rightarrow 2CaCO_3(s) + 2H_2O. \qquad (14.5)$$

One can stop at this point (*selective calcium softening*) or add enough further lime to precipitate the Mg^{2+} as $Mg(OH)_2$. In contrast to the calcium analogs, $MgCO_3$ is fairly soluble but $Mg(OH)_2$ is not (see Exercises 14.1 to 14.3):

$$Mg^{2+}(aq) + Ca(OH)_2(aq) \rightarrow Mg(OH)_2(s)_{\downarrow} + Ca^{2+}(aq). \qquad (14.6)$$

Finally, soda ash or washing soda is added to remove the Ca^{2+} put *into* the solution by reaction 14.6, as well as any "noncarbonate" calcium:

$$Ca^{2+}(aq) + Na_2CO_3(aq) \rightarrow CaCO_3(s)_{\downarrow} + 2Na^{+}(aq). \qquad (14.7)$$

So, we need to add $Ca(OH)_2$ equivalent to the temporary hardness *plus* the magnesium hardness (which is just the total hardness, if noncarbonate Ca^{2+} is absent), and Na_2CO_3 equivalent to the permanent (i.e., total minus temporary) hardness. Clearly, if lime–soda softening is to be effective, accurate analyses for Ca^{2+}, Mg^{2+}, and temporary hardness are needed, and the lime must be accurately weighed out accordingly.

There are several disadvantages to lime–soda softening. It leaves the water still saturated with 31 ppm $CaCO_3$, unless extra Na_2CO_3 is added. Even if no extra soda is added, the solution will be quite alkaline, causing dissolution of silica when hot and also caustic embrittlement of metals in nooks and crannies of pipes or boilers where water so treated may tend to become concentrated by evaporation. Finally, although colloidal $CaCO_3$ and $Mg(OH)_2$ carry opposite charges (negative and positive, respectively) and so tend to coagulate of their own accord, precipitation may need to be helped along by addition of sodium aluminate, powdered magnesium oxide, or recycled sludge, but, even then, filtration may be necessary to remove turbidity. Nevertheless, lime–soda softening is widely used in industry, especially where large volumes of water have to be treated.

Complexation. Complexation involves the addition of reagents that give *water-soluble complexes* with Ca^{2+} and Mg^{2+}. Sodium triphosphate (marketed as Calgon, etc.; the complexing anion is $P_3O_{10}{}^{5-}$), $Na_3(NTA)$, or $Na_2H_2(EDTA)$ may be used. Nitrilotriacetate ion and the less thermally stable $EDTA^{4-}$ ion are used to suppress boiler sludge formation, and so reduce blowdown frequency, but they should not be used if fittings of nickel, copper, or their alloys (other than stainless steel) are present, since Ni^{2+} and Cu^{2+} are strongly complexed by NTA^{3-} or $EDTA^{4-}$ and corrosion of these metals will be facilitated (Section 16.5).

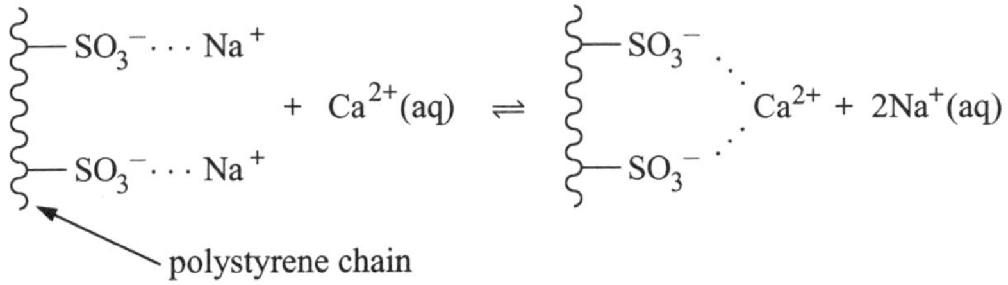

$$\left\{\begin{matrix} —SO_3^- \cdots Na^+ \\ —SO_3^- \cdots Na^+ \end{matrix}\right. + Ca^{2+}(aq) \rightleftharpoons \left\{\begin{matrix} —SO_3^- \\ —SO_3^- \end{matrix}\right\} Ca^{2+} + 2Na^+(aq)$$

polystyrene chain

Figure 14.1 Exchange of aqueous Ca^{2+} for Na^+ within a bead of a cation-exchange resin in hard water.

Sodium ion exchange. Sodium ion exchange on zeolites (Section 7.3) or on synthetic organic cation-exchange resins such as Dowex-50 (a sulfonated polystyrene; Fig. 14.1), in most circumstances, is superior to the above softening methods.[13] The exchange process favors binding of Ca^{2+} or Mg^{2+} over Na^+ in the solid resin phase:

$$RNa_2(s) + M^{2+}(aq) \rightleftharpoons RM(s) + 2Na^+(aq). \qquad (14.8)$$

Thus, if hard water is passed down a column of Na^+-form resin or zeolite, essentially all the Ca^{2+} and Mg^{2+} ions in the water are replaced by Na^+ until the ion exchange capacity of the resin or zeolite is used up. Furthermore, the resulting softened water is of neutral pH (unless it was alkaline to start with). Reaction 14.8 is reversible, however, and the ion exchanger can be restored to its Na^+ form by backflushing with sufficiently concentrated brine (NaCl solution). Any iron present in the hard water is likely to be oxidized by the air to Fe^{3+}, which is very strongly absorbed by ion-exchange resins or zeolites and cannot easily be removed. Accordingly, unless one can afford to throw away some ion exchanger periodically, any iron should be removed from the water before ion-exchange softening (described later, under *aeration*).

Other problems with ion exchangers include coating of the resin beads or zeolite particles with suspended matter from turbid water (pretreatment with a coagulant may be necessary) or algal growths (chlorination of the water may be required). Zeolites may cause significant silica carryover and should not be used to treat boiler water for steam turbines. Finally, although Ca^{2+} and Mg^{2+} are objectionable in boiler or laundry operations, they are necessary nutrients in the human diet. Furthermore, excessive consumption of Na^+ can contribute to hypertension and other blood circulatory problems.* In Canada, for example, the incidence of heart disease and related health problems is lower in areas where the water supply is

*With hypertensive individuals in mind, the European Community recommends that the sodium concentration in drinking water not exceed 20 mg L^{-1}.

hard. Accordingly, in homes where a sodium ion exchanger is installed to provide soft water for washing purposes, a separate supply of hard water should be available for drinking and cooking.

Demineralization. Demineralization (deionization) is possible if the hard water is first passed down a column of *cation* exchanger in the H^+ form ("RH_2")

$$RH_2(s) + M^{2+}(aq) \rightleftharpoons RM(s) + 2H^+(aq) \tag{14.9}$$

and then through an *anion* exchanger (usually a hydrocarbon polymer containing quaternary ammonium groups, R_4N^+) in the OH^- form,

$$R_4N^+OH^-(s) + X^-(aq) \rightleftharpoons R_4N^+X^-(s) + OH^-(aq). \tag{14.10}$$

There are, of course, equal numbers of anionic and cationic charges in the solution, and hence equal numbers of H^+ and OH^- ions are released, forming water:

$$H^+(aq) + OH^-(aq) \rightarrow H_2O(l). \tag{14.11}$$

Since the predominant anion in hard water is HCO_3^-, anion-exchange capacity can be conserved by purging the acidic water coming out of the cation exchanger with air or by vacuum degassing, so sweeping out most of the HCO_3^- as CO_2:

$$H^+(aq) + HCO_3^-(aq) \rightleftharpoons H_2CO_3(aq) \rightarrow CO_2(g) + H_2O(l). \tag{14.12}$$

Anion exchange is an effective way of removing nitrate and nitrite ions, should these be present at objectionable levels. If a strongly basic anion-exchange resin $R_4N^+OH^-$ is used, silica will also be removed as $(HO)_3SiO^-$ (see Section 7.6), leading to water of low total dissolved solids content:

$$R_4N^+OH^-(s) + Si(OH)_4(aq) \rightleftharpoons R_4N^+(HO)_3SiO^- + H_2O. \tag{14.13}$$

Hydrogen sulfide is similarly removed:

$$R_4N^+OH^-(s) + H_2S(aq) \rightleftharpoons R_4N^+SH^-(s) + H_2O. \tag{14.14}$$

The chief drawbacks to demineralization are the rather high cost of the resins (especially the anion exchanger) and the need to regenerate the cation and anion exchangers with relatively expensive H_2SO_4 and NaOH solutions, respectively.

Reverse osmosis. Reverse osmosis involves manipulation of the *osmotic pressure* of a solution. If a solution containing n moles of solute particles in a volume V (in m^3) of solution is separated from pure solvent by a semipermeable membrane (i.e., a membrane through which solvent molecules, but not solute particles, can pass), an osmotic pressure Π pascals develops across the membrane, the pure solution being the high pressure side (Fig. 14.2a).

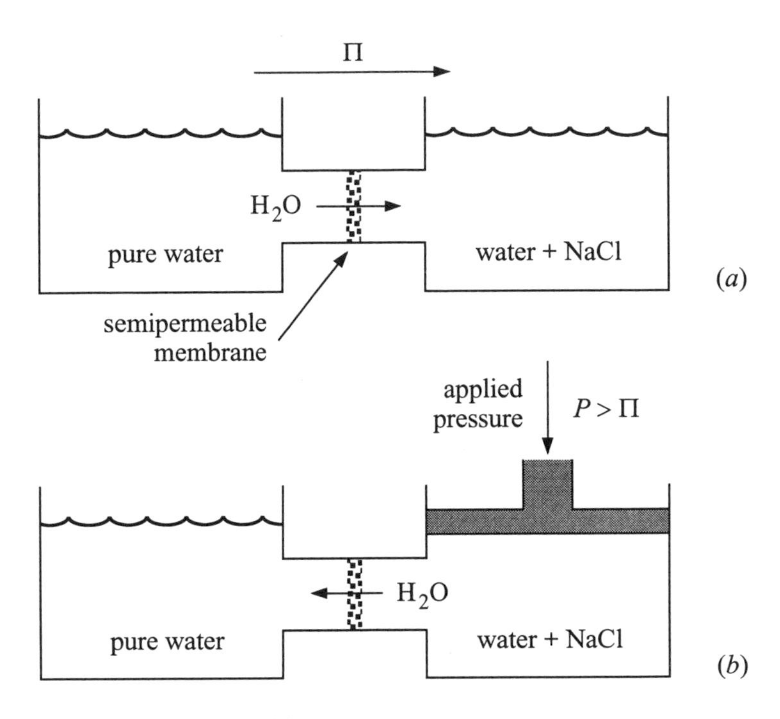

Figure 14.2 (*a*) Osmosis. (*b*) Reverse osmosis.

This is because the pure solvent tends to dilute the solution; the solute, however, cannot pass through the membrane to dilute itself. Ideally, the osmotic pressure is given by

$$\Pi = \frac{nRT}{V} = cRT \tag{14.15}$$

where T is the temperature in kelvins, Π is in pascals, the gas constant R = 8.3145 J K^{-1} mol^{-1}, and c is the solute particle concentration in moles per *cubic meter*. Note that, if the solute is an electrolyte M_mX_x, there will be $(m + x)$ moles of particles per mole of M_mX_x.

The net flow of solvent into the solution will continue unless a pressure equal to Π is applied in the reverse direction (for example, as a result of the buildup of a hydrostatic head on the solution side of the membrane as the dilution process proceeds). Indeed, if a pressure greater than Π is deliberately applied on the solution side, the net flow of solvent will be out of the solution and into the pure solvent (Fig. 14.2b). This is *reverse osmosis*, which obviously can be used to derive pure water from solutions. The semipermeable membrane must be able to withstand a substantial pressure differential, as well as to resist passage of ions and other solutes effectively. Membranes based on cellulose acetate are adequate for use with brackish water, but polyamides are preferred with seawater. Note that the pressurized brine is further concentrated and hence Π is increased by the reverse osmosis process. In practice, then, the plant is typically run at a pressure of about 2Π (6.0 MPa; cf. Exercise 14.4), and the brine is run off after only about 20% of the water content has been extracted.

Reverse osmosis is now extensively used to reduce salt concentrations in brackish waters and to treat industrial waste water, for example, from pulp mills. Reverse osmosis has also proved economical (the cost can be as low as about $1 per 1000 liters) for large-scale desalination of seawater, a proposition of major interest in the Middle East, where almost all potable water is now obtained by various means from seawater or from brackish wells. Thus, at Ras Abu Janjur, Bahrain, a reverse osmosis plant converts brackish feedwater containing 19,000 ppm dissolved solids to potable water with 260 ppm dissolved solids at a rate of over 55,000 m^3 per day, with an electricity consumption of 4.8 kilowatt hours per cubic meter of product. On a 1000-fold smaller scale, the resort community on Heron Island, Great Barrier Reef, Australia, obtains most of its fresh water from seawater (36,000 ppm dissolved salts) directly by reverse osmosis, at a cost of about $10 per 1000 liters.

Nonelectrolytes including dissolved silica can be also be removed by reverse osmosis, which can reduce colloidal silica levels to 0.1 ppm.

Electrodialysis. Electrodialysis[3,10] utilizes the fact that cations, but not anions, can readily pass through a cation-exchange membrane, while the reverse is true of anion-exchange membranes. If, for example, salt water is made to flow through a cell that is divided longitudinally into narrow sections by alternating cation- and anion-exchange membranes and is subjected to a transverse electric potential gradient, the Na^+ and Cl^- will become concentrated in alternate compartments and depleted in the others (Fig. 14.3).

Note that, for n pairs of anion and cation membranes, the passage of one faraday causes transfer of n moles of NaCl, but only one mole NaCl is discharged at the electrodes. Thus, by using a large number of membrane pairs, electrolytic production of unwanted NaOH, Cl_2, and H_2 can be minimized. Ordinary polystyrene ion-exchange resins (described earlier) will perform satisfactorily in electrodialysis, although fluorocarbon membrane materials such as Nafion (Section 11.4.1) are preferred if the budget permits. For best effect, the process is normally repeated in several stages. Electrodialysis can help recover process chemicals from dilute wastes, while at the same time purifying the excess water for discharge; for example, NaOH and H_2SO_4 are regenerated in the cathode and anode compartments when waste Na_2SO_4 solution is electrodialyzed. (This is called *salt splitting.*)

Adsorption onto suspended solids affords an inexpensive way of remov-

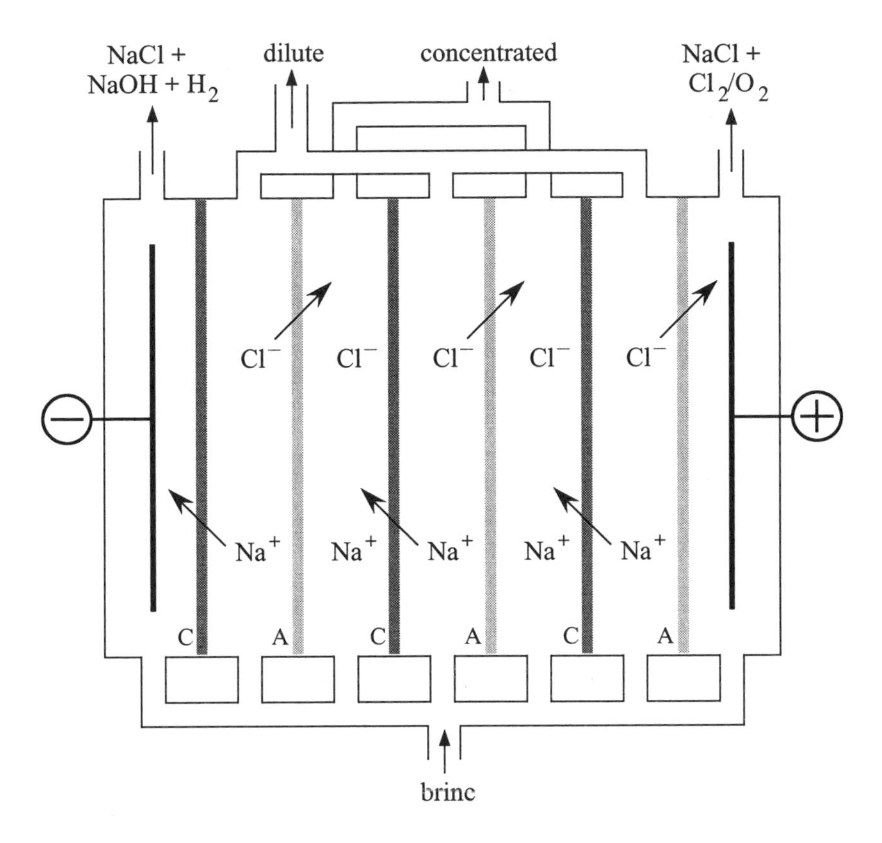

Figure 14.3 Schematic drawing of an electrodialysis cell. A, Anion-exchange membranes; C, cation-exchange membranes.

ing dissolved or colloidal silica from boiler feedwater. Silica may be adsorbed on solid magnesium oxide or hydroxide, or on crushed dolomite. This is *not* a stoichiometric reaction involving the whole solid phase, but occurs to a variable extent on the surface only. The extent of adsorption is adequately expressed by the *Freundlich adsorption isotherm**:

$$\frac{\text{mass SiO}_2\text{ adsorbed}}{\text{mass MgO (etc.)}} = k[\text{SiO}_2(\text{aq})]^{1/n} \quad (14.16)$$

where k and n are empirical constants at a given temperature. This proce-

*Although the Freundlich adsorption isotherm is empirical, it implies that the (negative) heat of sorption decreases exponentially with increasing coverage. Alternatives include the Langmuir adsorption isotherm, which assumes all sorption sites are equivalent, regardless of the extent of coverage, and the Frumkin isotherm, which assumes that it becomes linearly more difficult to adsorb molecules as surface coverage increases. See Bodek et al.[6]

dure can be used in conjunction with lime–soda softening (described earlier), since this results in a precipitate of $Mg(OH)_2$. Effective adsorption of silica onto the $Mg(OH)_2$ requires a temperature of at least 50 °C and preferably 100–110 °C (the *hot* lime–soda process), and it may be necessary to add $MgSO_4$ to get enough $Mg(OH)_2$.

Aeration. Aeration will remove iron(II) from most natural waters. It sweeps excess CO_2 out (so increasing the pH), and the Fe^{2+} then reacts relatively rapidly with excess oxygen to precipitate as $Fe(OH)_3$:

$$Fe^{2+}(aq) \xrightarrow[pH > 6]{O_2} Fe(OH)_{3\downarrow}. \qquad (14.17)$$

Iron(II) reacts only slowly with oxygen in more acidic conditions, and in any event the product iron(III) may then remain in solution. To ensure a high pH, lime may be added:

$$Fe^{2+}(aq) + 2OH^-(aq) \rightleftharpoons Fe(OH)_2(s) \xrightarrow[rapid]{O_2} Fe(OH)_3(s). \qquad (14.18)$$

If the iron is present as humic acid complexes,[11] these can be coagulated with alum (Section 14.2). Instead of trying to precipitate the iron, it may be better to keep it in solution, in which case it can be complexed with a chelating agent such as NTA^{3-} or $EDTA^{4-}$. As a last resort, Fe^{2+} or Fe^{3+} can be removed by cation exchange, but the absorption on the zeolite or resin is usually irreversible.

14.4.2 Removal of Organic Solutes

Organic matter in unpolluted freshwater is usually in the form of humic acids (Section 13.6), which may discolor drinking water, foul ion-exchange resins, transport toxic metal ions, or generate carcinogenic chlorocarbons if the water is chlorinated (see below). Humic compounds may be removed by coagulation. Low levels of organic solutes can be removed by sorption on activated charcoal.

In searching for chemical methods of destroying organic pollutants, chemists tend to look to *oxidative* methods, since the preferred end product is usually CO_2, but, as discussed below, special methods are usually necessary to bring about oxidation to acceptable products. In 1995, however, it was shown that elemental (zero-valent) metals such as ordinary iron filings can be effective in the *reductive* conversion of halocarbon pollutants such as trichloroethene in wastewater to innocuous hydrocarbons over a few days[14]:

$$2RCl + Fe(s) + 2H^+(aq) \rightarrow 2RH + Fe^{2+}(aq) + 2Cl^-(aq). \qquad (14.19)$$

As often happens, this is actually a rediscovery rather than new science, since reaction 14.19 has been known for some time in connection with the corrosion of iron. Alternative modes of action of the iron are corrosion to give hydrogen, which may then reduce the pollutant if a suitable catalyst is present, and reduction of the pollutant by the Fe^{2+}(aq) ion that is produced in either case. Wastewater containing chlorocarbons can be "remediated" by slow percolation through a bed of sand containing iron filings. The rise in pH implicit in reaction 14.19 may cause problems with precipitation of $Fe(OH)_2$, etc., but these are not insuperable. Some noxious inorganic pollutants are also reduced by iron; for example, carcinogenic chromium(VI) compounds (chromates or dichromates) from leather tanning or electroplating works are reduced to chromium(III), which, because of the rise in pH that accompanies the reduction, precipitates as the innocuous green gel $Cr(OH)_3$ (cf. Eq. 2.55).

Hoffmann and coworkers[15] developed some striking new techniques for the *oxidative* destruction of trace quantities of organic solutes in water. One is irradiation ("sonication") with *ultrasound*,[16] that is, sound of frequencies higher than can be heard by humans (>14 kHz). For example, sonication of water at 20 kHz and 75 W cm^{-2} creates small cavities in the water (diameter 0.17 mm) which, when they collapse, generate transient temperatures of about 5000 °C and pressures of around 50 MPa. At such temperatures, the supercritical water breaks down into ·H and ·OH radicals. The ·OH radicals are estimated to enter the solution at the rate of 3.5 μmol L^{-1} per minute; as discussed in Section 8.3.2, ·OH radicals are highly effective at oxidizing pollutants. As long as a cavity survives, organic solutes will tend to volatilize into it; consequently, when it collapses, the organic molecules are torn apart (pyrolyzed) by the extreme conditions and destroyed by the ·OH radicals. Higher frequencies generate more radicals but cause less pyrolysis.

Sonochemical processes are not very efficient in terms of moles of organic material reacting per unit time, but we are concerned only with low levels of objectionable pollutants such as CCl_4 and parathion, which have been shown in the laboratory to be completely destroyed after sonication for a few minutes. Ultrasonics can be combined with ozonolysis to eliminate organic pollutants. Pentachlorophenol, for example, is completely converted to CO_2 and HCl after 3–4 minutes of ozonolysis with ultrasound. Hoffmann also investigated destruction of aqueous organic pollutants by high voltage discharges, which give more intense shocks and higher temperatures than ultrasound (the *pulsed power plasma* process). For example, a 1 GW pulse of 300 kA destroyed more than half the aqueous dichloroaniline in a sample on a millisecond timescale; again, combination of this technique with ozonolysis destroyed all intermediates completely in about 0.2 ms.

We saw in Section 6.5.2 that ultraviolet irradiation of titanium dioxide promotes electrons into its conduction band and leaves reactive "holes" in the valence band with a band gap of 3 V (more, for very small TiO_2

particles). In aerated water, this potential is enough for the holes to oxidize water to $\cdot$OH radicals at the crystal surfaces, while the electrons, which in the short term are trapped as Ti^{3+} ions, go on to reduce O_2 to O_2^- or HO_2^- ions. All these species are highly reactive oxidants, and the $\cdot$OH radicals can attack otherwise unreactive pollutants such as chloroform or pentachlorophenol to start a sequence that results in oxidation to innocuous products[15,17]:

$$2CHCl_3 + O_2 + 2H_2O \xrightarrow[TiO_2]{UV} 2CO_2 + 6HCl, \qquad (14.20)$$

$$2C_6Cl_5OH + 7O_2 + 8H_2O \xrightarrow[TiO_2]{UV} 4HCO_2H + 8CO_2 + 10HCl. \qquad (14.21)$$

Hoffmann developed an ingenious fiber optic device that can bring light into otherwise inaccessible (e.g., underground) bodies of polluted water and photolyze organic contaminants on a TiO_2 coating on the surface of the quartz optical fibers.[15e]

14.5 Sewage Treatment

The processing of sewage is mainly microbiological, and the details lie outside the scope of this book. In outline, there are three levels to which sewage may be treated.

(*a*) *Primary treatment* involves mechanical removal of suspended solids and removal of colloidal matter by coagulation.

(*b*) *Secondary treatment* involves one of two biological processes that decompose (biodegrade) organic matter.[3,4,11] In the *trickling filter* method, the sewage percolates through a pebble bed on which the microorganisms that feed on the organic content accumulate. In the *activated sludge process*, which is preferred nowadays, the sewage is mixed with some sludge from a previous batch (to provide a source of aerobic bacteria) and is oxygenated with air or pure oxygen to the degree required by the measured *biochemical oxygen demand* (BOD). The solids are then separated, dewatered, and incinerated or otherwise disposed of. Ammonia and other nitrogen species tend to become oxidized to nitrite or nitrate by the aerobic bacteria; if the discharge of nitrate ions is unacceptable, an additional denitrification step, using anoxic bacteria, may be necessary. In any event, the issuing water is often chlorinated (Section 14.7) to kill escaping microorganisms before it is released into rivers or lakes.

In principle, the dewatered sludge is a valuable fertilizer, but it usually contains too many heavy metal ions to be safe for use in growing foodstuffs (it is often used to fertilize golf courses and the like). Methane is a by-product of sewage treatment that can be used to fuel generators for the sewage plant's electrical requirements. The use of oxygen, rather than air, in the activated sludge process keeps the working volume down, and the treatment can then be carried out odor-free in closed tanks.

(*c*) *Tertiary treatment* refers to removal of specific pollutants from the discharge of secondary treatment plants. Most commonly, this means removal of phosphates, which can be done by adding lime to precipitate hydroxyapatite (cf. Sections 7.7 and 9.6):

$$5Ca(OH)_2 + 3HPO_4{}^{2-} \rightarrow Ca_5(PO_4)_3OH_{\downarrow} + 6OH^- + 3H_2O. \quad (14.22)$$

Alternatively, addition of alum precipitates $AlPO_4$, while $FeCl_3$ brings down $FePO_4$. Another possibility is to rely on bacteria to consume the excess phosphate and any ammonia in the treated sewage.

14.6 Dissolved Gases

Henry's law states that the equilibrium solubility of a gas X in a liquid is directly proportional to the partial pressure P_X of X:

$$[X] = K_H P_X. \quad (14.23)$$

Furthermore, the solubility of gases in liquids decreases with rising temperature. Accordingly, the concentration of a dissolved gas in water can be reduced simply by vacuum degassing or by heating. Alternatively, one can resort to specific chemical methods of removal.[18, 19]

Carbon dioxide is quite soluble in cold water, largely because it hydrates to form carbonic acid

$$CO_2(aq) + H_2O(l) \rightleftharpoons H_2CO_3(aq) \quad (14.24)$$

and mineral springs often contain much more dissolved CO_2 than would be present if the water were in equilibrium with air (0.5 ppm). Carbon dioxide concentrations in excess of 10 ppm can cause greatly accelerated corrosion of metals (Chapter 16), but thorough aeration will reduce the CO_2 content to acceptable levels. Carbon dioxide in steam can be neutralized with organic amines.

Oxygen is soluble in water to the extent of 9.4 ppm from air at 100 kPa and 20 °C, and O_2 is the oxidant responsible for most metallic corrosion. Consequently, deaeration of water by purging with nitrogen or vacuum degassing may be desirable in some circumstances; this should not be undertaken without circumspection, since deoxygenation may cause activation of otherwise passive metals or cause cathodic areas to become anodic (Chapter 16). At high temperatures, aqueous oxygen is consumed quite rapidly by hydrazine or sodium sulfite (Section 16.7).

Hydrogen sulfide enters natural waters from decay of organic matter (e.g., in swamps), bacterial reduction of sulfate ion, or underground sour natural gas deposits. It can be removed by aeration, anion exchange (Eq. 14.14), or oxidation by chlorine to elemental sulfur:

$$Cl_2(g) + H_2S(aq) \rightarrow S(s) + 2HCl(aq). \qquad (14.25)$$

14.7 Bacteria and Algae

Bacteria in water are usually thought of in terms of human disease. Indeed, until quite late in the nineteenth century, disastrous outbreaks of water-borne diseases such as cholera, dysentery, and typhoid fever were common in the major cities of the world. The last outbreak of typhoid in the United Kingdom occurred in Croydon in 1937. Serious cholera epidemics still occur in some parts of the world; one that began in Peru in 1991 spread to several countries in the Americas, causing 391,000 cases of illness and 4000 deaths that year.

Bacteria are also responsible for destruction of wood, for example in cooling towers,[20] by breaking down the cellulose fibers. Certain bacteria derive their metabolic energy from the iron(II)–iron(III) redox cycle. These "iron bacteria" can proliferate to the extent that they block pipes. In any case, they will discolor water. In addition, objectionable growths of algae can occur in water tanks or circuits, given even minimal supplies of nutrients. Consequently, biocidal agents are widely used in the treatment of industrial, as well as municipal, water supplies.

Chlorination is the longest established and most economical means of disinfecting water on a large scale. Generally, less than 10 ppm Cl_2 will suffice to disinfect water. Chlorine is available at low cost as a by-product of caustic soda production (Section 11.4) and reacts quickly with water to give hypochlorous acid, HOCl (Section 12.2), which is the actual bactericidal agent. Alkaline solutions of sodium hypochlorite are less effective, inasmuch as HOCl penetrates bacterial cell walls much more readily than OCl^- ion, although the OCl^- will be hydrolyzed to HOCl when diluted with near-neutral water. Travelers in the tropics or other areas where drinking water is suspect are advised to filter the water, for example, through a wad of clean cloth, then to add 0.5 mL (~10 drops) of domestic bleach with 1% available chlorine* to each liter of filtered water, and finally to let the well-mixed water stand for about 30 minutes before drinking it. The water should smell slightly of chlorine after this time; if not, the process should be repeated. Alternatively, 0.2 mL of tincture of iodine (2.5%) can be substituted for the bleach, though bleach remains the safer choice.

*With higher bleach concentrations proportionally less is added, and twice the recommended amount should be used if the water remains cloudy after filtration.

Aqueous chlorine, however, reacts with other possible solutes such as H_2S (Eq. 14.25), NH_3 (giving chloramine, NH_2Cl), and organic matter, so the chlorination plant operator arranges for 0.2 to 0.5 ppm chlorine equivalent to remain in the water 5 minutes after treatment, as this is enough for continued bactericidal action en route to the user.

Chlorination can result in unacceptable taste intensification, where potable water is concerned. This often originates in the chlorination of phenols present in trace amounts from industrial pollution. If economics permit, use of chlorine dioxide (Section 12.2) or ozone (Section 8.3) in place of chlorine will minimize taste intensification and will also avoid formation of carcinogenic chlorocarbons, notably chloroform. These carcinogens may form from chlorination of contaminants such as acetone, a commonly used solvent that finds its way into water supplies:

$$(CH_3)_2CO + 3HOCl \rightarrow CH_3COOH + CHCl_3 + 2H_2O. \qquad (14.26)$$

However, the chief source of chloroform is probably chlorination of naturally formed humic acids, especially in the tropics and subtropics. The World Health Organization has set a limit of 30 $\mu g\ L^{-1}$ as the acceptable chloroform concentration in drinking water. Overzealous use of chlorine to sterilize sewage-plant effluent has also led to major fish kills in rivers.

Use of ozone for water sterilization is therefore gaining much favor, but ozone does not persist long enough to give continued protection of the water once it leaves the treatment plant. A compromise is to use ozone as the primary disinfectant and to add some chlorine for more lasting sterilization. Studies suggest, however, that even ozone may generate carcinogens indirectly: any bromide ion in the water will be oxidized by ozone to Br_2, which then can react with trace organic compounds in the water to produce brominated substances that are more carcinogenic than chloroform. A further alternative to chlorination is intense irradiation of the water with ultraviolet light, but this, like ozone, gives no continuing sterilization effect.

Other biocidal agents that can be used in closed industrial water systems include copper(II) salts (which, however, can cause corrosion of metals), chlorinated phenols such as sodium pentachlorophenate ($NaOC_6Cl_5$, which is toxic), and quaternary ammonium salts ($R_4N^+X^-$).

Exercises

14.1 (*a*) Suppose that the temporary and total hardnesses of municipal tap water were equivalent to 72.0 and 150.0 mg $CaCO_3$ per kilogram water (i.e., 72.0 and 150.0 ppm). Calculate the weights of lime (as CaO) and soda ash (as Na_2CO_3) needed to soften 1.000 m^3 of water by the lime–soda process, assuming that the permanent hardness is due to magnesium ion alone.

(*b*) How would the calculations be changed if 20 mg kg^{-1} $CaCO_3$ equivalent of the permanent hardness were actually due to noncarbonate calcium hardness?

[*Answers:* (*a*) 84.1 and 82.7 g, respectively.]

14.2 In water at ambient temperature, the solubilities of $MgCO_3$, $CaCO_3$, $Mg(OH)_2$, and $Ca(OH)_2$ are about 106, 15.3, 9.0, and 1600 mg kg^{-1}, respectively. Show that stoichiometric application of the lime–soda process should leave a residual hardness of some 31 ppm $CaCO_3$ equivalent.

14.3 Consider again the tap water described in Exercise 14.1a. Suppose, instead of lime–soda treatment, we add $EDTA^{4-}$ exactly equivalent to the total concentrations of Ca^{2+} and Mg^{2+}.

(*a*) What will be the residual $CaCO_3$ equivalent hardness, given that the stability constants β_1 of $CaEDTA^{2-}$ and $MgEDTA^{2-}$ are 5.0×10^{10} and 4.9×10^8 L mol^{-1}, respectively? (Ignore pH and ionic strength effects.)

(*b*) What would the residual hardness be, if *all* of the original 150 ppm $CaCO_3$ equivalent hardness were of the *temporary* type?

(*c*) Compare your answers with those for the lime–soda method.

[*Answers:* (*a*) 0.13 ppm; (*b*) 0.017 ppm $CaCO_3$ equivalent.]

14.4 Seawater typically contains 35 g kg^{-1} dissolved salts. Assuming this to be effectively all NaCl, and ignoring interionic interactions (Debye–Hückel, ion pairing, etc.), arrive at an order-of-magnitude estimate of the back-pressure necessary to obtain pure water from seawater by reverse osmosis.

[*Answer:* 3 MPa or more, i.e., 30 bars.]

14.5 Compare the advantages and disadvantages of zeolites with those of synthetic organic ion-exchange resins in the context of water treatment.

14.6 Distilled water is more corrosive than hard water toward ordinary glass. Why? (See Chapter 7.)

14.7 (*a*) From the information given in the text, calculate the molar concentrations of Ca^{2+}, Mg^{2+}, and Na^+ for typical Missouri River water samples at Great Falls, Montana.

(*b*) Colorado River water delivered to Los Angeles typically contains 220 ppm $CaCO_3$ equivalent sodium. What does this correspond to in terms of NaCl?

[*Answers:* (*a*) 1.0, 0.49, and 0.94 mmol L^{-1}; (*b*) 0.129 g L^{-1}.]

14.8 The published analysis of a well-known French brand of mineral water reads as follows (in ppm, i.e., mg L^{-1}): Ca, 78; Mg, 24; Na, 5; HCO_3,

357; SiO_2, 13; SO_4, 10; Cl, 3; NO_3, 1; F^-, 0.120; dissolved mineral salts, 310. The pH is 7.2. Show that these data are consistent with the origin of the Ca and Mg contents as carbonate and sulfate minerals. Do the total ionic contents tally? If not, suggest reasons.

14.9 For the dissolution of CO_2 in water at 25 °C, the proportionality constant K_H in Henry's law (Eq. 14.23) is 3.4×10^{-2} mol L^{-1} bar^{-1}. The first acid dissociation constant $K_{a(1)}$ of aqueous CO_2 (carbonic acid) is 4.2×10^{-7} mol L^{-1}. Calculate the pH of rain falling through air at 25 °C, assuming equilibrium with atmospheric CO_2 (see Section 8.1.1) and absence of pollutants.
[*Answer:* pH 5.64.]

References

1. W. H. Betz and L. D. Betz, "Handbook of Industrial Water Conditioning," 8th Ed. Betz Laboratories, Philadelphia, 1980.

2. F. N. Kemmer (ed.), "The Nalco Water Handbook." McGraw-Hill, New York, 1988.

3. N. Basta, "Shreve's Chemical Process Industries," 6th Ed. McGraw-Hill, New York, 1994.

4. J. A. Kent (ed.), "Riegel's Handbook of Industrial Chemistry," 9th Ed. Van Nostrand-Reinhold, New York, 1992.

5. R. M. Harrison (ed.), "Pollution: Causes, Effects and Control," 2nd Ed., Chapters 2–6. Royal Society of Chemistry, London, 1990.

6. I. Bodek, W. J. Lyman, W. F. Reehl, and D. H. Rosenblatt, "Environmental Inorganic Chemistry." Pergamon, New York, 1988. [Section 2.12.6 describes types of adsorption isotherms.]

7. N. Bunce, "Environmental Chemistry." Wuerz, Winnipeg, 1991.

8. S. E. Manahan, "Environmental Chemistry." Lewis Publishers, Chelsea, Michigan, 1991; S. E. Manahan, "Fundamentals of Environmental Chemistry." Lewis Publishers, Boca Raton, Florida, 1993.

9. L. A. Baker (ed.), "Environmental Chemistry of Lakes and Reservoirs," Advances in Chemistry Ser. 237. American Chemical Society, Washington, D.C., 1993.

10. D. Pletcher and F. C. Walsh, "Industrial Electrochemistry." Chapman & Hall, London, 1990.

11. J. E. Fergusson, "Inorganic Chemistry and the Earth." Pergamon, Oxford, 1982.

12. K. Goulding and P. Poulton, Unwanted nitrate. *Chem. Br.* **28**, 1100–1102 (1992).

13. K. Dorfner, "Ion Exchangers: Properties and Applications," 3rd Ed. Ann Arbor Science Publishers, Ann Arbor, Michigan, 1972; M. J. Slater, "The Principles of Ion Exchange Technology." Butterworth-Heinemann, Oxford, 1991 (engineering principles); R. W. Grimshaw and C. E. Harland, "Ion-Exchange: Introduction to Theory and Practice." Chemical Society, London, 1975.

14. E. K. Wilson, Zero-valent metals provide possible solution to groundwater problems. *Chem. Eng. News* July 3, 19–22 (1995).

15. (*a*) A. Kotronarou and M. R. Hoffmann, The chemical effects of collapsing cavitation bubbles: Mathematical modeling. *In* "Aquatic Chemistry," Advances in Chemistry Ser. 244, pp. 233–251. American Chemical Society, Washington, D.C., 1995; (*b*) I. Hua, R. H. Hoechemer, and M. R. Hoffmann, Sonolytic hydrolysis of *p*-nitrophenyl acetate: The role of supercritical water. *J. Phys. Chem.* **99**, 2335–2342 (1995); (*c*) M. R. Hoffmann, S. T. Martin, W. Choi, and D. W. Bahnemann, Environmental applications of semiconductor photocatalysis. *Chem. Rev.*, **95**, 69–96 (1995); (*d*) S. T. Martin, A. T. Lee, and M. R. Hoffmann, Chemical mechanisms of inorganic oxidants in the TiO_2/UV process: Increased rates of degradation of chlorinated hydrocarbons. *Environ. Sci. Technol.* **29**, 2790–2796 (1995); (*e*) N. J. Peill and M. R. Hoffmann, Development and optimization of a TiO_2-coated fiber-optic cable reactor: Photocatalytic degradation of 4-chlorophenol. *Environ. Sci. Technol.* **29**, 2974–2981 (1995).

16. J. P. Lorimer and T. J. Mason, "Sonochemistry." Ellis Horwood. Chichester, U. K., 1988; P. D. Lickliss and V. E. McGrath, Breaking the sound barrier. *Chem. Br.* **32**, 47–50 (1996).

17. P. A. Christensen, A. Hamnett, R. He, C. R. Howarth, and K. E. Shaw, Photocatalytic Detoxification of Water. *In* "Electrochemistry and Clean Energy" (J. A. G. Drake, ed.), pp. 64–86. Royal Society of Chemistry, Cambridge, 1994.

18. W. Stumm and J. J. Morgan, "Aquatic Chemistry," 3rd Ed. Wiley, New York, 1995.

19. S. D. Faust and O. M. Aly, "The Chemistry of Water Treatment." Butterworth, London, 1983.

20. W. L. Marshall, Cooling water treatment in power plants. *Industrial Water Engineering* February–March, 38–42 (1972).

Chapter 15

Oxidation and Reduction in Solution

15.1 Galvanic Cells

GALVANIC CELLS are electrochemical cells in which an external electric current is produced by an internal chemical reaction.[1–7] As an example, consider first the dissolution of zinc metal in a strong aqueous acid to give $Zn^{2+}(aq)$ and hydrogen gas:

$$Zn(s) + 2H^{+}(aq) \rightarrow Zn^{2+}(aq) + H_2(g). \tag{15.1}$$

The zinc is oxidized, that is, loses electrons to the hydrogen ions, which are reduced. If zinc metal is simply placed in direct contact with aqueous acid, the transfer of electrons takes place within the solution. If, however, zinc oxidation and hydrogen reduction are made to take place in different compartments (half-cells), the electrons can be transferred via an external circuit and become available for doing external electrical work. There are two other requisites to make this *galvanic cell* work (Fig. 15.1): a *salt bridge* or other means of allowing inert ions to flow between the half-cells so as to balance the flow of electrical charge through the external circuit, and an inert conducting surface (electrode) such as platinum metal to supply the electrons from the external circuit to the hydrogen ion reduction reaction. Thus, the complete galvanic cell

$$Zn(s) \mid Zn^{2+}(aq) \parallel 2H^{+}(aq) \mid H_2(g),Pt(s) \tag{15.2}$$

- first $\mid$: metal–solution interface
- $\parallel$: salt bridge or porous partition
- second $\mid$: solution–gas interface on Pt metal

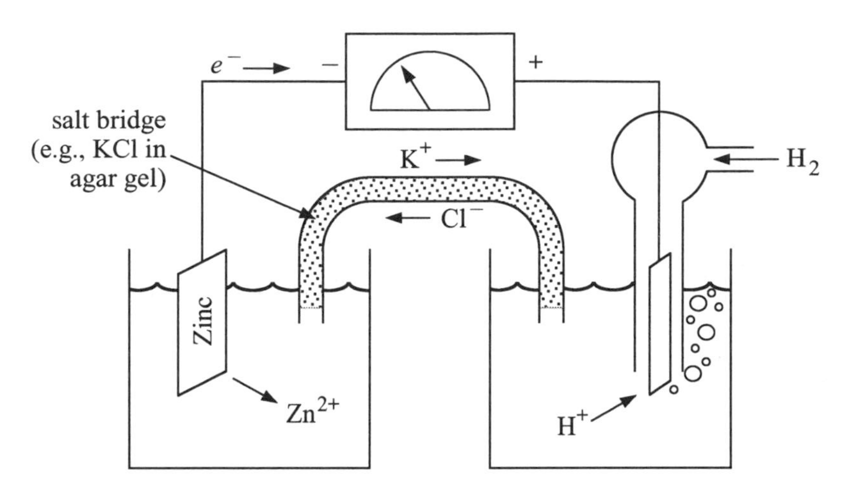

Figure 15.1 A galvanic cell, composed of Zn/Zn^{2+}(aq) and H_2/H^+(aq) electrodes.

is a combination of two half-cells

$$Zn(s)\,|\,Zn^{2+}(aq) + 2e^- \tag{15.3}$$

$$2e^- + 2H^+(aq)\,|\,H_2(g),Pt(s) \tag{15.4}$$

in which occur the two half-reactions:

$$Zn(s) \rightarrow Zn^{2+}(aq) + 2e^- \tag{15.5}$$

$$2e^- + 2H^+(aq) \rightarrow H_2(g). \tag{15.6}$$

The electrode in the half-cell in which oxidation is occurring is said to be the anode (here, the zinc metal), whereas the other is the cathode (here, the platinum). In principle, we could connect any pair of feasible half-cells to form a galvanic cell; the identity of the half-cells will determine which electrode will act as the anode, and which the cathode. The electromotive force (EMF, in volts) of the cell will depend on the identity of the half cells, the temperature and pressure, the activities of the reacting species, and the current drawn. An EMF will also be generated by a cell in which the two half cells are the chemically identical except for a difference in reactant activities (concentrations); this is called a *concentration cell.*

If we choose a set of standard conditions (cf. Section 2.3) and one convenient half-cell to serve as a reference for all others, then a set of standard half-cell EMFs or *standard electrode potentials* $E°$ (Appendix D)[1–9] can be measured while drawing a negligible electrical current, that is, with the cell working *reversibly* so that the equations of reversible thermodynamics

(Chapter 2) apply. The standard reference half-cell is reaction 15.6, the *standard hydrogen electrode* (SHE), and the standard conditions are those listed in Section 2.3, although for our purposes the molar concentration scale (mol L^{-1}) can generally be used without significant loss of precision. We will simplify matters further, for illustrative purposes, by *equating activities with molar concentrations;* our numerical results will therefore be only approximate, except where concentrations are very low. A thermodynamically acceptable treatment would require the calculation or measurement of ionic activities or, at the very least, maintenance of constant ionic strength, as outlined in Section 2.2.

By international agreement, the algebraic sign of $E°$ for a half-cell is chosen to be the same as its electrical sign relative to the SHE. This means, in effect, that we must write the half-reactions with the electrons on the left-hand side; in other words, $E°$ values are taken to be *reduction potentials.** Consequently, a reagent such as chlorine that is more oxidizing than aqueous H^+ ($\rightarrow \frac{1}{2}H_2$) under standard conditions will have a positive $E°$

$$\tfrac{1}{2}Cl_2(g) + e^- \rightarrow Cl^-(aq) \quad E° = +1.36\,V, \tag{15.7}$$

and a reagent such as chromium(II) ($\rightarrow Cr^{3+}$) that is more reducing than H_2 ($\rightarrow 2H^+$) will be associated with a negative $E°$:

$$Cr^{3+}(aq) + e^- \rightarrow Cr^{2+}(aq) \quad E° = -0.42\,V. \tag{15.8}$$

Note that *reducing reagents* (i.e., suppliers of electrons) appear on the right in a half-reaction.

Consideration of electrode potentials in the context of Born–Haber-type cycles (cf. Sections 4.7 and 13.1) leads to the following generalizations†:

(*a*) Strongly *reducing* reagents (very negative $E°$ values) result from low ionization potentials (e.g., alkali metals), low heats of sublimation (e.g., alkali metals again), and strongly negative heats of hydration (associated with small ionic radii; thus, Li is more reducing than Na);

(*b*) Strongly *oxidizing* reagents (very positive $E°$ values) are associated with thermodynamically negative electron affinities (e.g., halogens), low heats of dissociation (e.g., halogens again), and strongly negative heats of hydration (associated with small ionic radii; thus, fluorine is more oxidizing than chlorine).

---

*In some of the older American literature, such as the venerable and still useful books by Wendell M. Latimer,[9] the reverse convention (*oxidation potentials*) was used; the electrons appear on the right, and the sign of $E°$ is reversed.

†As an exercise, check these conclusions and compare them with actual tabulated $E°$ values (Appendix D).

As noted in Section 2.3, E° is a measure of the available free energy *per mole of electrons* to be transferred, under standard conditions:

$$\Delta G^\circ = -nFE^\circ \tag{15.9}$$

where n is the number of (moles of) electrons transferred and F is the faraday (i.e., 96,485 C mol^{-1}). The more positive E° is, the greater is $-\Delta G^\circ$ and hence the greater the tendency for that particular half-cell to go in the direction indicated, with respect to oxidizing H_2 to $H^+(aq)$. More generally, a half-reaction should proceed spontaneously, in the sense written, by *reversing* another half-reaction of less positive E°. Iron(III) ions, for example, should (and do) oxidize aqueous iodide to iodine:

$$Fe^{3+}(aq) + e^- \rightarrow Fe^{2+} \qquad E^\circ = +0.771\,\text{V} \tag{15.10}$$

$$\tfrac{1}{2}I_2(s) + e^- \rightarrow I^- \qquad E^\circ = +0.536\,\text{V} \tag{15.11}$$

$$\text{Subtract:} \quad Fe^{3+} + I^- \rightarrow Fe^{2+} + \tfrac{1}{2}I_2 \qquad \Delta E^\circ = +0.235\,\text{V}. \tag{15.12}$$

We can extract the equilibrium constant K for reaction 15.12 from this information (cf. Section 2.3).

$$nF(\Delta E^\circ) = -\Delta G^\circ = RT\ln K \tag{15.13}$$

$$\begin{aligned} K &= \exp\frac{nF(\Delta E^\circ)}{RT} \\ &= \exp\frac{1 \times 96485 \times 0.235}{8.3145 \times 298.15} \\ &= 9.4 \times 10^3 \\ &= \frac{[Fe^{2+}][I_2]^{1/2}}{[Fe^{3+}][I^-]}. \end{aligned} \tag{15.14}$$

For nonstandard conditions, we need the *Nernst equation*:

$$\begin{aligned} E &= E^\circ - \frac{RT}{nF}\ln Q \\ &= E^\circ - \frac{0.05916}{n}\log Q \end{aligned} \tag{15.15}$$

for a half-reaction at 25 °C, or

$$\Delta E = \Delta E^\circ - \frac{RT}{nF}\ln Q \tag{15.16}$$

for a balanced reaction, where

$$Q = \frac{[\text{products}]^p}{[\text{reactants}]^r} \tag{15.17}$$

which, in the case of reaction 15.12, means

$$Q = \frac{[Fe^{2+}][I_2]^{1/2}}{[Fe^{3+}][I^-]}. \tag{15.18}$$

When reaction 15.12 is allowed to come to equilibrium, ΔE (the EMF of a galvanic cell based on reaction 15.12) runs down to zero, all concentrations assume their equilibrium values, and so Q becomes K. To put it another way, if all the reagents are present at standard-state concentrations (unity, by definition), Q becomes unity and E is just E°.

There are some complications. As we know from Section 13.6, iron(III) ions tend to hydrolyze in aqueous solution unless the pH is very low. Accordingly, it is understood that, unless otherwise stated, E° values refer to measurements in *1.0 molal acid* solution, even if the hydrogen ions do not explicitly appear in the balanced redox equation (e.g., reaction 15.12). Second, iodide ion actually reacts with iodine in water to give brown I_3^-:

$$I_2(s) + I^-(aq) \rightarrow I_3^-(aq). \tag{15.19}$$

Since the equilibrium constant of reaction 15.19 is 0.93, purple solid iodine is significantly more soluble in aqueous iodide solutions than in pure water. For dilute solutions, however, this can be ignored, as can a similar reaction involving chlorine and chloride in connection with reaction 15.7. Finally, we should remind ourselves that we have assumed that all the activity coefficients are unity (although here again the discrepancy so introduced can be ignored for simplicity).

Because electrode potentials are defined with reference to the H^+/H_2 electrode under standard conditions, E° values apply implicitly to (hypothetically ideal) acidic solutions in which the hydrogen ion concentration is 1 mol kg^{-1}. Such E° values are therefore tabulated in Appendix D under the heading *Acidic Solutions.* Appendix D also lists electrode potentials for *basic solutions*, meaning solutions in which the hydroxide ion concentration is 1.0 mol kg^{-1}. The conversion of E° values to those appropriate for basic solutions is effected with the Nernst equation (Eq. 15.15), in which the hydrogen ion concentration (if it appears) is set to 1.0×10^{-14} mol kg^{-1} and the identity and concentrations of other solute species are adjusted for pH 14. For example, for the $Fe^{3+/2+}$ couple in a basic medium, the relevant forms of iron(III) and iron(II) are the solid hydroxides, and the concentrations of $Fe^{3+}(aq)$ and $Fe^{2+}(aq)$ to be inserted into the Nernst equation are those determined for pH 14 by the solubility products of $Fe(OH)_3(s)$ and $Fe(OH)_2(s)$, respectively. Examples of calculations of electrode potentials for nonstandard pH values are given in Sections 15.2 and 15.3.

15.2 Manipulation and Use of Electrode Potentials

15.2.1 Example: Analysis of Brass

As an illustration of the use of electrode potentials, consider the classical method of analysis of copper in brass, which involves dissolving the weighed sample in nitric acid to obtain Cu^{2+}(aq), adjusting the pH to a weakly acidic level, allowing the Cu^{2+} to react completely with excess potassium iodide to form iodine and the poorly soluble CuI, and then titrating the iodine with sodium thiosulfate solution that has been standardized against pure copper by the same procedure:

$$Cu^{2+} + I^- + e^- \rightarrow CuI(s) \qquad E^\circ = +0.86\,V \tag{15.20}$$

$$S_4O_6{}^{2-} + 2e^- \rightarrow 2S_2O_3{}^{2-} \qquad E^\circ = +0.08\,V \tag{15.21}$$

$$NO_3{}^- + 2H^+ + 2e^- \rightarrow NO_2{}^- + H_2O \qquad E^\circ = +0.94\,V. \tag{15.22}$$

It follows from reactions 15.11 and 15.20 that copper(II) ion can oxidize iodide to iodine, and from reactions 15.11 and 15.21 that thiosulfate can reduce I_2 back to I^- while forming the tetrathionate ion (Section 3.4). Because the relevant ΔE° values are several hundred millivolts, the equilibrium constants are large, and these reactions go to essential completion.

We must, however, ensure that there is no contaminant such as free Fe^{3+} in the solution before the potassium iodide is added; otherwise, more I^- will be oxidized than there is Cu^{2+} in the sample (reaction 15.12). The effective concentration of Fe^{3+} can be reduced to negligible levels by adding sodium fluoride to complex it. (The divalent copper ion is little affected.)

From the E° of half-reaction 15.22, it would seem that the nitrate ion present from the dissolution of the brass should also oxidize iodide ion. This E° value, however, refers to *standard* conditions, which implies 1 mol L^{-1} H^+, whereas we have adjusted the pH to near neutrality. Suppose the pH is adjusted to 7.0, that is, $[H^+] = 1 \times 10^{-7}$ mol L^{-1}, while $[NO_2{}^-]$ and $[NO_3{}^-]$ retain their standard-state values of unity; the corresponding EMF for the half-reaction 15.22 is then

$$E_{\text{pH }7} = E^\circ - \frac{0.05916}{2}\log\frac{[NO_2{}^-]}{[NO_3{}^-][H^+]^2} \tag{15.23}$$

$$= +0.94 - \frac{0.05916}{2}(2 \times 7)$$

$$= +0.53\,V$$

so that the oxidation of I^- by $NO_3{}^-$ is no longer favored. Even in slightly acidic solutions, oxidation of I^- by $NO_3{}^-$, though marginally favored thermodynamically, is so *slow* that it does not occur, in practice. This illustrates an important limitation to the use of E° data to predict the outcome of an electrochemical reaction; we can say definitely what the equilibrium position is, but *not* whether the reaction will proceed to that equilibrium at a detectable rate.

15.2.2 Example: Oxidation States of Manganese

Manganese can exist in an unusually large number of oxidation states in 1.0 mol L^{-1} aqueous acid, as summarized in Eq. 15.24, and these afford a colorful illustration of how E° data can be manipulated. Some examples follow.

	$MnO_4^-(aq)$	permanganate ion	(intensely purple)	Mn^{VII}
$+e^-$	$\downarrow$ +0.56 V			
	$MnO_4^{2-}(aq)$	manganate ion	(dark bottle-green)	Mn^{VI}
$+2e^-$ $+4H^+$	$\downarrow$ +2.26 V			
	$MnO_2(s)$	pyrolusite	(brown-black solid)	Mn^{IV}
$+e^-$ $+4H^+$	$\downarrow$ +0.95 V			
	$Mn^{3+}(aq)$	manganic ion	(red-violet)	Mn^{III}
$+e^-$	$\downarrow$ +1.51 V			
	$Mn^{2+}(aq)$	manganous ion	(pale pink)	Mn^{II}
$+2e^-$	$\downarrow$ −1.18 V			
	$Mn(s)$	manganese metal		Mn^0

(15.24)

Calculation of E° for multielectron half-reactions. Equation 15.9 makes it clear that E° is proportional to the (negative) free energy change *per mole of electrons transferred.* Consequently, we can obtain EMFs of cells by simply adding the E° values of the two constituent half-cells, and we can predict if a given half-reaction will go by reversing another by direct comparison of the E° data (Appendix D), since each of these is on a "per electron" basis. We *cannot,* however, obtain E° for the direct reduction of $MnO_4^-(aq)$ to MnO_2 simply by adding E° for the MnO_4^-/MnO_4^{2-} couple to that for the MnO_4^{2-}/MnO_2 pair, since different numbers of electrons are transferred. It is the *free energies,* and not the E° values, that are additive. So, to evaluate E° for

$$MnO_4^- + 4H^+ + 3e^- \rightarrow MnO_2(s) + 2H_2O \qquad (15.25)$$

we can convert E° to the appropriate free energies, add these, and then convert back to E°. To take a short cut, the procedure reduces to multiplying E° for each step by the number of electrons in that step, adding all these products algebraically, and dividing the sum by the total number of electrons. For the specific half-reaction 15.25:

$$E^\circ = \frac{(1 \times 0.56) + (2 \times 2.26)}{3} = +1.69\text{ V.} \tag{15.26}$$

Identification of species likely to disproportionate. According to thermodynamics, the half-reaction

$$\text{Mn}^{3+} + e^- \rightarrow \text{Mn}^{2+} \quad E^\circ = +1.51\,\text{V} \tag{15.27}$$

can proceed spontaneously by reversing some other reaction with $E^\circ <$ $+1.51\,\text{V}$. One such reaction is

$$\text{MnO}_2(\text{s}) + 4\text{H}^+ + e^- \rightarrow \text{Mn}^{3+} + 2\text{H}_2\text{O} \quad E^\circ = +0.95\,\text{V} \tag{15.28}$$

which means that Mn^{3+} is capable of oxidizing (or reducing) *itself,* to Mn^{2+} and MnO_2 (disproportionation):

$$2\text{Mn}^{3+} + 2\text{H}_2\text{O} \rightarrow \text{Mn}^{2+} + \text{MnO}_2 + 4\text{H}^+ \quad \Delta E^\circ = +0.56\,\text{V.} \tag{15.29}$$

Indeed, manganese(III) solutions do decompose slowly, putting down brown MnO_2 on the vessel walls. However, reaction 15.29 tells us that this disproportionation reaction is less thermodynamically favored at high $[H^+]$, and the *kinetics* of decomposition are also slower in highly acidic media, so that $Mn^{3+}(aq)$ chemistry *can* be studied at very low pH, so long as the timescale is kept short. In a reduction-sequence display such as sequence 15.24, any species that has a more positive E° following than preceding it will be thermodynamically unstable with respect to disproportionation.*

pH effects. Whenever hydrogen ions appear in a half-reaction, E° must be pH dependent. There is no *direct* pH effect on the $MnO_4{}^-/\ MnO_4{}^{2-}$ couple, since reaction 15.30 is balanced as it is:

$$\text{MnO}_4{}^- + e^- \rightarrow \text{MnO}_4{}^{2-}. \tag{15.30}$$

The $MnO_4{}^{2-}/MnO_2$ couple, however, is much less oxidizing in 1.0 mol L^{-1} aqueous alkali ($[H^+] = 1 \times 10^{-14}$ mol L^{-1}) than in aqueous acid:

$$\text{MnO}_4{}^{2-} + 4\text{H}^+ + 2e^- \rightarrow \text{MnO}_2(\text{s}) + 2\text{H}_2\text{O} \quad E^\circ = +2.26\,\text{V} \tag{15.31}$$

$$E_{\text{pH 14}} = E^\circ - \frac{0.05916}{2}\log\frac{[\text{MnO}_2]}{[\text{MnO}_4{}^{2-}][\text{H}^+]^4}. \tag{15.32}$$

*As an exercise, you may check this and note that manganate ion is a case in point.

which works out to $2.26 - [(0.05916/2) \times 4 \times 14] = +0.60\,\mathrm{V}$, since MnO_2 is a solid phase of unit activity and $[MnO_4{}^{2-}]$ in the standard state is set to unity. It is therefore possible to minimize the tendency of manganate ion to disproportionate by working in strongly alkaline solution.

The Mn^{3+}/Mn^{2+} couple is ostensibly pH independent, but we must bear in mind that manganese(III), in particular, will hydrolyze unless the acidity is very high, and both $Mn(OH)_3$ and $Mn(OH)_2$ will come out of solution in alkaline media. Small degrees of hydrolysis in solution have little impact on E°, but precipitation of hydroxides (and all metal hydroxides except those of the alkali metals and Ca, Sr, Ba, and Ra are poorly soluble in water) affects E° profoundly. Thus, in alkaline media, the electrode potentials (often called E_h by geologists, wherever $[H^+]$ is not the standard value) of Mn^{2+} and Mn^{3+} are controlled by the solubility products (K_{sp}) of $Mn(OH)_2$ and $Mn(OH)_3$, respectively. In practice, $Mn(OH)_3$ tends to dehydrate to $MnO(OH)$, so we consider the $Mn^{2+}(aq)/Mn(s)$ couple:

$$\mathrm{Mn^{2+}(aq)} + 2e^- \rightarrow \mathrm{Mn(s)} \quad E^\circ{}_{\mathrm{pH}\ 0} = -1.18\,\mathrm{V}. \tag{15.33}$$

In 1 mol L^{-1} NaOH solution (pH 14.0),

$$\mathrm{Mn(OH)_2} \rightleftharpoons \mathrm{Mn^{2+}(aq)} + 2\mathrm{OH^-(aq)} \quad K_{\mathrm{sp}} = 3 \times 10^{-13} \tag{15.34}$$

so

$$[\mathrm{Mn^{2+}}]_{\mathrm{pH}\ 14} = \frac{3 \times 10^{-13}}{[\mathrm{OH^-}]^2} = 3 \times 10^{-13} \tag{15.35}$$

$$E_{\mathrm{h(pH\ 14)}} = -1.18 - \frac{0.05916}{2} \log \frac{1}{[\mathrm{Mn^{2+}}]} = -1.55\,\mathrm{V}. \tag{15.36}$$

Effects of complexation. Just as K_{sp} of an insoluble phase can control the concentration of free metal ion and hence E_h, so control by complexing the metal ion can alter E°. For example, E° for the half-reaction

$$\mathrm{Mn^{III}(EDTA)^-} + e^- \rightarrow \mathrm{Mn^{II}(EDTA)^{2-}} \tag{15.37}$$

can be calculated from

$$\mathrm{Mn^{3+}} + e^- \rightleftharpoons \mathrm{Mn^{2+}} \quad E^\circ = +1.51\,\mathrm{V} \tag{15.38}$$

given that the stability constants β^{III} and β^{II} for $Mn^{III}(EDTA)^-$ and $Mn^{II}(EDTA)^{2-}$ are $10^{25.3}$ and $10^{13.8}$, respectively. Ignoring pH effects, we can write

$$[\mathrm{Mn^{3+}}] = \frac{[\mathrm{Mn^{III}(EDTA)^-}]}{\beta^{\mathrm{III}}[\mathrm{EDTA^{4-}}]} \tag{15.39}$$

and

$$[\mathrm{Mn}^{2+}] = \frac{[\mathrm{Mn}^{\mathrm{II}}(\mathrm{EDTA})^{-}]}{\beta^{\mathrm{II}}[\mathrm{EDTA}^{4-}]} \tag{15.40}$$

from which $[\mathrm{EDTA}^{4-}]$ can be eliminated and, for the standard-state conditions relevant to reaction 15.37, the concentrations of the EDTA complexes can be set to unity (the degree of dissociation is very small):

$$\begin{aligned} \frac{[\mathrm{Mn}^{2+}]}{[\mathrm{Mn}^{3+}]} &= \frac{[\mathrm{Mn}^{\mathrm{II}}(\mathrm{EDTA})^{2-}]\beta^{\mathrm{III}}}{[\mathrm{Mn}^{\mathrm{III}}(\mathrm{EDTA})^{-}]\beta^{\mathrm{II}}} \\ &= \frac{\beta^{\mathrm{III}}}{\beta^{\mathrm{II}}}. \end{aligned} \tag{15.41}$$

The standard electrode potential $E°_{\mathrm{EDTA}}$ for half-reaction 15.37 is then derived from the Nernst equation for reaction 15.38:

$$\begin{aligned} E°_{\mathrm{EDTA}} &= +1.51 - \frac{0.05916}{1}\log\frac{[\mathrm{Mn}^{2+}]}{[\mathrm{Mn}^{3+}]} \\ &= +1.51 - 0.05916 \times (25.3 - 13.8) \\ &= +0.83\,\mathrm{V}. \end{aligned} \tag{15.42}$$

Notice that since complexation stabilizes $\mathrm{Mn}^{\mathrm{III}}$ much more than $\mathrm{Mn}^{\mathrm{II}}$, $E°$ is reduced to a more moderate value. Note the standard electrode potential for the couple

$$\mathrm{Mn(CN)_6}^{3-} + e^{-} \rightarrow \mathrm{Mn(CN)_6}^{4-} \quad E° = -0.24\,\mathrm{V}; \tag{15.43}$$

$\mathrm{Mn}^{\mathrm{III}}$ is thus strongly stabilized by complexing with cyanide.

Redox of solvent water. Any half-reaction with a negative electrode potential can, in principle, be reversed by

$$2\mathrm{H}^{+}(1\ \mathrm{mol\ L^{-1}}) + 2e^{-} \rightarrow \mathrm{H_2(g)} \quad E° = 0.000\,\mathrm{V} \tag{15.44}$$

and, if $E°$ is more negative than $-0.414\,\mathrm{V}$, by neutral water (pH 7.0):

$$2\mathrm{H}^{+}(1 \times 10^{-7}\ \mathrm{mol\ L^{-1}}) + 2e^{-} \rightarrow \mathrm{H_2(g)} \quad E_{\mathrm{h}} = -0.414\,\mathrm{V} \tag{15.45}$$

or, if $E°$ is more negative than $-0.828\,\mathrm{V}$, by 1.0 mol $\mathrm{L^{-1}}$ aqueous alkali:

$$2\mathrm{H}^{+}(1 \times 10^{-14}\ \mathrm{mol\ L^{-1}}) + 2e^{-} \rightarrow \mathrm{H_2(g)} \quad E_{\mathrm{h}} = -0.828\,\mathrm{V}. \tag{15.46}$$

Metallic manganese certainly dissolves readily in 1 mol $\mathrm{L^{-1}}$ aqueous acid, as expected, but in aqueous base a coating of insoluble $\mathrm{Mn(OH)_2}$ would form immediately, stopping the predicted reaction. In cases such as aluminum [$E°$ for $\mathrm{Al^{3+}/Al(s)} = -1.67\,\mathrm{V}$], the product $\mathrm{Al^{III}}$ is soluble in alkali as the aluminate ion $\mathrm{Al(OH)_4}^{-}$, but not in neutral water, in which the aluminum oxide/hydroxide layer is protective. A further factor that may prevent expected H_2 evolution is *overpotential*, which is discussed in detail in Section 15.4; in essence, overpotential is the result of the kinetic slowness of one or more steps in a thermodynamically feasible electrochemical reaction.

Similarly, any half-reaction with E° greater than +1.229 V should proceed by oxidizing water to O_2, under standard conditions (1.0 mol L^{-1} H^+):

$$O_2(g) + 4H^+(aq) + 4e^- \rightarrow 2H_2O(l) \quad E^\circ = +1.229\,\text{V}. \tag{15.47}$$

We have shown (Eq. 15.26) that E° is +1.69 V for the MnO_4^-/MnO_2 couple, and therefore we may expect aqueous acidic $KMnO_4$ solution to evolve oxygen and deposit MnO_2. Indeed, it does, though slowly, and solutions of $KMnO_4$ used for volumetric analysis must be restandardized if kept for more than a few hours.

According to reaction 15.47, for which the Nernst equation gives E_h = +0.401 V at pH 14.0, the oxidation of water is clearly favored by alkaline conditions (see Exercise 15.3). At the same time, however, many oxidation half-reactions also have lower E_h values in basic media. For example, E_h for the manganate/manganese dioxide couple

$$MnO_4^{2-} + 2H_2O + 2e^- \rightarrow MnO_2(s) + 4OH^- \quad E^\circ = +0.61\,\text{V} \tag{15.48}$$

is only 0.21 V more positive than that required for water oxidation in 1.0 mol L^{-1} aqueous alkali, as against 1.03 V more positive in aqueous acid (cf. reactions 15.24 and 15.47), because *two* H^+ ions are consumed (two OH^- liberated) per electron transferred in reaction 15.48 as compared with *one* in reaction 15.47. So, quite apart from the problem of disproportionation discussed above, aqueous MnO_4^{2-} is expected to be tractable only in highly alkaline solutions. Once again, any such predictions may be subject to modification by overpotential effects.

15.3 Pourbaix (E_h–pH) Diagrams

The ranges of E_h and pH over which a particular chemical species is thermodynamically expected to be dominant in a given aqueous system can be displayed graphically as *stability fields* in a *Pourbaix diagram.*[10–14] These are constructed with the aid of the Nernst equation, together with the solubility products of any solid phases involved, for certain specified activities of the reactants. For example, the stability field of liquid water under standard conditions (partial pressures of H_2 and O_2 of 1 bar, at 25 °C) is delineated in Fig. 15.2 by

$$E_h = 1.229 - 0.05916\,\text{pH V (for } O_2 \text{ evolution)} \tag{15.49}$$

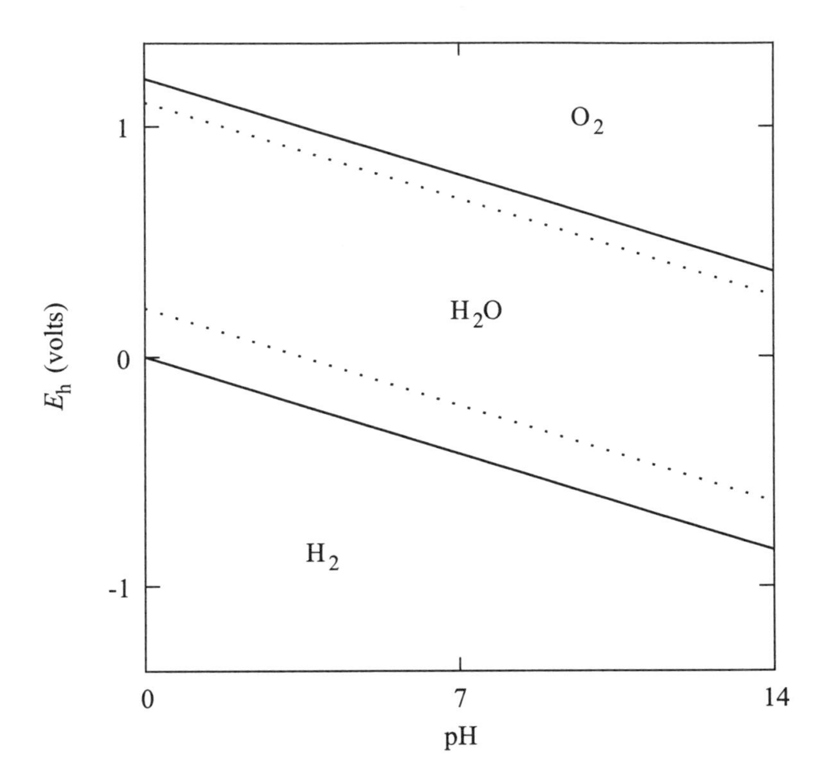

Figure 15.2 Redox stability field of liquid water at 25 °C. Solid lines refer to gas partial pressures of 1 bar (standard state), broken lines to 10^{-6} bar.

and

$$E_h = 0.000 - 0.05916 \text{ pH V (for } H_2 \text{ evolution)}. \tag{15.50}$$

For nonstandard partial pressures of the gases, these boundaries will be slightly displaced, but their slopes will remain the same. In Fig. 15.2, the stability field of water is only slightly narrowed by considering gas pressures a millionfold lower.

The E_h–pH relations for the important iron–water system at 25 °C are summarized in Fig. 15.3 with some simplifications. First, it is assumed that no elements other than Fe, O, and H are involved; in a natural water system, the presence of CO_2 would oblige us to include $FeCO_3$ (siderite), and sulfur compounds could lead to precipitation of iron sulfides in certain E_h–pH regimes. As it is, the only Fe–O–H solids we have considered are Fe metal, $Fe(OH)_2$, and $Fe(OH)_3$, whereas in practice magnetite (Fe_3O_4), hematite (α-Fe_2O_3), goethite [α-FeO(OH)], and other Fe–O–H phases could be present. Indeed, our choice of solubility products for $Fe(OH)_2$ and

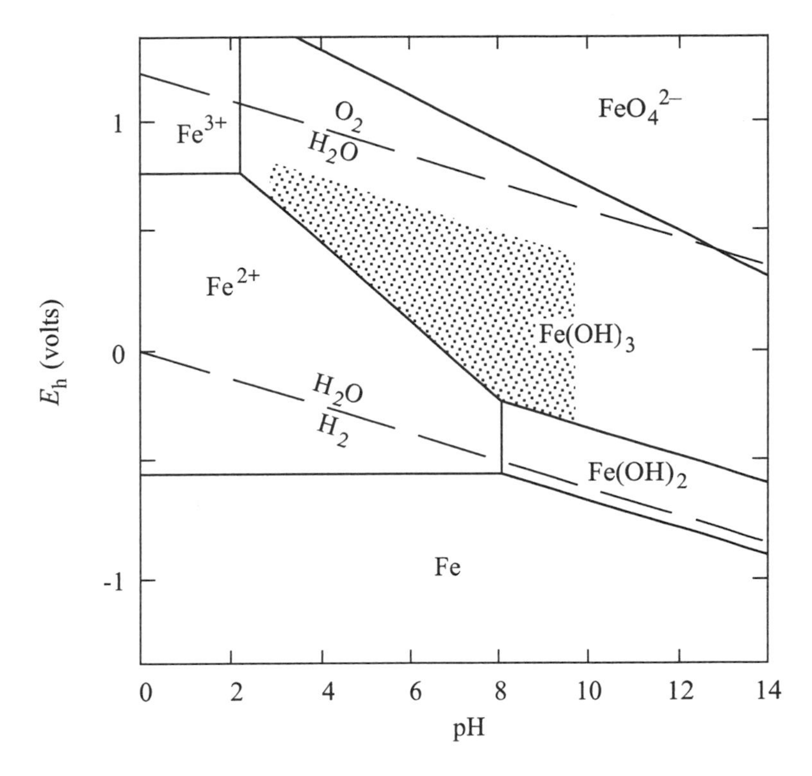

Figure 15.3 E_h–pH (Pourbaix) diagram for the iron–water system at 25 °C for a maximum concentration of 0.001 mol L^{-1} dissolved iron. The shaded area shows the E_h–pH range of natural waters.

$Fe(OH)_3$ is somewhat arbitrary, since these materials become less soluble as they "age," that is, become better crystallized as time passes; here, mid-range values are chosen. Furthermore, the hydrolytic sequence given in Fig. 13.6 for iron(III) has been ignored. A small modification[12] is needed to take this into account, but for simplicity we may represent the hydrolysis of Fe^{3+}(aq) as leading directly to $Fe(OH)_3$(s). Finally, certain of the boundary lines in Fig. 15.3 refer to a specific activity of aqueous iron ions. This is arbitrarily set at 10^{-3} mol L^{-1} (and, for convenience, activity is equated with concentration), but comparison with Fergusson[14] for an iron ion activity of 10^{-5} shows that the qualitative features of the diagram are not very sensitive to this variable.

When interpreting a Pourbaix diagram, one must not infer that a particular species cannot exist outside its stability field. The field boundaries represent contours at which that species and an alternative are equally important; outside its stability field, that species is no longer dominant, and it becomes less and less important the further outside the field we go.

Figure 15.3 is constructed as follows. The boundary between the stability fields of solid iron and 1×10^{-3} mol L^{-1} $Fe^{2+}(aq)$ is given by the Nernst equation (15.52) for

$$Fe^{2+}(aq) + 2e^- \rightarrow Fe(s) \quad E^\circ = -0.44\,V \tag{15.51}$$

$$E_h = -0.44 - \frac{0.05916}{2} \log [Fe^{2+}]^{-1} = -0.53\,V. \tag{15.52}$$

Since this is independent of pH, it is simply a horizontal line at the lower left of Fig. 15.3. The value of E° used in Eq. 15.52 is somewhat more negative than the widely cited -0.409 V derived from the U.S. National Bureau of Standards (now the National Institute of Standards and Technology) thermodynamic data tables (cf. Appendix C). In fact, since various values between -0.409 and -0.475 V have been reported by experienced experimentalists using several different methods, the value selected here is intended to be a conservative compromise.[8] The moral is that one should not place blind faith in tabulated data, even from the most authoritative sources.

The boundary between the $Fe^{2+}(aq)$ and $Fe^{3+}(aq)$ stability fields will be simply the pH-independent E° value for reaction 15.53, if hydrolysis is ignored, since (by definition) the concentrations of the two ions will be equal at the boundary:

$$Fe^{3+}(aq) + e^- \rightarrow Fe^{2+}(aq) \quad E^\circ = +0.77\,V \tag{15.53}$$

$$E_h = E^\circ - 0.05916 \log \frac{[Fe^{2+}]}{[Fe^{3+}]} = +0.77\,V. \tag{15.54}$$

The vertical $Fe^{3+}(aq)/Fe(OH)_3(s)$ boundary at top left involves pH through Eqs. 15.55 and 15.56, but not E_h, since no electrons are transferred in reaction 15.55.

$$Fe(OH)_3(s) \rightleftharpoons Fe^{3+}(aq) + 3OH^-(aq) \quad K_{sp} = 10^{-38.1} \tag{15.55}$$

$$[H^+][OH^-] = K_w = 1 \times 10^{-14} \tag{15.56a}$$

$$\log K_{sp} = \log [Fe^{3+}] - 3(14.0 - pH) \tag{15.56b}$$

$$pH = 14.0 + \frac{-38.1 + 3.0}{3} = 2.3. \tag{15.56c}$$

The $Fe^{2+}(aq)/Fe(OH)_2(s)$ dividing line is similarly calculated to stand at pH 7.95 for 1.0×10^{-3} mol L^{-1} Fe^{2+}, given $\log K_{sp} = -15.1$ for $Fe(OH)_2$:

$$Fe(OH)_2(s) \rightleftharpoons Fe^{2+}(aq) + 2OH^-(aq) \quad K_{sp} = 10^{-15.1}. \tag{15.57}$$

The boundary for the Fe(s) and $Fe(OH)_2(s)$ stability fields, however, is given by a combination of reactions 15.51 and 15.57:

$$Fe(OH)_2(s) + 2e^- \rightarrow Fe(s) + 2OH^-(aq). \qquad (15.58)$$

That is, reaction 15.57 controls $[Fe^{2+}]$ in the half-reaction 15.51:

$$\begin{aligned} E_h &= -0.44 - \frac{0.05916}{2} \log \left(\frac{K_{sp}}{[OH^-]^2} \right)^{-1} \qquad (15.59) \\ &= -0.44 - 0.02958[15.1 - 2(14.0 - pH)] \\ &= -0.06 - 0.059\,pH. \end{aligned}$$

This boundary line therefore depends on both E_h and pH and has a slope of $-(\ln 10)(RT/F)$ (bottom right of Fig. 15.3). Parallel to this and above it is the $Fe(OH)_2/Fe(OH)_3$ boundary, which is defined by a combination of reactions 15.53, 15.55, and 15.57; the concentrations of $Fe^{2+}(aq)$ and $Fe^{3+}(aq)$ in reaction 15.53 are controlled by the pH-dependent solubilities of $Fe(OH)_2$ and $Fe(OH)_3$, so that what appears to be a solid–solid reaction

$$Fe(OH)_3(s) + e^- \rightarrow Fe(OH)_2(s) + OH^-(aq) \qquad (15.60)$$

is actually mediated by a solution-phase redox process in an aqueous environment. Development as above leads to

$$E_h = +0.24 - 0.05916\,pH. \qquad (15.61)$$

The $Fe^{2+}(aq)/Fe(OH)_3(s)$ boundary, which cuts through the center of Fig. 15.3, represents the situation in which the $Fe^{3+}(aq)$ concentration in half-reaction 15.53 is controlled by the solubility of $Fe(OH)_3$, given that $[Fe^{2+}]$ is 1×10^{-3} mol L^{-1}. Proceeding as before, we obtain

$$\begin{aligned} E_h &= 0.77 - 0.05916 \log \frac{[Fe^{2+}][OH^-]^3}{K_{sp}} \qquad (15.62) \\ &= 1.00 - 0.177\,pH - 0.05916 \log [Fe^{2+}] \\ &= 1.18 - 0.177\,pH. \end{aligned}$$

As a final embellishment, the oxidation of iron(III) to the aqueous ferrate(VI) ion, ${FeO_4}^{2-}$, is included:

$${FeO_4}^{2-}(aq) + 8H^+(aq) + 3e^- \rightarrow Fe^{3+}(aq) + 4H_2O \quad E^\circ = +1.9\,V. \qquad (15.63)$$

As the finished diagram (Fig. 15.3) shows, we need only consider the ferrate(VI) ion in alkaline environments, in which the iron(III) concentration is controlled by the poor solubility of $Fe(OH)_3$, as in reaction 15.55,

$$\begin{aligned} E_h &= 1.9 - \frac{0.05916}{3} \log \frac{K_{sp}}{[OH^-]^3[{FeO_4}^{2-}][H^+]^8} \qquad (15.64) \\ &= 1.7_9 + 0.0197 \log [{FeO_4}^{2-}] - 0.0986\,pH \\ &= 1.7_3 - 0.0986\,pH \end{aligned}$$

where again the total dissolved iron concentration has been set at 1×10^{-3} mol L^{-1}.

In addition to the above information, Fig. 15.3 includes the standard-state stability field of liquid water itself (broken lines) from Fig. 15.2, as well as the E_h–pH ranges found in natural waters (shaded area). The usefulness of the Pourbaix diagram now becomes apparent:

(*a*) We see that solid $Fe(OH)_2$ will be oxidized by O_2 in aqueous media, no matter what the pH, and, of course, bare iron metal will always be susceptible to corrosion by aqueous oxygen but may become coated with potentially protective oxide/hydroxide films at pH greater than about 9.

(*b*) Aqueous Fe^{2+} (1×10^{-3} mol L^{-1}) will be oxidized by O_2 to aqueous Fe^{3+} below pH 2.3, and to solid $Fe(OH)_3$ above pH 2.3. Mechanistically, this process is complex, but since it is strongly retarded by hydrogen ions, it is possible to handle aqueous iron(II) salt solutions in the open air for limited periods without significant oxidation, if they kept acidic. Thus, kinetic factors may limit the utility of Pourbaix diagrams, like any other thermodynamic tool.

(*c*) Iron metal can, in principle, reduce water to hydrogen at any pH, although the margin in E_h is narrow and so the rate of H_2 evolution is likely to be negligible because of the high overpotential (Section 15.4), except in acidic media.

(*d*) Oxidation of iron(III) to iron(VI) is unlikely to be significant except in strongly alkaline media, since the solvent water will tend to be oxidized instead (to oxygen). Overpotential effects (Section 15.4), however, may suppress O_2 evolution kinetically. It is possible to produce purple solutions of $FeO_4{}^{2-}$ by anodic oxidation of iron in concentrated aqueous alkali.

(*e*) The $Fe^{2+}(aq)/Fe(OH)_3(s)$ couple provides an E_h–pH buffer in natural waters. Except for some extreme examples associated with acid rain (Sections 8.4 and 8.5), or drainage from coal mines in which oxidation of pyrite (FeS_2) in the exposed coal by aerated water can produce sulfuric acid, the pH of natural waters ranges from an upper limit of pH 9 to 10, set by $CO_3{}^{2-}/HCO_3{}^-$ and $H_4SiO_4/H_3SiO_4{}^-$ buffers, down to around pH 3. The stability field of water is a somewhat narrower band than is shown in Fig. 15.2, since the relevant partial pressures of O_2 and H_2 may be quite low and other factors may intervene,[12] but, even so, the effect of the ubiquitous and abundant iron in the Earth's crust (cf. Table 1.1) is clear. The upper limit of dissolved iron in natural waters is around 1×10^{-3} mol L^{-1}, the value chosen in constructing Fig. 15.3. The lower limit of pH will be buffered by Fe^{III}

hydrolysis,[12] but the details of this have been omitted for simplicity in Fig. 15.3.

One final comment on the Fe/H_2O system is needed: if Fig. 15.3 were extended beyond pH 14, we would have to include the stability fields of $Fe(OH)_4^-(aq)$ and $Fe(OH)_3^-(aq)$.[12] In other words, iron, like aluminum, chromium, zinc, and many other metals, exhibits *amphoteric behavior* (i.e., has both acidic and base like properties), but only if a sufficiently wide range of pH is considered. Amphoteric behavior, like many other chemical properties, is not so much something that a given element does or does not exhibit, but rather is a trait that different elements display to different extents.

15.4 Kinetic Aspects of Electrochemistry: Overpotential

The foregoing considerations are based on the concepts of *reversible thermodynamics*: the electrochemical cells are considered to be operating reversibly, which means in effect that no net current is drawn. Real cell EMFs, however, can differ substantially from the predictions of the Nernst equation because of electrochemical *kinetic* factors that emerge when a nonnegligible current is drawn. An electrical current represents electrons transferred per unit of time, that is, it is proportional to the extent of electrochemical reaction per unit of time, or *reaction rate.* The major factors that can influence the cell EMF through the current drawn are

(*a*) the ohmic voltage drop across the cell itself, since cells necessarily have some internal electrical resistance (this loss in EMF will be proportional to the current drawn, if the internal resistance remains constant, per Ohm's law);

(*b*) depletion of reactants or buildup of products near the electrode surfaces (the so-called *concentration polarization*—diffusion of molecules or ions to or from these surfaces through the solutions is not instantaneous and may slow down the electrochemical reaction significantly);

(*c*) slowness of one or more steps in the electrochemical reactions themselves (*activation polarization*), which is a matter of chemical reaction mechanism and manifests itself as an *overpotential* η, which reduces the galvanic EMF predicted by the Nernst equation.[1,3–7,15–17]

Factor (*a*) can be minimized by making the internal resistance of the cell as small as possible, for example, by having a high concentration of an inert electrolyte in the cell. Factor (*b*) can be reduced by stirring the cell contents vigorously. Factor (*c*), since it originates in chemical reaction kinetics, can often be modified by catalyzing one or more of the electrochemical reactions or half-reactions that make up the cell (see later).

Contributions to overpotential for the $H_2(g)/H^+(aq)$ electrode include the following:

(*i*) A layer of water molecules is inevitably adsorbed on the electrified metal surface (see Chapter 6), so creating an electrical double layer, since the water molecules are dipolar. Furthermore, the $H^+(aq)$ ion itself is solvated by several water molecules in dilute solution (see Chapter 13) and can be regarded as having two tightly coordinated water molecules to which it forms a linear hydrogen bond

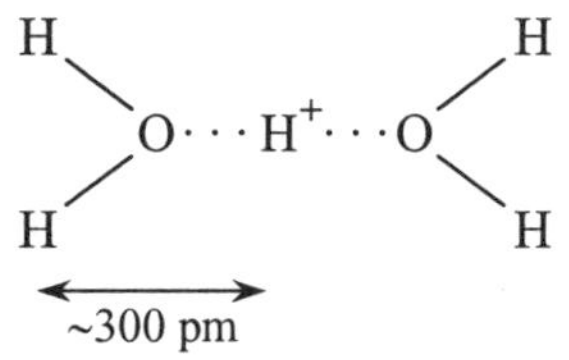

$$H_2O \cdots H^+ \cdots OH_2$$

~300 pm

plus several further water molecules hydrogen bonded to the four outer protons in turn. Consequently, even if, in the approach of $H^+(aq)$ to the metal, substantial desolvation of both the metal surface and the proton were to occur (and that itself would cost energy and so contribute to η), the electron would still have to jump some 300 pm through a potential energy barrier in order to reduce $H^+(aq)$ to the hydrated hydrogen atom, $H\cdot(aq)$.

(*ii*) The hydrated hydrogen atom is higher in energy than the parent $H^+(aq)$ and the electron residing on the metal. In quantum mechanical parlance, the empty $1s$ orbital on the $H^+(aq)$ ion is high in energy relative to the electronic conduction bands on the metal, so the electrical potential of the electrons on the metal must be increased to match the $1s$ energy.

(*iii*) Even if electron transfer to $H^+(aq)$ does occur, the newly created $H\cdot(aq)$ species must find another $H\cdot(aq)$, and desolvate, to form H_2 gas. If it does not, and is not scavenged by some other reactive species, then either the electron returns to the electrode or the $H\cdot$ atom diffuses into the metal causing *hydriding* (and usually embrittlement) of the metal, as explained in Section 5.7.

These considerations refer to the formation of $H_2(g)$ from $H^+(aq)$, but any forward pathway for a reaction is necessarily a reverse pathway, too (cf. the principle of microscopic reversibility; Section 2.5), so the same factors create an overpotential for the oxidation of $H_2(g)$ to $H^+(aq)$. In particular, other reactions involving diatomic gases will be subject to constraints similar to those for H_2. Indeed, high overpotentials are associated with most electrodes involving diatomic gases.

Quantitative treatment of overpotential and related phenomena goes back to 1905, when Tafel showed empirically that, for an electrochemical half-cell from which a net electrical current I is being drawn, an excess potential ΔE away from the equilibrium potential will inevitably exist, and ΔE will be a linear function of the logarithm of the *current density* i ($i = I$/area of interface):

$$\Delta E = a + b \log i. \tag{15.65}$$

Figure 15.4 illustrates the Tafel plot for the hydrogen electrode, operating in both the conventional forward direction (current density i_f, b negative)

$$2H^+(aq) + 2e^- \rightarrow H_2(g) \tag{15.66}$$

and the reverse direction (reversed current density i_r):

$$H_2(g) \rightarrow 2H^+(aq) + 2e^- \tag{15.67}$$

$$\Delta E = a' + b' \log i_r \tag{15.68}$$

where b' is opposite in sign but often very similar in magnitude to b. This is illustrated in Fig. 15.4.

As we see later, the Tafel equations for the forward and reverse electrode reactions are simply limiting cases for relatively high currents of the comprehensive, theoretically based Butler–Volmer relationship (Eq. 15.70). Tafel plots are widely used for practical purposes, however, and provide the least mathematical mode of entry into electrode kinetics and the phenomenon of overpotential. For an electrode process involving only a single electrochemical step, the Tafel slope b for the forward reaction is given by $-2.303RT/\beta nF$, where β is a symmetry factor of value between 0 and 1 but usually close to 0.5, and the other symbols have the same meaning as in the Nernst equation (Eq. 15.15). One can regard β as being the ratio of the distance along the reaction coordinate to the top of the potential energy barrier for electron transfer from the metal surface, to the total distance between the metal surface and the electron acceptor in the solution [$H^+(aq)$, in the present discussion]. For electron transfer in the *reverse* direction, the corresponding factor is therefore $(1 - \beta)$, so that b' is given by 2.303 $RT/(1-\beta)nF$. If, however, *several* electrochemical steps are involved [and we have seen that this is the case for the $H_2(g)/H^+(aq)$ electrode reaction], we must write $b = -2.303RT/\alpha_f nF$ and $b = -2.303RT/\alpha_r nF$ where α_f and α_r are the *transfer coefficients* for the forward and reverse reactions, respectively.*

*The relationship between transfer coefficients and reaction mechanism is explained by Bockris and Reddy[1]; for our purposes, it is enough to note that this theoretical foundation exists and to take experimental values of b and b'.

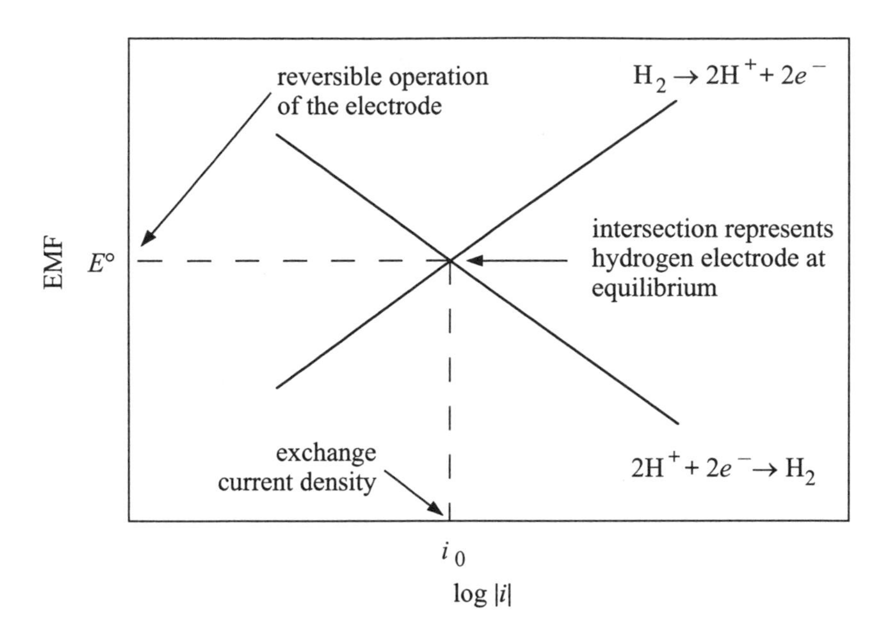

Figure 15.4 Tafel plot for the two half-reactions of the reversible hydrogen electrode.

In any operating condition of the hydrogen electrode, there will in principle be some reduction of H^+ going on together with some oxidation of H_2. The predominance of the first would mean that the electrode is working as a cathode, while predominance of the second would correspond to operation as an anode. If, however, the electrode is at equilibrium (the situation for which $E°$ applies), the forward and reverse reaction rates are exactly equal, and the corresponding values of i_f and i_r are therefore also equal but opposite in direction, so that no net current flows. The value of i_f or i_r that corresponds to this is called the *exchange current density,* i_0, and the corresponding ΔE value relative to SHE is just $E°$, the standard electrode potential, if standard conditions apply. (If not, this voltage is given by the Nernst equation.) If the constants in the Tafel equations (Eqs. 15.65 and 15.68) are known, $\log i_0$ is given by a/b or a'/b'.

Figure 15.5 shows the net current density i as a curve defined by the sum of i_f and i_r, which goes to zero when $i_f = i_r = i_0$. This curve merges asymptotically with the two Tafel lines when substantial currents are drawn in either direction, so that the intersection point of these lines, which defines $E°$ and i_0, is obtainable experimentally by extrapolation of the linear (i.e., Tafel) portions of EMF versus log i plots. We can now define the overpotential η quantitatively. It is the excess electrical potential of the electrode relative to the reversible value $E°$, for a particular value of the

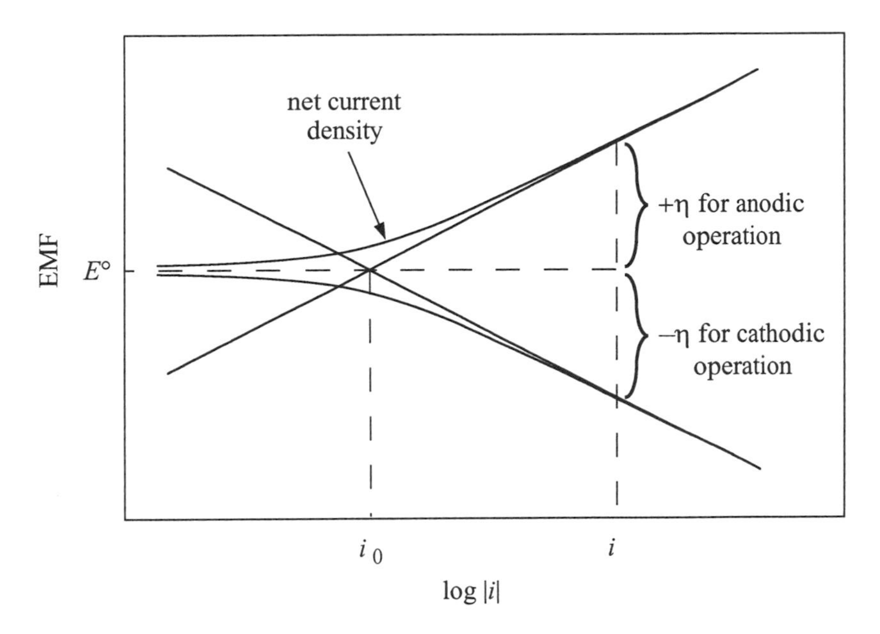

Figure 15.5 Overpotential η associated with a particular current density i. The net current density curve is given by the Butler–Volmer equation.

current density:

$$\eta = b \log \frac{i}{i_0}. \tag{15.69}$$

For the forward reaction, the sign of b is negative, so η reduces the EMF. Equations 15.65, 15.68, and 15.69 can be combined and rearranged to give the *Butler–Volmer equation* (Eq. 15.70) for the net current density, i, of an electrode process involving a single electrochemical step:

$$i = i_0 \left[\exp \frac{(1-\beta)\eta nF}{RT} - \exp \frac{-\beta \eta nF}{RT} \right]. \tag{15.70}$$

Derivation of the Butler–Volmer equation in terms of electrode reaction rate constants is given in most electrochemical texts.[1, 3–7, 15]

If i is not too large, the overpotential η can be made negligible by making i_0 as large as possible. This is achieved by catalyzing the electrode reactions. Catalysis affects only the reaction rate and does not affect the position of equilibrium (see Section 6.1). In other words, catalysis changes i_f and i_r without changing the EMF of the intersection point, so that i_0 increases while $E°$ remains constant. This serves to emphasize a general chemical principle: catalysis affects *both* the forward *and* the reverse rates of a chemical reaction to the same extent. For example, the metallic contact surface chosen for the hydrogen electrode in cells such as that shown

TABLE 15.1
Exchange Current Densities for the Half-reaction $2H^+ + 2e^- \rightleftharpoons H_2$

Electrode	i_0 *(A cm^{-2})*
Pb	2×10^{-13}
Zn	5×10^{-11}
Cu	2×10^{-7}
Fe	1×10^{-6}
Pt	1×10^{-3}
platinized Pt	1×10^{-2}

in Fig. 15.1 is usually platinized platinum, that is, platinum foil on which finely divided platinum has been deposited to obtain the maximum effective surface area for catalysis of the H_2/H^+ reaction. In chloralkali cells (Section 11.4), it is now usual to coat the titanium anodes with ruthenium dioxide (RuO_2), as this material has a lower overpotential for chlorine evolution than does the graphite formerly used.

Since different surfaces will catalyze a given reaction to different degrees, it follows that i_0 values are specific for a particular electrode surface. The exchange current densities for the evolution of hydrogen from 1 mol L^{-1} HCl, for example, range over some 11 powers of 10 (Table 15.1).

To illustrate the significance of this, let us consider the dissolution of zinc and, separately, iron in 1 mol L^{-1} acid. We might naively expect the zinc to dissolve (corrode) more rapidly than the iron, since the corresponding $E°$ values are −0.76 and −0.44 V, against 0.00 V for hydrogen evolution. In fact, the iron dissolves faster than the zinc. Figure 15.6 shows that $i(\mathrm{Zn})$ for the dissolution of zinc

$$\mathrm{Zn} \rightarrow \mathrm{Zn}^{2+} + 2e^- \tag{15.71}$$

will match $i(\mathrm{H_2})$ for the evolution of hydrogen

$$2\mathrm{H}^+ + 2e^- \rightarrow \mathrm{H}_2 \tag{15.72}$$

at the point $i(\mathrm{H_2}) = i(\mathrm{Zn}) = i_{\mathrm{corr}}$ (the corrosion current density) and $E = E_{\mathrm{corr}}$ (the corrosion potential). This point corresponds to the dissolution reaction

$$\mathrm{Zn} + 2\mathrm{H}^+ \rightarrow \mathrm{Zn}^{2+} + \mathrm{H}_2 \tag{15.73}$$

if the anodic and cathodic areas are equal (*general corrosion*, Section 16.5), since the rate of delivery of electrons (the corrosion current) by half-reaction 15.71 must equal the rate of electron consumption by half-reaction 15.72.

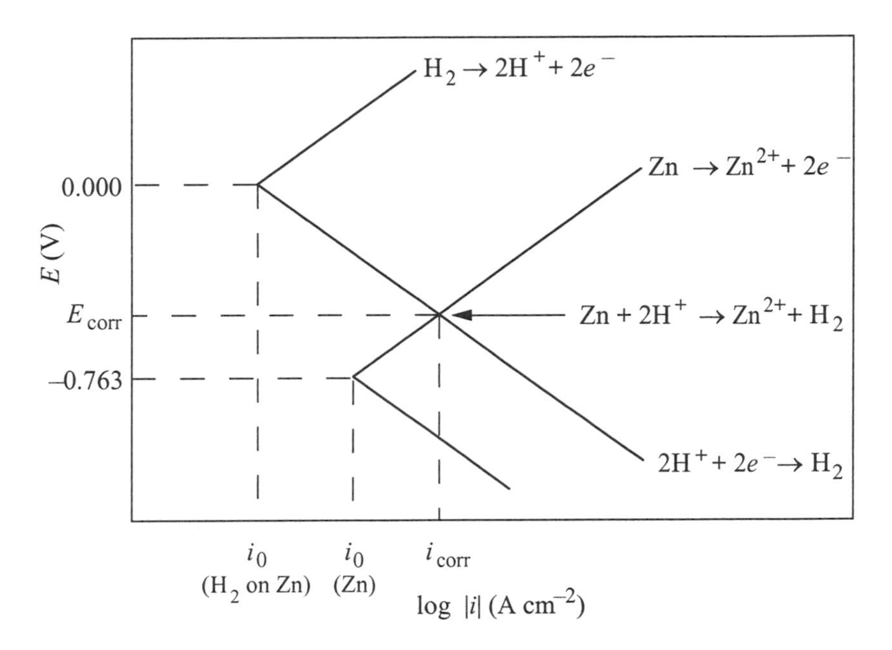

Figure 15.6 Mixed-potential representation of the dissolution of zinc metal in 1.0 mol L^{-1} aqueous acid.

The corrosion current is a direct measure of the rate of reaction 15.71. Figure 15.7 shows the same kind of *mixed-potential* plot for the dissolution of iron in 1 mol L^{-1} acid, superimposed on that for zinc. It is seen that the corrosion current density, and hence the dissolution rate, of the iron is somewhat higher than for the zinc and that this is a consequence of a much higher exchange current density for H_2 evolution on iron relative to zinc.

15.5 Fuel Cells

A conventional power plant fired by fossil fuels converts the chemical energy of combustion of the fuel first to heat, which is used to raise steam, which in turn is used to drive the turbines that turn the electrical generators. Quite apart from the mechanical and thermal energy losses in this sequence, the *maximum* thermodynamic efficiency ε for any heat engine is limited by the relative temperatures of the heat source (T_{hot}) and heat sink (T_{cold}):

$$\varepsilon = \left(\frac{T_{\text{hot}} - T_{\text{cold}}}{T_{\text{hot}}} \times 100 \right) \%. \tag{15.74}$$

Even with an optimistic heat sink temperature of 300 K, an almost uncontainable heat source temperature of 3000 K would be needed to have a

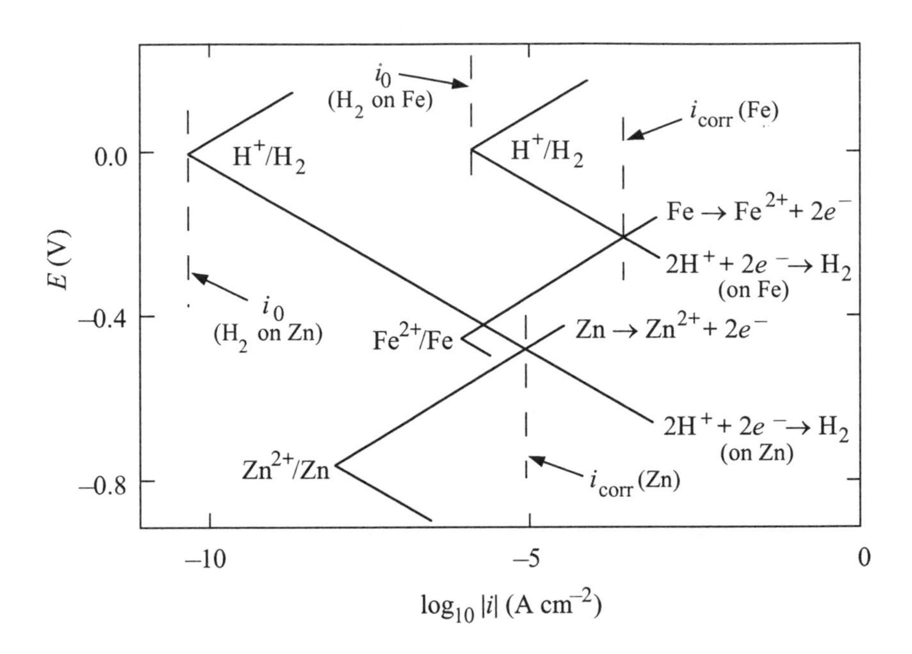

Figure 15.7 Influence of exchange current density for different surfaces on the rate of metal dissolution by hydrogen evolution.

maximum theoretical efficiency of 90%. This refers to the conversion of the *enthalpy* of combustion, ΔH.

If, on the other hand, we could convert the *free energy change* of combustion ΔG *directly* to electricity in a suitable galvanic cell, the maximum extractable energy would be

$$-\Delta G^\circ = nF\Delta E^\circ \tag{15.75}$$

and the efficiency, in terms of what one could extract by simply burning the fuel (ΔH), would be

$$\varepsilon = \left(\frac{\Delta G^\circ}{\Delta H^\circ} \times 100\right) \% \tag{15.76}$$

which, because ΔH° is usually the major component of ΔG°, can be close to 100%. Indeed, consideration of Eq. 2.12

$$\Delta G^\circ = \Delta H^\circ - T\Delta S^\circ$$

shows that there will be some reactions for which ε (defined as above) could exceed 100%!

Such cells are called *fuel cells* and were discovered as long ago as 1839 by Sir William Grove, who noticed that, when water had been electrolyzed and the hydrogen and oxygen products were left in contact with the electrodes, the cell could run "backward," producing a current in an external circuit. Yet it is only relatively recently that fuel cells have attracted major industrial interest. For reasons that will emerge later, the reaction usually chosen is that discovered by Grove:

$$H_2 + \tfrac{1}{2}O_2 \rightarrow H_2O. \tag{15.77}$$

15.5.1 Hydrogen as a Fuel

In the future, hydrogen gas may be distributed widely for domestic and industrial use from a central facility where water is electrolyzed (using the surplus generating capacity of a nuclear, hydroelectric, or geothermal power plant at times when consumption of electricity is low) or possibly one in which water is photolyzed. The gas would be piped to consumers much as methane is today. Hydrogen is especially attractive as a nonpolluting fuel for use in cities, as it burns to give no pollutants or greenhouse gases other than pure water; this is a particular asset in manned spacecraft, where H_2 is used in fuel cells to provide drinking water as well as electrical energy.

Early visions of a full-scale "hydrogen economy," in which piped H_2 would be the universal fuel for industrial and domestic use as well as a feedstock for production of several metals and of chemicals such as methanol and ammonia, have faded to some extent, largely because petroleum fuels and natural gas are again inexpensive and readily available. Hydrogen has a low energy density relative to petroleum and must be compressed (with consumption of 35% of its energy equivalent), liquefied (losing 60–70% of the available energy), or stored as metal hydrides (Section 5.7) in order to compete with liquid fossil fuels. Production of H_2 is relatively expensive, especially if CO_2-producing methods such as the water–gas reaction (reactions 9.5–9.7) or steam reforming of methane (reaction 9.8) are ruled out because of greenhouse considerations; moreover, the use of solar energy for water photolysis or electrolysis is not feasible, on the scale required, during winter in higher latitudes when fuel demands are greatest.

That said, the future for fuel cells as such has become much brighter because of environmental, rather than economic, exigencies. In particular, the State of California has ruled that at least 10% of all new vehicles sold in that state must be "zero emission vehicles" (ZEVs)* by the year 2003,

*This means zero emissions from the running vehicle; in practice, there will be some emissions at central locations where the energy source for the ZEVs is produced, be it hydrogen from the reforming of hydrocarbons or electricity generated to recharge battery-powered cars. Such emissions, however, will be lower and more readily controlled than those of cars with internal combustion engines.

and Ballard Power Systems of North Vancouver, British Columbia, has demonstrated the practicality of on-board fuel cells to power electric buses. We may be sure that other jurisdictions will introduce similar legislation to encourage the automotive industry to develop and produce ZEVs (in effect, electric vehicles) despite contrary market pressures.

In any event, the existence of the infrastructure of a "hydrogen economy" is not a prerequisite for the widespread use of fuel cells for stationary electric power generation. At present, hydrogen for fuel cells is usually generated on-site by reforming propane or natural gas. This is an excellent way of providing electricity in remote communities where bottled propane is available. The cost of long-distance transmission of electricity from a conventional power plant large enough to be economical is eliminated, and the silent, nonpolluting fuel cell has a better response to load fluctuations than the noisy, polluting diesel or gasoline generators presently used. Fuel cells require almost no supervision or maintenance (there are few, if any, moving parts). The advantage in efficiency (>40%) over conventional electrical generating systems is most evident for fuel cell installations in the range 10 kW to 10 MW, and even in large cities there are advantages to generating electricity for, say, a hospital complex on-site with an efficient, reliable fuel cell system. The electricity delivered by a single cell is low voltage direct current, but higher voltages can be generated using batteries of fuel cells in series, and an inverter can be installed to give alternating current.

15.5.2 General Principles of Fuel Cells

The efficiency ε of a fuel cell operating at standard conditions, in which the hydrogen is oxidized to liquid water,

$$\mathrm{H_2} + \tfrac{1}{2}\mathrm{O_2} \rightarrow \mathrm{H_2O(l)} \qquad \Delta H = -285.83\ \mathrm{kJ\ mol^{-1}} \qquad \Delta E^\circ = +1.229\,\mathrm{V}, \tag{15.78}$$

is given by

$$\varepsilon = \frac{\Delta G^\circ}{\Delta H^\circ} = \frac{-2 \times 96485 \times 1.229}{-285830} = 0.830 = 83.0\%. \tag{15.79}$$

If the product is water *vapor,* ε becomes 98.1%. Such theoretical efficiencies would require intractably high temperatures in a heat engine.

In practice, there are several factors that reduce ε, but even in the worst case ε is higher than the 25 to 30% efficiency realized by conventional or nuclear power stations. These factors include the following:

(*a*) Parasitic reactions (e.g., of hydrocarbons in the fuel supply) may reduce the energy output of the cell.

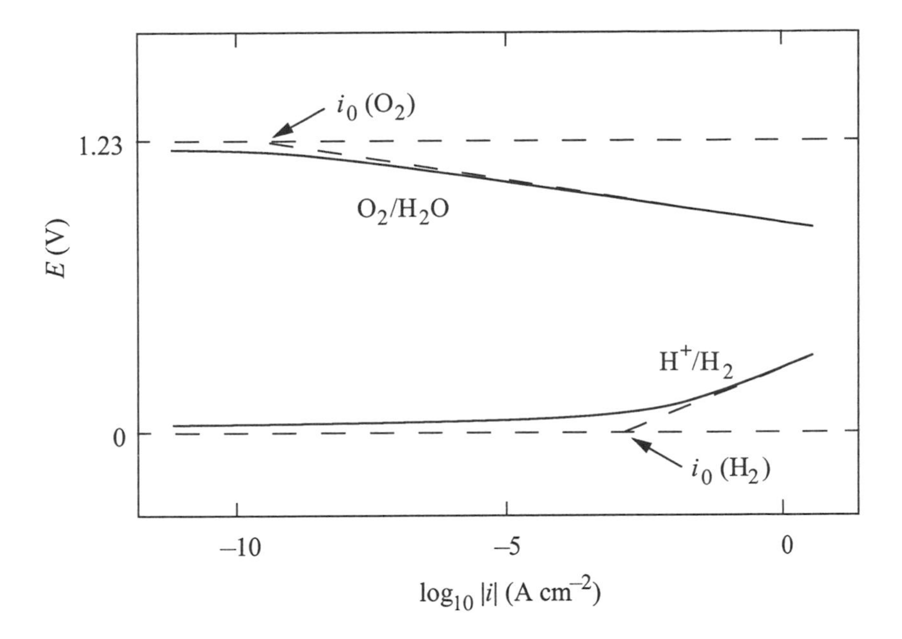

Figure 15.8 Oxygen reduction and hydrogen oxidation electrode reactions (standard conditions, on shiny platinum).

(*b*) As with all galvanic cells, the internal electrical resistance will reduce the EMF, and so ε.

(*c*) Concentration polarization (Section 15.4) may reduce the EMF. Porous electrodes are used to improve gas–electrolyte–electrode contacts, and some of the energy output of the cell can be used to stir the electrolyte.

(*d*) Activation polarization (overpotential) is usually significant, especially at the cathode, since the exchange current density for O_2 reduction is even lower than i_0 for H_2 oxidation and must be catalyzed (Fig. 15.8, cf. Section 15.4).

To minimize overpotential effects, cathodes are usually made of finely divided platinum on a porous support, for aqueous electrolytes. The catalytic surfaces of the anodes are particularly susceptible to poisoning by CO, olefins, sulfur compounds, and other impurities in the fuel. These lie above H_2 in the chemisorption series (Eq. 6.3).

Consequently, the *direct* use of hydrocarbon gases as fuel is usually considered to be impractical, although Whitesides and co-workers[18] describe an aqueous fuel cell in which methane reacts with aqueous iron(III) ions over a platinum black catalyst to form CO_2 and iron(II); the Fe^{2+} solution

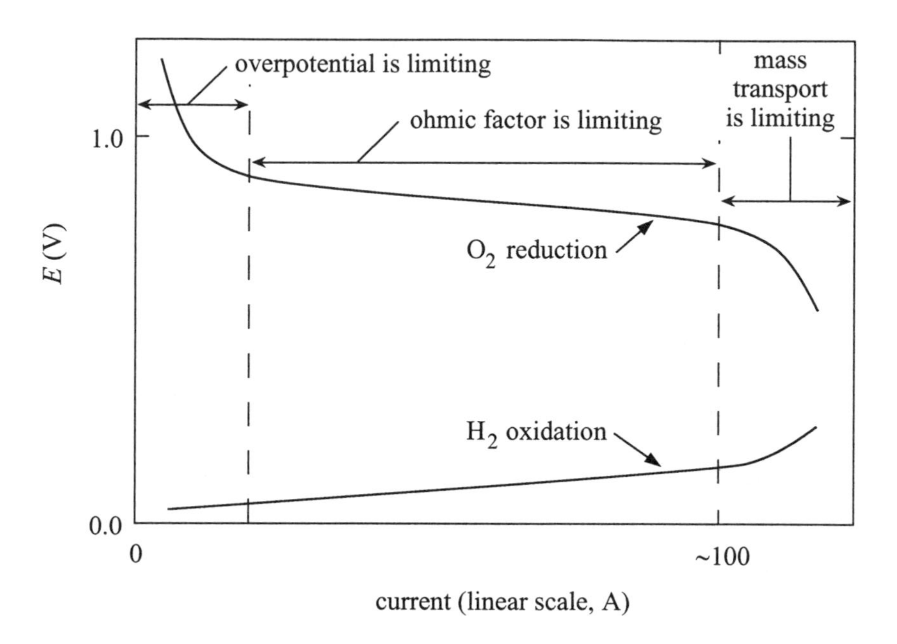

Figure 15.9 Voltage–current relationship for a typical hydrogen–oxygen fuel cell. The vertical separation between the anode and cathode curves represents the cell output voltage.

is pumped to the anode compartment of the cell, where it is reoxidized to Fe^{3+}, while O_2 is reduced in the cathode compartment in presence of a $VO_2^+/VO^{2+}/HNO_3$ catalyst:

$$CH_4 + 8Fe^{3+} + 2H_2O \xrightarrow{\text{platinized Pt}} CO_2 + 8Fe^{2+} + 8H^+ \qquad (15.80)$$

$$2O_2 + 8H^+ + 8e^- \xrightarrow{\text{at cathode}} 4H_2O \qquad (15.81)$$

$$8Fe^{2+} \xrightarrow{\text{at anode}} 8Fe^{3+} + 8e^-. \qquad (15.82)$$

The anode and cathode compartments are separated by a Nafion membrane in H^+ form, through which the H^+ ions flow to complete the circuit. The cell operates at a modest 120 °C.

Figure 15.9 shows how the standard potentials of the anode and cathode in a practical aqueous-medium H_2/O_2 fuel cell change with the current drawn. The horizontal scale is linear in current, not logarithmic in current density as in Figs. 15.4 to 15.8, so even the overpotential-limited Tafel region is nonlinear. Note that most of the voltage loss occurs at the oxygen electrode. At higher currents, the internal resistance of the cell is the chief limiting factor, and ultimately mass transport (i.e., the rate of supply of

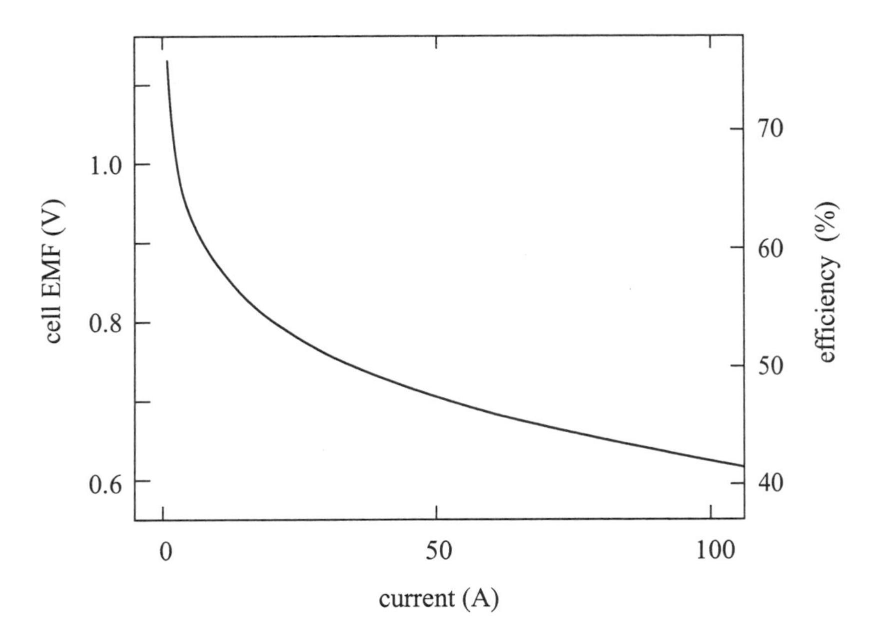

Figure 15.10 Hydrogen/oxygen fuel cell voltage and efficiency as a function of the current drawn.

reactants and ions to the electrodes) becomes limiting. Figure 15.10 gives the same information in terms of the net EMF of the cell (anode versus cathode) and the efficiency ε.

It is usual to operate an aqueous-medium fuel cell under pressure at temperatures well in excess of the normal boiling point, as this gives higher reactant activities and lower kinetic barriers (overpotential and reactant diffusion rates). An alternative to reliance on catalytic reduction of overpotential is use of molten salt or solid electrolytes that can operate at much higher temperatures than can be reached with aqueous cells. The ultimate limitations of any fuel cell are the thermal and electrochemical stabilities of the electrode materials. Metals tend to dissolve in the electrolyte or to form electrically insulating oxide layers on the anode. Platinum is a good choice for aqueous acidic media, but it is expensive and subject to poisoning.

15.5.3 Commercial Fuel Cells

As of 1996, there are five main types of fuel cells that have been developed to the commercial level.[19,20]

(*a*) *Phosphoric acid fuel cells* (PAFCs) work very well and are the most widely used of the aqueous-acid fuel cells (indeed, of any commercial cell at present). Despite much research and development, however, they are still more costly to build than conventional electrical generators of the same capacity, largely because of the need for catalytic platinum for the electrodes. They use concentrated H_3PO_4 as the electrolyte, usually on a porous solid support, and are usually operated at 160–220 °C. The catalysts are susceptible to poisoning by CO, but sufficiently pure hydrogen can be generated by reforming natural gas or naphtha.

(*b*) In *alkaline fuel cells* (AFCs), the electrolyte is 34–46% KOH, immobilized on a porous support, and the operating temperature is 60–120 °C. Because the environment is alkaline, Raney nickel (a finely divided form of nickel) can be used in place of expensive platinum. However, the alkali will be neutralized by any CO_2 in the hydrogen fuel, so AFCs are not suitable for operation with reformed hydrocarbons but can be fueled with alcohols or hydrazine. AFCs were used successfully on the Apollo space missions.

(*c*) In *solid polymer membrane fuel cells* (SPMFCs), robust cation-exchanging membranes such as Du Pont's Nafion (Section 11.4.1) in H^+ form can transport H^+ freely, so long as they are kept moist. They can therefore serve as a solid electrolyte layer between the anode and the cathode in acid-type fuel cells. Operating temperatures are usually in the range 60–120 °C. Costly platinum is again required for the catalytic electrodes, but otherwise SPMFCs hold much commercial promise, having proved their practicality in the Gemini space flights and in the Ballard Power Systems bus. The Ballard bus runs on compressed hydrogen (stored at 25 MPa and delivered to the cell stacks at 0.3 MPa absolute pressure); the 20 fuel cell stacks operate at 90 °C and deliver 205 kW direct current at 650 V.

(*d*) *Molten carbonate fuel cells* (MCFCs) exploit the fact that molten salt media permit fuel cell operation at temperatures high enough to obviate the need for catalysts that are expensive and prone to poisoning. Molten carbonates are the favored electrolytes in this case, since CO_2 from reformed hydrocarbon fuels (commonly methane or propane) is then actually an active participant in the fuel cell chemistry. The operating temperature of a MCFC with a molten K_2CO_3/Li_2CO_3 electrolyte is 600–700 °C, and the necessary heat can be obtained from an internal hydrocarbon reformer, which delivers hydrogen at some 0.3 MPa absolute pressure:

$$C_3H_8 \xrightarrow{\text{limited air}} 3CO + 4H_2. \tag{15.83}$$

The anode reaction is then

$$H_2 + {CO_3}^{2-}(\text{solvent}) \xrightarrow{700\,^\circ\text{C}} H_2O + CO_2 + 2e^-. \tag{15.84}$$

The issuing gases contain some carbon monoxide, which is burned to CO_2 before the gases are fed to the cathode compartment:

$$CO_2 + \tfrac{1}{2}O_2 + 2e^- \rightarrow CO_3{}^{2-}(\text{solvent}). \qquad (15.85)$$

Thus, the net reaction is just Eq. 15.77. Methane-fueled MCFCs are leading candidates for local power plants in the range 1–3 MW, as their cost and efficiency are more attractive than those of PAFCs. The relatively high temperature of the exhaust gases means that they can be used effectively to drive conventional gas or steam turbines to generate additional ("topping") electricity for increased overall efficiency. Less CO_2 per kilowatt-hour is produced than by other fossil fuel-fired power generation methods, and emissions of NO_x, SO_x, and unburned hydrocarbons are minor.

(*e*) In *solid oxide fuel cells* (SOFCs), just as the membrane in a SPMFC permits transport of H^+ from anode to cathode, a solid that permits rapid transport of O^{2-} (a *fast ion conductor*)[21,22] from cathode to anode separates the electrode compartments. Although β-alumina (Section 5.5) might be used as such a solid electrolyte, the usual choice is a Y_2O_3/ZrO_2 mixed oxide ("yttria-stabilized zirconia"), which becomes a sufficiently good ionic conductor at 900–1000 °C. In this material, which is a solid solution of Y_2O_3 in ZrO_2, replacement of some Zr^{4+} ions by Y^{3+} results in there being vacancies in the oxide ion array, giving rise to O^{2-} mobility. A typical composition might be $Zr_{0.9}Y_{0.1}O_{1.95}$. In a typical SOFC, hydrogen (which may contain CO) is introduced over a Ni–ZrO_2 "cermet" (metal–ceramic composite) anode, air enters over a $LaMnO_3$ cathode, and oxide transport between the two electrodes takes place through a layer of $Zr_{0.9}Y_{0.1}O_{1.95}$.

Although still under development, SOFCs are promising in that they present no electrolyte evaporation or catalyst-poisoning problems. The high operating temperature can limit the choice of structural materials, and SOFCs that operate at a moderate 800 °C with $Ce_{0.9}Gd_{0.1}O_{1.95}$ as the electrolyte have been built. On the other hand, high temperature operation permits the hot exhaust gases to be used efficiently to drive turbines for topping power or to run the gas compressors.

15.6 Electrochemical Energy Storage Cells

The term "fuel cell" is usually reserved for electrochemical cells to which the reagents are fed from an external source. Storage cells (often called "batteries," although this term really refers to a collection of interconnected cells) already contain all the necessary reagents, and they may be classed into two groups, *rechargeable* and *disposable* cells. In this section, we survey briefly the chemical characteristics of some common storage cells.[16a] As with fuel cells, the maximum theoretical (*reversible* or *no-load*) EMF of a storage cell is reduced in practice by the voltage drop across the internal resistance of the cell and by the overpotentials developed by the anode and cathode reactions, and the cell design therefore aims largely to minimize these effects.

The Leclanché cell, the inexpensive disposable flashlight-type cell, has been on the market for over 100 years, yet its chemistry is not completely understood. The cell consists of an outer zinc shell that acts as the anode (seen by the external circuit as the source of electrons and hence the *negative* terminal) and oxidizes away during operation of the cell, a carbon rod or disk that serves as the cathodic current collector (*positive* terminal), and a moist paste of manganese dioxide, ammonium chloride, and zinc chloride that fills the cell and acts as both the electrolyte and the source of the cathodic reaction (reduction of Mn^{IV}). Usually, graphite in the form of carbon black is added to the paste to increase the electrical conductivity. The basic reactions are

Cathode:

$$2MnO_2(s) + 2NH_4{}^+(aq) + 2e^- \rightarrow 2MnO(OH)(s) + 2NH_3(aq) \qquad (15.86)$$

Anode:

$$Zn(s) \rightarrow Zn^{2+}(aq) + 2e^- \qquad (15.87)$$

$$Zn^{2+}(aq) + 2NH_3(aq) + 2Cl^-(aq) \rightarrow Zn(NH_3)_2Cl_2(aq).$$

Ammonium chloride plays a key role in formation of a soluble complex of zinc(II), which would otherwise precipitate as $Zn(OH)_2$ on the anode. The cell EMF, which ideally is 1.55 V, may fall by several tenths of a volt because of concentration polarization if large currents are drawn continuously, but it tends to recover (though slowly and incompletely) on breaking the circuit, as reaction products diffuse into the bulk paste. Leclanché cells cannot be recharged. The small 9 V batteries used in transistor radios, etc., typically consist of six shallow Leclanché cells stacked and connected in series.

An alternative to the Leclanché cell is the *alkaline manganese cell,* in which the electrolyte is a strongly alkaline paste and the anode and cathode reactions in their respective compartments are

$$Zn(s) + 4OH^-(aq) \rightarrow Zn(OH)_4{}^{2-}(aq) + 2e^- \qquad (15.88)$$

$$2MnO_2(s) + 2H_2O + 2e^- \rightarrow 2MnO(OH)(s) + 2OH^-. \qquad (15.89)$$

The alkaline cell has a longer operating life than a Leclanché cell, but, because only high-grade electrolytically manufactured MnO_2 can be used, the cost of manufacture is higher.

The design of alkaline *nickel-cadmium* (Ni/Cd, "nicad") cells varies according to the intended use, ranging from large rechargeable storage batteries for power system backup to disposable button cells for miniature electronic devices. However, the essential features are in all cases a cadmium anode (negative terminal in the external circuit) and a nickel current collector, which are in contact with the respective metal hydroxides and the electrolyte, aqueous KOH paste. A separator (usually of porous polyethylene) keeps the anode and cathode regions apart. The chemistry is complicated, especially at the anode (where various oxides/hydroxides of Ni^{II}, Ni^{III}, and Ni^{IV} are involved), but can be simplified for illustrative purposes as follows:

Cathode:

$$2NiO(OH)(s) + 2H_2O + 2e^- \rightarrow 2Ni(OH)_2(s) + 2OH^-(aq) \quad (15.90)$$

Anode:

$$Cd(s) + 2OH^-(aq) \rightarrow Cd(OH)_2(s) + 2e^-. \quad (15.91)$$

The reaction products remain in place near the current collectors, and the cell is rechargeable, if suitably designed. The maximum EMF is 1.48 V.

Although more expensive, the nickel-cadmium cell is superior to the Leclanché cell in almost all respects, except that the toxicity of cadmium places some restrictions on the disposal of defunct nicad cells. Even the rechargeable Ni/Cd cell has a limited life, due to a "memory effect" after discharge (i.e., it is not quite fully rechargeable), and consideration must be given to proper disposal or, better, recycling. Peugeot's entry in the ZEV field, the Model 106 electric car, uses 20 liquid-cooled 6 V Ni/Cd cells to deliver 120 V, and the supplier undertakes to recycle the battery at the end of its useful life.

The *silver oxide* and *mercuric oxide button cells* used in cameras and other devices requiring a miniature source of EMF consist of a zinc disk, which serves as the anode, and, on the other side of a porous separator, a paste of Ag_2O or HgO. The reaction products are zinc hydroxide and metallic silver or mercury. Inert metal caps serve as the current collectors.

The *lead/acid battery* used in conventional gasoline-fueled automobiles consists of six 2.05 V cells connected in series (for a 12 V electrical system). The current collectors are lead grids filled, when in the charged condition, with powdered lead (anode) and a lead/lead(IV) oxide mixture (cathode), and the electrolyte is aqueous sulfuric acid. During discharge, the following reactions occur:

Cathode:

$$PbO_2(s) + 2H^+(aq) + H_2SO_4(aq) + 2e^- \rightarrow PbSO_4(s) + 2H_2O \quad (15.92)$$

Anode:

$$Pb(s) + H_2SO_4(aq) \rightarrow PbSO_4(s) + 2H^+(aq) + 2e^-. \qquad (15.93)$$

The solubility of lead(II) sulfate in water is low and, as a result of the common ion effect, is virtually nil in aqueous sulfuric acid, so the newly formed $PbSO_4$ remains inside the lead grids and the electrolyte becomes more dilute as the sulfuric acid is consumed. Thus, the state of discharge of the battery may be monitored through the falling density (specific gravity) of the liquid phase. Conversely, since all the electroactive materials are solids and remain localized within the grids, the battery is readily rechargeable; there is always some powdered lead remaining in the electrodes, so that their conductivity remains high. Care should be taken not to overcharge lead/acid batteries, especially those of the sealed "maintenance-free" type, since hydrogen and oxygen will then be liberated at the electrodes, with an attendant explosion hazard. Lead/acid batteries are well suited to deliver currents on the order of 100 A over several seconds, such as are needed to start an automobile engine.

Lead/acid batteries, with their long history and proven reliability, are an obvious choice to power electric cars (ZEVs). For practical purposes, however, a car should have a range of about 150 km, which implies a battery capacity of about 75 kWh (at 40% efficiency) coupled with a power delivery of 35–40 kW to the drive train. Since the energy density of a conventional lead/acid battery is only about 0.04 kWh kg^{-1} when the masses of all the components are added together, this implies a daunting 1.8 metric tons of battery. Nevertheless, in late 1995, General Motors introduced the Impact lead/acid battery-powered car, which has a range of up to 145 km with a regulated maximum speed of 130 km $hour^{-1}$ and an acceleration of 0–100 km $hour^{-1}$ in less than 9 s. As with the Peugeot 106 Ni/Cd car, however, the batteries must be replaced triennially. Concern has been expressed that processing of so much toxic lead could create a serious new pollution problem while attempting to solve the old ones (primarily NO_x and ozone pollution), but this view has not gained wide acceptance.

Sodium cells[20–23] operate at fairly high temperatures (300–400 °C) and require an inert atmosphere (argon) in a sealed, corrosion-resistant vessel (e.g., Cr-coated steel). Furthermore, leakage of liquid Na could obviously have dire consequences. Nevertheless, sodium–sulfur cells have received serious consideration as rechargeable batteries:

$$Na(l) \xrightarrow{\text{anode}} Na^+ + e^- \qquad (15.94)$$

$$S(l) + 2e^- \xrightarrow{\text{porous graphite cathode}} S^{2-}. \qquad (15.95)$$

The sulfide ion is probably present as polysulfides, for example, S_3^{2-} (Section 3.4), in the molten sulfur solvent. The anode and cathode compartments are separated by a refractory fast ion conductor through which Na^+

can pass freely; β-alumina (actually $Na_{1+2x}Al_{11}O_{17+x}$, Section 5.5) is the usual choice. As an alternative to a sulfur cell, the rechargeable ZEBRA battery (Beta Research & Development, Derby, England)[20] uses the net reaction

$$2Na(l) + NiCl_2/Na[AlCl_4](l) \rightleftharpoons Ni(s) + 2NaCl. \tag{15.96}$$

The electrolyte surrounding the central positive electrode is sodium tetrachloroaluminate, which melts at 157 °C and acts as a solvent for the nickel(II) chloride, and this is separated from the molten sodium in the outer compartment by a β-alumina tube which, again, serves as a fast ion conductor for Na^+. The cell operates at 300 °C and delivers up to 2.58 V.

Lithium cells[20–22] with anode chemistry analogous to reaction 15.94 have been developed, but several alternative cathodic half-cells can be substituted for reaction 15.95. Lithium cells can offer special advantages over other candidates for ZEV rechargeable batteries in that the energy density in the charged cell is high because Li is very light. Furthermore, the Li^+ ion has a very small radius together with a low charge, so that it can move very freely through defect sites in a suitably chosen fast ion conductor without seriously distorting the lattice (most other migrating ions impose severe structural strains on solid electrolytes). For example, in the lithium/manganese dioxide cell

$$Li \xrightarrow{\text{anode}} Li^+ + e^- \tag{15.97}$$

$$xLi^+ + MnO_2 \xrightarrow{\text{cathode}} Li_xMnO_2, \tag{15.98}$$

the Li^+ ion causes only very small, readily reversible distortions of the rutile structure of the MnO_2 as it migrates into it (an *insertion cathode*, Eq. 15.98), and the energy density of the cell is about six times that of a lead/acid cell if only the actual reactants are considered. In general, Li cells recharge by inserting Li^+ into a solid electrolyte at the positive electrode to form a solid solution and plating out Li metal at the negative electrode, so that there is no buildup of Li^+ in the electrolyte. In so-called *lithium ion batteries*, the cell depends entirely on insertion and extraction reactions of Li^+ at relatively low potentials, and no lithium metal is involved.

15.7 Electrolysis, Electroplating, and Electroforming

Crude hydrogen for industry is most economically made today by cracking hydrocarbons (Section 7.3), steam reforming or partial oxidation of methane, or the water–gas reaction (Section 9.3), but if very pure hydrogen and/or oxygen are required, or if electricity is inexpensive, water may be electrolyzed instead. As noted in Section 15.5, electrolytic hydrogen generation offers a means of converting surplus electrical generating capacity during periods of low demand to an alternative, readily stored energy source.

A water electrolysis cell is like a hydrogen/oxygen fuel cell run in reverse, and factors, such as overpotential, that limit fuel cell efficiency are also important in electrolysis. The electrolyte, which must be unchanged by electrolysis, is usually aqueous NaOH or KOH, as these are not aggressively corrosive toward metals such as the nickel-on-steel commonly used for the electrodes. Platinum would be the best electrode material, in terms of both corrosion resistance and lowered overpotentials for gas evolution, but it is very expensive. Overpotentials on nickel are satisfactory if the Ni is deposited so as to present a high surface area. The operating temperature is typically 80 °C (again to minimize overpotential effects), but temperatures up to 150 °C can be used if the cell is pressurized—this has the additional advantage that the gas bubbles which tend to coat the electrodes and increase the effective electrical resistance are kept small. The anode and cathode compartments need to be kept separated by a diaphragm of some sort (cf. chloralkali diaphragm cells, Section 11.4), although in H_2/O_2 production the purpose is just to keep the gaseous products separated, as the electrolyte undergoes no net change.

Other commercially important inorganic chemicals that can be made electrolytically include caustic soda and chlorine, chlorate and perchlorate salts (Chapter 12), potassium dichromate ($K_2Cr_2O_7$), manganese dioxide, and potassium permanganate.[16]

Electrolytic purification of metals is considered at length in Chapter 17. In essence, metals can be deposited in high purity from solution on a cathodic surface, by careful control of the voltage and other parameters. The anode can be a billet of the impure metal, and the impurities will either stay in solution or form an insoluble "anode slime"; here, both dissolution and reprecipitation of the desired metal are accomplished in a single electrolytic step. Alternatively, a crude solution of the metal ion might be prepared by some other means, and the pure metal deposited on a cathode with an anode of some inert material; the product of electrolysis at the anode will normally be oxygen gas.

Electroplating of one metal onto another is widely used for protection against corrosion and wear or for cosmetic purposes.[16a] Again, the source of metal for deposition could be anodic dissolution or a prepared solution with an inert anode. In contrast to electrolytic refining, only a very thin layer (typically on the order of 1 to 10 μm) of the plating metal is wanted, but usually this layer must be uniform, cohesive, and nonporous, and often a shiny appearance is desired. To understand the roles of some of the variables in electroplating, it is useful to consider the electrodeposition process as having at least three steps:

(*a*) diffusion of a metal ion to the cathodic surface;

(*b*) reduction of the metal ion to form an *adatom;* and

(*c*) migration of the adatom to its final site in the metal lattice.

After initial deposition of a few atomic layers of the plating metal on the substrate, the quality of the thickening film will depend largely on whether process (*c*) is completed before more adatoms are created. For example, if the adatoms have time to migrate to join the step of a screw dislocation (Fig. 5.3), the step will grow laterally and rotate helically about the screw axis to create a continuous layer of new metal atoms. This kind of situation requires that the reduction step (*b*) be slow compared to migration (*c*)—that is, the current density must be *low.* High current densities may lead to formation of local irregularities, culminating in the extreme case with growth of whiskers and feathery structures called *dendrites.*

High concentrations of electrolyte are needed to ensure high conductivity of the solution and hence a more uniform deposition potential over the surface of the specimen. At the same time, however, adatom formation (*b*) may be too rapid unless the *free* metal ion concentration is kept low by complexing it with appropriate ligands, such as silver ion with cyanide:

$$Ag^{+} + 2CN^{-} \rightleftharpoons Ag(CN)_2{}^{-} \quad \beta_2 \approx 1 \times 10^{20}. \tag{15.99}$$

Another factor that may prevent the occurrence of excessive local potentials (and hence excessive deposition rates) on the metal surface is incursion of a different electrochemical reaction, such as hydrogen evolution, at such places as an alternative to the metal deposition that predominates at lower potentials.

Finally, metal objects can sometimes be fabricated in their entirety by electrodeposition (*electroforming*), with much the same considerations as electroplating. Conversely, portions of a metal specimen can be selectively electrolyzed away (*electrochemical machining*). This technique is especially useful where the metal to be shaped is too hard or the shape to be cut is too difficult for conventional machining. The sample is made the anode, a specially shaped tool the cathode, and electrolyte solution (e.g., aqueous NaCl) is fed rapidly but uniformly over the surface to be machined. Current densities may reach several hundred amperes per square centimeter across the electrolyte gap of a millimeter or so. Excellent tolerances can be achieved in favorable circumstances.[16]

Exercises

15.1 The standard pressure for thermodynamic calculations was changed in

1982 from one atmosphere (101.325 kPa) to one bar (100 kPa exactly). What correction should be applied to old $E°$ values for half-reactions *not* involving gases, relative to the H^+/H_2 standard electrode, to bring them to the new scale?
[*Answer:* Subtract 0.169 mV. Note that this is less than the typical uncertainty of ±1 mV in $E°$.]

15.2 Calculate the stability constant β_2 of $Au(CN)_2^-$ at 25 °C from the following standard electrode potentials:

$$Au^+ + e^- \rightleftharpoons Au \qquad E° = +1.68\,V$$
$$Au(CN)_2^- + e^- \rightleftharpoons Au + 2CN^- \qquad E° = -0.6\,V$$

[*Answer:* 3.3×10^{38}.]

15.3 Given that $E°= +1.229$ V for the half-reaction

$$O_2 + 4H^+ + 4e^- \rightleftharpoons 2H_2O$$

in acid solution (1.00 mol L^{-1}), calculate the EMF of the same half-cell in neutral water (pH 7.00) and in aqueous NaOH (1.00 mol L^{-1}), under standard conditions.
[*Answers:* +0.815, +0.401 volt.]

15.4 From the table of standard electrode potentials given in Appendix D, select reagents that should be thermodynamically capable (in 1 mol L^{-1} acid, standard conditions) of (*a*) oxidizing HCl to chlorine gas, (*b*) reducing aqueous chromium(III) to chromium(II), (*c*) reducing solid silver chloride to metallic silver, (*d*) oxidizing water, and (*e*) reducing water. (*f*) Waste aqueous cyanide is often destroyed with sodium hypochlorite solution (domestic chlorine bleach); what are the expected products of the reaction, according to thermodynamics?

15.5 Consider the following electrode potentials (in volts, for 1 mol L^{-1} H^+, at 25 °C):

$$UO_2^{2+} \xrightarrow{+0.05} UO_2^+ \xrightarrow{+0.62} U^{4+} \xrightarrow{-0.61} U^{3+} \xrightarrow{-1.85} U$$
$$PuO_2^{2+} \xrightarrow{+0.91} PuO_2^+ \xrightarrow{+1.17} Pu^{4+} \xrightarrow{+0.98} Pu^{3+} \xrightarrow{-2.03} Pu$$

(*a*) Identify (with rationale) any of these species that would be unstable with respect to disproportionation under these conditions. (*b*) Would a change in pH affect your answer to (*a*)? (*c*) Calculate $E°$ for the following half-reactions:

$$U^{VI} + 3e^- \rightarrow U^{III}$$
$$U^{VI} + 2e^- \rightarrow U^{IV}$$
$$U^{VI} + 6e^- \rightarrow U^0$$
$$Pu^{VI} + 3e^- \rightarrow Pu^{III}.$$

(*d*) With the aid of the standard electrode potentials listed in Appendix D, select a reagent that should reduce plutonium(VI) and plutonium(IV) to plutonium(III) in 1 mol L^{-1} acidic solution, but leave uranium(VI) untouched, assuming standard state conditions.

[*Answers:* (*c*) +0.02, +0.335, −0.915, +1.01 V; (*d*) Any reagent on the right-hand side of a reaction with $E°$ lying between 0.335 and 0.91 volts—iron(II) sulfate solution, perhaps. Considerations such as these can be developed as a basis for separating uranium and plutonium in the reprocessing of spent nuclear fuel elements.]

15.6 Which of the stability field boundaries in Fig. 15.3 are dependent on the arbitrary choice of [Fe^{n+}(aq)] (here, 1×10^{-3} mol L^{-1})?

15.7 Operation of the Ni/Cd alkaline flashlight cell involves the higher oxidation states of nickel. Assuming (for simplicity) that the only such material present is the insoluble solid NiO_2, construct a Pourbaix (E_h–pH) diagram for Ni in aqueous systems, showing the Ni(s), Ni^{2+}(aq), $Ni(OH)_2$(s), NiO_2(s), and H_2O (i.e., $O_2/H_2O/H_2$) stability fields for $[Ni^{2+}] = 1 \times 10^{-3}$ mol L^{-1} (where this is not limited by solubility, etc.). Relevant data are as follows:

$$\begin{array}{rl} Ni^{2+} + 2e^- \rightleftharpoons Ni(s) & E° = -0.257\,V \\ Ni(OH)_2(s) \rightleftharpoons Ni^{2+} + 2OH^- & K_{sp} = 1.6 \times 10^{-16} \\ NiO_2(s) + 4H^+ + 2e^- \rightleftharpoons Ni^{2+} + 2H_2O & E° = +1.593\,V \\ O_2(g) + 4H^+ + 4e^- \rightleftharpoons 2H_2O & E° = +1.229\,V \\ 2H^+ + 2e^- \rightleftharpoons H_2(g) & E° = 0.000\,V. \end{array}$$

Under what circumstances is breakdown of the water theoretically possible? What could prevent this from occurring?

15.8 C. G. Vayenas and R. D. Farr, *Science* **208**, 593 (1980), describe a solid electrolyte fuel cell in which ammonia is the fuel and is catalytically converted at 1000 K with oxygen (or air) to nitric oxide. The idea is that the energy released in this step in industrial nitric acid production (Section 9.4) could be recovered directly as electricity.

Given that $\Delta G°$ for the oxidation of one mole of NH_3 exclusively to NO is −269.9 kJ mol^{-1} at 1000 K and 1 bar, calculate (*a*) the maximum no-load EMF that the cell could (in theory) deliver and (*b*) the maximum theoretical efficiency of the cell, given the following data (25 °C, 0.1 MPa) and assuming $\Delta C_P°$ to be independent of temperature:

	$\Delta H_f°$ (kJ mol^{-1})	$\Delta C_P°$ (J K^{-1} mol^{-1})
O_2	0	29.35
NO	90.25	29.84
NH_3	−41.66	35.06
H_2O(g)	−241.82	33.58

[*Answers:* (*a*) 0.56 V; (*b*) 120%.]

15.9 A hydrogen/oxygen fuel cell, operating at near-standard conditions to produce *liquid* water as the reaction product, delivers 50 A at 0.70 V. What is the efficiency of the cell?
[*Answer:* 47%.]

15.10 Consideration of the relevant E° data (Appendix D) might suggest that electrolysis of aqueous NaCl should give O_2 rather than Cl_2 at the anode. In fact, O_2 is produced only in minor quantities. Suggest possible reasons for this.

15.11 Under what circumstances could the theoretical efficiency of a fuel cell (as defined in this chapter) exceed 100%? Suggest an appropriate combination of fuel and oxidant to achieve this.

15.12 From the data in Appendix C, calculate the theoretical maximum EMF of a methane/oxygen fuel cell with an acidic electrolyte under standard conditions. Assume the products to be liquid water and aqueous CO_2. [*Hint:* You need to know the number of electrons transferred per mole CH_4 consumed. Write a balanced equation for the net reaction, and obtain the number of electrons from Eq. 15.47.]
[*Answer:* 1.05 V.]

15.13 Why is concentrated phosphoric acid preferred over other aqueous acid media as the electrolyte in aqueous acid fuel cells?

15.14 Aluminum might seem to offer a convenient and inexpensive substitute for lithium in lightweight rechargeable cells. Why is this not a practical proposition?

15.15 Why is a Leclanché cell not rechargeable?

References

1. J. O'M. Bockris and A. K. N. Reddy, "Modern Electrochemistry," Vols. 1 and 2. Plenum, New York, 1970.

2. R. A. Robinson and R. H. Stokes, "Electrolyte Solutions," 2nd Ed. (revised). Butterworth, London, 1965.

3. P. H. Rieger, "Electrochemistry," 2nd Ed. Prentice-Hall, Englewood Cliffs, New Jersey, 1994.

4. K. B. Oldham and J. C. Myland, "Fundamentals of Electrochemical Science." Academic Press, San Diego, 1994.

5. D. Pletcher, "A First Course in Electrode Processes." Electrochemical Consultancy, Romsey, Herts., U. K., 1991.

6. J. O'M. Bockris and S. U. M. Khan, "Surface Electrochemistry: A Molecular Level Approach." Plenum, New York, 1993.

7. D. R. Crow, "Principles and Applications of Electrochemistry," 3rd Ed. Chapman & Hall, London, 1988.

8. A. J. Bard, R. Parsons, and J. Jordan (eds.), "Standard Potentials in Aqueous Solution." Dekker, New York, 1985.

9. W. M. Latimer, "Oxidation Potentials," 2nd Ed. Prentice-Hall, Englewood Cliffs, New Jersey, 1952; W. M. Latimer and J. H. Hildebrand, "Reference Book of Inorganic Chemistry," 3rd Ed. Macmillan, New York, 1951.

10. M. Pourbaix, "Atlas of Electrochemical Equilibria in Aqueous Solution." Pergamon, Oxford, 1966.

11. R. M. Garrels and C. L. Christ, "Solutions, Minerals and Equilibria," Chapter 7. Harper & Row, New York, 1965.

12. D. W. Barnum, Potential-pH diagrams. *J. Chem. Educ.* **59,** 809–812 (1882).

13. I. Bodek, W. J. Lyman, W. F. Reehl and D. H. Rosenblatt, "Environmental Inorganic Chemistry." Pergamon, New York, 1988.

14. J. E. Fergusson, "Inorganic Chemistry and the Earth," pp. 53–61. Pergamon, Oxford, 1982.

15. E. Gileadi, "Electrode Kinetics for Chemists, Engineers and Materials Scientists." VCH, New York, 1993.

16. (*a*) D. Pletcher and F. C. Walsh, "Industrial Electrochemistry." Chapman & Hall, London, 1990; (*b*) F. C. Walsh, "A First Course in Electrochemical Engineering." Electrochemical Consultancy, Romsey, Herts., U. K., 1991.

17. M. G. Fontana, "Corrosion Engineering," 3rd Ed., Chapter 9. McGraw-Hill, New York, 1986.

18. S. H. Bergens, C. B. Gorman, G. T. R. Palmore and G. M. Whitesides, A redox fuel cell that operates with methane as fuel at 120 °C. *Science* **265**, 1418–1420 (1994).

19. K. V. Kordesch and G. R. Simader, Environmental impact of fuel cell technology. *Chem. Rev.* **95**, 191–207 (1995).

20. J. A. G. Drake (ed.), "Electrochemistry and Clean Energy." Royal Society of Chemistry, Cambridge, 1994.

21. R. C. T. Slade, Fast ion conductors. *In* "Insights into Speciality Inorganic Chemicals" (D. Thompson, ed.), pp. 169–195. Royal Society of Chemistry, Cambridge, 1995.

22. P. G. Bruce (ed.), "Solid State Electrochemistry." Cambridge Univ. Press, Cambridge, 1995.

23. R. P. Tischer (ed.), "The Sulfur Electrode: Fused Salts and Solid Electrolytes." Academic Press, New York, 1983.

Chapter 16

Corrosion of Metals

METALLIC CORROSION is a major engineering and economic problem. In North America and Europe, the cost of corrosion is on the order of 3–4% of gross domestic product (GDP). Not surprisingly, there is an abundance of good books on this topic.[1–15] This chapter focuses mainly on corrosion of metals in aqueous systems, as this is the most commonly encountered problem and is a natural corollary of the material of the two preceding chapters.

We saw in Section 5.6 that the "dry" oxidation of metals by oxygen or air can be viewed as an electrochemical process in which the *electrolyte* of the cell is the developing solid oxide layer itself. If liquid water is present, diffusion of the ions and molecules involved in the electrochemical corrosion process is greatly facilitated, and consequently aqueous corrosion of metals is much more important than dry oxidation at near-ambient temperatures. Although most corrosion problems encountered in practice involve only a single metal, aqueous electrochemical corrosion can be especially severe, and its principles most clearly illustrated, in cases where two different metals are in electrical contact with one another.

16.1 Bimetallic Corrosion

Consider a zinc strip immersed in water. At equilibrium, a small number of Zn^{2+} ions will pass into solution per unit time, leaving twice as many electrons behind, while an equal number of Zn^{2+} ions already in the water will be redeposited as elemental zinc (reaction 16.1). The rate of this process, in terms of the electrons transferred per unit surface area of the metal, is the exchange current density i_0 for equilibrium 16.1, as explained in Section 15.4:

$$Zn^{2+}(aq) + 2e^- \rightleftharpoons Zn(s). \tag{16.1}$$

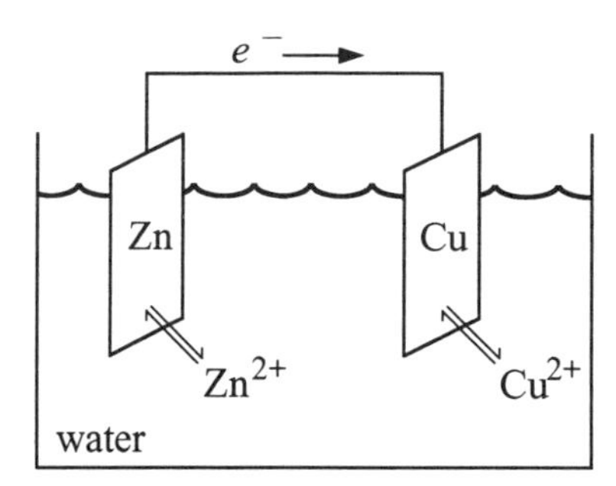

Figure 16.1 The basis of the galvanic (bimetallic) corrosion cell. Zinc, which is more readily oxidized than copper, becomes the anode.

For a strip of copper immersed in water, a similar equilibrium will be set up:

$$Cu^{2+}(aq) + 2e^- \rightleftharpoons Cu(s). \tag{16.2}$$

The driving force for copper deposition (reaction 16.2) is much greater than for zinc precipitation (reaction 16.1), as the corresponding standard electrode potentials of +0.340 and −0.763 V indicate. Consequently, if the copper and zinc strips are placed in the same aqueous medium and are in electrical contact with each other as in Fig. 16.1, reaction 16.2 will proceed in the direction left to right by driving reaction 16.1 in the sense right to left. The zinc will therefore dissolve while the copper ions in the aqueous phase are redeposited.

This situation cannot persist for long: the concentration of copper(II) ions in water will initially be extremely small, unless some other source is also involved, and will quickly be depleted. The important point is that, as soon as electrical contact is made, the zinc becomes an anodic electrode, and the copper a cathode. If another cathodic reaction besides reaction 16.2 is possible, however, then dissolution (i.e., corrosion) of the zinc will continue, while the copper will serve merely as an electrically conducting surface to deliver electrons for the alternative cathodic reaction. In pure water, the obvious alternative reaction is hydrogen evolution (reaction 16.3) for which E_h is −0.414 V at pH 7:

$$2H^+(aq) + 2e^- \rightarrow H_2(g). \tag{16.3}$$

In practice, the overpotential for hydrogen evolution on copper is moderately high (i.e., i_0 is fairly low; Table 15.1), though not as high as on zinc itself. Furthermore, the zinc will tend to become coated with zinc hydroxide as the pH begins to rise as a result of progress of reaction 16.3. The zinc–copper couple will, however, produce hydrogen from hot water, although zinc alone will not, because of the greater overpotential for hydrogen evolution on zinc. Clearly, if the medium is acidic, H_2 will be evolved much more readily—from the copper surface, and not from the zinc, even though it is the zinc that is corroded away. In general, the same considerations apply to other metals of moderately negative $E°$ values: corrosion with hydrogen evolution as the cathodic process is important only at low pH.

16.1.1 Corrosion by Oxygen

If, however, the water contains dissolved oxygen, as it usually does if it is or has been in contact with air, the oxygen absorption reaction 16.4 becomes a likely cathode reaction:

$$O_2(g) + 4H^+(aq) + 4e^- \rightarrow 2H_2O(l) \quad E° = +1.229\,V. \tag{16.4}$$

In neutral water or even 1 mol L^{-1} alkali, E_h for reaction 16.4, better written for basic media as reaction 16.5, is still quite positive (+0.815 and +0.401 V for 0.1 MPa O_2 at pH 7 and 14, respectively, or about 0.01 V less for air at 0.1 MPa). Thus, oxygen absorption remains an important cause of aqueous metallic corrosion, even in alkaline media, although it is clear from reactions 16.4 and 16.5 that it is favored by low pH.

$$O_2(g) + 2H_2O(l) + 4e^- \rightarrow 4OH^-(aq) \tag{16.5}$$

Oxygen absorption, then, is the usual complementary cathodic reaction for the anodic dissolution of metals other than the most reactive ones (Li, Na, K, Ca, etc.), in near-neutral aerated water. Regardless of whether the cathodic reaction is 16.3 or 16.5, however, the following observations regarding bimetallic corrosion apply:

(*a*) The cathodic metal is *protected* from corrosion. It has a *surfeit* of electrons (for delivery to reaction 16.3 or 16.5) and hence is not going to be oxidized so long as the metal with which it is in electrical contact is corroding and so supplying the electrons.

(*b*) The electrical current flowing between the two metals (the corrosion current) is a direct measure of the corrosion rate, since each mole of zinc that dissolves away releases two moles ($2F$) of electrons for consumption at a relatively remote site, i.e., the cathode.

(*c*) Either cathodic reaction involves a diatomic gas, and hence high overpotentials (Section 15.4); therefore, the cathodic reaction is usually the major factor in controlling the corrosion *rate*. The rate is determined by the corrosion current, that is, by the corrosion current *density*, i_{corr} of Fig. 15.6, multiplied by the cathodic surface area. Thus, *the larger the cathodic surface area, the faster the corrosion occurs*, in terms of moles of metal lost per unit time. Furthermore, the same rate of metal loss will be borne by the anode regardless of its size. A small anode will soon be completely dissolved away, with possible catastrophic consequences in engineering applications. One must therefore never allow small metallic structural components to remain in contact with a less anodic metal; otherwise, they will corrode away very rapidly. Iron nails or rivets must not be used with copper or stainless steel sheets, for example. If dissimilar metals must be joined, they should be separated from one another by an electrically insulating layer such as a plastic or rubber grommet.

(*d*) Lowered oxygen concentrations (*deaeration*) or increased pH (alkaline conditions) reduce the EMF of the cathodic half-reactions 16.3 and 16.4 (or 16.5) and so will ordinarily help reduce the corrosion rate.

16.1.2 Bimetallic Corrosion of Iron

In the corrosion of iron by neutral aerated water, in a bimetallic couple with, for example, copper or platinum as the cathode (as in Fig. 16.2), the initial product at the anode is $Fe^{2+}(aq)$:

$$Fe^{2+}(aq) + 2e^{-} \rightarrow Fe(s) \quad E^{\circ} = -0.44\,V. \tag{16.6}$$

Iron(II) is only slowly oxidized further to iron(III), partly because E_h for oxygen reduction at pH 7 (+0.815 V) is only marginally more positive than that for the iron(II)/iron(III) couple

$$Fe^{3+}(aq) + e^{-} \rightarrow Fe^{2+}(aq) \quad E^{\circ} = +0.771\,V, \tag{16.7}$$

but mainly because the rate equation

$$-\frac{d[Fe^{2+}]}{dt} = k[Fe^{2+}][O_2][OH^{-}]^2 \tag{16.8}$$

(where $k = 5 \times 10^{13}$ L^3 mol^{-3} s^{-1} at 25 °C) shows that the half-period for the oxidation of $Fe^{2+}(aq)$ to Fe^{III} would be over an hour at pH 7 and about 9 ppm dissolved O_2. Consequently, the Fe^{2+} ion has ample time to diffuse away from the anode before it is oxidized further.

The example just given assumed a constant pH of 7. Simultaneously with $Fe^{2+}(aq)$ release at the anode, however, hydroxide ion is formed by reduction of oxygen at the cathode (reaction 16.5), increasing the local pH, and diffuses out to mix with Fe^{2+} in the bulk aqueous phase. The solubility product of $Fe(OH)_2$ is about 8×10^{-16}, which means that concentrations of several micromoles Fe^{II} per liter could build up before $Fe(OH)_2$ is precipitated. At the same time, increased pH will accelerate the oxidation

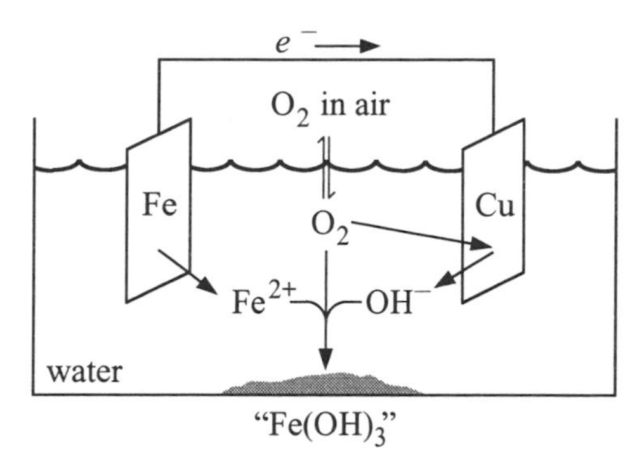

Figure 16.2 Bimetallic corrosion of iron. The immediate anode product is Fe^{2+}(aq), which reacts with dissolved oxygen and OH^- (from the cathode) in bulk solution to form iron(III) oxide/hydroxide which precipitates away from the anodic surface.

by O_2 of Fe^{2+} to Fe^{3+}, which is therefore precipitated as the very insoluble hydroxide—nominally $Fe(OH)_3$, although in practice the brown deposit familiar to us as rust is the partially dehydrated form, γ-FeO(OH) (lepidocrocite). This substance will be deposited from bulk solution rather than on the anodic or cathodic surfaces, and consequently is not protective, in marked contrast to the protective film of iron oxides that forms in the dry oxidation of iron (Section 5.6). The E_h–pH relationships that govern this process are summarized in the Pourbaix diagram for the iron–water system (Fig. 15.3). Under some circumstances, the main product of corrosion of iron can be goethite [α-FeO(OH)]; the "rusticles" (iciclelike growths of rust) that have grown on the wreck of the S. S. *Titanic*, which has rested under 3800 m of cold ocean water off Newfoundland since 1912, consist of cores of needlelike crystals of α-FeO(OH) with a thin surface coat of γ-FeO(OH).[16]

The products of corrosion of iron by well-oxygenated *hot* water are, in fact, usually goethite and ultimately hematite (α-Fe_2O_3). If the oxygen concentration is low, however, the product is usually magnetite (Fe_3O_4) or some partially oxidized nonstoichiometric oxide Fe_xO_4 ($2.667 \leq x \leq 3.000$; see Section 5.6).

16.2 Single-Metal Corrosion

In bimetallic corrosion, the anodic and cathodic surfaces are well defined, being different metals, and are established instantly on placing the metals in electrical contact. The corrosion that results can be very vigorous, but one can usually arrange either to avoid using dissimilar metals together or

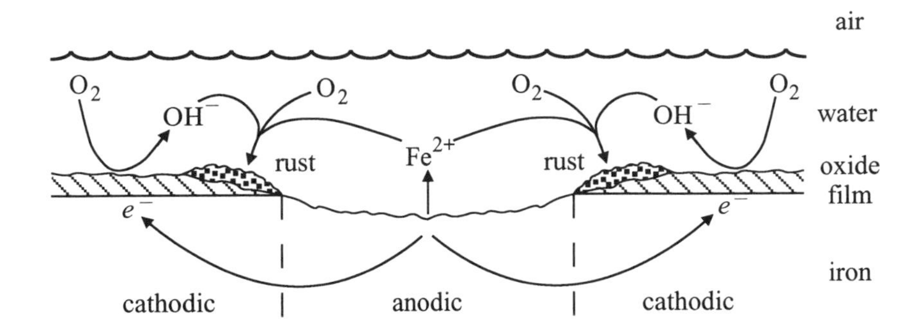

Figure 16.3 Single-metal corrosion of iron by aerated water.

else to place an electrically insulating gasket between them.

Even single metals, however, are subject to aqueous corrosion by essentially the same electrochemical process as for bimetallic corrosion. The metal surface is virtually never completely uniform; even if there is no preexisting oxide film, there will be lattice defects (Chapter 5), local concentrations of impurities, and, often, stress-induced imperfections or cracks, any of which could create a local region of abnormally high (or low) free energy that could serve as an anodic (or cathodic) spot. This electrochemical differentiation of the surface means that local galvanic corrosion cells will develop when the metal is immersed in water, especially aerated water.

A more effective source of anodic and cathodic regions on a single metal is variation in the thickness, or even local absence, of a protective oxide film. The oxide film on an iron specimen, for example, is usually quite thin everywhere (Section 5.6), but there will usually be some spots where it is particularly thin or has been abraded mechanically. These spots become anodes when the specimen is placed in neutral aerated water, and the areas with a more protective oxide film become cathodic. Iron(II) release (reverse of reaction 16.6) then starts up at the anodic sites, and the rate-controlling oxygen reduction reaction 16.5 commences at the relatively large cathodic areas. Iron metal at the cathodic areas remains negatively charged and thus is protected from corrosion, while metal loss at the anodic sites causes them to deepen (Fig. 16.3). The anodic–cathodic differentiation is therefore self-perpetuating and, worse, tends to cause the loss of metal to be locally severe (another point of contrast to "dry" oxidation; Section 5.6).

As in bimetallic corrosion, the Fe^{2+} and the OH^- that constitute the immediate corrosion products are created at quite separate places on the metal surface and must diffuse together over substantial distances in the water phase to complete the electrochemical circuit. Where they come

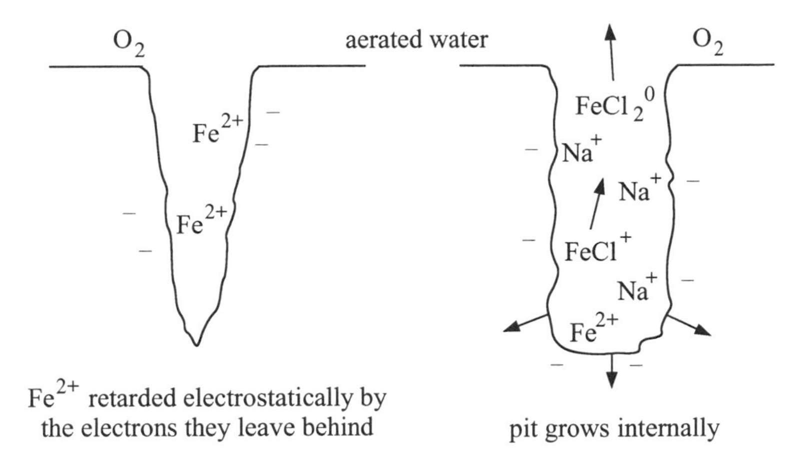

Figure 16.4 Acceleration of corrosion in a crack or crevice (left) by a dissolved electrolyte, NaCl (right).

together, oxidation by dissolved O_2 to iron(III) occurs, and rust is precipitated in a ring around the anodic spot. The diffusion of the ions toward one another is greatly facilitated if other "inert" ions such as Na^+ and Cl^- are present in the aqueous phase and can move to correct the electrical imbalance caused by the movement of the Fe^{2+} or OH^-. To look at the same problem a little differently, the flow of electrons from the anodic site to the cathodic areas requires that an electrical current flow through the aqueous phase to match it and to complete the electrical circuit. If there are substantial concentrations of ions in addition to Fe^{2+} and OH^- already in the water, they can carry most of this current and so increase the corrosion rate (i.e., the electrical corrosion current). Dissolved electrolytes therefore accelerate corrosion (often greatly), even though they usually do not cause the corrosion themselves (the cause is usually reaction 16.5).

Newly formed Fe^{2+} ions, then, must escape from the electrical field of the electrons they are leaving behind on the metal. In addition to the simple current-carrying function of an "inert" electrolyte in the water (described above), some anions such as chloride can form complexes with the Fe^{2+}, facilitating its departure by reducing its effective charge, that is, by forming $FeCl^+$(aq) or even $FeCl_2{}^0$(aq). The transport of iron(III) as $FeCl^{2+}$(aq) and $FeCl_2{}^+$(aq) is similarly enhanced; all these species are more soluble than the hydroxide counterparts. The consequence is that dissolved ionic chlorides such as road salt (NaCl, or $CaCl_2$ from Solvay wastes) are especially aggressive in promoting corrosion. Worse yet, they are particularly effective in accelerating crevice corrosion or pitting, as shown in Fig. 16.4. The result can be the failure of a steel component, weakened by deep pits or crevices, even though the traditional corrosion rate expressed in *mils* per year (1 mil = 0.001 inch or 0.0254 mm), that is, thickness of metal lost per year if corrosion were uniform across the surface, may have seemed insignificant.

16.3 Role of Oxide Films

The oxide film which helped create the cathodic areas in the first place may be destroyed by *reductive dissolution,* at least in the case of iron, because of the excess of electrons present at cathodic sites. This is possible for iron because of the accessibility of the iron(II) oxidation state. The extreme insolubility (and hence protective capacity) of the oxide film derives from its iron(III) content, but iron(II) hydroxide has a significant solubility in neutral water, as noted above, and of course is readily soluble in acidic media.

$$\begin{aligned} Fe_2O_3(s) + 3H_2O + 2e^- &\rightleftharpoons 2Fe(OH)_2(s) + 2OH^-(aq) \\ &\rightleftharpoons 2Fe^{2+}(aq) + 6OH^-(aq) \end{aligned} \qquad (16.9)$$

Equation 16.9 tells us that the protective oxide film on iron will be preserved in alkaline media, weakened in neutral water, and lost in acidic environments. Indeed, in very acidic solutions, the distinction between *extended* anodic and cathodic sites will be lost along with the oxide film, although local anodic and cathodic spots will persist, and so dissolution of iron with accompanying hydrogen evolution becomes general across the surface of the specimen.

The oxide films on the alkali metals and on calcium, strontium, and barium are soluble in water at virtually any pH, and these metals, which also have very negative $E°$ values for reduction of the aqueous ions to the metal, react vigorously—in fact, violently—with water at any pH to liberate hydrogen gas. Sodium, for example, melts from the heat of the reaction and, being less dense than water, rushes around the water surface on a cushion of hydrogen until the gas ignites with a resounding explosion and the characteristic yellow flame associated with vaporized sodium. In Section 5.6 it was noted that the oxide film on these metals is also unprotective in "dry" oxidation, although for an entirely different reason.

At the other extreme, the oxide layers on aluminum, beryllium, titanium, vanadium, chromium, nickel, and tantalum are very insoluble in water at intermediate pH values and do not have easily accessible reduced states with higher solubility. The oxide films on those metals are therefore highly protective against aqueous corrosion.

Aluminum, for example, is a very reactive metal if it is freed of its oxide film:

$$Al^{3+}(aq) + 3e^- \rightarrow Al(s) \quad E° = -1.67\,V, \qquad (16.10)$$

and, if the oxide film is broken under aerated water, Al^{3+}(aq) forms instantly along with OH^-. Thus, a protective film of aluminum oxides or hydroxides immediately forms again on the fresh anodic site, sealing it off and stopping the corrosion reaction at once unless the water is acidic, when Al^{3+}(aq) can escape into bulk solution, or very alkaline, in which case the soluble aluminate ion, $Al(OH)_4^-$(aq), forms:

$$2Al(s) + 2OH^-(aq) + 6H_2O \rightarrow 2Al(OH)_4^-(aq) + 3H_2. \qquad (16.11)$$

There is *no* stable entity Al^{2+}(aq) to compare with Fe^{2+}(aq); consequently, the mechanism that causes rust to be nonprotective because of migration of Fe^{2+}(aq) through the water before precipitation as FeO(OH) does not apply to aluminum, on which $Al(OH)_3$ or AlO(OH) forms, at once, on the anodic site. Conversely, removal of the protective aluminum oxide film cannot occur by the reductive dissolution mechanism described for iron.

The oxide film on chromium, presumably Cr_2O_3, is similarly highly insoluble and not easily reduced to more soluble chromium(II) oxide/hydroxide. Unlike Al^{3+}(aq), Cr^{3+}(aq) *can* be reduced to the sky-blue Cr^{2+}(aq) ion in water, for example, by treatment with zinc metal ($E°$ for Zn^{2+}/Zn is -0.76 V):

$$Cr^{3+}(aq) + e^- \rightarrow Cr^{2+}(aq) \quad E° = -0.42\,V. \qquad (16.12)$$

However, it is obvious from the $E°$ of reaction 16.12 that Cr^{2+}(aq) is a powerful reductant and is very rapidly reoxidized to various green chromium(III) species by aerated water. As long as there is *some* oxygen present (more accurately, as long as the E_h of the system is more positive than about -0.4 V), reductive dissolution will not affect the protective Cr_2O_3 film, and the chromium will be very resistant to corrosion. The metal in this condition is said to be *passive*. If, however, the E_h of the environment becomes so negative that reductive dissolution sets in (read Cr for Fe in reaction 16.9), the passivity of the chromium will be lost, and the metal will then become *active* in corrosion.

Stainless steels, all of which contain at least 11% Cr, owe their usual passivity to an oxide layer that can be approximately formulated as $FeCr_2O_4$, or (if nickel is also present) as $(Ni,Fe)Cr_2O_4$. These spinel-type oxides are similar to the mineral chromite (Section 4.6.2), which is extremely insoluble in aqueous media. Again, however, the oxide film may be weakened or lost under strongly reducing conditions, and the "stainless" steel may then become active. This problem is analyzed in electrochemical terms in Section 16.6.

16.4 Crevice and Intergranular Corrosion

In the rusting of iron, loss of metal occurs at the anodic sites *only*. Oxygen reduction occurs elsewhere, where the metal is cathodic and therefore protected by the excess of electrons (whether the oxide film survives there or not). Paradoxically, then, locally *high* concentrations of dissolved oxygen tend to create cathodic areas on the metal and so protect it. This phenomenon is referred to as the *principle of differential aeration*. Conversely, anodic sites tend to form in places where the oxygen concentration is locally low, for example, in cracks or crevices where the initially available dissolved oxygen becomes depleted by the first, general oxidation of the iron and cannot be replaced rapidly enough by diffusion.

While some crevices occur naturally, others may either be induced by stress or be created by poor engineering practice. For example, sloppy riveting of iron plates with an iron rivet may leave confined spaces within the joint (Fig. 16.5) in which the oxygen concentration is soon depleted to well below that in the water outside. The metal inside the joint therefore becomes anodic and, worse yet, the entire outer surface of the riveted plates becomes an extended cathode where the rate-determining oxygen reduction step in the overall corrosion process occurs. The small anodic region at the rivet therefore supports essentially the whole of a substantial corrosion current, and rapid metal loss at the rivet will lead to failure of the joint. The deposit of rust will tend to form just outside the crevices where corrosion is occurring, since the $Fe^{2+}(aq)$ will have to diffuse out to where the oxygen concentration and the pH are relatively high to be oxidized to γ-FeO(OH). As noted earlier in Sections 16.2 and 16.3, the corrosion rate will be even greater if dissolved salts are present in the aqueous phase (e.g., seawater), and worse still if the rivet is of a more reactive metal than the plates.

Local accumulation of dirt on a steel structure in a damp environment is enough to set up an anodic area underneath it by excluding air. Similarly, chipped paintwork results in lateral spreading of anodic areas under the paintwork, radially outward from the chips. At the chipped site, air has relatively free access to the metal, but under the paint the oxygen is excluded and anodic activity becomes intense, spreading under the paint and leaving a trail of rust behind where air has slowly diffused in to oxidize the $Fe^{2+}(aq)$.

Such rusting phenomena as these are distressingly familiar in marine environments or in moderately cold climates where salts are used to deice roads. Acceleration of corrosion by seawater or sea spray, or by road salt, has several origins:

(*a*) Salts increase the electrical conductivity of the aqueous phase and so increase the corrosion current (Section 16.2).

(*b*) Some anions such as chloride complex Fe^{2+} and Fe^{3+} and facilitate

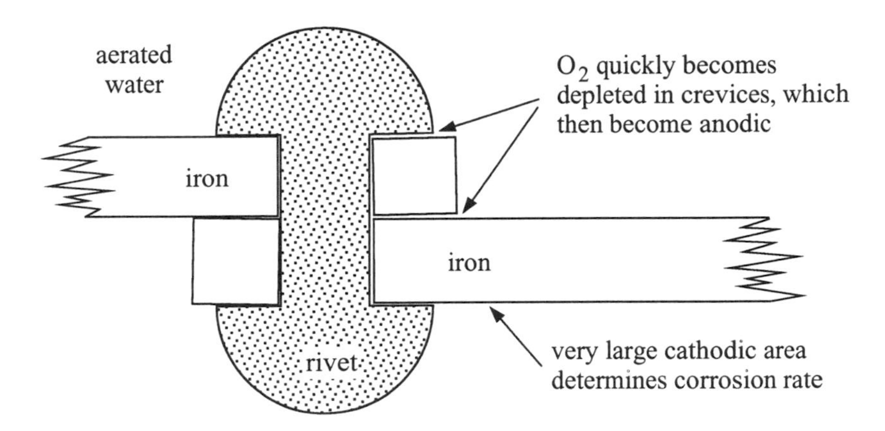

Figure 16.5 Differential aeration around a poorly fitted rivet leads to intense local corrosion.

their dispersal (Section 16.2).

(*c*) Complexing anions like chloride may act as bridging ligands between, say, Fe^{2+} in solution and Fe^{3+} in the oxide film, so facilitating electron transfer into the oxide film (Fe^{III} becoming Fe^{II}) and expediting its breakup:

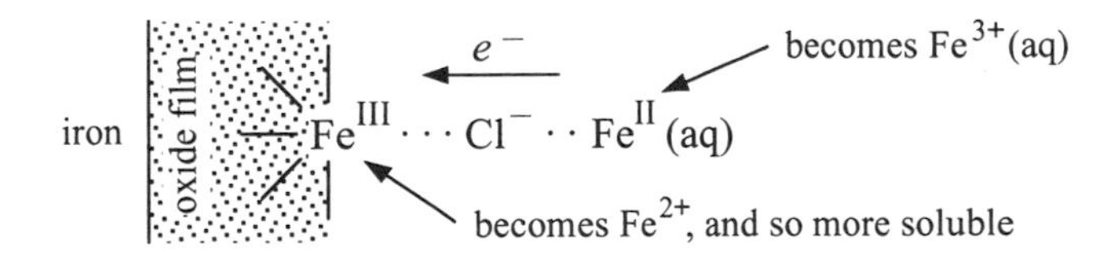

(*d*) Dissolved salts depress the freezing point of water, so aqueous corrosion continues below 0 °C.

(*e*) Dissolved salts depress the vapor pressure of liquid water [the ultimate cause of (*d*)] and so retard its evaporation; the metal stays wet longer.

The worst aspect of crevice corrosion is that it weakens the specimen locally, often unseen in a casual inspection, and may lead to failure of the specimen even though only a very small amount of metal has been lost. This type of corrosion can originate where stress has induced defects in the atomic lattice of the metal, creating spots with locally high free energy and hence higher chemical reactivity, or where stress caused cracking (***stress cracking corrosion***).

Even in unstressed metals, however, the boundaries of the grains in the metal structure (Fig. 5.4) have relatively high free energy and thus tend to

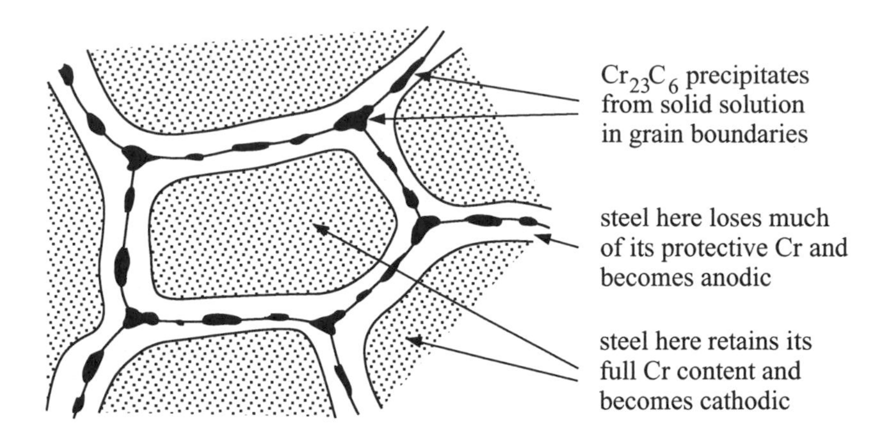

Figure 16.6 Creation of microscopic anodic areas in the weld decay of a typical stainless steel.

become anodic, leading to *intergranular corrosion*. The intergranular regions also tend to be the places where impurities, precipitated from solid solution, tend to collect, and this is illustrated by the *weld decay* of austenitic stainless steels, which typically contain 18% Cr and 8% Ni as well as iron and 0.1 to 0.2% C. At 500 to 800 °C, a chromium carbide phase $Cr_{23}C_6$ separates from the chemically homogeneous stainless steel and precipitates in the grain boundaries, leaving a Cr-depleted zone around the boundaries (Fig. 16.6). The steel depends on a high Cr content to remain corrosion-resistant (Section 16.3), so that the regions around the grain boundaries become anodic when the cold specimen is left in contact with aerated water, the rest of the steel serving as a much larger cathode. As a result, intense intergranular corrosion will occur in those parts on either side of a weld in a stainless steel specimen where the temperature has been 500 to 800 °C for an extended time during welding (Fig. 16.7).

There are several ways in which weld decay can be avoided:

(*a*) High temperature arc welding helps get the job completed quickly, before much $Cr_{23}C_6$ is precipitated.

(*b*) The sample might be reheated to about 1120 °C, whereupon the $Cr_{23}C_6$ will go back into solid solution. The sample should then be quenched in water.

(*c*) Elements that form more stable carbides than does chromium can be alloyed into the stainless steel when it is made. The idea is to prevent the chromium from being consumed locally by the carbon by tying up the carbon in another carbide. From the discussion in Section 5.7 it follows that transition metals to the left of chromium

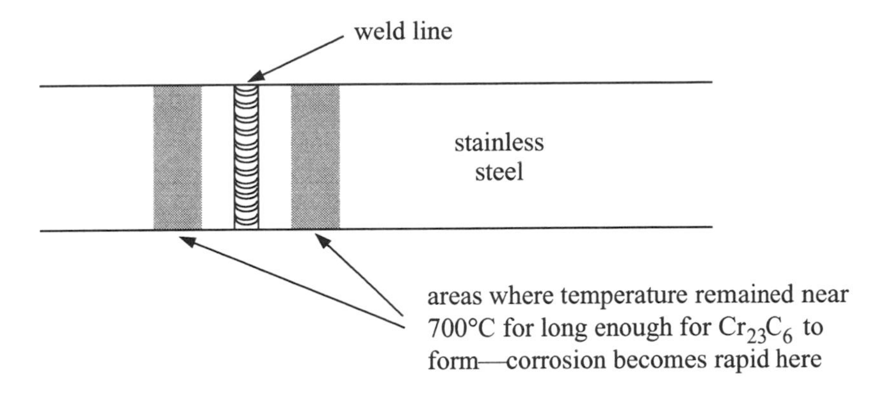

Figure 16.7 Areas susceptible to weld decay (*heat affected zones* or HAZ) in a welded stainless steel article.

in the periodic table should be suitable. In practice, titanium is used in AISI type 321 stainless steel,* and niobium (columbium, in U.S. industrial usage) and tantalum in AISI type 347.

(*d*) A special low-carbon (<0.03 weight %) stainless steel can be prepared, for example, AISI types 304L and 316L, where "L" designates low carbon content. This may, of course, mean some loss of desirable mechanical properties.

16.5 Corrosion by Acids and with Complexing Agents

In general, the susceptibility of metals M to aqueous corrosion is expected to correlate inversely with the E° values for the reduction of M^{m+}(aq) to M(s): the less positive E° is, the greater is the tendency of M to corrode in aerated water. Factors that can upset predictions based on E° include the presence of protective films (***passivation***), overpotential effects (Sections 15.4 and 16.6), effects of complexing agents, and incursion of a cathode reaction other than O_2 reduction or H_2 evolution.

Complexing agents usually promote corrosion, either by stripping away protective films or metal hydroxide deposits (cf. Section 13.5) or by changing the E° or E_h value itself (Section 15.2). In Exercise 15.2, we saw that E° for the gold(I)/gold metal couple went from +1.68 V for the unstable ion Au^+(aq) to −0.6 V if gold(I) were complexed with cyanide ligands. The latter figure shows that gold can be oxidized by *aerated* aqueous NaCN solu-

*The composition of stainless steels is discussed in Section 16.8. AISI stands for American Iron and Steel Institute.

tion, which dissolves it as $Na[Au(CN)_2]$, and in fact this reaction is used in the hydrometallurgical extraction of elemental gold from crushed ores (Section 17.2.1). [Note that pK_a for HCN is about 9, so NaCN solutions are very alkaline, and the lowered E_h for H^+/H_2 plus overpotential effects will prevent the $Au(CN)_2^-/Au$ couple from reducing the solvent to hydrogen.]

Stabilization of nickel(II) and copper(II) by complexing with $EDTA^{4-}$ or NTA^{3-} similarly increases the vulnerability of these normally corrosion-resistant metals to corrosion by aerated waters. For this reason, $EDTA^{4-}$ or NTA^{3-} solutions should not be used in copper or high nickel alloy equipment.

Corrosion of metals by aqueous acids with hydrogen evolution is usually rapid and fairly uniform across the surface (*general corrosion*), since the reductive dissolution of the oxide film that helps maintain the distinction between anodic and cathodic sites is favored by low pH (reaction 16.9). Thus, although local anodic and cathodic areas persist, pitting becomes less important than overall loss of metal. If the oxide film is sufficiently insoluble in acids and is also resistant to reductive dissolution, as with titanium or stainless (>11% Cr) steels, the metal may remain unaffected by aqueous acids, except at quite negative E_h values. In cases where the cathodic discharge of hydrogen ions

$$H^+(aq) + e^- \rightarrow H\cdot(aq) \tag{16.13}$$

does occur but the formation of diatomic hydrogen gas is inhibited (e.g., by the presence of H_2S), the hydrogen atoms may penetrate the metal, giving an interstitial hydride (Sections 5.7 and 15.4). Indeed, since H_2S is quite acidic, it can itself attack steel with the formation of iron sulfides and hydrogen and so is particularly problematic in sour natural gas wells and associated equipment.

Hydrogen atom penetration may lead to *embrittlement* of the metal, or, if the hydrogen atoms diffuse together through grain boundaries or voids to form gaseous H_2 *within* the metal, *hydrogen blistering* of the metal may result. In either case, it is a highly undesirable development from the engineering standpoint. For example, a combination of hydrogen embrittlement, sulfide corrosion, and stress cracking corrosion has been held responsible for several natural gas pipeline ruptures in Saskatchewan, Manitoba, and Ontario throughout the 1990s that caused spectacular fires locally and severely disrupted gas deliveries to eastern Canada and United States. In at least some of these cases, the hydrogen was evidently generated by bacterial action at breaks in the protective outer coating of the pipeline. The crack-inducing stress was the high operating pressure of the pipeline (5–10 MPa).

If oxidizing anions are present, the main cathodic reaction in aqueous corrosion may be neither oxygen absorption nor hydrogen evolution but *reduction of the anion.* Reduction of these anions, which are almost inevitably oxoanions, usually involves consumption of several hydrogen ions, so that this type of cathodic reaction is usually encountered only under very acidic conditions. Aqueous nitric acid or concentrated sulfuric acid provide familiar examples:

$$4H^+ + NO_3^- + 3e^- \rightarrow NO(g) + 2H_2O \quad E^\circ = +0.96\,V \quad (16.14)$$

$$3H^+ + HSO_4^- + 2e^- \rightarrow SO_2(g) + 2H_2O \quad E^\circ = +0.17\,V. \quad (16.15)$$

Copper metal therefore dissolves readily in relatively dilute aqueous nitric acid (see Section 15.2) to give sky-blue $Cu^{2+}(aq)$ and nitric oxide gas:

$$Cu^{2+}(aq) + 2e^- \rightarrow Cu(s) \quad E^\circ = +0.34\,V \quad (16.16)$$

$$3Cu(s) + 8HNO_3(aq) \rightarrow 3Cu(NO_3)_2(aq) + 4H_2O + 2NO(g) \quad \Delta E^\circ = +0.62\,V. \quad (16.17)$$

However, in concentrated (70%) nitric acid the gaseous product is mainly nitrogen dioxide:

$$Cu(s) + 4HNO_3(aq) \rightarrow Cu(NO_3)_2(aq) + 2H_2O + 2NO_2(g). \quad (16.18)$$

Copper does not dissolve in 1 mol L^{-1} sulfuric acid, since E° for the half-reaction 16.16 is too positive relative to sulfate reduction (reaction 16.15) or hydrogen evolution. It will, however, dissolve in hot, concentrated (18 mol L^{-1}) sulfuric acid, in which the activities of H^+ and HSO_4^- are much higher and the activity of water (now a minor constituent) much lower than in the standard-state conditions for which the E° of half-reaction 16.15 applies:

$$Cu(s) + 2H_2SO_4(\text{hot, concentrated}) \rightarrow CuSO_4\ \text{solution} + 2H_2O + SO_2(g). \quad (16.19)$$

Sometimes it happens that incursion of oxoanion reduction in place of hydrogen evolution as the cathodic reaction in the corrosion of iron leads, not to an increased rate of corrosion, but to a drastic retardation. This is because strongly oxidizing conditions (e.g., in concentrated nitric acid) can force the immediate oxidation of iron to iron(III), rather than via the persistent iron(II) intermediate (as described in Sections 16.1 and 16.2), so that an insoluble iron(III) oxide layer forms at once on the anodic and cathodic surfaces alike and the iron becomes passivated (Section 16.3). Michael Faraday's demonstration of this phenomenon is instructive:

(*a*) An iron specimen is placed in dilute nitric acid. It is seen to begin to dissolve with a steady evolution of gas.

(*b*) The specimen is transferred to concentrated nitric acid. A few bubbles of gas may escape at first, then no more, and the now passive iron is not attacked further.

(*c*) The passivated iron may be transferred back to the dilute nitric acid. If this is done carefully, the iron remains passive because of its oxide film, and no gas at all is evolved.

(*d*) The specimen, still under dilute HNO_3, is struck lightly. The chemically inert but mechanically fragile film is broken, and gas evolution recommences vigorously, initially where the iron was struck, but subsequently over the whole surface of the specimen as the remnants of the oxide film undergo reductive dissolution. Obviously, such a delicate oxide film has no value in corrosion-proofing iron (contrast the tough "$FeCr_2O_4$" film on stainless steels).

16.6 Role of Overpotential in Corrosion

Consider first the *polarization curve* (i.e., Tafel plot) for the anodic half-reaction occurring in corrosion of stainless steels (Fig. 16.8). The diagram for the active region is much the same as has been seen for other anodes (Figs. 15.4 to 15.7). As E_h is increased to a certain specific value, however, a sudden and dramatic drop in the anodic current density i occurs, corresponding to formation of an oxide film. At higher E_h, i remains constant at a very low level (the horizontal scale in Fig. 16.8 is logarithmic), and the metal has become passive, that is, effectively immune from corrosion.

The oxide film may be regarded as consisting of chromite ($FeCr_2O_4$), the protective capacity of which is due to the extreme insolubility conferred by the cationic chromium(III) content. Under sufficiently oxidizing conditions, however, chromium(III) is oxidized to chromium(VI), which characteristically forms the soluble oxoanions CrO_4^{2-} (chromate, yellow), $HCrO_4^-$ (hydrogen chromate, orange, $pK_a \approx 6$), and $Cr_2O_7^{2-}$ (dichromate, orange, from condensation of $HCrO_4^-$). There is therefore an upper limit in E_h to the passivity of a stainless steel, beyond which active corrosion again sets in with release of orange or yellow chromium(VI) into the water phase. This transpassive region is not normally entered in aqueous systems at ordinary temperatures, although in highly oxygenated water far above its normal boiling point (under pressure) chromium(VI) is sometimes seen to form.

The passive range of typical stainless steels conveniently spans most of the E_h stability field of neutral water. This can be appreciated by examination of the E° or E_h values for reactions 16.20 to 16.25, with the caveat that these refer to pure iron and chromium metals rather than to stainless steels and that the conditions are standard ones rather than, for example, the very low $[Cr^{3+}]$ in equilibrium with the "$FeCr_2O_4$" film. The formation

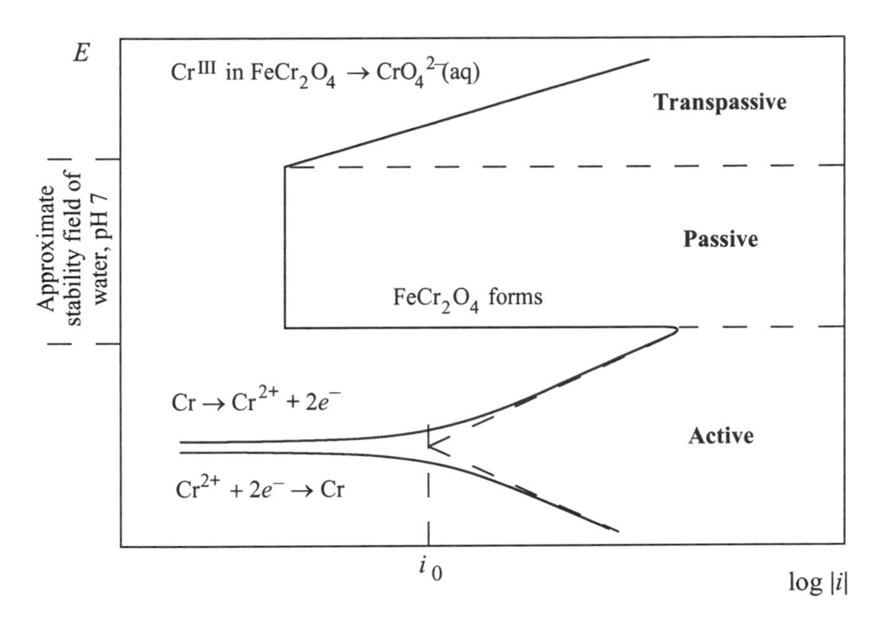

Figure 16.8 Polarization (voltage versus log current density) curve for a typical stainless steel in water (cf. Fig. 15.5).

of the $FeCr_2O_4$ film becomes possible when Cr^{II} can be oxidized to Cr^{III}, and when iron metal can be oxidized to Fe^{II}.

$$Cr^{2+}(aq) + 2e^- \rightarrow Cr(s) \qquad E^\circ = -0.91\,V \quad (16.20)$$

$$Fe^{2+}(aq) + e^- \rightarrow Fe(s) \qquad E^\circ = -0.44\,V \quad (16.21)$$

$$Cr^{3+}(aq) + e^- \rightarrow Cr^{2+}(aq) \qquad E^\circ = -0.42\,V \quad (16.22)$$

$$Cr_2O_7{}^{2-} + 14H^+ + 6e^- \rightarrow 2Cr^{3+} + 7H_2O \qquad E^\circ = +1.36\,V, \quad E_{h(pH\ 7)} = +0.39\,V \quad (16.23)$$

$$2H^+(aq) + 2e^- \rightarrow H_2(g) \qquad E^\circ = \ 0.00\,V, \quad E_{h(pH\ 7)} = -0.41\,V \quad (16.24)$$

$$O_2(g) + 4H^+(aq) + 4e^- \rightarrow 2H_2O \qquad E^\circ = +1.229\,V, \quad E_{h(pH\ 7)} = +0.81\,V \quad (16.25)$$

A diagram similar to Fig. 16.8 could be constructed for ordinary iron or steel, but the onset of passivity, corresponding to the formation of an Fe_2O_3 film directly over the entire surface of the metal, would occur at a much higher E_h value than for stainless steel [E° for $Fe^{3+}(aq)/Fe^{2+}(aq)$ is +0.77 V]. The transpassive region, corresponding to anodic oxidation of the

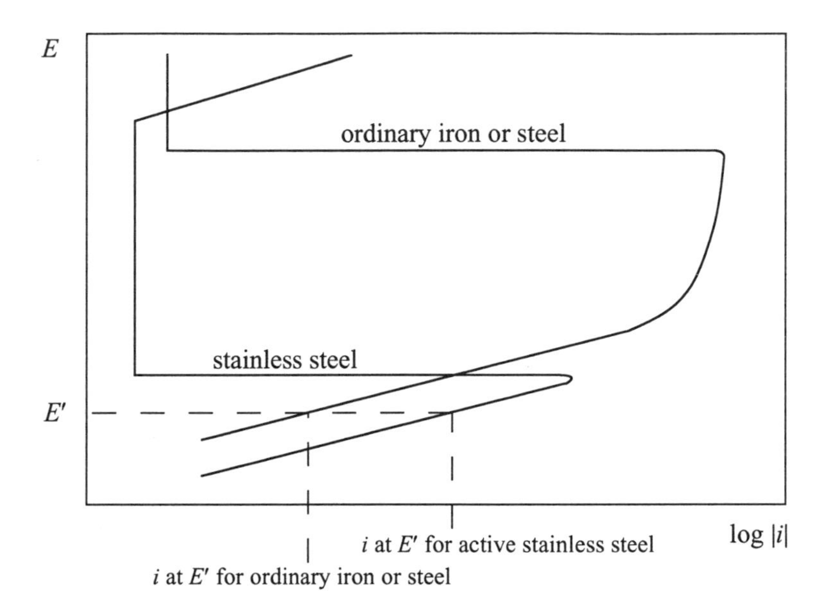

Figure 16.9 Anodic polarization curves for typical stainless and ordinary steels. At low E_h, stainless steel may become active, in which case it corrodes more rapidly than ordinary iron or steels.

Fe_2O_3 film to soluble $FeO_4{}^{2-}$, requires a high impressed EMF and is only feasible in alkaline conditions (cf. the Pourbaix diagram of the iron–water system, Fig. 15.3).

In Fig. 16.9, the operationally significant parts of qualitative polarization curves for a typical steel and a stainless steel are superimposed. It is seen that, for a given E_h value E' in the active range of the stainless steel, the current density will be higher for dissolution of the stainless steel than for corrosion of the iron. It is therefore very important that stainless steels be prevented from becoming active in service, because, if they do, they corrode rapidly, more than ordinary iron would.

Figures 16.8 and 16.9 show only the anodic polarization curves for corrosion cells. The important question is, where do these curves intersect with the polarization curves for likely cathodic reactions, such as hydrogen evolution or oxygen absorption? The intersection point defines the corrosion current density i_{corr} and hence the corrosion rate per unit surface area. As an example, let us consider the corrosion of titanium (which passivates at negative E_h) by aqueous acid. In Fig. 16.10, the polarization curves for H_2 evolution on Ti and for the Ti/Ti^{3+} couple intersect in the active region of the Ti anode. To make the intersection occur in the passive region (as in Fig. 16.11), we must either move the H^+/H_2 polarization curve bodily

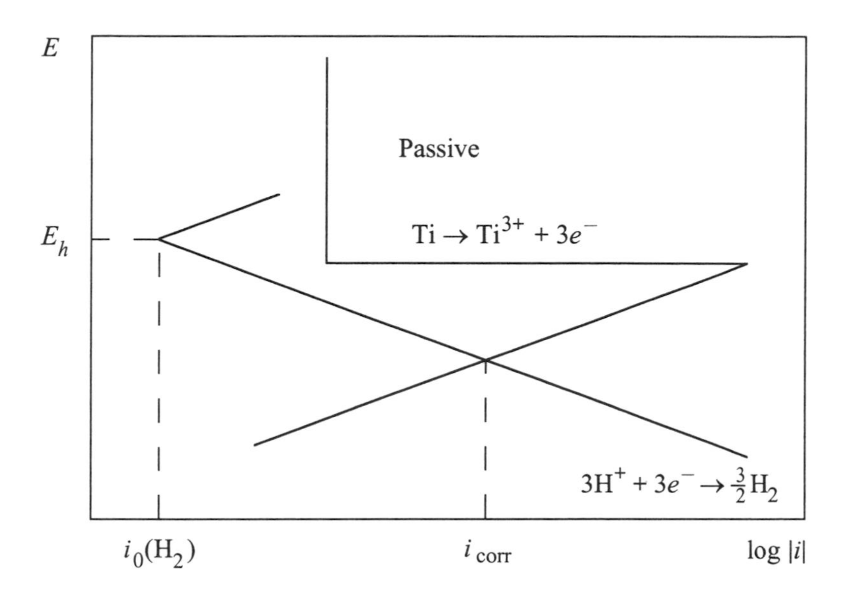

Figure 16.10 Active corrosion of titanium by aqueous acid.

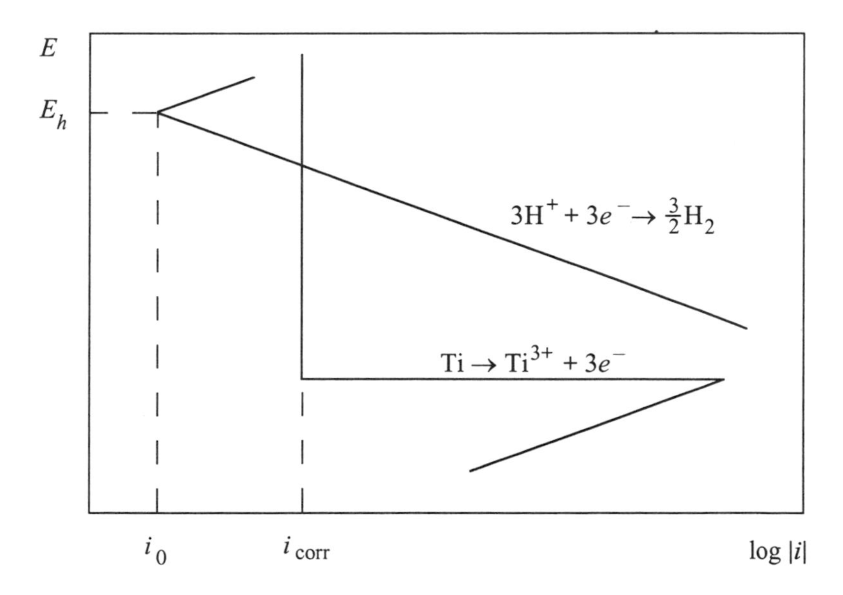

Figure 16.11 Passive titanium–0.2% palladium alloy in aqueous acid. Here, i_0 refers to catalyzed H_2 evolution.

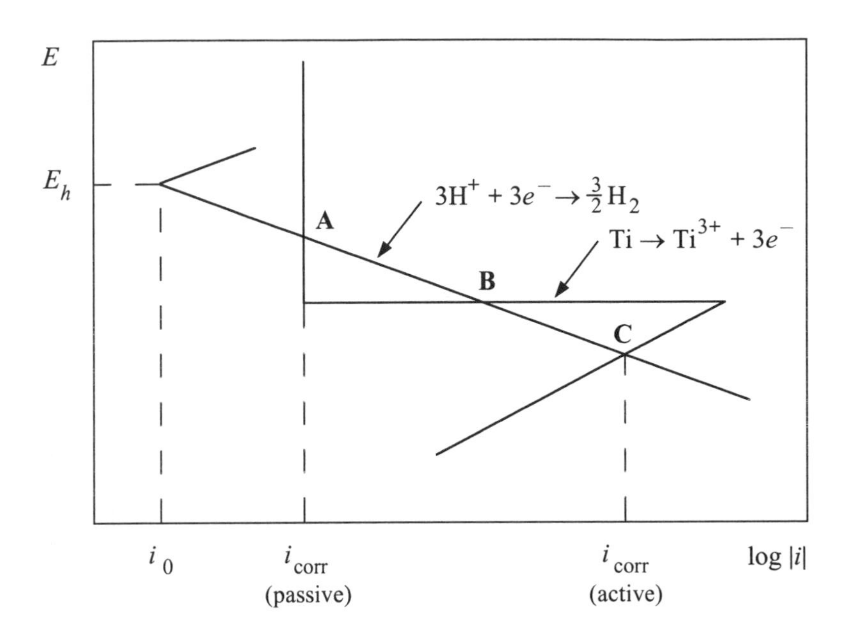

Figure 16.12 Titanium with metastable passivity. If E_h or i_0 are not quite high enough, the metal may go from passive to active behavior without warning. If this happens, the corrosion current will be high (cf. normal active corrosion, Fig. 16.10).

upward, which means, paradoxically, increasing the hydrogen ion concentration and hence E_h for that electrode, or slide the curve to the right by increasing the exchange current density i_0 for the evolution of hydrogen on titanium, or both. The exchange current density can be increased by catalyzing hydrogen evolution with 0.1 to 0.2% palladium, alloyed into the titanium. The resulting alloy costs at least twice as much as ordinary commercial Ti alloys but may be worth the extra cost if service in acids at low E_h values is required.

We must, however, avoid the situation shown in Fig. 16.12, where the cathodic polarization curve intersects the anodic curve not only in the passive region at point A, but also on the active "nose" of the curve at B and C. The intersection at B is of no real significance, as it is electrically unstable (a slight perturbation in E_h will send the system to situation A or C). Intersection C, however, represents a high rate of corrosion, and although we may seem to have a passive, noncorroding system corresponding to A, this serene situation may change suddenly and without warning to that represented by C. It is therefore very important that E_h and i_0 for the cathode reaction be set high enough to place A well into the passive zone, so that the cathode polarization curve easily clears the nose on the anodic polarization curve. The same considerations apply to any metals or alloys (notably stainless steels) that rely on an oxide film for corrosion resistance.

16.7 Control of Corrosion

Many ploys to minimize corrosion will suggest themselves from the foregoing; for example, steel structures should be kept dry and free of dirt and salts, and water could be deaerated. It must be kept in mind, however, that steps taken to reduce one technological problem may create another; for example, softening of water may increase corrosion, while the use of toxic chromates as anodic inhibitors (see Section 16.7.3) has obvious environmental limitations. In corrosion control, as in all applications of chemistry, one has to consider the broader consequences before taking action, and often one perceived remedy for corrosion (e.g., removing excess CO_2 from water by thorough aeration) may accelerate corrosion in another way (in this case, by saturating the water with oxygen at its atmospheric partial pressure).

16.7.1 Cathodic Protection

Cathodic protection involves simply making the metal object electrically negative with respect to the surroundings. This can be done by connecting the object electrically to a more reactive metal or "sacrificial anode," which is allowed to corrode away while supplying protective electrons to the metal object. Typical examples are protection of steel power pylons with magnesium stakes sunk into the ground, use of magnesium bars inside hot-water tanks, and attachment of magnesium or zinc ingots to the hulls of steel ships. In all cases, the sacrificial anode will have to be replaced from time to time. Steel used for tubs or outdoor structures can be coated with zinc (*galvanized iron*); the idea is the same, except that the lost coating cannot easily be replaced. Alternatively, the structure to be protected may be connected to the negative terminal of a DC source such as a fuel cell or a rectified low voltage tap through a transformer from a nearby AC transmission line; the former is an attractive proposition for natural gas pipelines in remote locations, since a miniscule portion of the gas can be bled off to operate the unattended fuel cell.

16.7.2 Protective Coatings

Protective coatings of paint or plastic resins are a familiar means of corrosion control by cutting off access of water, air, and electrolytes to the structure. The coating must, however, be complete and remain intact to be protective. Local penetration of the coating will generally create an active corrosion site that may be driven by a cathodic reaction involving exposed areas elsewhere or slow diffusion of oxygen through the large area of surviving coating. Likewise, *plating* with a corrosion-resistant metal such as chromium, tin, or gold is effective only as long as the plate remains *completely* intact. If the underlying metal (usually steel) is exposed anywhere, it will immediately become anodic and will have to support a large corrosion current controlled by the much greater cathodic area of intact plate. The intense localized corrosion so caused will be considerably worse than in the case of chipped paint or resin.

If the metal to be corrosion-proofed is one that normally forms a protective oxide film of low electrical conductivity (e.g., aluminum, beryllium, titanium, or zirconium), the film can be forcibly thickened by making the metal object the anode of an electrolytic cell until the current it can pass becomes negligible. This process is known as *anodizing* and is extensively applied to aluminum products. Typically, the anodizing of aluminum in 3 mol L^{-1} aqueous sulfuric acid produces an oxide coat 10 μm thick, whereas in air at 400 °C the oxide film thickness would reach only about 2 nm in the same time. The outer layers of oxide produced in anodizing are porous to some degree and can adsorb dyes while they form, giving a lustrous, brightly colored product.

16.7.3 Corrosion Inhibitors

Modification of the corrosive environment may be feasible if one is dealing with a closed system. For example, if oxygen absorption is the main cathodic process driving corrosion, deoxygenation of the water phase by purging with nitrogen or vacuum degassing may be effective in suppressing corrosion. Alternatively, the oxygen in a hot aqueous system (such as boiler water) may be removed chemically by addition of sodium sulfite, which will leave possibly undesirable sodium sulfate in solution, or hydrazine ($H_2N{-}NH_2$, Section 2.10), which goes to nitrogen and water and so is preferred for "zero dissolved solids" operation. Such chemical oxygen absorbers are sometimes classed as corrosion inhibitors, but this term is properly reserved for additives that are not consumed (see below).

It should be remembered that, even with ordinary steels, high oxygen concentrations are locally protective, so that deoxygenation is not recommended if the metal to be protected is already cathodic. With stainless steel equipment, the E_h of the system must not be taken too low, otherwise the metal may become active (Section 16.6).

Corrosion inhibitors are solutes that blanket the electrochemically active surfaces of the corrosion-prone metal and suppress corrosion either by physically blocking the flow of ions or molecules to or from these surfaces or by altering the electrical double layer at the metal surface in such a way as to retard the anodic or cathodic corrosion processes (they may, for example, displace aggressive adsorbed ions such as chloride).[17–19] Many *anodic inhibitors* operate by causing the freshly formed $Fe^{2+}(aq)$ ions at the anodic surfaces of corroding iron to be precipitated immediately on the anode itself (cf. passivation of iron by concentrated HNO_3, which, however, affects *all* surfaces of the iron). *Alkaline solutions* achieve this by causing $Fe(OH)_2$ to precipitate at once, rather than after diffusion of the $Fe^{2+}(aq)$. If dissolved oxygen is present, the $Fe(OH)_2$ oxidizes to insoluble Fe^{III} oxides or hydroxides, while at high temperatures and low oxygen partial pressures (e.g., in deoxygenated boilers or in heat transfer circuits of water-cooled nuclear power reactors), $Fe(OH)_2$ decomposes spontaneously to hydrogen and a black, protective film of magnetite (the *Schikorr reaction*):

$$3Fe(OH)_2(s) \xrightarrow{\text{heat}} Fe_3O_4(s) + 2H_2O(l) + H_2(g). \tag{16.26}$$

Hydrogen could also be generated *at high temperatures* by direct reaction of iron with water to form $Fe(OH)_2$; in principle, this is marginally possible at high pH even at 25 °C, according to Fig. 15.3, but is inhibited by overpotential. Furthermore, in *very* concentrated alkali, the product $Fe(OH)_2$ may form soluble green hydroxo complexes (Section 15.3) and so may not assist in formation of protective layers. Since hydrogen can cause hydriding of the metal (Sections 5.7 and 16.5), *caustic embrittlement* of the iron may become locally intense in a boiler or other heat transfer system wherever hot aqueous alkali becomes highly concentrated (e.g., in steam-forming pockets, in stress-induced cracks, or in leaky seams or rivets; cf. Section 16.4), with potentially disastrous consequences. This must be borne in mind whenever alkali is used for corrosion suppression in aqueous systems at high temperatures.

Phosphates, *molybdates*, and (at high pH) *silicates* act as anodic inhibitors much as do alkalis, except that the iron oxides/hydroxides formed on anodic sites then contain some $PO_4{}^{3-}$, $MoO_4{}^{2-}$, or $SiO_4{}^{4-}$ ("basic" iron phosphates, etc.). These inhibitors require the presence of O_2 to produce basic iron(III) phosphate, molybdate, or silicate films, whereas oxidizing anions such as *chromates* and *nitrites* oxidize $Fe^{2+}(aq)$ rapidly to insoluble iron(III) oxides on anodic sites. *Dianodic inhibitors* combine complementary inhibition mechanisms; for example, sodium triphosphate may be used with sodium chromate, or sodium molybdate with $NaNO_2$.

A different approach to blanketing anodic areas involves dissolving organic anions with appropriate hydrophobic (i.e., water-repellent) substituents. The anionic heads of the molecules specifically seek out a positively charged anodic spot, and the hydrophobic tails serve to isolate it from the aqueous solution and so block the ionic part of the corrosion circuit or, at least, modify the electrical double layer (Fig. 16.13). Sodium benzoate and especially sodium cinnamate are particularly effective in this regard.

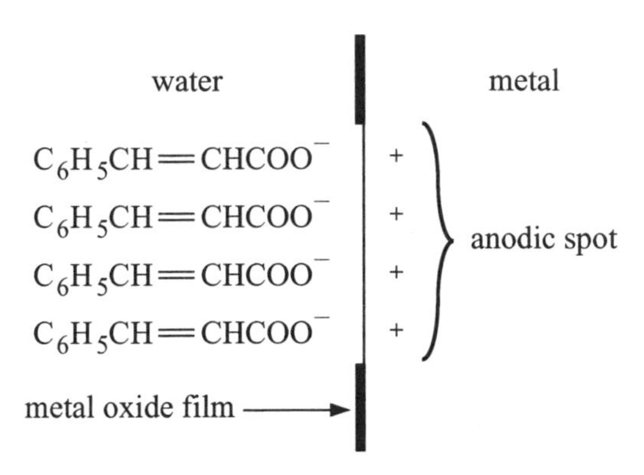

Figure 16.13 Blanketing action of an organic anodic inhibitor—in this case, sodium cinnamate.

Anodic inhibitors should *not*, however, be used on galvanized iron, in which protection of the iron depends upon the zinc coating being anodic; an anodic inhibitor would coat the zinc, reverse the polarity of the corrosion cell, and cause the iron to corrode merrily away.

A serious limitation of the use of anodic inhibitors is that they must be used in sufficiently high concentration to eliminate *all* the anodic sites, otherwise the anodic area that remains will carry the whole corrosion current, which is usually *cathodically* controlled. Intense local corrosion may then result, possibly leading to failure of the specimen. *Cathodic inhibitors*, on the contrary, are helpful in *any* concentrations; for example, the blanketing of only half the cathodic surface will still roughly halve the corrosion rate. The presence of temporary hardness or magnesium ions can help reduce corrosion through deposition of $CaCO_3$ or $Mg(OH)_2$, specifically on the cathodic surfaces where OH^- is produced in the oxygen absorption reaction:

$$O_2 + 2H_2O + 4e^- \rightarrow 4OH^- \tag{16.27}$$

$$Ca^{2+} + HCO_3{}^- + OH^- \rightarrow CaCO_3(s) + H_2O \tag{16.28}$$

$$Mg^{2+} + 2OH^- \rightarrow Mg(OH)_2(s). \tag{16.29}$$

It follows that hard water is paradoxically less corrosive than soft, deionized, or distilled water, other things (notably the O_2 concentrations) being equal. Zinc ions precipitate $Zn(OH)_2(s)$ on cathodic surfaces, but even better protection is possible if phosphate is also present, giving a basic zinc phosphate coating. Of course, phosphate ions can also act as anodic inhibitors, but it should be recognized that only *one* of the electrode reactions—the anodic *or* the cathodic—can be rate controlling at a given time. In practice, the inhibition of corrosion by Zn^{2+} and phosphate ions acting together is greater than the sum of the separate contributions; this *synergism* is typical of many mixed inhibitors.

In *acidic* solutions, organic amines protonate to form cations with hydrophobic tails. These ions will seek out and blanket cathodic surfaces, much as carboxylate anions seek out and cover anodic spots in neutral or basic media. The usual choices include amylamine ($C_5H_{11}NH_2$), cyclohexylamine ($C_6H_{11}NH_2$), pyridine (C_5H_5N), and morpholine [$O(CH_2CH_2)_2NH$]. Metallic iron behaves as a *soft acid* in terms of the HSAB classification (Section 2.9), despite the indubitable "hard" behavior of its trivalent ion, and so molecules with soft donor atoms adsorb more strongly than hard bases ($S > N > O$). This principle can be applied to the design of inhibitors.[17, 19]

16.7.4 Atmospheric Corrosion

Aqueous corrosion can occur even when the metallic object to be protected is ostensibly not immersed in water, if the relative humidity of the atmosphere exceeds 60%. In that case, a film of water will in fact be present on the metal surface. Further, if sulfur dioxide is present in the air, corrosion in the thin film of water will be greatly accelerated, partly because the acidity of the dissolved SO_2 facilitates the oxygen absorption reaction

$$SO_2(g) + H_2O(l) \rightleftharpoons H_2SO_3(aq) \qquad (16.30)$$

$$\rightleftharpoons H^+(aq) + HSO_3^-(aq) \text{ (etc.)}$$

$$O_2(g) + 4H^+ + 4e^- \rightarrow 2H_2O, \qquad (16.31)$$

but also because SO_2 will favor reductive dissolution of partially protective iron(III) oxide films:

$$Fe_2O_3(s) + H_2SO_3(aq) + 2H^+(aq) \rightarrow 2Fe^{2+}(aq) + SO_4^{2-}(aq) + 2H_2O. \qquad (16.32)$$

Dust (especially from industrial activities) and salt spray will also exacerbate atmospheric corrosion (Section 16.4). In enclosed industrial premises, atmospheric corrosion could be minimized by preventing noxious emissions, filtering the air to remove particulate matter, and scrubbing the air with water to remove SO_2 and other objectionable gases, although the humidity should itself be kept as low as possible (e.g., steam leaks should not be tolerated). On the global scale, however, the cost to the public of atmospheric corrosion could be substantially reduced by sharply limiting SO_2 and, to a lesser extent, NO_x emissions from power plants, smelters, automobiles, and other industrial functions. This is an aspect of the acid rain threat (Chapter 8) that is usually overlooked.

16.7.5 Corrosion-Resistant Metals

Modification of the metal itself, by alloying for corrosion resistance, or substitution of a more corrosion-resistant metal, is often worth the increased capital cost. *Titanium* has excellent corrosion resistance, even when not alloyed, because of its tough natural oxide film, but it is presently rather expensive for routine use (e.g., in chemical process equipment), unless the increased capital cost is a secondary consideration. *Iron* is almost twice as dense as titanium, which may influence the choice of metal on structural grounds, but it can be alloyed with 11% or more *chromium* for corrosion resistance (stainless steels, Section 16.8) or, for resistance to acid attack, with an element such as *silicon* or *molybdenum* that will give a film of an acidic oxide (SiO_2 and MoO_3, the anhydrides of silicic and molybdic acids) on the metal surface. Silicon, however, tends to make steel brittle. Nevertheless, the proprietary alloys *Duriron* (14.5% Si, 0.95% C) and *Durichlor* (14.5% Si, 3% Mo) are very serviceable for chemical engineering operations involving acids. Molybdenum also confers special acid and chloride resistant properties on type 316 stainless steel. Metals that rely on oxide films for corrosion resistance should, of course, be used only in E_h conditions under which passivity can be maintained.

Nickel confers excellent resistance to alkaline attack and to stress-cracking corrosion and is also quite resistant to nonoxidizing acids. It is widely used to the extent of a few percent in stainless steels (300 series, see Section 16.8), and in several high nickel alloys, including the following proprietary metals:

Monel 400: 66% Ni, 31% Cu, 1% Fe
Inconel 600: 76% Ni, 16% Cr, 8% Fe
Hastelloy C: 56% Ni, 15% Cr, 17% Mo, 5% Fe, 1% Si, 4% W.

Hastelloy C has superb corrosion resistance over wide ranges of pH and E_h. The main disadvantage of such non-ferrous alloys over stainless steels is usually cost, although the vulnerability of high-nickel and copper alloys to attack by chelating agents (Sections 14.4.1 and 16.5) should be borne in mind.

Alloying to modify the overpotential of the metal surface for H_2 evolution or O_2 absorption can help control corrosion, although it is not always obvious whether these cathodic processes should be suppressed (i_0 lowered) or stimulated to produce the desired corrosion resistance. In the case of titanium (see Section 16.6), for example, palladium was alloyed in to *catalyze* H_2 evolution and to force the metal into a passive condition.

16.8 Stainless Steels

The common feature of all stainless steels is the presence of at least 11%, and more usually 18%, chromium. This confers corrosion protection through formation of a particularly insoluble and reduction-resistant oxide film. The steels are heat-treatable for improved mechanical properties (Section 5.7) if they can undergo the α-Fe/γ-Fe transition at elevated temperatures. It is useful to distinguish four groups of commonly used stainless steels:

(*a*) *Martensitic chromium steels* (AISI 400 series) contain no (or very little) nickel, and the chromium content is typically about 12%. These steels can undergo the α-Fe/γ-Fe transition at about 1050 °C and so can be heat-treated for improved mechanical properties, much as can ordinary carbon steels. Since they have the α-Fe structure at ambient temperatures, they are ferromagnetic in ordinary service. Examples are type 410 (11.5–13.5% Cr), which is used for turbine blades, and type 416 (12–14% Cr with minor amounts of Se, Mo or Zr), which has good machinability.

(*b*) *Ferritic chromium steels* (also 400 series) again contain no nickel, but the chromium content typically ranges from 12 to 20%, which prevents the α-Fe/γ-Fe transition at any temperature. However, since the structure is basically the α-Fe type, these alloys are ferromagnetic. They have good resistance to oxidizing acids, by virtue of the high Cr content. In an early application of such alloys in the chemical industry, type 430 (14–18% Cr) has been used for fabricating nitric acid tank cars.

(*c*) *Austenitic stainless steels* (300 series) usually contain about 18% Cr, and 8 to 12% nickel, which suppresses stress-cracking corrosion and improves general chemical resistance. The nickel content preserves the γ-Fe (austentite) structure at all temperatures, so these austenitic steels are *not* heat-hardenable nor ferromagnetic. Type 304 is the widely used "18-8" stainless steel, while type 316, with 2 to 3% molybdenum, has improved resistance to acids and is an excellent choice for chemical process equipment. To avoid weld decay (Section 16.4), types 304L or 316L, with carbon contents less than 0.03%, or types 321 (% Ti not less than 4 × % C) or 347 [%(Nb + Ta) at least 10 × % C] may be used. For better high temperature performance, higher nonferrous metal contents are required, for example, 25% Cr and approximately 20% Ni in type 310 stainless steel.

(*d*) The *200 series stainless steels* are again austenitic, like the 300 series, but usually manganese is used in place of nickel.

Exercises

16.1 With reference to the discussion of the bimetallic corrosion of iron given in Section 16.1, confirm (*a*) that the solubility of iron(II) hydroxide is 5.8 μmol L^{-1} if OH^- is produced along with iron(II) according to reaction 16.5 and the reverse of reaction 16.6 and (*b*) that the timescale of the oxidation of iron(II) in solution at pH 7 is as given following Eq. 16.8. *Hint:* Incorporate $[O_2]$ and pH into a rate constant k_1 for a first-order process, then calculate the half-period $t_{1/2} = (\ln 2)/k_1$.

16.2 Express the "mils per year" corrosion rate in terms of electrical current density for iron (density 7.86 g cm^{-3}; one mil = 0.001 inch = 0.00254 cm).

16.3 Provide explanations for the following:

(*a*) Extremely pure metals (e.g., zinc) often corrode less rapidly than the same metals in impure form.

(*b*) If an aluminum sheet ($E°$ for $Al^{3+}/Al = -1.67$ V) is in contact with one of iron ($E°$ for $Fe^{2+}/Fe = -0.44$ V), it is the iron that corrodes (i.e., becomes anodic).

(*c*) A rubber band was slipped over a mild steel plate, which was then left lying in a damp place for several days. A rust deposit formed on the exposed metal (especially on either side of the rubber band), but the metal under the band was rust free (though markedly etched).

(*d*) Corrosion of steel by a single drop of water produces a ring of rust, the outer diameter of which is somewhat less than the diameter of the drop.

(*e*) A steel pile, partly immersed in still water, tends to corrode mainly at the lower end; there is a protected area under water near the meniscus.

(*f*) Health authorities have warned against the use of copper or stainless steel fittings in contact with the lead piping that is still found in domestic plumbing in some older houses in Britain.

(*g*) Water containing high concentrations of dissolved carbon dioxide is very corrosive toward iron.

16.4 Compare and contrast the mechanisms of oxidation of iron by dry air ("scaling"; see Chapter 5) and by aqueous environments.

References

1. U. R. Evans, "An Introduction to Metallic Corrosion," 3rd Ed. Arnold, London, 1981.

2. U. R. Evans, "The Corrosion and Oxidation of Metals." Arnold, London, 1960, and Supplements 1 (1968) and 2 (1978).

3. M. D. Fontana, "Corrosion Engineering," 3rd Ed. McGraw-Hill, New York, 1986.

4. J. C. Scully, "The Fundamentals of Corrosion," 3rd Ed. Pergamon, Oxford, 1990.

5. National Association of Corrosion Engineers, "Corrosion Basics." Houston, Texas, 1984.

6. K. R. Trewethey and J. Chamberlain, "Corrosion." Longman, Harlow, U. K., 1988.

7. H. H. Uhlig and R. W. Revie, "Corrosion and Corrosion Control." Wiley, New York, 1985.

8. L. L. Shreir, R. A. Jarman, and G. T. Burstein, "Corrosion," Vols. 1 and 2. Butterworth–Heinemann, London, 1994.

9. G. Wranglén, "An Introduction to Corrosion and Protection of Metals." Institut för Metallskydd, Stockholm, 1972.

10. P. A. Schweitzer (ed.), "Corrosion and Corrosion Protection Handbook," 2nd Ed. Dekker, New York, 1989.

11. C. P. Dillon, "Corrosion Control in the Chemical Process Industries." McGraw–Hill, New York, 1986; C. P. Dillon, "Materials Selection for the Chemical Process Industries." McGraw–Hill, New York, 1992.

12. National Association of Petroleum Engineers, "Corrosion Control in Petroleum Production." Houston, Texas, 1979.

13. K. A. Chandler, "Marine and Offshore Corrosion." Butterworth, London, 1985.

14. D. J. De Renzo and I. Mellan, "Corrosion Resistant Materials Handbook." Noyes Data, Park Ridge, New Jersey, 1985.

15. J. O'M. Bockris and A. K. N. Reddy, "Modern Electrochemistry," Vol. 2, especially Chapter 11. Plenum, New York, 1970.

16. P. Stoffyn-Egli and D. E. Buckley, The *Titanic*: From metals to minerals. *Can. Chem. News* October, 26–28 (1993); P. Stoffyn-Egli and D. E. Buckley, The micro-world of the *Titanic*. *Chem. Br.* **31**, 551–553 (1995).

17. B. G. Clubley (ed.), "Chemical Inhibitors for Corrosion Control." Royal Society of Chemistry, Cambridge, 1990.

18. P. Boden and D. Kingerley, Inorganic chemicals as metallic corrosion inhibitors. *In* "Insights into Speciality Inorganic Chemicals" (D. Thompson, ed.). Royal Society of Chemistry, Cambridge, 1995.

19. F. H. Walters, Design of corrosion inhibitors. *J. Chem. Educ.* **68**, 29–31 (1991)

Chapter 17

Extractive Metallurgy

EXTRACTION OF metals from minerals[1–7] has many points in common with corrosion control. Redox chemistry is applied to wrest metals from Nature, just as it is used to prevent her reclaiming them through corrosion. There is an additional factor to be considered, however. Most ores contain only minor amounts of the metal of interest, and, before final reduction or refining, the desired metal or its compounds must be concentrated and separated from other metals that may be present.

17.1 Gravity and Flotation Methods of Ore Concentration

The simplest concentration technique is the use of *gravity* to separate dense metal or ore particles from the much less dense silicate and other rock-forming minerals by suspending the latter, finely divided, in swirling water. Placer gold (density 19.3 g cm^{-3}, cf. about 2.5 g cm^{-3} for silicates) is the time-honored example. The gold dust that has accumulated in some riverbed silts was concentrated there by the same process, having weathered out of the surrounding rock in which it was present in very small amounts.

Froth flotation[1] is widely used to concentrate ores, particularly sulfides such as galena (PbS), although it is by no means restricted in use to metallic sulfides (e.g., the technique is also used to separate KCl from NaCl, Section 9.7). In the case of metal sulfides, the crushed rock is suspended in water, and the particles of metal sulfide, which may be denser than the unwanted siliceous *gangue*, are nevertheless caught up in a froth generated by blowing air through the mixture after addition of a frothing agent such as pine oil. The froth can then be skimmed off the top and the metal sulfide recovered. This will only work if the metal sulfide particles specifically can be made water repellent, so that these unwetted particles seek the air–

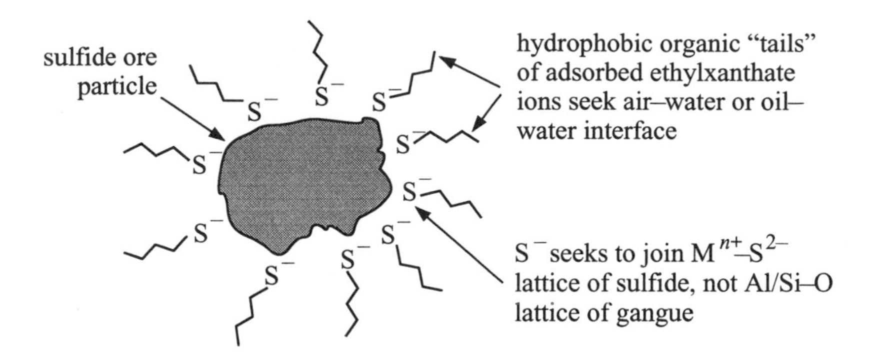

Figure 17.1 Mode of action of sodium ethylxanthate as a froth flotation agent for sulfide minerals.

water interface of the froth while the water-wetted gangue particles sink in the aqueous phase. This is achieved by adding to the water a water-soluble soaplike salt in which the negative charge of the organic anion is borne by sulfur atoms, which will therefore tend to adsorb on the metal sulfide surfaces but not on the silicate or aluminosilicate gangue surfaces, which have negligible affinity for sulfur. The salt most commonly used with metal sulfides is sodium ethylxanthate ($C_2H_5OCS_2{}^-Na^+$), which is a sulfur-containing analog of organic carbonate salts (Fig. 17.1; compare and contrast Fig. 7.7). Nonsulfur surface-active agents that are used as collectors include hydroxamates and oleic acid.

17.2 Hydrometallurgical Concentration and Separation

Hydrometallurgical methods[4, 5] use reactions in aqueous solution (often involving metal complex formation) to concentrate and/or separate the metal ions of interest. A commercially important example is the heap leaching of low-grade copper ores with acid.

17.2.1 Cyanide Leaching

Gold particles in crushed rock can be brought into solution as $Au(CN)_2{}^-$ ion by leaching the rock with aerated sodium cyanide solution (see Exercise 15.2); the aluminosilicates remain undissolved:

$$2Au(s) + O_2(g) + 4CN^-(aq) + 2H_2O(l) \rightarrow 2Au(CN)_2{}^-(aq) + H_2O_2(aq) + 2OH^-(aq). \quad (17.1)$$

The reaction is rather slow, requiring several days, after which the gold may be recovered by electrolysis[6] (Section 15.7) or by reaction with zinc[7] ("cementation"):

$$Zn(s) + 2Au(CN)_2^-(aq) \rightarrow Zn(CN)_4^{2-}(aq) + 2Au(s). \qquad (17.2)$$

When zinc cementation is to be used, the gold-bearing cyanide solution must first be deoxygenated to prevent oxidation of Zn(s) to $Zn(CN)_4^{2-}$ by O_2 rather than $Au(CN)_2^-$.*

17.2.2 Ammonia Leaching

The effectiveness of aqueous ammonia as a complexing agent for Ni^{2+}, Cu^{2+}, and Co^{3+}, but not for Fe^{2+}, Fe^{3+}, or Mn ions, is the basis of the *Sherritt Gordon ammonia leach process*, developed in the 1950s. Crushed ore containing small amounts of nickel (~10%), copper (<2%), and cobalt (~0.3%) sulfides is treated with aqueous $NH_3/(NH_4)_2SO_4$ buffer solution at 85 °C under 10 bars pressure of air, whereupon $Ni(NH_3)_6^{2+}$, $Cu(NH_3)_4^{2+}$, and $Co(NH_3)_5OH_2^{3+}$ form. [The cobalt is present as cobalt(II) originally, but aqueous $Co(NH_3)_6^{2+}$ reacts with oxygen to give primarily the aquapentaamminecobalt(III) ion; Section 13.2.] Simultaneously, the sulfide ions are oxidized to sulfate, thiosulfate, and polythionates (Section 3.4), and all these pass into solution, leaving behind a gangue of iron(III) oxides/hydroxides and aluminosilicate minerals. The excess NH_3 is then distilled off, so that the pH falls and the sulfur anions disproportionate, giving H_2S. The H_2S causes CuS ($K_{sp} \approx 10^{-36}$) to reprecipitate, leaving the Ni (for NiS $K_{sp} = 1 \times 10^{-22}$) in solution along with $Co(NH_3)_5OH_2^{3+}$ which is kinetically inert and so does not part with the ammonia ligands at this stage.

The nickel in the aqueous phase is recovered by reduction with H_2 (Section 17.6) until about as much nickel as cobalt remains in rather dilute solution. Both are then precipitated with added H_2S as NiS and CoS, which are redissolved in aqueous H_2SO_4 at pH 2 by oxidation with air and again converted with ammonia and air to $Ni(NH_3)_6^{2+}$ and $Co(NH_3)_5OH_2^{3+}$. Now, if the concentrated solution is acidified with sulfuric acid, the nickel ammine, being labile (i.e., able to react rapidly), goes to $Ni(H_2O)_6^{2+}$, which comes out of solution as $NiSO_4 \cdot (NH_4)_2SO_4 \cdot 6H_2O$, a rather poorly soluble "double salt" (cf. alum, Section 14.2). Cobalt would do the same if it were present

*Gold can also be extracted from crushed rock by dissolving it in mercury, with which it forms an amalgam, and then distilling off the mercury.[8] Amateur gold miners must be aware of the toxicity of cyanides and of mercury vapor to themselves and to the ecosystem before attempting either technique. The release of mercury into the Amazon River by miners working alluvial gold is causing serious pollution problems reminiscent of the Minamata disaster (see Section 11.4.2).

as $Co(H_2O)_6{}^{2+}$, but it is not, because $Co(NH_3)_5OH_2{}^{3+}$ is kinetically inert and remains in solution as such.

Thus, after a good deal of complicated aqueous chemistry, we have separated Cu, Ni, and Co ions from one another. Reduction of the ions to the metals is described in Section 17.6. A more recent version of the ammonia leach process uses an ammonia/ammonium carbonate, rather than sulfate, buffer solution.

17.2.3 Acid and Microbial Leach Processes

Acid leaching of metal sulfides such as FeS, CuS, or ZnS with, say, aqueous sulfuric acid might be expected to give H_2S gas and the corresponding aqueous metal sulfates. In practice, though, many such sulfides are just too insoluble. Treatment with hot aqueous acid under a few bars pressure of air or oxygen, however, can oxidize the sulfide ion to elemental sulfur, polythionates, and eventually sulfate ion (cf. the ammonia leach process), so that the desired metal ions pass into solution. Such hydrothermal processes have an important advantage over sulfide ore roasting (Section 17.8) in that emissions of the notorious air pollutant SO_2 (Section 8.5) can be completely avoided[4]:

$$ZnS(s) + H_2SO_4(aq) + \tfrac{1}{2}O_2(g) \rightarrow ZnSO_4(aq) + S(s) + H_2O. \qquad (17.3)$$

Oxidation of insoluble mineral sulfides to the usually water-soluble sulfates ($PbSO_4$ is an exception) can also be carried out in many cases by *microbial leaching*, that is, by the use of bacteria such as *Thiobacillus ferrooxidans* which can use the sulfide–sulfate redox cycle to drive metabolic processes. The overall reaction still consumes oxygen

$$MS(s) + 2O_2(g) \xrightarrow[\text{in water}]{\text{bacteria}} M^{2+}(aq) + SO_4{}^{2-}(aq), \qquad (17.4)$$

and any unwanted iron(II) is usually also oxidized to iron(III) which may then separate as a hydrolytic precipitate (this may, however, be a disadvantage if it impairs the bacterial activity).

Heap (dump) acid leaching of copper sulfide ores is possible with the aid of microbial oxidation. Not all copper minerals are sulfidic, however—malachite, azurite, and chrysocolla are basic copper carbonates—and sulfuric acid heap leaching of low-grade copper carbonate ores can give solutions from which the Cu^{2+} ion can be separated by solvent extraction (Section 17.3) and copper metal obtained by electrowinning.

Other sulfide-oxidizing processes include the use of nitric acid, a strong oxidant, in place of sulfuric acid:

$$3CuS(s) + 8HNO_3(aq) \rightarrow 8NO + 3CuSO_4(aq) + 4H_2O. \qquad (17.5)$$

Better yet, the metal sulfide may be made the anode of an electrolytic cell, in which the electrolyte is $H_2SO_4(aq)$ and hydrogen evolution is the cathodic reaction:

$$NiS(s) \rightarrow Ni^{2+}(aq) + S(s) + 2e^- \quad (17.6)$$

$$2H^+(aq) + 2e^- \rightarrow H_2(g). \quad (17.7)$$

17.2.4 Alkali Leaching

Alkali leach methods are exemplified by the *Bayer process* for the preparation of pure α-Al_2O_3 for electrolysis (Section 17.5) from the mineral *bauxite.* Bauxite consists mainly of α-AlO(OH) (diaspore) and/or γ-AlO(OH) (boehmite), the difference between these being essentially that the oxygen atoms form hcp and ccp arrays, respectively. The chief contaminants are silica, some clay minerals, and iron(III) oxides/hydroxides, which impart a red-brown color to the mineral. Aluminum(III) is much more soluble than iron(III) or aluminosilicates in alkali, so that it can be leached out with aqueous NaOH (initially 10–15 mol L^{-1}) at 165 °C under approximately 0.6 MPa pressure, leaving a "red mud" of iron (and other transition metal) oxides/hydroxides and aluminosilicates:

$$AlO(OH)(s) + OH^-(aq) + H_2O \rightarrow Al(OH)_4{}^-(aq). \quad (17.8)$$

The silica content of the bauxite reacts with the aqueous aluminate to form insoluble sodium aluminosilicates, typically zeolite-like solids such as sodalite and particularly cancrinite. The hot solution is decanted from the red mud, cooled, and diluted about threefold, whereupon gibbsite [α-$Al(OH)_3$] will crystallize out if seeded with crystals from a previous batch. The gibbsite is then dehydrated to α-Al_2O_3 by heating to 1250 °C; lower temperatures would give γ-Al_2O_3, which tends to carry absorbed water that would give hydrogen instead of aluminum in the electrolytic reduction to Al that follows. The alkali solution is reconcentrated by evaporation and is recycled, but some NaOH is consumed in precipitating the sodium aluminosilicates, adding significantly to the cost of the Bayer process.

17.3 Solvent Extraction and Ion-Exchange Separations

Formation of mineral deposits occurs through geochemical reactions which often concentrate and precipitate specific compounds quite locally and separately from others of sufficiently different chemical properties. In such cases, natural processes do much of the work of concentration and separation for us. If, however, the elements of interest have closely similar chemistries, they will tend to occur together and thus will require highly selective chemical techniques to separate them after concentration. Examples include the lanthanides, also "*rare earths*" even though this is a misnomer, as they are not so much rare as dispersed. Even when lanthanides are found as a relatively rich deposit such as the phosphate mineral monazite, they are difficult to separate from one another. The chemical properties of lanthanides generally change very slightly but progressively, following trends set by the lanthanide contraction in ionic radii (Section 2.6), as we go from element 57 (lanthanum) to element 71 (lutetium).

A secondary consequence of the lanthanide contraction is that hafnium (element 72), which follows the lanthanides in the periodic table, has almost identical atomic and ionic radii and ionization potentials to zirconium, its predecessor in the titanium group, instead of having bigger and more easily ionized atoms as one might otherwise expect. Indeed, it was not until 1923 that it was shown that the then unknown element 72 was present in samples of zirconium and its compounds prepared up to that time. Yet Zr and Hf differ markedly in *nuclear* properties; in particular, Zr has a conveniently low "*cross section*" for neutron absorption, whereas that for Hf is very high. Consequently, in preparing the highly corrosion-resistant zirconium alloy cladding for water-cooled nuclear power reactor fuel elements, it is essential that Hf be eliminated as far as possible.

17.3.1 Solvent Extraction

Difficult separations can often be effected by *liquid liquid solvent extraction*, which depends on differences in the distribution of solute species between two immiscible or partially immiscible phases.[4,9] For a solute species A, this distribution is governed by the *Nernst partition law*

$$\frac{[\mathrm{A}]_\mathrm{e}}{[\mathrm{A}]_\mathrm{r}} = \lambda_\mathrm{A} \tag{17.9}$$

where $[\mathrm{A}]_\mathrm{e}$ represents the concentration of A in the preferred solvent (the *extract*, usually in an organic solvent) and $[\mathrm{A}]_\mathrm{r}$ the concentration in the depleted solvent (the *raffinate*, usually the aqueous phase) after the two phases have been thoroughly mixed and allowed to separate at equilibrium. The *distribution coefficient* λ_A is constant for a particular solvent pair and solute at a given temperature. A similar equation

$$\frac{[\mathrm{B}]_\mathrm{e}}{[\mathrm{B}]_\mathrm{r}} = \lambda_\mathrm{B} \tag{17.10}$$

can be written for some other solute B, from which A is to be separated, and so on for solutes C, D, etc. The separation factor

$$\alpha_\mathrm{AB} = \frac{\lambda_\mathrm{A}}{\lambda_\mathrm{B}} = \frac{[\mathrm{A}]_\mathrm{e}[\mathrm{B}]_\mathrm{r}}{[\mathrm{A}]_\mathrm{r}[\mathrm{B}]_\mathrm{e}} \tag{17.11}$$

should be as large as possible if we want A to be concentrated in the extract and B in the raffinate; if $\alpha_{AB} = 1$, no separation is possible. Often, as in the case of the lanthanide or the zirconium–hafnium separation, the difference between A and B (etc.) in any one chemical property may be small, so we seek an extraction system that compounds differences in several chemical factors, as discussed later. In addition, we rely on multiple equilibrations of extract and raffinate with fresh solvent portions; it is more effective to make several extractions with small amounts of extractant, and then combine the extracts, than to equilibrate the solution just once with the same total volume of extractant. In practice, one designs a multistage, countercurrent extractor to work in the continuous flow mode. Highly efficient separations are possible, but loss of extractant through dissolution or emulsification in the waste raffinate may limit the cost-effectiveness of solvent extraction technology in some circumstances.

Selective extraction of a metal cation from water into an organic extractant will be facilitated if (*a*) the organic solvent has a significant tendency to solvate that particular metal ion, (*b*) the metal ion can be induced to form a neutral species (e.g., by complexing with the anions that are necessarily present in the solution), and/or (*c*) the metal ion can be made to form a complex of zero net charge with the anion of an acidic organic chelating agent when the pH of the aqueous phase is adjusted to an appropriate value. For example, the aqueous Co^{2+} ion has little tendency to pass into the commonly used extractant tributyl phosphate [TBP, $(C_4H_9O)_3PO$] but is readily extracted as the chargeless $Co(BTA)_2(TBP)_2$ complex (BTA^- is the anion of the β-diketone 3-benzoyl-1,1,1-trifluoroacetone, C_6H_5CO-CH_2COCF_3); furthermore, since this extraction into the nonaqueous phase drives the equilibrium 17.12 to the right, the efficiency of the extraction process is compounded:

$$Co^{2+}(aq) + 2HBTA \rightleftharpoons Co(BTA)_2(OH_2)_2 + 2H^+. \tag{17.12}$$

Thus, HBTA and TBP act *synergistically* to extract the cobalt(II).

Application to nuclear fuel reprocessing. Factors (*a*) and (*b*) are illustrated by the use of TBP in some nuclear fuel reprocessing cycles. In most nuclear fission power reactors, uranium-235 nuclei undergo fission following absorption of moderated neutron (i.e., a neutron of thermal energy), releasing fission products, much energy, and several neutrons to perpetuate the chain reaction. Since civilian nuclear fuels are usually UO_2 with either the natural (0.72%) or slightly enriched (~3%) abundance of ^{235}U, large amounts of the nonfissile isotope ^{238}U are exposed to high neutron fluxes in the reactor core. The ^{238}U can absorb a neutron to form ^{239}U, which undergoes two successive β^- decays to form first neptunium-239 and then plutonium-239. The latter is alpha-radioactive with a long half-life (24,000 years),* and is itself fissile by fast neutrons. Recovery of ^{239}Pu from spent UO_2 fuel elements may be contemplated for use in nuclear power plants or weapons.

A further possible reason for separating plutonium from uranium and the fission products relates to the extreme toxicity of Pu. Plutonium(IV) mimics iron(III) (the aqueous E° and charge-to-radius ratios of the two ions are very similar), so that cancers are likely to result from the absorption of even microgram amounts of ingested radioactive Pu into organs of the human body (bone marrow, spleen, liver) that store iron(III). It may therefore be considered desirable to remove ^{239}Pu, a long-lived health hazard, from spent nuclear fuels before disposal of the latter in repositories that may not remain inviolate for thousands of years into the uncertain future (most of the fission products decay away to negligible levels of activity in an acceptable time).

The spent fuel element is still mainly UO_2 and is dissolved in aqueous nitric acid, which is oxidizing enough to take the uranium to the VI oxidation state as $UO_2^{2+}(aq)$ and Pu to $Pu^{4+}(aq)$ (the uranyl ion UO_2^{2+} can be regarded as hydrolyzed U^{6+}; see Section 13.6). Treatment of the solution of uranyl and plutonium(IV) nitrates with either an iron(II) salt or SO_2 will reduce all the Pu to $Pu^{3+}(aq)$, which is not extractable with TBP, but will leave the uranium(VI) untouched (see Exercise 15.5). The solution is then equilibrated with TBP (which is immiscible with water) or TBP in an alkane solvent. The UO_2^{2+} forms a neutral complex containing both TBP and the nitrate ions, which are present in large excess:

$$UO_2^{2+} + 2NO_3^- + 2TBP \rightleftharpoons UO_2(NO_3)_2(TBP)_2. \tag{17.13}$$

The U complex, having no net charge and already containing TBP solvent ligands, passes preferentially into the TBP solvent phase, leaving the Pu^{3+} and almost all the fission products such as Sr^{2+}, I^-, and Cs^+ behind in the aqueous phase.

Separation of hafnium from zirconium. In a zirconium–hafnium separation process developed by Eldorado Nuclear, a mixture of sodium zirconate and hafnate can be obtained by fusing zircon sand with NaOH and dissolving the product in water. Acidification with nitric acid then gives aqueous zirconyl (ZrO^{2+}, or hydrolyzed "Zr^{4+}") and hafnyl (HfO^{2+}) nitrates. Extraction with TBP then gives an extract containing mainly $[HfO(NO_3)_2(TBP)_n]$, but most of the ZrO^{2+} remains in the aqueous phase and can be recovered on evaporation as essentially Hf-free $ZrO(NO_3)_2(s)$. The effectiveness of this process can be ascribed to compounding of factors

*Radioactive decay follows the same rate equation as first-order chemical kinetics (Section 2.5); the half-life $t_{1/2}$, the time required for one-half of the sample to decay, is given by (ln 2)/(rate constant). Decay of a sample is considered arbitrarily to be practically complete after $10t_{1/2}$, which is 240,000 years for ^{239}Pu.

(*a*) (affinity for the organic solvent) and (*b*) (complexing by anions to form a chargeless species) mentioned above to favor extraction of the hafnyl ion.

Extraction of transition metals from low grade ores. Factor (*c*) (formation of a chargeless chelate complex) can be illustrated by considering the formation of complexes between 8-hydroxyquinoline (HQ) and a mixture of metal ions, say, $M^{2+} = Fe^{2+}$, Co^{2+}, Ni^{2+}, and Cu^{2+}. This is in fact the order of increasing stability constants of the complexes MQ_2 (equilibrium constants β_2 for Eq. 17.14): $\log \beta_2 = 15.0$, 17.2, 18.7, and 23.4, respectively, in dilute solution at 20 °C. This commonly encountered sequence for complex formation by the divalent Fe, Co, Ni and Cu ions is known as the *Irving–Williams order* (cf. the susceptibility of Ni^{2+} and Cu^{2+} to complexing by NTA^{3-}, noted in Sections 14.4 and 16.5).

$$2HQ + M^{2+} \rightleftharpoons MQ_2 + 2H^+ \tag{17.14}$$

Since, according to Eq. 17.14, M^{2+} must compete with $2H^+$ for Q^-, we can set the pH low enough to ensure that *only* copper(II) forms major amounts of an uncharged di(hydroxyquinolinato) complex, which can then be extracted selectively into, say, chloroform. When this is complete, the pH could be increased enough to permit selective extraction of NiQ_2, etc. The metal ions can be stripped from the extractant by the reverse process, i.e., by equilibration with sufficiently concentrated aqueous acid.

Several such schemes have been developed to recover copper from dilute solutions obtained (usually) by acid heap leaching of low-grade copper ores.[9] The required properties of a complexing agent L in the extractant are:

(*i*) low solubility in water;

(*ii*) high selectivity for the metal ion M^{m+} of interest;

(*iii*) chemical stability under the reaction conditions;

(*iv*) sufficiently high distribution coefficient λ_{ML} under the conditions of the extraction process; and

(*v*) sufficiently low distribution coefficient under the conditions of the reverse (stripping) process to allow efficient recovery of M.

Usually, the composition of commercial extractants is proprietary information, but a typical example is an α-hydroxyoxime carried in a hydrocarbon solvent such as kerosene.

Anionic solutes X^- can be solvent extracted from aqueous solution into organic phases containing trialkylammonium (R_3NH^+) or similar cations:

$$R_3N(\text{org}) + H^+(\text{aq}) + X^-(\text{aq}) \rightleftharpoons \{R_3NH^+, X^-\}(\text{org}). \tag{17.15}$$

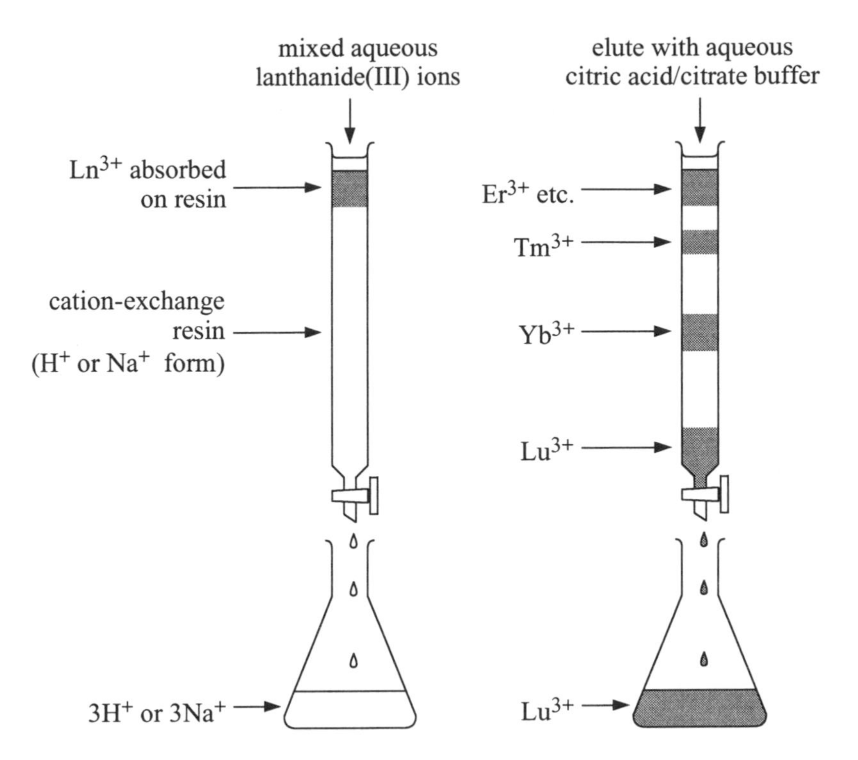

Figure 17.2 Ion-exchange chromatographic separation of mixed aqueous lanthanide(III) ion.

17.3.2 Ion-Exchange Separations

Compounds of the individual lanthanides find various "high technology" applications, including color video phosphors, lasers (e.g., neodymium in yttrium aluminum garnet), materials with special electrical or magnetic properties for electronic applications (e.g., yttrium iron garnet), superconducting ceramics such as $YBa_2Cu_3O_{7-x}$, and catalysts. Liquid–liquid solvent extraction is actually the method most commonly used for separation of the lanthanide(III) ions from one another on an industrial scale, but, where the highest purity is required, separation by ion-exchange chromatography is preferable.[10]

The mixed lanthanide(III) ions (Ln^{3+}) can be absorbed from solution at the top of a column of a cation-exchange resin RSO_3Na (Section 14.4) and then selectively *eluted* from it, that is, swept down the column in bands by a stream of a solution of a substance that competes with the Ln^{3+} for sites in the resin. An acidic tribasic chelating agent H_3X (usually citric acid/sodium citrate buffer) is used so that, as in reaction 17.16, the tendency for a specific Ln^{3+} to form neutral LnX and so escape the electrostatic attractions holding it within the resin can be fine-tuned by adjusting the pH, that is, the buffer composition:

$$Ln^{3+}(aq) + H_3X(aq) \rightleftharpoons LnX(aq) + 3H^+(aq) \tag{17.16}$$

in competition with

$$Ln^{3+}(aq) + 3RSO_3Na(s) \rightleftharpoons (RSO_3)_3Ln(s) + 3Na^+(aq). \tag{17.17}$$

The lanthanide contraction determines the order of ease of elution to be $Lu^{3+} > Yb^{3+} > Tm^{3+} >$ etc. (Fig. 17.2), for at least two reasons:

(*a*) The ions with the largest *hydrated* radii are the least firmly held by electrostatic attractions in the resin, and the extent of *hydration* of Ln^{3+} increases as the core ion becomes smaller. So, paradoxically, $Lu^{3+}(aq)$ is the largest hydrated ion, followed by $Yb^{3+}(aq)$, etc.

(*b*) The smaller the naked Ln^{3+} ion is, the higher is the stability constant for LnX.

17.4 Separations Utilizing Special Properties

The list of special chemical traits that might be exploited in extractive metallurgy is long, but two cases deserve special mention. The only widespread mineral in the Earth's crust that is ferromagnetic (meaning, in effect, that it can be picked up by a powerful magnet) is magnetite (Fe_3O_4). (Maghemite, "γ-Fe_2O_3," can be included with magnetite for the present purpose; Section 4.6.2.) This means that deposits of this high Fe mineral are readily located in the field through magnetic anomalies, and also that Fe_3O_4 can be picked out selectively from the crushed ore by passing it between the poles of a large magnet.

Most transition metals form complexes known as *carbonyls* with carbon monoxide as ligands. Examples include $Fe(CO)_5$, $Fe_2(CO)_9$, $Cr(CO)_6$, and $Rh_6(CO)_{16}$, in all of which the metal is ostensibly in the oxidation state of zero, and many mixed-ligand carbonyls such as $Mn(CO)_5I$, $CH_3Mn(CO)_5$, and $(C_6H_6)Mo(CO)_3$ are known. Such compounds have an organiclike chemistry, being essentially covalent (see Section 8.2 and Chapter 18), and the simple carbonyls such as $Ni(CO)_4$ are volatile liquids that can be purified by fractional distillation. Of all these, however, only $Ni(CO)_4$ (bp 43 °C) forms *rapidly* (and reversibly) from the elemental metal and CO gas *at low pressures and temperatures**:

$$Ni(s) + 4CO(g) \underset{\sim 200\,^\circ C}{\overset{40\text{–}90\,^\circ C}{\rightleftharpoons}} Ni(CO)_4(g), \quad \Delta H^\circ = -160.8 \text{ kJ mol}^{-1}. \tag{17.18}$$

Nickel can therefore be separated very specifically from impurities by treating the crude metal with CO gas and pyrolyzing (i.e., decomposing at high temperatures) the $Ni(CO)_4$ vapor so formed to obtain nickel in 99.97% or better purity. This reaction was discovered in 1889 by Mond, and in 1902 the Mond Nickel Company began producing pure nickel by the carbonyl process at Clydach, Wales, using crude nickel obtained by reduction of nickel oxide with hydrogen.[9] This plant, now greatly modernized, still operates. (Mond amalgamated with Inco in 1929.)

In the 1970s, Inco brought on stream a pressure carbonyl plant at Copper Cliff, near Sudbury, Ontario, in which crude nickel from the smelter (72% Ni, 18% Cu, 3% Fe, 1% Co, 5% S) is treated with CO at 70 bars and temperatures up to 170 °C. High pressures favor $Ni(CO)_4$ formation because of the large negative molar volume change associated with reaction 17.18, but they also favor formation of small amounts of $Fe(CO)_5$ (bp 103 °C) and $Co_2(CO)_8$ (mp with decomposition, 51 °C) which, fortunately, can easily be removed by fractional distillation. The purified $Ni(CO)_4$ can then be pyrolyzed at 200 °C at *low* pressure.

Tetracarbonylnickel is colorless and odorless but extremely toxic (carcinogenic), and no leakage, down to the part-per-billion detection limit, can be tolerated in the plant. This has undoubtedly discouraged more widespread use of carbonyl refining in place of the usual electrolytic processes (see Section 17.5).

17.5 Electrolytic Reduction of Concentrate

Since chemical reduction means gain of electrons, *electrolysis* is the most direct way of recovering a metal from its ores, as long as these can be handled in a fluid state.[6] Consideration of E° values for reactive metal half-cells such as $Na^+(aq)/Na(s)$, $Mg^{2+}(aq)/Mg(s)$, and $Al^{3+}(aq)/Al(s)$ (−2.71, −2.36, and −1.67 V, respectively) shows that these metals can never be obtained by electrolysis of aqueous solutions of their salts, as H_2 would be produced instead, but they can often be obtained by electrolysis of suitable *molten salts* such as NaCl and $MgCl_2$:

$$2NaCl(l) \xrightarrow[\text{graphite anode}]{\text{steel cathode}} 2Na(l) + Cl_2(g). \tag{17.19}$$

*Iron pentacarbonyl forms readily from CO and elemental iron only at ~150°C and ~10 MPa CO pressure, although long exposure of iron, e.g., steel gas cylinder walls, to pressurized CO can lead to the formation of small amounts of $Fe(CO)_5$.

It is usual to use a mixture of salts (in the case of sodium, 60% NaCl + 40% $CaCl_2$) so as to depress the melting point and permit electrolysis at a lower temperature (600 °C, instead of 801 °C for pure NaCl), but obviously the second metal ion (Ca^{2+}) must be less readily reduced than the one of interest.

In the case of aluminum, the product of the Bayer prepurification process (Section 17.2.4) is high-melting α-Al_2O_3 (mp ~2015 °C), direct electrolysis of which is impractical. (In any case, the melt is a non-electrolyte.) It can, however, be dissolved (2 to 8%) in a much lower melting mixture of cryolite (Na_3AlF_6) with about 10% fluorite; here again, the purpose of the CaF_2 is simply to depress the melting point without itself being preferentially electrolyzed. The electrolysis is carried out at about 970 °C with graphite electrodes (at ~4.5 V and current density ~1 A cm^{-2}). The graphite anodes, at which the O^{2-} ions are discharged, react to form CO_2 gas and so are consumed as the electrolysis proceeds (the *Hall–Héroult process*):

$$2Al_2O_3(\text{in cryolite}) + 3C(s) \xrightarrow{\text{electrolyze}} 4Al(l) + 3CO_2(g). \qquad (17.20)$$

Consumption of the graphite is quite acceptable; inert anode materials for these conditions are hard to find, but in any event involvement of C in reaction 17.20 reduces the electricity requirement of the energy-intensive electrolysis by nearly half. Other, less commonly used processes that involve lower temperature electrolysis of molten $AlCl_3$ (mp 183 °C, performed in a closed vessel to prevent sublimation) have been developed. In these, the aluminum is formed as a solid (mp 660 °C), and the anodes are not consumed. Of course, one has to convert Al_2O_3 to $AlCl_3$ first, and this is usually done using graphite (coke) anyway:

$$Al_2O_3 + 3C + 3Cl_2 \xrightarrow{900\,^\circ C} 2AlCl_3 + 3CO \qquad (17.21a)$$

$$2AlCl_3(l) \xrightarrow{\text{electrolyze}} 2Al(s) + 3Cl_2(g). \qquad (17.21b)$$

Electrolysis of aqueous solutions may be used to obtain less reactive metals, such as Cu, Ni, Zn, and Cr, in high purity, either from aqueous concentrates of the metal salts themselves (*electrowinning*) or from anodes of the crude metal prepared by other, usually pyrometallurgical, techniques (*electrorefining*). For example, crude copper from a smelter can be made the anode in a bath of dilute sulfuric acid; it dissolves when the electric current is passed and is redeposited on the refined Cu cathode in high purity, if the operating conditions of the cell are correctly adjusted (Fig. 17.3). The acid-soluble impurities such as iron also go into solution and remain there as long as the cathode potential is not too negative. Since each metal ion/metal couple has a characteristic E° value, electrorefining can be highly selective, giving very pure products.

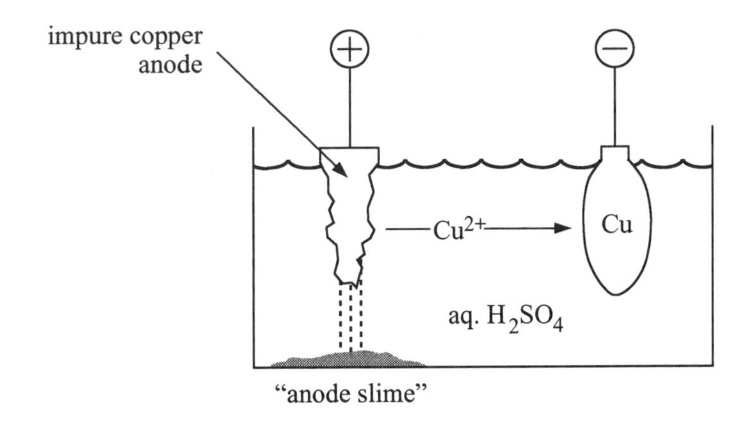

Figure 17.3 Principle of electrorefining of smelter copper.

Insoluble impurities fall to the floor of the cell as "anode slime." Despite the derogatory name, this material contains precious metals such as gold, silver, and platinum. Anode slime from the electrorefining of nickel[11] at Sudbury, Ontario, is a significant source of platinum and palladium as by-products (~0.34 g Pt and 0.36 g Pd per metric ton of ore), whereas deposits in the Bushveld complex (Transvaal, South Africa) are so rich in platinum-group metals (Ru, Os, Rh, Ir, Pd, Pt) that the associated Co, Ni, and Cu recovered are considered to be by-products of the lucrative platinum production (4.78 g Pt and 2.03 g Pd per metric ton of ore).[7]

17.6 Chemical Reduction of Concentrate

There has been much discussion of the possibility of a future hydrogen-based economy, as noted in Section 15.5.1, and advocates of this scenario point out that *gaseous hydrogen* could serve as the reductant in the extraction of some metals. The hydrometallurgical applicability of gaseous hydrogen might seem to be limited, since many economically important metals have standard electrode potentials E° for $M^{m+}(aq)/M(s)$ that are negative, whereas that for $H^+(aq)/H_2(g)$ is zero by definition. The conventional values of E°, however, refer to standard conditions (in practical terms, 1 mol L^{-1} M^{m+}, 1 mol L^{-1} H^+, 1 bar H_2, 25 °C, etc.), and hydrogen reduction of several metal ions can often be made thermodynamically feasible if these conditions are adjusted appropriately—notably to higher hydrogen pressures, higher pH, and, if the reduction is endothermic, higher temperatures.

Nickel(II), for example, would seem an unlikely candidate for H_2 reduction in aqueous solution:

$$Ni^{2+}(aq) + H_2(g) \rightleftharpoons Ni(s) + 2H^+(aq) \quad \Delta E = -0.26\,V, \quad \Delta H^\circ = +54 \text{ kJ mol}^{-1}. \tag{17.22}$$

In the Sherritt Gordon ammonia leach process, however, aqueous Ni^{2+} is reduced to nickel metal by hydrogen by using a high pressure (30 bars) of H_2, low acidity (NH_4^+/NH_3 buffer; the pK_a of NH_4^+ is 9.2 at 25 °C and 5.8 at 200 °C), and high temperature (175 to 200 °C). The kinetics of the reduction are prohibitively slow unless precipitation of the nickel is nucleated (seeded) with pure nickel powder from a previous batch. Cobalt, copper, and several other metals can be produced in this way. The integration of metallurgical operations and ammonia fertilizer production in the Sherritt plant at Fort Saskatchewan, Alberta, is noteworthy: hydrogen from the reforming of locally produced natural gas is also used to make the ammonia needed to concentrate and separate Co, Ni, and Cu by the ammonia leach process (Section 17.2.2), while sulfuric acid (a by-product of natural gas production) is used in making ammonium sulfate fertilizer as well as in some steps in the ammonia leach sequence.

In *pyrometallurgy*[1,7,12,13] (high temperature dry smelting), however, hydrogen is much less effective as a reductant, but *carbon* becomes important. For reasons we discuss in Section 17.8, the concentrate to be reduced is usually in the form of a metal oxide, so the reductant has to be something that has a greater affinity for the available oxygen atoms than has the

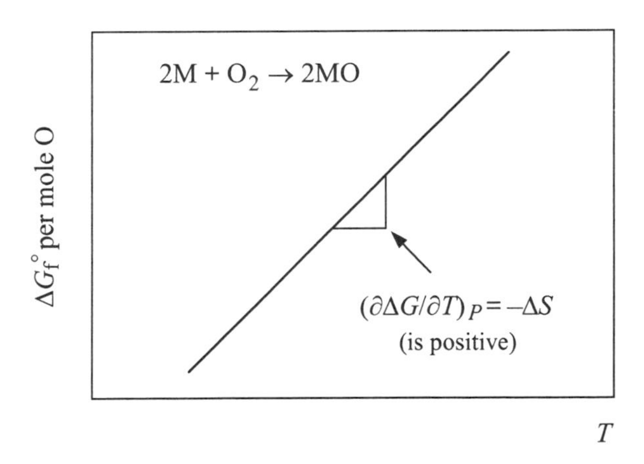

Figure 17.4 Ellingham plot (free energy of formation per mole O atoms versus temperature) for the reaction (metal) + (O_2 gas) → (metal oxide); the metal is not necessarily divalent. The plot is rectilinear if the heat capacity change of the reaction is negligible.

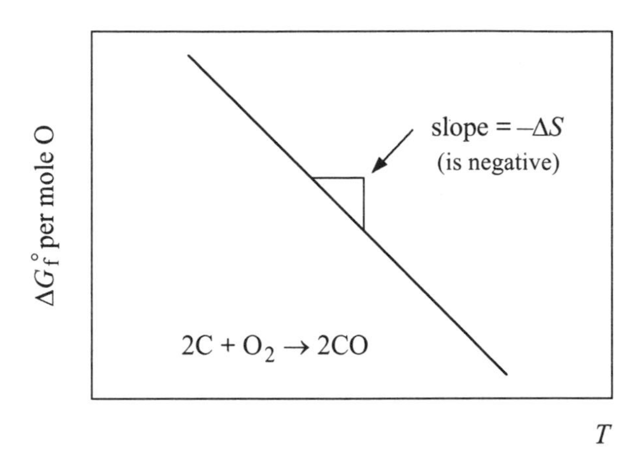

Figure 17.5 Ellingham plot for combustion of carbon to CO gas.

metal itself. In more precise terms, the oxide of the reductant must have a more negative free energy of formation ΔG_f° *per mole O atoms* than does the metal. The relevant information is usefully summarized in *Ellingham diagrams*, which are plots of ΔG_f° per mole O against temperature.[12] These plots usually appear almost linear because, although the standard heat capacities C_P° of the individual reactants will be temperature dependent, the net change ΔC_P° in heat capacity on reaction is usually negligible (Section 2.3).

The standard entropies S° of gases are much larger than those of solids and liquids (Section 2.3). This may be understood by the somewhat simplistic view of S° as a measure of disorder at the molecular level. The molecules of gases have much greater freedom of translational motion, and hence are less ordered, than those of liquids and especially solids. Consequently, for oxidation of a *solid* metal to a *solid* oxide with consumption of *gaseous* oxygen

$$M(s) + \tfrac{1}{2}O_2(g) \rightarrow MO(s), \tag{17.23}$$

the entropy change ΔS_f° will be *negative.* The dependence of ΔG_f° for reaction 17.23 on temperature, which may be presumed to be linear ($\Delta C_P^\circ \approx 0$), will therefore have a *positive* slope, as in Fig. 17.4:

$$\Delta G_f^\circ = \Delta H_f^\circ - T\Delta S_f^\circ. \tag{17.24}$$

For oxidation of solid carbon to gaseous carbon monoxide by O_2, however, there is an *increase* in the number of gas molecules, hence an *increase* in ΔS_f° with rising temperature, and so the Ellingham plot has a negative slope (Fig. 17.5):

$$C(s) + \tfrac{1}{2}O_2(g) \rightarrow CO(g). \tag{17.25}$$

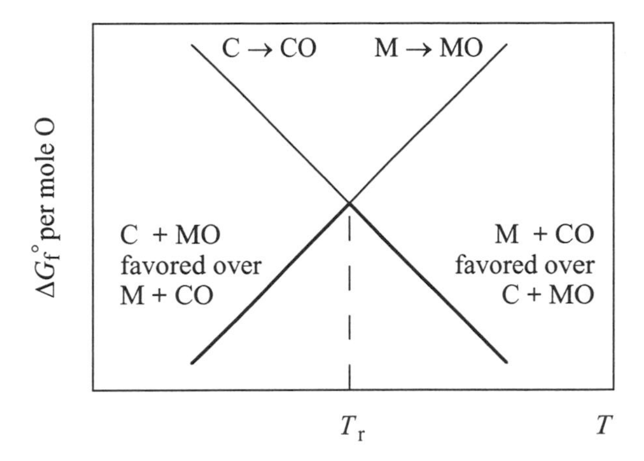

Figure 17.6 Combination of Figs. 17.4 and 17.5, representing the system M + C + O.

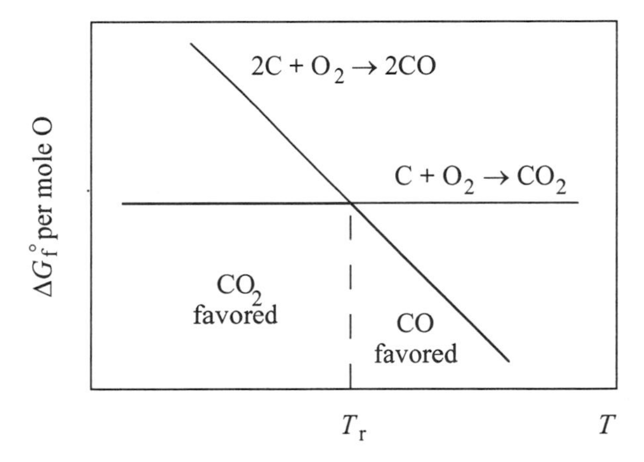

Figure 17.7 Relative stability of carbon monoxide and carbon dioxide in the presence of excess carbon.

Combining Figs. 17.4 and 17.5 into Fig. 17.6, we see that, although MO and C are favored over M and CO at low temperatures, there is a temperature T_r above which the reverse becomes true, for the diagram as drawn. In other words, above T_r, carbon (usually supplied as coke, i.e., metallurgical-grade coal from which the volatiles have been distilled) is thermodynamically capable of reducing the metal oxide to the metal.

There are, of course, some complications. First, there may or may not be an intersection as in Fig. 17.6 in the accessible temperature range (in practice, up to about 2000 °C, above which many refractory materials melt

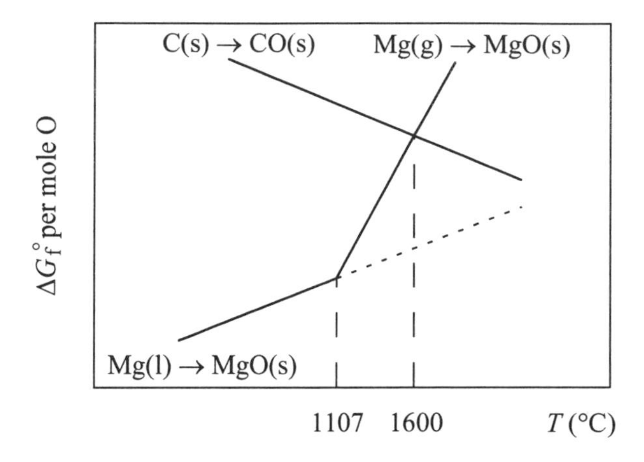

Figure 17.8 Effect of vaporization of magnesium on the reducibility of magnesium oxide with coke.

or are themselves reduced). Second, the activities of the reactants (particularly the gases) will in general not be the standard ones, so that simple calculations of the reduction temperature T_r based on standard values of ΔH_f° and ΔS_f° should be regarded only as guidelines for real smelting conditions. Third, the reaction *rate* may be prohibitively low at moderate temperatures, even where the reaction is thermodynamically strongly favored; reactions between coarsely mixed solids (C+MO) tend to be inherently slow unless mediated by some fluid component (cf. iron smelting, Section 17.7.2). Fourth, the predominant product of coke oxidation will be CO_2 at lower temperatures, and it continues to be so at higher temperatures, if the air supply is liberal. Under pyrometallurgical conditions (excess C, limited O_2), however, CO becomes stable with respect to disproportionation above about 700 °C (see Fig. 17.7):

$$2CO(g) \rightleftharpoons C(s) + CO_2(g). \tag{17.26}$$

The entropies of liquids are not strikingly large, compared to those of solids, while those of gases are. Consequently, the Ellingham plot for metal oxides is little affected as we pass the melting point of the metal. However, if the metal *boils* (e.g., magnesium, at 1107 °C), the slope of the plot increases sharply since the entropy loss on solid oxide formation is much greater for gaseous than for solid metal:

$$\underset{\text{1 mole gas}}{Mg(g)} + \underset{\frac{1}{2}\text{ mole gas}}{\tfrac{1}{2}O_2(g)} \rightarrow \underset{\text{no gas}}{MgO(s)}. \tag{17.27}$$

It is therefore possible to reduce magnesium oxide to the metal with coke above about 1600 °C (Fig. 17.8), even though there would be no intersection of the Ellingham lines for the oxidations of *solid* Mg and coke in an accessible temperature range. The practicability of producing magnesium in this way depends on rapid quenching of the Mg vapor below temperatures at which the reverse reaction

$$Mg(g) + CO(g) \rightarrow MgO(s) + C(s) \tag{17.28}$$

proceeds at a significant rate, since reaction 17.28 again becomes the preferred direction of reaction below 1600 °C. Most magnesium is in fact made by electrolysis of molten $MgCl_2$, as this is currently more economical.

17.7 Pyrometallurgy of Oxides

17.7.1 Use of the Ellingham Diagram for Oxides

The combined Ellingham diagram for a selection of oxides is shown in Fig. 17.9. From this it follows that most common metals, even calcium, could, at least in principle, be made by high temperature reduction of the oxide with coke. In practice, however, this is not always feasible or desirable. For example, research on reduction of alumina to aluminum with coke inspired by the high cost of electricity in Japan has revealed problems with the volatility of the suboxide Al_2O at the very high reduction temperatures dictated by Fig. 17.9. Titanium dioxide could be reduced by coke, but the very stable compound TiC (Section 5.7) and possibly TiN from the nitrogen of the air would result from any attempt to make titanium by direct coke reduction. Titanium metal is therefore made by the *Kroll process* (Section 17.8). Copper oxide could be reduced easily by coke, but it turns out that copper can be obtained by roasting copper sulfide ores without recourse to an added reductant, as described in Section 17.8.3. On the other hand, nickel sulfides are usually converted to the oxide by roasting, and the oxide is reduced with coke to give crude nickel for hydrometallurgical electrorefining.

Figure 17.9 also indicates that hydrogen gas can be an effective reductant for Cu_2O, PbO, NiO, and CoO but not for FeO, ZnO, etc. Indeed, as noted in Section 17.4, NiO can be reduced with hydrogen (at 400 °C) to give crude nickel for the Mond process; the gas used is actually water–gas (Section 9.3), and the CO content is used in making the tetracarbonylnickel.

The most powerful reductants for the oxides shown in Fig. 17.9 are aluminum, magnesium, and calcium. Scrap aluminum can be used in the *thermite reaction* to reduce metal oxides, for example, Cr_2O_3:

$$Cr_2O_3 + 2Al \rightarrow Al_2O_3 + 2Cr. \tag{17.29}$$

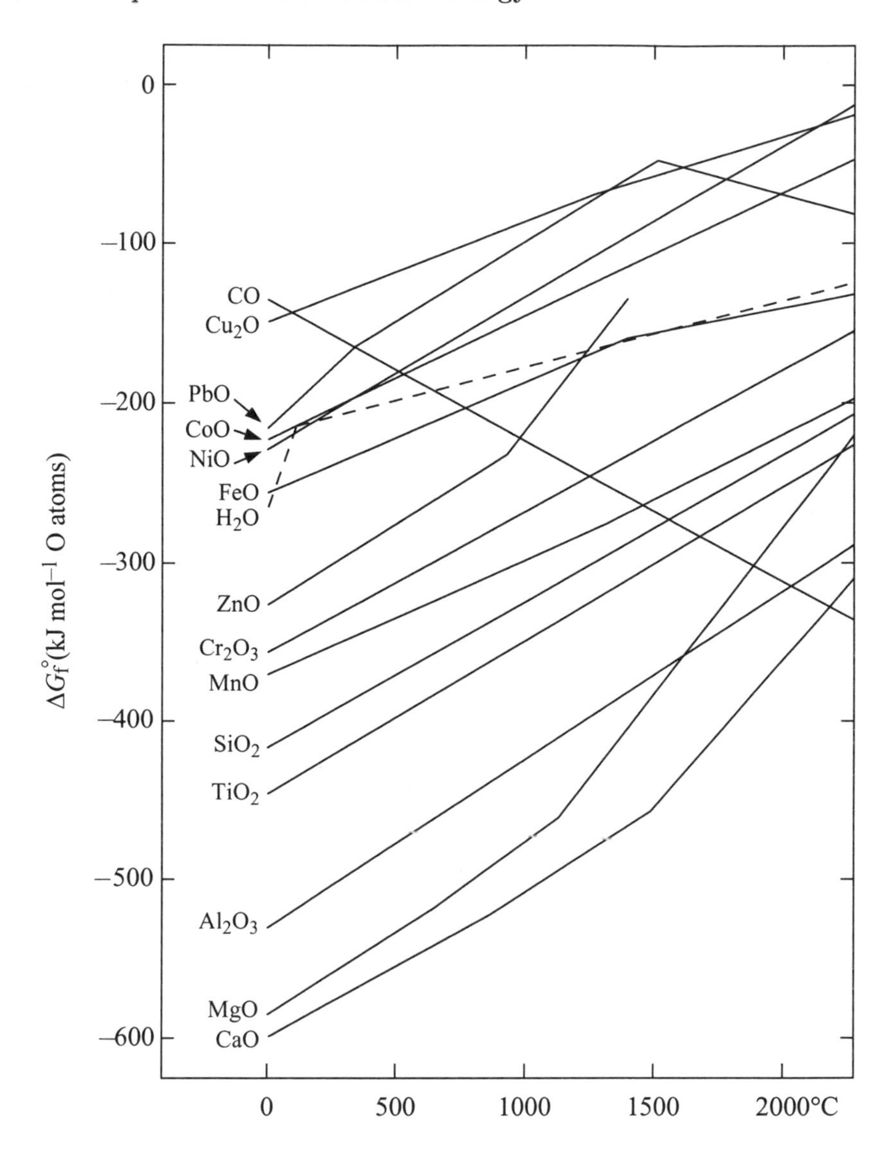

Figure 17.9 Ellingham diagram for formation of several oxides.

Much heat is generated, and the mixture becomes white-hot. Indeed, thermite reactions have been used to generate intense heat as well as the reduced metal for certain welding applications. Chromium, however, is usually made by coke reduction of chromite ($FeCr_2O_4$) in the electric furnace. The product is *ferrochrome*, an Fe–Cr alloy, but this is acceptable if, as is often the case, the intention is to make stainless steel.

17.7.2 Iron Production

Iron production is the most important application of coke pyrometallurgy in economic terms. Figure 17.10 shows the chemical reactions that occur in the various temperature regimes of the traditional blast furnace. The furnace is charged from the top with a mixture of iron ore (nominally Fe_2O_3), coke (C), and limestone ($CaCO_3$). If the charge contains fine ore, coke, or limestone particles, recycled blast furnace flue dust, etc., it is first sintered at an elevated temperature into a porous mass. Air is injected into the bottom of the blast furnace through pipes known as *tuyères*, and combustion of coke in this air blast raises the temperature to about 2000 °C, at which the predominant carbon oxide is CO (Fig. 17.7). Thus, the Fe_2O_3 charge at the top of the furnace meets hot CO coming upward and is reduced, first

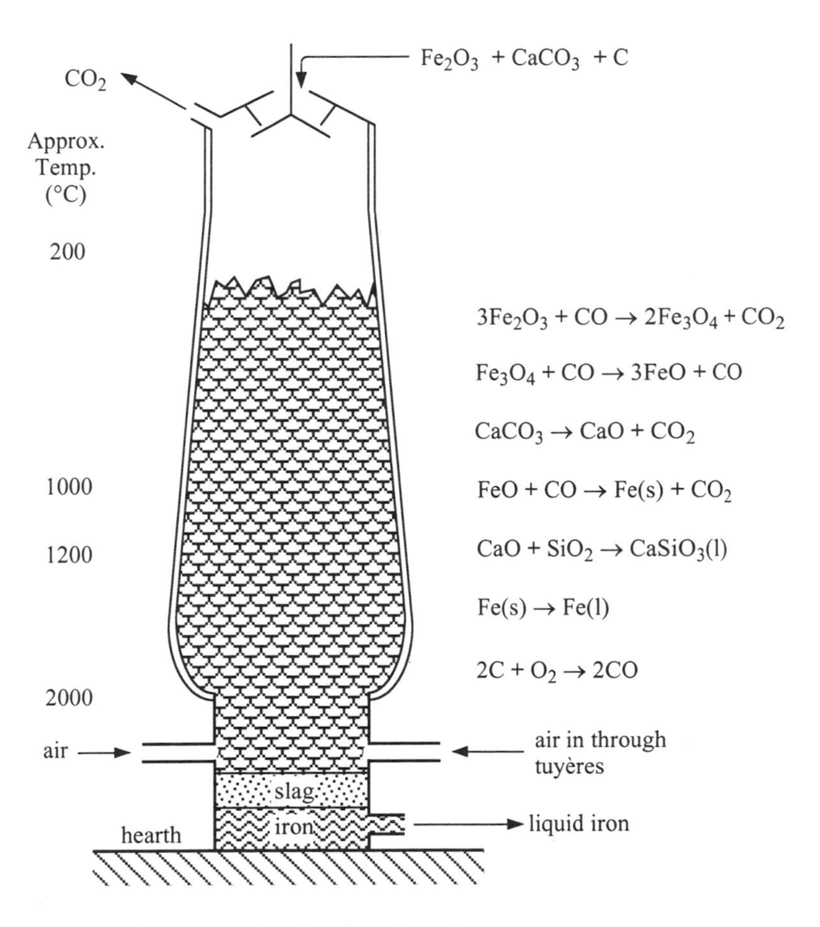

Figure 17.10 Iron ore reduction in a blast furnace.

to magnetite and then to wüstite (nominally FeO), while the CO goes to CO_2. At the same time, the limestone is converted to lime, adding more CO_2 to the exhaust gases. Any remaining CO will tend to disproportionate to CO_2 and carbon in the cooler, upper part of the furnace (cf. Fig. 17.7); however, this process will be relatively slow and incomplete, and in practice CO in the exit gases is burned in heat exchangers to preheat the air blast before it enters the tuyères.

The critical reduction step involves wüstite and CO at 1000 to 1200 °C

$$FeO(s) + CO(g) \rightarrow Fe(s) + CO_2(g), \tag{17.30}$$

with the CO arising from the reaction of CO_2 with coke

$$CO_2(g) + C(s) \rightarrow 2CO(g), \tag{17.31}$$

which is obviously entropy driven. The net reaction is therefore the one expected on the basis of the Ellingham diagram:

$$FeO(s) + C(s) \rightarrow Fe(s) + CO(g). \tag{17.32}$$

However, direct reaction of two solids as in reaction 17.32 is almost always slow because of the limited contact between the reactants, whereas in the combined reactions 17.30 and 17.31 the fluid CO/CO_2 phase provides a mechanism for relatively rapid reaction.

As the coke is consumed, allowing the iron and lime to fall toward the hottest part of the furnace, the lime reacts with the silica or silicates that were inevitably present in the iron ore, and a molten calcium silicate *slag* is formed. The iron also melts as much as 400 °C below the melting point of pure iron (1539 °C) because of dissolved carbon (see Fig. 5.8), and molten iron collects on the floor of the hearth under a pool of the lighter molten slag. From time to time, molten iron is run out from the hearth. In former times, it was made to flow along narrow channels in sand on the foundry floor into pools called *pigs*, where it would solidify to form *pig iron*. In modern foundries, the molten iron is taken directly to steelmaking vessels in special railcars (torpedo cars).

Most of the pollution associated with an iron foundry comes, not from the blast furnace (the exhaust gases of which are freed of recyclable particulates and then used for heating the air intake), but from the coking plant and the sintering machine. In addition to particulates and gaseous emissions containing SO_2, coke ovens release phenols, ammonia, and cyanides into the water used to quench the hot coke. Sintering the furnace feed reduces emissions in the actual smelting process, but some SO_2 that would otherwise be caught in the slag in the blast furnace is released in the sintering process. Large volumes of water are needed to cool the exterior of the blast furnace, but pollution in that case is mostly thermal.

A large fraction of the iron and steel produced today is recycled scrap. Since scrap does not require reduction, it can be melted down directly in an electric arc furnace, in which the charge is heated through its own electrical resistance to arcs struck from graphite electrodes above it. The main problem with this process is the presence of "tramps" (i.e., copper from electrical wiring, chromium, nickel, and various other metals) that accompany scrap steel such as crushed automobile bodies and that lead to brittleness in the product. Tin in combination with sulfur is the most troublesome tramp. Only the highest quality recycled steel—specifically, steel with no more than 0.13% tramps—can be used for new automobile bodies, and usually reprocessed scrap has to be mixed with new steel to meet these requirements.

17.7.3 Steelmaking

Consideration of the Ellingham diagram for oxides (Fig. 17.9) shows that, at the high temperatures prevailing near the tuyères, coke may reduce silica to silicon (mp 1410 °C):

$$SiO_2(s) + 2C(s) \rightarrow Si(l) + 2CO(g) \tag{17.33}$$

which will dissolve in the iron. This is undesirable, as silicon causes iron to be brittle, as do the other major impurities in pig iron—C, P, Mn, and a minor amount of S. The limestone in the blast furnace minimizes this occurrence by converting the SiO_2 to slag, but inevitably the concentrations of impurities, including Si, in pig iron must be reduced to acceptable levels in a separate conversion process, namely, by oxidation with air or pure oxygen in the presence of lime or limestone, to give serviceable steels. Usually one aims for carbon contents below 1.5% for hard (tool) steels and not more than 0.3% for mild steels, with Si, P, S, and Mn contents lower still.

The reactions occurring in the molten iron/lime mixture are

$$2C + O_2 \rightarrow 2CO \tag{17.34}$$
$$Si + O_2 + CaO \rightarrow CaSiO_3(l) \tag{17.35}$$
$$2P + \tfrac{5}{2}O_2 + 3CaO \rightarrow Ca_3(PO_4)_2(\text{in slag}) \tag{17.36}$$
$$2Mn + O_2 \rightarrow 2MnO(\text{in slag}) \tag{17.37}$$
$$Mn + S + CaO \rightarrow CaS(\text{in slag}) + MnO(\text{in slag}). \tag{17.38}$$

The CO gas is easily removed, while the calcium phosphate and manganese oxide go into the molten slag, which has value in fertilizer manufacture because of the phosphate content (Section 9.6). Figure 17.9 shows that Mn, Si, and C will react preferentially with O_2 before oxidation of the molten iron "solvent" begins, but SO_2 and P_2O_5 have less negative ΔG_f°

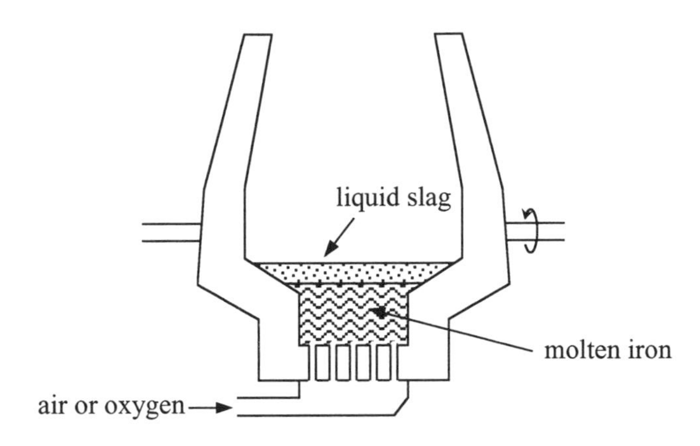

Figure 17.11 Bessemer converter.

values per mole oxygen than does FeO, and it is only because of extraction of $Ca_3(PO_4)_2$ and CaS into the molten slag that P and S can be removed without oxidation of the liquid iron. The slag, then, has a crucial role to play.

Numerous different steelmaking processes involving the same basic principles have been used at various times.[14–16] The bottom-blown *Bessemer converter* (Fig. 17.11), which originated in England in mid-nineteenth century, was largely superseded by the top-blown *Siemens regenerative open hearth furnace*—a long, shallow bath containing molten iron that is heated radiatively by burning fuel oil or gas injected through nozzles in the top of the furnace. Since the late 1950s, this in turn has been largely replaced by the *basic oxygen* or *Linz–Donawitz process* (BOP or LD process, Fig. 17.12), in which a powerful jet of oxygen is blown into the molten slag layer from the top through a water-cooled lance, so as to provide better stirring than is possible with the open hearth furnace. Agitation can be improved by rotating the vessel (*KALDO* and *Rotor* processes), in which case a lower pressure oxygen jet that does not part the slag layer can be used.

Bottom blowing of BOP converters is possible but technically difficult, although modern Japanese BOP furnaces are top and bottom blown simultaneously for best mixing.[16] BOP steelmaking is much faster than the open-hearth and other methods; it is fast enough to feed *continuous casting* machines that have revolutionized steelmaking by supplying a continuous slab of hot steel to the rolling mill instead of ingots that have to be separately poured, solidified, transported, and reheated. In modern foundries, the whole steelmaking process is automated to provide computer control of steel quality and synchronization of the various steps.

The use of oxygen instead of air is now widespread in the steel industry.

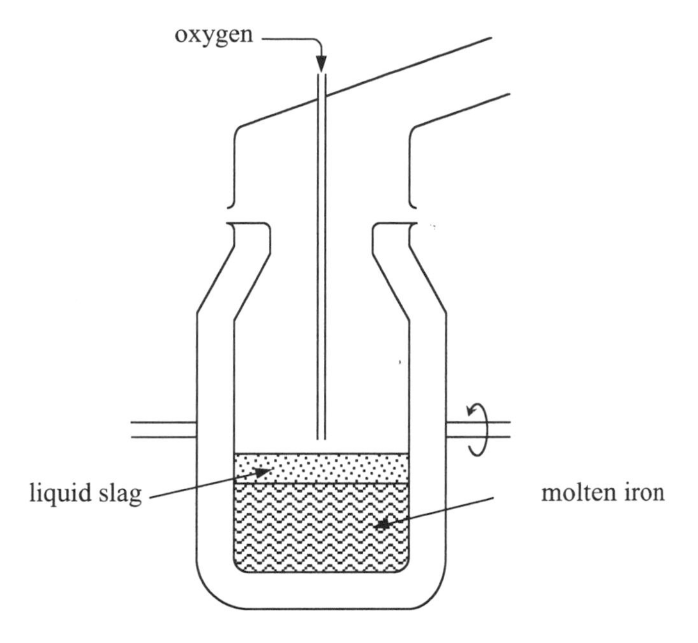

Figure 17.12 The basic oxygen (LD) process.

Some 3,000,000 metric tons of O_2 are used annually for steelmaking in the United States alone. This practice has the following advantages which more than compensate for the extra cost:

(*a*) The higher partial pressure of O_2 means that reactions 17.34 through 17.38 proceed more rapidly, so reducing the operating time.

(*b*) The heat loss due to having to heat the nitrogen (79%) and other air components besides O_2 is eliminated. In fact, in most major processes other than the open hearth method, the heat generated by reactions 17.34 through 17.38 is great enough to keep the charge molten, if 100% O_2 is used and the P content is about 2% or more.

(*c*) Nitriding of the steel (Section 5.7.2) by the N_2 of the air to form small amounts of Fe_4N is eliminated.

On the other hand, the exhaust gases from BOP converters contain relatively large amounts of red α-Fe_2O_3 dust; this, however, can be trapped by electrostatic precipitation or water scrubbing and recycled, and the exit gases can be scrubbed with water to reduce air pollution.

Whatever the process, the steelmaking vessel must be lined with a suitable refractory material, usually bricks of calcined dolomite, (Mg,Ca)O. Silicate firebricks cannot be used in the presence of lime.

17.8 Pyrometallurgy of Halides and Sulfides

Figure 17.13 shows immediately that carbon is useless as a reductant for chlorides, since CCl_4 is barely thermodynamically stable even at room temperature and becomes endergonic above about 400 °C. Similarly, CF_4 is the least exergonically formed (per F atom) of all the fluorides covered by Fig. 17.13, and so carbon cannot reduce them. In principle, hydrogen could reduce titanium tetrachloride to titanium, but in reality titanium hydrides (Section 5.7) would be obtained. The only satisfactory reductants for metal chlorides and fluorides are magnesium and (less practically, on account of their extreme reactivity) sodium or calcium.

17.8.1 Titanium and Titanium Dioxide

Titanium (and also zirconium and tantalum) are made industrially by the *Kroll process*, which involves conversion of TiO_2 to $TiCl_4$ by coke reduction in the presence of Cl_2

$$TiO_2(s) + C(s) + 2Cl_2(g) \xrightarrow{500\,^\circ C} TiCl_4(g) + CO_2(g), \tag{17.39}$$

after which the titanium tetrachloride, which is a liquid at ambient temperature (bp 140 °C), is condensed, purified by fractional distillation, and then reduced with molten magnesium (mp 649 °C, bp 1107 °C) to solid Ti (mp 1660 °C), giving liquid magnesium chloride (mp 714 °C):

$$TiCl_4(g) + 2Mg(l) \xrightarrow[850\,^\circ C]{Ar} Ti(s) + 2MgCl_2(l). \tag{17.40}$$

It is necessary to use argon as the inert atmosphere for reaction 17.40 because even nitrogen will react with Ti to give an interstitial compound (Section 5.7). For the same reason, reduction of TiO_2 by Mg, though theoretically possible according to Fig. 17.9, is not feasible because of retention of interstitial oxygen in the solid titanium. Besides, the other product MgO is extremely refractory (mp 2852 °C) and would be difficult to remove.

The high reactivity of titanium metal is disguised in normal use by its unusually corrosion-resistant protective oxide film (Chapter 16), but it must always be borne in mind by users of Ti process or laboratory equipment. In particular, titanium should not be used with pressurized oxygen: pressures of oxygen of as little as 0.4 MPa have been reported to lead to ignition of Ti at ambient temperature if the oxide film is penetrated.

In passing, we note that reaction 17.39 is also used in the conversion of natural rutile ore, which is usually red-brown because of its iron(III) content, to pure TiO_2, which is used as a filler for paper, as a white pigment in paints, and as a photocatalyst (Sections 10.4 and 14.4.2). The high optical refractive index gives TiO_2 the best available "covering power" in

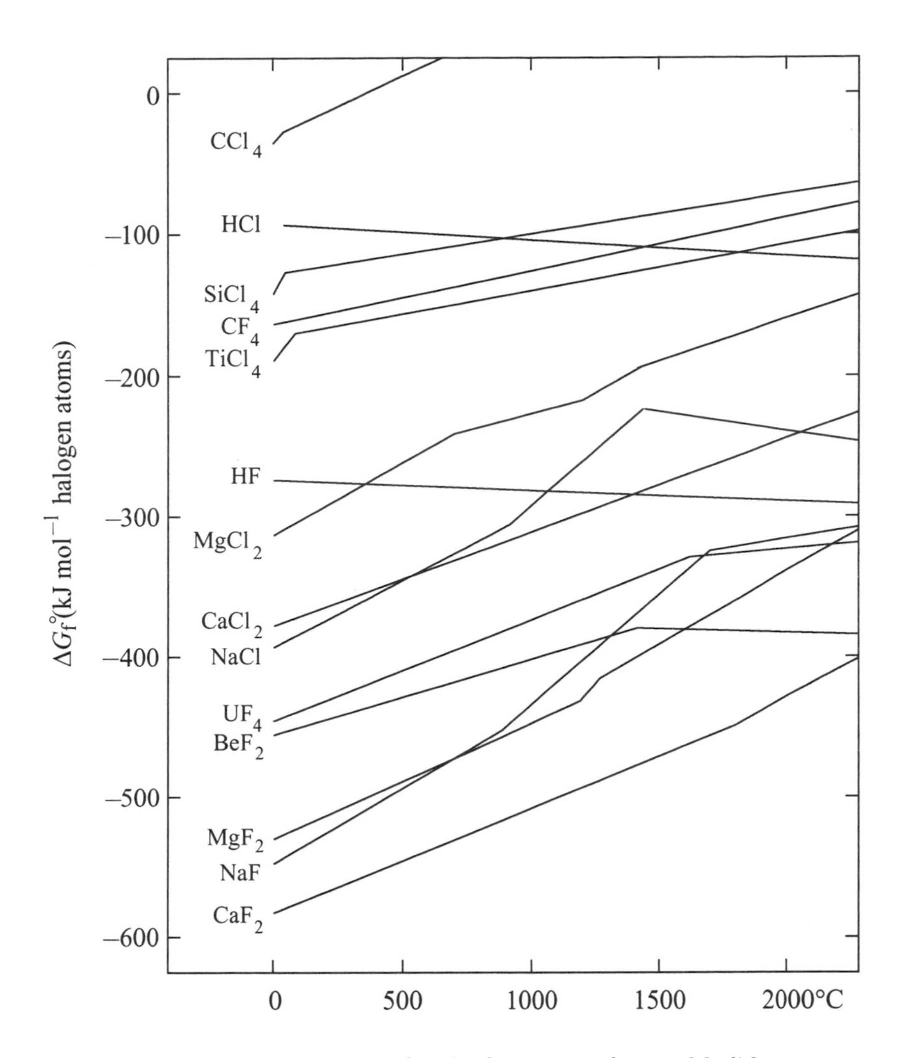

Figure 17.13 Ellingham diagram for the formation of several halides.

paints, and, unlike the lead carbonate formerly used, TiO_2 is very inert and nontoxic. The purified $TiCl_4$ from reaction 17.39 is converted to pure TiO_2 (rutile form) by high temperature oxidation:

$$TiCl_4(g) + O_2(g) \rightarrow TiO_2(s) + 2Cl_2(g). \quad (17.41)$$

Titanium dioxide, in the form of either rutile or anatase, can also be obtained from ilmenite by treatment with sulfuric acid (the *sulfate process*), followed by hydrolysis and then thermal dehydration of the resulting $TiO(OH)_2$. The chloride process, however, is now more widely used than the sulfate process that was dominant until about 1990, largely because, as a continuous process, it is more economical than the sulfate method. The sulfate process also produces waste sulfuric acid, which is usually neutralized with limestone; the resulting gypsum ($CaSO_4{\cdot}2H_2O$) can be used in making plasterboard.

17.8.2 Silicon

Crude elemental silicon can be obtained by reduction of silica sand with coke in the electric furnace (reaction 17.33) and may be adequate for making ferrosilicon alloys (Section 16.7.5) or silicones (Section 3.5). The high purity silicon used for electronic "chips" can be made from silica via silicon tetrachloride, which, like $TiCl_4$, is a volatile liquid (bp 57 °C) susceptible to hydrolysis but readily purifiable by fractional distillation. Indeed, the procedure for silicon resembles the Kroll process for titanium, except that an argon atmosphere is not necessary:

$$SiO_2(g) + 2C(s) + 2Cl_2(g) \xrightarrow{\text{red heat}} SiCl_4(g) + 2CO(g) \quad (17.42)$$

$$SiCl_4(\text{redistilled}) + 2Mg(\text{pure}) \xrightarrow{\text{heat}} Si(s) + 2MgCl_2. \quad (17.43)$$

The magnesium chloride and any excess Mg are then vaporized off, and the remaining silicon is further purified, usually by *Czochralski crystal growth* (whereby a spinning crystal of pure Si is drawn slowly from a silicon melt, solidifying as it goes; Section 19.2). Increasingly, though, electronics-grade silicon is being made via H_2 reduction of $HSiCl_3$. The Ellingham diagram for halides (Fig. 17.13) shows that H_2 can reduce $SiCl_4$ to Si and HCl(g) at elevated temperatures, and the analogous reaction of $HSiCl_3$, which has a less negative ΔG_f° than $SiCl_4$, is even more favorable. The trichlorosilane is made from crude silicon by the *Siemens process*:

$$Si(s) + 3HCl(g) \rightarrow HSiCl_3(g) + H_2(g). \quad (17.44)$$

Redistilled $SiCl_4$ is also used in the preparation of pure SiO_2 for fiber optics (Section 7.5).

17.8.3 Metal Sulfides

Figure 17.14 shows that coke cannot serve as a reductant for sulfides, either. Carbon disulfide is seen to be barely stable thermodynamically, while CS, the analog of carbon monoxide which is so important in oxide pyrometallurgy, is highly unstable and was first characterized in the laboratory by Klabunde as recently as 1984. The strongly negative ΔG_f° of SO_2, however, means that oxygen can remove S from many metal sulfides, although the product is usually not the metal but rather an oxide of the metal. This is

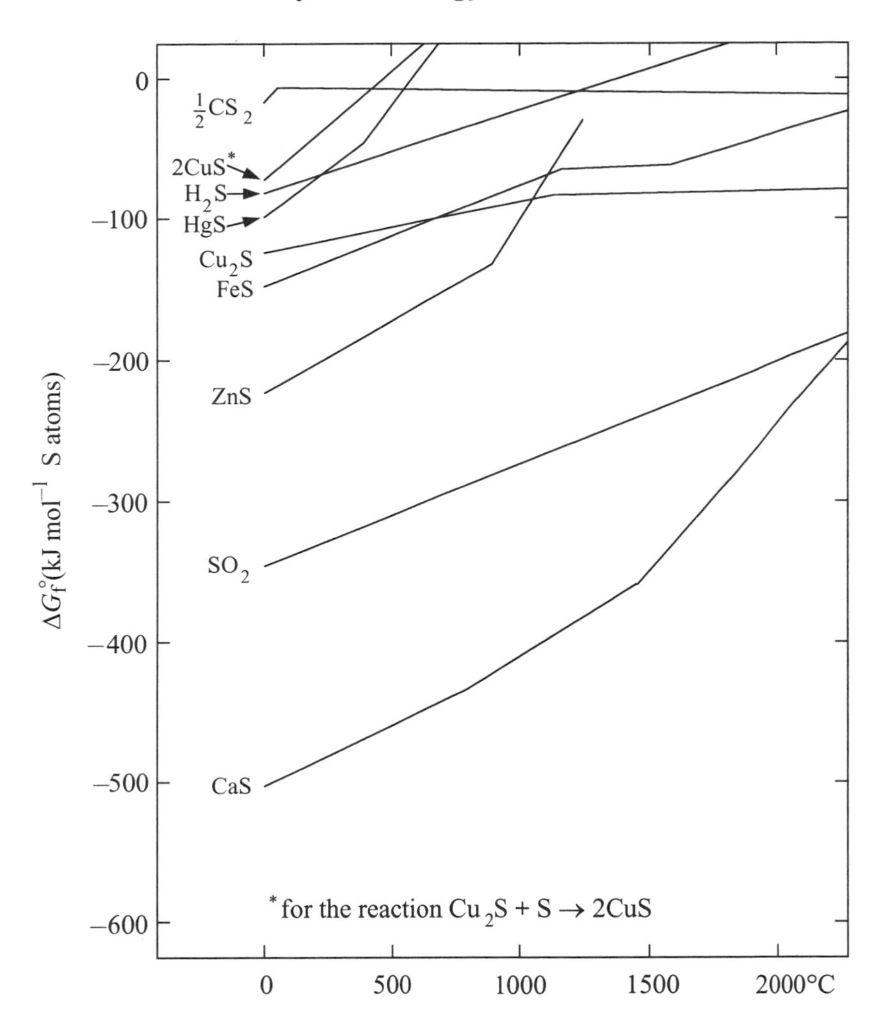

Figure 17.14 Ellingham diagram for formation of some sulfides.

helpful because metal oxides are often reducible with coke. Thus, coke pyrometallurgy of several sulfides such as ZnS (zinc blende, wurtzite) becomes possible if the ore concentrate is first "roasted" in air:

$$2ZnS + \text{excess } 3O_2 \xrightarrow{800\text{–}900\,^\circ C} 2ZnO + 2SO_2 \tag{17.45}$$

$$ZnO + C \xrightarrow{\text{blast furnace}} Zn + CO. \tag{17.46}$$

Roasting at a lower temperature in a limited air supply can give the metal sulfate, rather than the oxide,

$$ZnS + 2O_2 \xrightarrow{600\text{–}700\,^\circ C} ZnSO_4, \tag{17.47}$$

and the cooled sulfate can be dissolved in water for electrolytic reduction to the metal (a *sulfatizing roast*). The normal commercial process for electrolytic zinc, however, involves roasting ZnS to ZnO, followed by dissolution of the latter in sulfuric acid (made on-site from SO_2 liberated in reaction 17.45) and finally electrolysis to produce zinc metal. For lead, these methods are inapplicable, since $PbSO_4$ has very low solubility in water. On the other hand, if the oxygen and SO_2 partial pressures and the temperature are correctly controlled, it is possible to obtain molten lead directly from PbS by the *roast reduction* reaction:

$$\mathrm{PbS} + \mathrm{O_2} \rightarrow \mathrm{Pb} + \mathrm{SO_2}. \tag{17.48}$$

Copper, too, can be recovered from copper sulfides without resort to a reductant such as coke. The concentrate usually contains iron sulfides (in fact, the most widely occurring ore of copper is $CuFeS_2$, chalcopyrite), as well as silicates. The chemistry of copper smelting is subtle, but the commonly employed two-stage process can be summarized as follows:

(*a*) *Matte smelting* involves roasting the concentrate with lime and recycled converter slag at about 1300 °C to form a slag of molten oxides of iron, silicon, and other impurities, and an immiscible *matte* or molten sulfide layer, which contains Cu_2S along with some FeS.

(*b*) The molten matte layer is transferred to a converter, where air is blown through it and sand is added so that the FeS forms a silicate slag

$$\mathrm{FeS} \xrightarrow{\mathrm{O_2} + \mathrm{SiO_2}} \mathrm{FeSiO_3(l)} + \mathrm{SO_2}, \tag{17.49}$$

and then, above 1250 °C, the Cu_2S is converted to copper:

$$\mathrm{Cu_2S} + \mathrm{O_2} \rightarrow 2\mathrm{Cu} + \mathrm{SO_2}. \tag{17.50}$$

Copper made by reaction 17.50 (*blister copper*) is about 98% pure but may be refined electrolytically. As far as the copper itself is concerned, the two stages could be represented as

$$2\mathrm{CuS} + \mathrm{O_2} \rightarrow \mathrm{Cu_2S} + \mathrm{SO_2} \tag{17.51}$$

and

$$\mathrm{Cu_2S} + \mathrm{O_2} \rightarrow 2\mathrm{Cu} + \mathrm{SO_2}. \tag{17.52}$$

The role of the slag in what is in effect liquid–liquid solvent extraction is, once again, crucial in the removal of unwanted material such as iron compounds.

Exercises

17.1 At what pH is hydrogen gas at 1 bar pressure theoretically capable of reducing aqueous Ni^{2+} ion ($E° = -0.25$ V, assumed 1 mol kg^{-1}) to nickel metal (*a*) at 25 °C and (*b*) at 150 °C (if $\Delta H = 54$ kJ mol^{-1} and $\Delta C_P^\circ = 0$)? (*c*) In practice, what factors may prevent the reduction from taking place? (*d*) How might these be overcome?
[*Answers:* (*a*) pH 4.23; (*b*) pH 2.83.]

17.2 From the following thermodynamic data, with the assumptions that the heat capacities of reaction are negligible and that standard conditions (other than temperature) prevail, calculate the temperatures above which (*a*) carbon monoxide becomes the more stable oxide of carbon, in the presence of excess C; (*b*) carbon is thermodynamically capable of reducing chromia (Cr_2O_3) to chromium metal; (*c*) carbon might, in principle, be used to reduce rutile to titanium metal; and (*d*) silica (taken to be α-quartz) may be reduced to silicon in a blast furnace.

Substance	ΔH_f° *(kJ mol^{-1})*	$S°$ *(J K^{-1} mol^{-1})*
C (graphite)	0	5.740
CO(g)	−110.525	197.674
CO_2(g)	−393.509	213.74
Ti(s)	0	30.63
TiO_2 (rutile)	−944.7	50.33
Cr(s)	0	23.77
Cr_2O_3(s)	−1139.7	81.2
Si(s)	0	18.83
SiO_2 (α-quartz)	−910.94	41.84

[*Answers:* (*a*) 707 °C; (*b*) 1217 °C; (*c*) 1714 °C; (*d*) 1639 °C.]

17.3 On the basis of the following thermochemical data, (*a*) calculate the equilibrium constant $K°$ for the formation of *gaseous* nickel tetracarbonyl from Ni and Co under standard conditions; (*b*) estimate the temperature at which $K°$ becomes unity; and (*c*) explain how this and other related information can be applied in the refining of nickel.

Substance	ΔH_f° *(kJ mol^{-1})*	$S°$ *(J K^{-1} mol^{-1})*
Ni(s)	0	29.87
CO(g)	−110.525	197.674
$Ni(CO)_4$(g)	−602.91	410.6

Note: The heat capacity of reaction is small and may be neglected.
[*Answers:* (*a*) 5.7×10^6; (*b*) 119 °C.]

17.4 Magnesium metal can be made industrially by electrolysis of magnesium chloride. Magnesium oxide (which can be obtained from mineral $MgCO_3$ by firing it in a kiln) can be converted to magnesium chloride by passing chlorine gas through a bed of MgO mixed with coke at 600 to 1000 °C:

$$MgO(s) + Cl_2(g) + C(s) \rightleftharpoons MgCl_2(\text{mp } 714\,^\circ\text{C}) + CO(g). \qquad (1)$$

Concern has been expressed that the chlorine may react with the carbon monoxide to form the highly toxic gas *phosgene* (carbonyl chloride, $COCl_2$):

$$CO + Cl_2 \rightleftharpoons COCl_2. \qquad (2)$$

Are such fears justified? In formulating your answer, you may wish to consider, first, that for reaction 2 at 25 °C, $\Delta H^\circ = -108.3$ kJ mol^{-1}, $\Delta S^\circ = -137.2$ J K^{-1} mol^{-1}, and $\Delta C_P^\circ = -5.39$ J K^{-1} mol^{-1}, and, second, that chlorine will never be present at the exit from the hot reactant bed unless the supply of one of the other reactants in reaction 1 runs out.

17.5 Trace the steps in the production of high purity nickel metal from crude sulfide ore (*a*) by the Sherritt Gordon ammonia leach process, (*b*) by the Mond process, and (*c*) by acid leaching.

17.6 Suggest a procedure for the industrial production of hafnium-free zirconium metal from baddeleyite (ZrO_2), on the basis of information given in this chapter and in Appendix C.

References

1. R. H. Parker, "An Introduction to Chemical Metallurgy," 2nd Ed. Pergamon, Oxford, 1978.

2. J. E. Fergusson, "Inorganic Chemistry and the Earth." Pergamon, Oxford, 1982.

3. J. C. Bailar, H. J. Emeléus, R. S. Nyholm, and A. F. Trotman-Dickenson (eds.), "Comprehensive Inorganic Chemistry," Vols. 1–5. Pergamon, Oxford, 1973.

4. A. R. Burkin, "The Chemistry of Hydrometallurgical Processes." E. and F. N. Spon, London, 1966.

5. F. Habashi, Hydrometallurgy. *Chem. Eng. News* February 8, 46–58 (1982).

6. D. Pletcher and F. C. Walsh, "Industrial Electrochemistry." Chapman & Hall, London, 1990.

7. J. Hill, Refining processes relevant to the production of speciality inorganic chemicals. *In* "Insights into Speciality Inorganic Chemicals" (D. Thompson, ed.), pp. 5–34. Royal Society of Chemistry, Cambridge, 1995.

8. R. Hoffmann, Winning gold. *Am. Sci.* **82**, 15–17 (1994).

9. G. M. Ritcey and A. W. Ashbrook, "Solvent Extraction." Elsevier, New York, 1979; Y. Marcus and A. S. Kertes, "Ion Exchange and Solvent Extraction." Wiley (Interscience), New York, 1969; T. Sekine and Y. Hasegawa, "Solvent Extraction Chemistry." Dekker, New York, 1977.

10. K. Dorfner, "Ion Exchangers: Properties and Applications," 3rd ed. Ann Arbor Science Publishers, Ann Arbor, Michigan, 1972; R. W. Grimshaw and C. E. Harland, "Ion Exchange: Introduction to Theory and Practice." Chemical Society, London, 1975; M. J. Slater, "The Principles of Ion Exchange Technology." Butterworth–Heinemann: Oxford, 1991 (mainly engineering principles).

11. J. R. Boldt, Jr., "The Winning of Nickel" Longmans, Toronto, 1967.

12. C. B. Alcock, "Principles of Pyrometallurgy." Academic Press, London, 1976.

13. O. Kubaschewski and C. B. Alcock, "Metallurgical Thermochemistry," 5th Ed. Pergamon, Oxford, 1979.

14. C. S. Russell and W. J. Vaughan, "Steel Production: Processes, Products, and Residuals." Johns Hopkins Univ. Press, Baltimore, 1976.

15. F. N. Kemmer (ed.), "The Nalco Water Handbook," 32.6ff. McGraw-Hill, New York, 1988.

16. N. Ohashi, "Modern steelmaking." *Am. Sci.* **80**, 540–555 (1992).

Chapter 18

Organometallics

ORGANOMETALLIC COMPOUNDS ("organometallics") are usually defined as those that have at least one direct metal-to-carbon bond. This definition includes metal alkyl compounds such as diethylzinc [$Zn(C_2H_5)_2$] and metal arene (metallocene) complexes like ferrocene [$Fe(C_5H_5)_2$], as well as metal carbonyls such as $Ni(CO)_4$ (Sections 8.2 and 17.4) which do not have organic ligands in the usual sense but which behave much like organic compounds in terms of volatility, solubility in nonpolar organic solvents, and general properties. On the other hand, lead(IV) acetate [$Pb(OOCCH_3)_4$], a nonpolar liquid with organic ligands, is excluded because it contains no Pb—C bonds. Interstitial metal carbides (Section 5.7) might be said to contain metal–carbon bonds, but these refractory solids bear no resemblance to organometallics.

The chemistry of organometallics[1-11] is extraordinarily rich and often complicated, because the variety of coordination geometries and oxidation states that characterize the chemistry of metals (especially transition metals) is added to the diversity of carbon frameworks of organic compounds, together with some bonding modes not encountered in other kinds of compounds. Much research effort has gone into the development of organometallics as reagents for synthetic organic and pharmaceutical chemistry, which is beyond the scope of this book. Accordingly, only a brief introduction to the subject is given here, with emphasis on a few aspects of technological interest, namely

(*a*) use of specific organometallics as biocides, stabilizers for plastics, etc.

(*b*) homogeneous catalysis of industrial organic reactions

(*c*) catalysis of alkene polymerization in the plastics industry

(*d*) organometallic precursors for metal film deposition in the electronics industry (taken up in Chapter 19).

18.1 Alkyl Compounds of Some Main Group Metals

Most truly metallic elements M have electronegativities in the range 0.8 to 1.8 (cf. 2.55 for alkyl carbon, Table 2.2), so that M—C bonds can be expected to be strongly polar with a substantial partial negative charge on the carbon. Consequently, we can expect the M—C bond to be highly vulnerable to attack at C by *electrophiles* (i.e., species that seek negative charge) such as H^+, and at M by *nucleophiles* such as OH^-, water, and ammonia. The M—C bond is therefore expected to be highly reactive with respect to polar reagents, even though it is often thermochemically quite strong.

18.1.1 Group 12 Organometallics

Diethylzinc, one of the first organometallic compounds to be isolated (Frankland, 1849), epitomizes these characteristics. The compound is easily prepared by heating powdered zinc or zinc–copper alloy with ethyl iodide under an atmosphere of dry nitrogen or CO_2 (the initial product is actually C_2H_5ZnI, which disproportionates on distillation to zinc iodide and diethylzinc):

$$Zn(s) + 2C_2H_5I(l) \rightarrow Zn(C_2H_5)_2(l) + ZnI_2(s). \tag{18.1}$$

Despite a respectable Zn—C bond energy of 148 kJ mol^{-1}, however, diethylzinc is immediately hydrolyzed by moisture to form $Zn(OH)_2$ and ethane, catches fire spontaneously in the air to form a ZnO smoke, and reacts readily with many organic compounds as an alkylating or reducing reagent. The hydrolysis reaction is finding application in the preservation of books and documents made with acidic paper (Section 10.4.2); the item to be protected against acid embrittlement is exposed to diethylzinc (DEZ) vapor in a previously evacuated chamber, whereupon hydrogen ions in the paper are mopped up to form ethane and zinc ions, while the hydrolytic deposition of small amounts of $Zn(OH)_2$ within the paper gives lasting protection against reacidification. The procedure cannot, however, reverse any embrittlement that has already set in.

In the same periodic group, cadmium forms alkyls much as does zinc, whereas mercury is readily methylated biologically to form the very toxic methylmercury ion CH_3Hg^+, as discussed in Section 11.4.2. It should be noted that CH_3Hg^+ is a metal alkyl species that resists hydrolysis and is insensitive to the oxygen of the air. In the past, organomercury compounds [(hydroxymercuri)chlorophenols, Semesan] were used by farmers as fungicides on seed grain, but this practice resulted in accidental poisonings, and even fatalities, and has been discontinued.

18.1.2 Organometallics of Groups 1 and 2

In general, metal alkyls conform to the zinc stereotype of reactivity, and indeed it is the reactivity of metal alkyls that makes them so important as reagents. As might be expected, alkali metal alkyls are particularly reactive, perhaps too much so in the case of the reaction of sodium or potassium metals with alkyl or aryl halides RX, which tend to give the alkane R—R and NaX or KX (the *Wurtz–Fittig reaction*) rather than follow an analog of reaction 18.1. On the other hand, lithium metal will react with RX such as n-butyl chloride in a suitable solvent such as diethylether or hexane under a nitrogen or (better) argon atmosphere to give synthetically useful organolithium compounds:

$$2\mathrm{Li(s)} + \mathrm{RX(in\ ether)} \xrightarrow[\mathrm{N_2}]{\mathrm{Ar\ or}} \mathrm{LiR(in\ ether)} + \mathrm{LiX(s)}. \qquad (18.2)$$

For example, phenyllithium can be used to make tetraphenyltin:

$$4\mathrm{C_6H_5Li} + \mathrm{SnCl_4} \rightarrow \mathrm{(C_6H_5)_4Sn} + 4\mathrm{LiCl}. \qquad (18.3)$$

For many syntheses that do not require a reagent as aggressive as an organolithium compound, *Grignard reagents* (RMgX) may be used. These were discovered in 1900 by Victor Grignard, who was awarded the Nobel Prize in 1912 for his work in developing their chemistry. Grignard reagents are invariably prepared in solution in an ether [usually diethyl ether, $(C_2H_5)_2O$] which serves to stabilize them, probably as ether complexes $RMgX[O(C_2H_5)_2]_2$, although these exist in solution in equilibrium with other organomagnesium species such as the dimer $RMgX_2MgR$ and dialkylmagnesium R_2Mg (cf. the disproportionation of RZnX, Section 18.1.1). The reaction of small magnesium flakes with an organic halide in ether can be carried out in air (contrast the preparation of zinc alkyls), but water must be rigorously excluded (usually, the ether is predried over sodium wire):

$$\mathrm{Mg(s)} + \mathrm{RX(in\ ether)} \rightarrow \mathrm{RMgX(in\ ether)}. \qquad (18.4)$$

Solutions of Grignard reagents have largely replaced organozinc compounds in organic and organometallic syntheses. Unlike the zinc alkyls, Grignard reagents do not require exclusion of oxygen, but they do react readily with CO_2. In fact, this is a useful way to make unusual carboxylic acids:

$$\mathrm{RMgX} + \mathrm{CO_2} \longrightarrow \mathrm{RCOOMgX} \xrightarrow{\mathrm{HX(aq)}} \mathrm{RCOOH} + \mathrm{MgX_2}. \qquad (18.5)$$

18.1.3 Group 14 Organometallics

Grignard reagents can be used to make organosilicon, -germanium, -tin, and -lead compounds by halide abstraction from the corresponding halo-compounds (e.g., chlorosilanes, Section 3.5.1), much as in reaction 18.3:

$$n\mathrm{RMgX} + \mathrm{MX_4} \rightarrow \mathrm{MR}_n\mathrm{X}_{(4-n)} + n\mathrm{MgX_2}. \quad (18.6)$$

Industrially, however, organosilicon and organotin chlorides are usually obtained by the direct Rochow process (Section 3.5.2):

$$\mathrm{Sn} + 2\mathrm{RX} \rightarrow \mathrm{R_2SnX_2}\ (\text{etc.}). \quad (18.7)$$

Tin alkyls can also be made from the reaction of $SnCl_4$ with aluminum alkyls, highly reactive, oxygen-sensitive materials which are obtained industrially from olefins, hydrogen, and aluminum powder:

$$2\mathrm{Al(s)} + 3\mathrm{H_2(g)} + 6\mathrm{RCH{=}CH_2(g)} \xrightarrow[6\ \mathrm{MPa}]{100\,^\circ\mathrm{C}} 2\mathrm{Al(CH_2CH_2R)_3}. \quad (18.8)$$

Industrial organosilicon chemistry centers around silicone production (Section 3.5.2). Organotin compounds are widely used as stabilizers for polyvinyl chloride plastics (usually as R_2SnX_2, where R is typically n-octyl and X is a long chain carboxylate ligand such as laurate or maleate) and as curing agents for silicone rubbers (e.g., di-n-butyltin diacetate). Organotin compounds of the type R_3SnX such as n-Bu_3SnOH are important biocides and are used as antifouling agents in marine paints, in suppression of fungal growths in agriculture, and in slime control in the pulp and paper industry. They have the advantage of being very selective, and neither they nor their degradation products are particularly toxic to the higher animals, including humans. Nevertheless, release of organotin compounds into the environment should be minimized.

Organolead compounds, notably tetraethyllead, have been extensively used as antiknock additives for gasoline (see Section 7.3.3). The Pb—C bond is thermochemically weak, so that PbR_4 molecules decompose readily within the hot cylinders of an automobile engine to give radicals R· which terminate the chain reactions (Section 2.5) that cause explosive rather than smooth burning of the fuel vapor. Thus, it is possible to upgrade low-octane petroleum fractions. However, both tetraalkyllead compounds and their combustion products are dangerously toxic. Lead poisoning is particularly insidious in that the symptoms may not be recognized as such; for example, lead is known to inhibit the mental development of children, and it has been suggested that the phenomenon of steadily falling average scores in certain scholastic aptitude tests may be due in part to ingestion by young children of lead spread in the environment by automobile exhausts (and, to a lesser degree, by lead carbonate-based paints, which are now largely being phased out in favor of nontoxic TiO_2-containing paints).

Furthermore, lead compounds poison the platinum metal-based catalysts in catalytic converters, so that efforts to abate air pollution by automobiles through the use of catalytic converters (Section 8.4.2) are dependent on the use of lead-free fuels. Fuels of sufficiently high octane equivalents can be made from low octane fractions by shape-selective reforming or by use of additives such as MTBE (Sections 7.3.3 and 7.4) or organometallics of relatively low toxicity and catalyst poisoning tendencies such as methylpentacarbonylmanganese [$CH_3Mn(CO)_5$] or (methylcyclopentadienyl)tricarbonylmanganese [η^5-$CH_3C_5H_4Mn(CO)_3$, MMT].

Several main group organometallics have significant catalytic activity [e.g., dibutyltin compounds catalyze the formation of polyurethane from organic isocyanates (R—N═C═O) and alcohols], but perhaps the most important are the trialkylaluminums, which, following reaction with titanium chlorides, form the *Ziegler* and *Ziegler–Natta catalysts* for olefin polymerization. These are considered along with the more recently developed zirconocene olefin polymerization catalysts in Section 18.4.

18.2 Organotransition Metal Compounds

The transition metals often have partly filled *d* electron shells and so offer particularly rich opportunities to the organometallic chemist because (*a*) the oxidation states are more readily varied than those of the main group metals and (*b*) the empty *d* orbitals are able to accept electrons from the ligands (σ dative bonding) while the electrons in filled *d* orbitals can be donated back into low lying empty π^* antibonding molecular orbitals on the ligands (π "back-bonding"; cf. bonding in metal carbonyls, Section 8.2). The back-bonding mechanism stabilizes low oxidation states of metal atoms because it drains excess electron density away from them, and conversely the return of electron density to the ligand increases its basicity and hence its effectiveness as a σ donor ("synergic" or "push–pull" bonding). Back-bonding is especially favored by transition metals, as their *d* orbitals are often of just the right symmetry to mesh with the π^* antibonding molecular orbitals on a wide variety of organic ligands. Suitable π^* antibonding molecular orbitals are associated with carbon–carbon double bonds (alkenes), triple bonds (alkynes), and delocalized π-bond systems as in allylic ligands and aromatic hydrocarbons (arenes) like benzene.

The first organometallic compound of the transition metals to be characterized (1827) was *Zeise's salt*, $K[(C_2H_4)PtCl_3]\cdot H_2O$ (Fig. 18.1). It forms when $K_2[PtCl_4]$ in aqueous ethanol is exposed to ethylene (ethene); a dimeric Pt—C_2H_4 complex with Cl bridges is also formed. In both species, the ethylene is bonded *sideways* to the platinum(II) center so that the two carbon atoms are equidistant from the metal. This is called the *dihapto-* or η^2 mode. A ligand such as an allyl radical with three adjacent carbons directly bonded to a metal atom would be *trihapto-* or η^3, and so on.

The nature of the bonding in Zeise's anion and other η^2-olefin complexes is illustrated in Fig. 18.2. Without the push–pull mechanism, the π electrons of the olefin would have little or no tendency to allow themselves

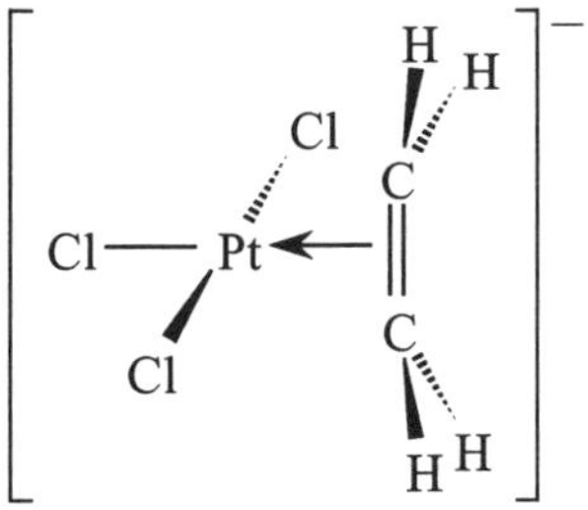

Figure 18.1 Anion of Zeise's salt.

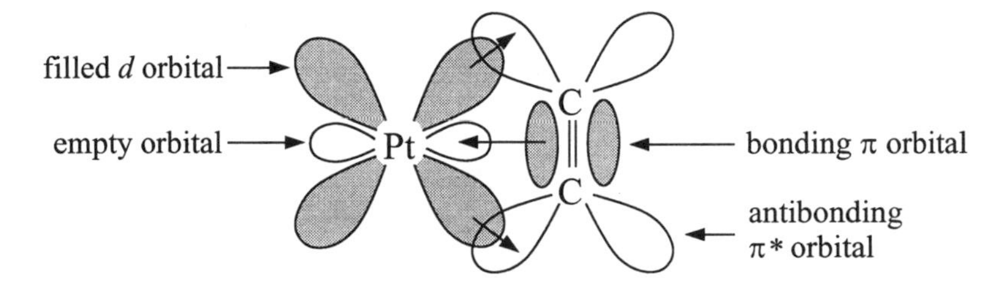

Figure 18.2 Bonding in Zeise's anion.

to be shared with the metal. The displacements of electron density in the π and π^* systems result in the substituents on the carbon atoms (H, in the case of ethylene) being bent away from the metal atom.

Aromatic hydrocarbons such as benzene (C_6H_6) and the anion $C_5H_5^-$ ("Cp^-") of cyclopentadiene (C_5H_6, "CpH") are "side-on bonders" par excellence. They bond to the metal atom perpendicularly to the plane of the aromatic ring by multiple interactions involving the several bonding and antibonding orbitals of the π-electron systems.[1–11] Thus, two cyclopentadienide units can form a *sandwich complex* with Fe^{II} known as *ferrocene*, $Fe(\eta^5\text{-}C_5H_5)_2$ (Fig. 18.3).

Ferrocene is an orange-brown solid (mp 173 °C) that is easily made by heating freshly distilled C_5H_6 with an iron(II) salt and solid KOH. Ferrocene is extraordinarily inert: it can be heated to 500 °C without decomposition, is stable indefinitely in the air or in contact with water, resists attempts to hydrogenate it, and generally requires rather brutal methods such as Friedel–Crafts syntheses to introduce substituents to the organic rings. It is oxidized by nitric acid or halogens, but even then the Fe—C bonds remain intact, giving the ferricenium ion $Fe^{III}(\eta^5\text{-}C_5H_5)_2^+$ or its derivatives. All this stands in sharp contrast to the traditional picture of organometallic reactivity that came from early experience with zinc alkyls.

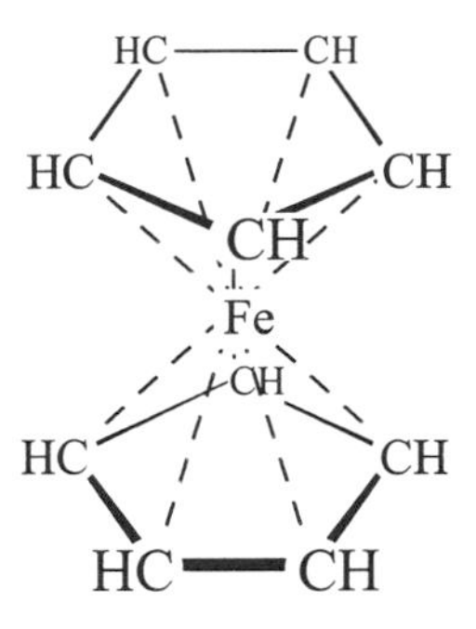

Figure 18.3 Ferrocene.

The inertness of ferrocene is a rather extreme example of the special properties conferred by arene ligands. Some other metal arene sandwich compounds such as $Cr(\eta^6\text{-}C_6H_6)_2$ and $Ni(\eta^5\text{-}C_5H_5)_2$ (nickelocene) are much more reactive than ferrocene and require special handling. Complexes containing just one aromatic ligand with side-on bonding ("open-faced sandwich" or "piano stool" complexes) are also well known. In all such *metallocene* complexes, the extensive π interactions of the ligand allow it to act as a "sponge" for excess electron density in the complex.

The draining away of excess electron density from the metal atom into the π^* system of olefin, aromatic, or carbonyl ligands helps stabilize low oxidation states of the metal. This is why metal carbonyls such as $Ni(CO)_4$ or $Fe(CO)_5$ are so stable despite the formal oxidation state of zero for the metal. Organophosphanes (i.e., organophosphines, organic derivatives PR_3 of phosphine, PH_3, now called phosphane for consistency with alkanes) are customarily viewed as behaving similarly to CO as back-bonding ligands, except that the π-acceptor orbitals are said to be the unoccupied $3d$ orbitals of the P atom.

Nitric oxide bonds to metals in the same way as does CO, forming *nitrosyl* complexes. These hardly qualify per se as organometallics, but in practice nitrosyl ligands are often associated with carbonyl or organic ligands because of the compatibility of the bonding modes, for example, in $Co(CO)_3NO$. The molecules N_2 and O_2 can also bond to transition metal centers using variants of the back-bonding mechanism. Even H_2 molecules can bond to metal centers in certain organotransition metal complexes in a dihapto mode; here, no back-bonding is involved, since H_2 has no π-acceptor system, and the bonding is presumably of the three-center, two-electron type well known in borane (boron hydride) chemistry. More importantly, the presence of π-acceptor ligands on the metal center stabilizes hydrido (H^-) complexes through the "electron sponge" action noted above, and so addition of H_2 to organotransition metal centers usually gives ultimately hydrido complexes, perhaps via the η^2-H_2 intermediate. In a similar way, alkyl derivatives of the transition metals, which might be expected to be extremely reactive on the basis of electronegativity, are substantially stabilized by the presence of π-acceptor ligands, as in $CH_3Mn(CO)_5$ (Section 18.1).

18.2.1 Eighteen-Electron Rule

It is commonly observed that organometallic, carbonyl, and nitrosyl complexes of the transition metals are usually most stable when the metal has a total of 18 electrons in the "valence shell" (i.e., beyond the preceding noble gas configuration), including electrons donated by the ligands. In effect, the *eighteen-electron rule* implies that *all* the orbitals of the valence shell of a transition metal atom—the five $(n-1)d$, the ns, and the three np atomic orbitals—are pressed into service, either to accept σ electrons from the ligands or to donate metal d electrons in back-bonding to the ligands. In the context of the eighteen-electron rule, ligands such as halide ions, hydride, η^2-alkenes, and CO count as two-electron donors, uncharged NO and the allyl ligand η^3-C_3H_5 are three-electron donors, and aromatic ligands with all six π electrons involved in bonding to the metal (e.g., η^5-$C_5H_5^-$, η^6-C_6H_6, and the tropylium ion η^7-$C_7H_7^+$) are six-electron donors.

Thus, the Cr in $Cr(CO)_6$ provides 6 electrons (Cr is the sixth element after Ar in the periodic table), and the six CO provide 12 electrons, totaling 18. The iron in ferrocene is formally iron(II), so the metal again provides 6 electrons and the two $C_5H_5^-$ ligands give 12 for a total of 18. The great majority of organometallic compounds follow the eighteen-electron rule, and those that do not usually show special reactivity. For example, the high reactivity of nickelocene, a 20-electron compound, was noted above. Furthermore, square-planar d^8 complexes of monovalent Group 9 or divalent Group 10 metals, although often very stable thermodynamically, are 16-electron molecules and often can add a fifth ligand or become involved reversibly in *oxidative addition reactions* (losing two electrons through oxidation but gaining two two-electron donor ligands) to form 18-electron species; this is the essence of their catalytic action, described in Section 18.3. Indeed, reactions of 18-electron organometallic compounds often involve 16-electron species, or, less commonly, intermediates with 14, 20, or odd numbers of electrons, but the transient nature of such compounds underlines the general validity of the eighteen-electron rule.

To summarize, the special bonding characteristics of π-acceptor ligands in organotransition metal compounds enable these complexes to coordinate small molecules such as ethylene, CO, and H_2 and also provide an electronic buffer system to facilitate changes of metal oxidation state and coordination number. Such complexes, and especially those of the 16-electron type, are therefore well suited to act as catalysts for homogeneous reactions involving these molecules, usually through cycles involving *oxidative addition* and *reductive elimination*, as follows.

18.3 Transition Metal Complexes as Homogeneous Catalysts

Homogeneous catalysts are those that are present in the same phase as the reaction that they facilitate (e.g., nitric oxide in the lead chamber or Deacon processes, where all the reactants and the catalyst are gaseous). Several transition metal complexes have gained industrial importance as homogeneous catalysts for reactions in solution. In Chapter 6, we considered the many advantages of *heterogeneous* catalysts. Homogeneous catalysts are in many respects less convenient—in particular, one is faced with the problem of separating the catalyst from the products at the end of the reaction—but, as molecular entities, they have precisely reproducible properties and are much more amenable to study and to systematic chemical modification than are catalytic surfaces. Thus, it is possible to *design* homogeneous catalysts for extremely high activity and selectivity through the techniques of synthetic chemistry.[12–14]

In essence, the functions of the catalyst are (*a*) to bring the reactant molecules together and (*b*) to facilitate rearrangement of their chemical bonds by acting as an electron "bank." Transition metal complexes can fulfill function (*a*) through their ability to accept and exchange various ligands, whereas the variability of the oxidation states of the central transition metal atoms, particularly where π-acceptor ligands are present, can provide very effectively for function (*b*). The specificity and effectiveness of the catalytic complex can, in principle at least, be modified at will by altering the "spectator" ligands or by choosing a different central metal atom.

In particular, if we have a complex that normally has n ligands when the oxidation state of the central metal is z, but prefers $(n+2)$ ligands when the oxidation state is increased to $(z+2)$, we have the prerequisites for facile *oxidative addition* of a polyatomic molecule such as H_2 to form two new ligands (here, hydrido ligands, H^-) by breaking a covalent bond within the molecule and taking two electrons from the metal atom M (reaction 18.9). The reverse process is called *reductive elimination.*

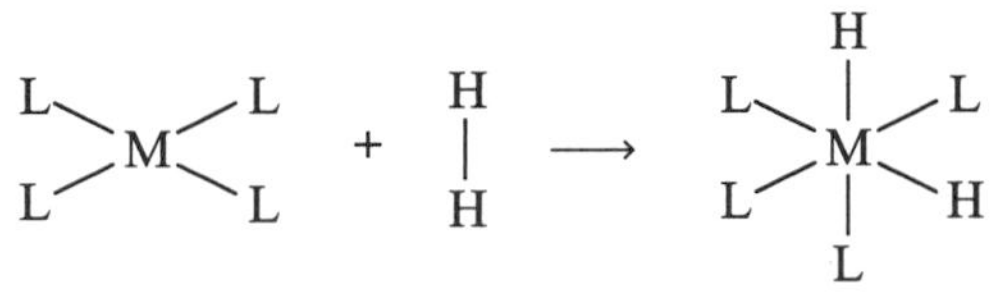

(18.9)

18.3.1 Homogeneous Hydrogenation

In reaction 18.9, the H—H bond is broken to form two hydrogen atoms in a reactive form (hydrido ligands), much as H_2 is activated by chemisorption on a nickel metal surface (Section 6.1); the hydrogens can attack appropriate ligands L intramolecularly (hydride transfer). Thus, if one of the ligands L is an olefin (alkene) coordinated to M as in Zeise's anion (Fig. 18.1), the hydrido ligands can attack the neighboring C=C function to form first an alkyl-metal complex and then a free alkane, with concomitant return of the two "borrowed" electrons back to M and resumption of four coordination in ML_4. The end result is therefore *hydrogenation* of the olefin by H_2 in homogeneous solution through the action of ML_4, which, as the definition of a catalyst requires, undergoes no net change.

This sequence of events may be illustrated by the homogeneous hydrogenation of ethylene in (say) benzene solution by *Wilkinson's catalyst*, $RhCl(PPh_3)_3$ (Ph = phenyl, C_6H_5; omitted for clarity in cycle 18.10). In that square-planar complex, the central rhodium atom is stabilized in the oxidation state I by acceptance of excess electron density into the 3*d* orbitals of the triphenylphosphane ligands but is readily oxidized to rhodium(III), which is preferentially six coordinate. Thus, we have a typical candidate for a catalytic cycle of oxidative addition and subsequent reductive elimination:

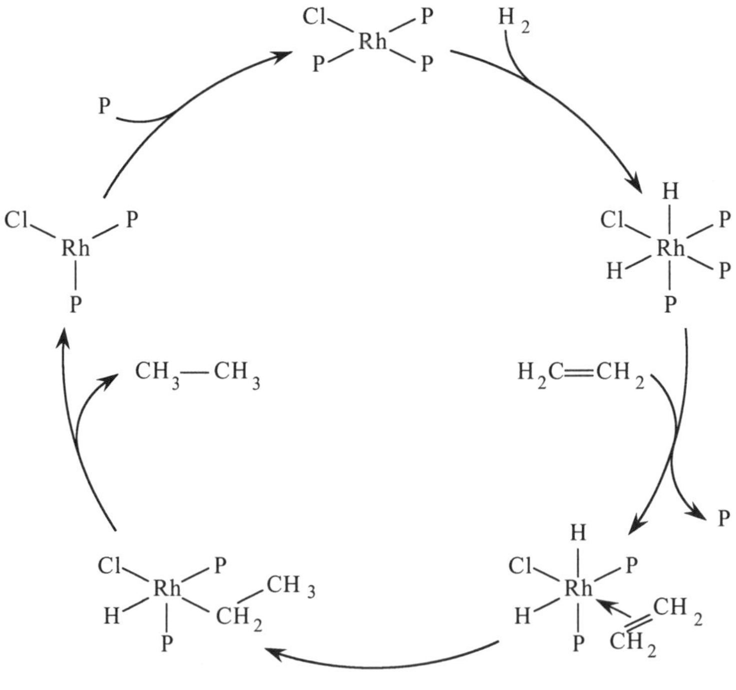

(18.10)

The mechanistic details of cycle 18.10 have been represented in a somewhat arbitrary fashion, but the essence of the mode of action of transition metal complexes (in particular, complexes of the Group 9 elements Co, Rh, and Ir in the I oxidation state) as homogeneous catalysts for hydrogenation reactions should be clear.

18.3.2 Hydroformylation

Carbon monoxide, like hydrogen, readily coordinates to transition metal centers. Consequently, cycles such as 18.10 can be used to catalyze the addition of CO (*carbonylation*) as well as hydrogen to a terminal $—CH{=}CH_2$ function, giving an aldehyde:

$$R—CH{=}CH_2 + CO + H_2 \xrightarrow{\text{catalyst}} R—CH_2—CH_2—CHO. \qquad (18.11)$$

This reaction is called *hydroformylation* and is typified by the OXO process (O. Roelen, 1938) in which the catalyst is $HCo(CO)_4$. In practice, almost any source of cobalt will serve, since it will be converted to $HCo(CO)_4$ in the presence of CO and H_2 under the operating conditions, which are rather severe: 110 to 180 °C and a total gas pressure of 20 to 35 MPa. Evidently, the alkene $R—CH{=}CH_2$ adds to the Co complex, as in Zeise's compounds, and is converted by coordinated H to an alkyl group as in cycle 18.10. This alkyl group then migrates to the C atom of one of the carbonyl ligands to form a $Co—CO—CH_2CH_2R$ grouping; this step is often referred to as "carbonyl insertion" into the metal alkyl bond, but in reality it is the alkyl group that migrates reversibly onto a bound CO ligand. The $Co—CO—CH_2CH_2R$ group in turn reacts with a further H ligand at the carbonyl C atom and is released as the aldehyde.

The use of the volatile $HCo(CO)_4$ is falling into disfavor because of catalyst losses, the need for high operating pressures and temperatures, and the loss of some alkene as alkane through a hydrogenation side reaction. The currently preferred Union Carbide hydroformylation process uses a rhodium triphenylphosphane complex, $HRh(CO)(PPh_3)_3$, that loses one PPh_3 under reaction conditions to form the 16-electron catalyst $HRh(CO)(PPh_3)_2$. This drives a reaction cycle analogous to the OXO hydroformylation reaction but at lower temperatures (~100 °C) and pressures (1–2 MPa), and with lower losses through hydrogenation.

Hydroformylation is clearly related to the *Fischer–Tropsch reactions*, in which CO/H_2 mixtures (in effect, water–gas; Section 9.3) react over *heterogeneous* catalysts to give organic compounds such as methanol.

18.3.3 The Wacker and Monsanto Processes

The previous examples involve *reduction* (hydrogenation) of organic molecules, but transition metal complexes can also catalyze *oxidation*. For example, the *Wacker process*, which has been widely used to convert ethylene to acetaldehyde, depends on catalysis by palladium(II) in the presence of copper(II) in aqueous HCl. The role of the copper chloride is to provide a means of using air to reoxidize the palladium to palladium(II). Once again, Zeise-type coordination of the ethylene to the metal center is believed to be involved:

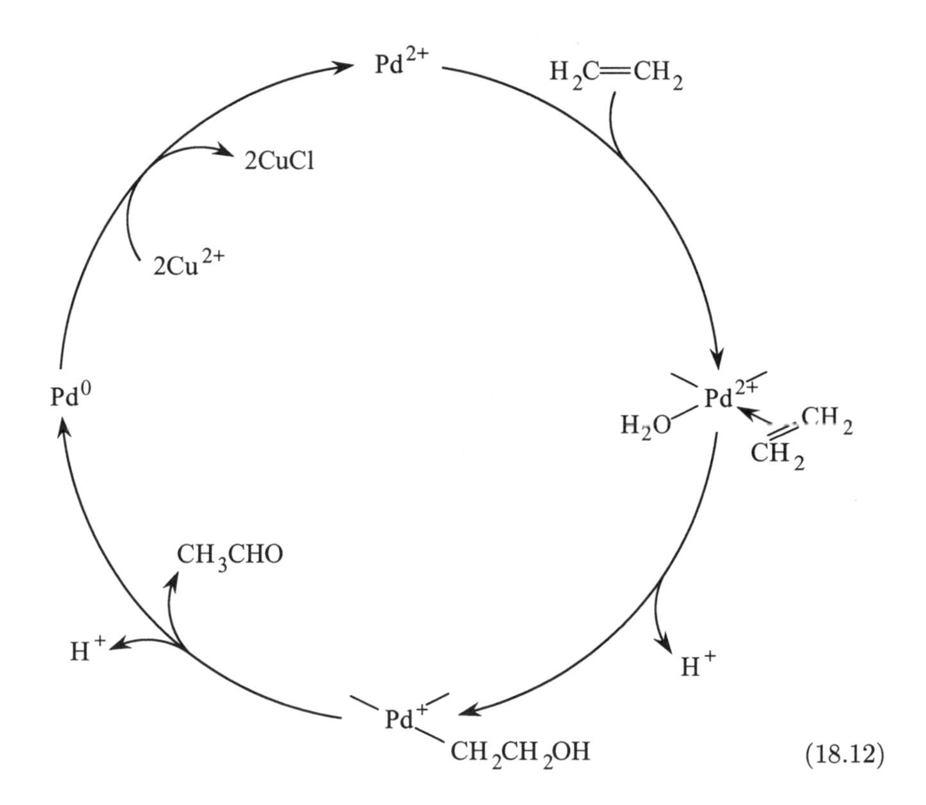

(18.12)

$$2CuCl + 2H^+ + \tfrac{1}{2}O_2 \rightarrow 2Cu^{2+} + 2Cl^- + H_2O. \qquad (18.13)$$

The Wacker process itself is becoming obsolete, largely because acetaldehyde is now more economically made by the reaction of methanol with CO via oxidative addition to the homogeneous catalyst $Rh^I(CO)_2I_2{}^-$ (the *Monsanto process*):

$$\mathrm{CH_3OH + HI \rightarrow CH_3I + H_2O} \tag{18.14}$$

$$\mathrm{Rh^{I}(CO)_2I_2^{-} + CH_3I \rightarrow CH_3Rh^{III}(CO)_2I_3^{-}} \tag{18.15}$$

$$\mathrm{CH_3Rh^{III}(CO)_2I_3^{-} + CO \rightarrow CH_3{-}CO{-}Rh^{III}(CO)_2I_3^{-}} \tag{18.16}$$

$$\mathrm{CH_3CORh^{III}(CO)_2I_3^{-} \rightarrow Rh^{I}(CO)_2I_2^{-} + CH_3COI} \tag{18.17}$$

$$\mathrm{CH_3COI + H_2O \rightarrow CH_3COOH + HI.} \tag{18.18}$$

Nevertheless, the Wacker process has been seminal in the development of palladium complexes as oxidation catalysts.

18.4 Olefin Polymerization Catalysts

18.4.1 Ziegler–Natta Catalysts

Polymerized olefins such as polyethylene were originally produced from the monomeric olefins only by processes involving elevated temperatures and very high pressures (typically 1000 bars), with the attendant risks and capital expenses. K. Ziegler found that triethylaluminum reacts with $TiCl_4$ in inert hydrocarbon solvents to give a brown suspension that causes ethylene to polymerize even at room temperature and pressure; furthermore, the product is of significantly higher density than polyethylene from the high pressure processes and is ideally suited to molding. Ziegler high density polyethylene has taken over a large share of the plastics market, although low density polyethylene from the high pressure process is still needed for making plastic film.

The exact nature of the Ziegler catalysts is somewhat obscure. The essential features seem to be that the $TiCl_4$ (a covalent liquid, soluble in hydrocarbon solvents) is alkylated by the trialkylaluminum and also at least partly reduced to titanium(III). Titanium(III) chloride is an ionic solid, insoluble in organic solvents, so it is not surprising that the material so formed is not readily soluble in the reaction medium. This material, which is apparently somewhat variable in structure and composition but which contains Al^{III} as well as Ti^{III}, can add an olefin molecule at a Ti center, presumably in much the same way that ethylene adds to platinum(II) chloride to form Zeise's salts (Section 18.2), and an alkyl group already present on the Ti can then migrate onto the olefin to form a new, longer alkyl group:

$$\mathrm{{>}Ti{<}^{R}_{Cl}} \xrightarrow{H_2C=CHR'} \overset{H_2C=CHR'\ \downarrow}{\mathrm{{>}Ti{<}^{R}_{Cl}}} \longrightarrow \mathrm{{>}Ti{<}^{CH_2-CHR'R}_{Cl}}$$

$$\xrightarrow{H_2C=CHR'} \overset{H_2C=CHR'\ \downarrow}{\mathrm{{>}Ti{<}^{CH_2-CHR'R}_{Cl}}} \qquad \text{etc.} \tag{18.19}$$

Thus, polymer chains $-(CH_2CHR')_n-$ are created. This is an important route to ***polypropylene*** (polypropene, in the newer nomenclature: $R' = CH_3$), which is difficult to make by other methods. Furthermore, it was shown by G. Natta that the heterogeneous (*Ziegler–Natta*) catalysts prepared using $TiCl_3$ in place of $TiCl_4$ produced polymers in which the side groups R' were arranged in a regular manner in relation to the central carbon chain (Fig. 18.4).

When the side chains are all on the same side of the central carbon backbone, the polymer is said to be *isotactic;* when they alternate regularly on either side, it is called *syndiotactic;* and when there is no regular placement of the side chains, the polymer is *atactic.* Isotactic polypropylene is a crystalline thermoplastic (mp ~165 °C) with excellent mechanical properties, and it is widely used in injection molding and fiber manufacture. In contrast, atactic polypropylene is an amorphous gum elastomer with low strength and little practical value. If, however, the polypropylene chain can be built of alternating stereoregular (isotactic or syndiotactic) and atactic segments, the mechanical properties of the resulting thermoplastic can be adjusted to strike a balance between toughness and stretchability; in Section 18.4.2, we see how this might be done.

The production of polyolefins in 1995 totaled 5.4×10^7 metric tons

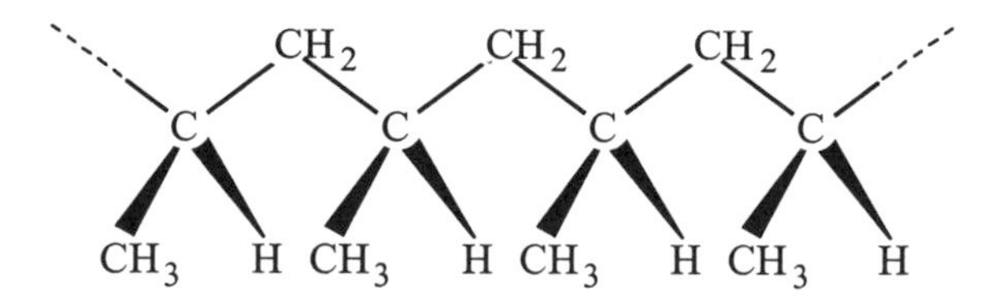

Figure 18.4 An isotactic polymer.

worldwide, mostly with use of Ziegler–Natta-type catalysts.[15] Thus, the work of Ziegler and Natta has assumed major commercial importance. In 1963, they were awarded the Nobel Prize for their contributions.

18.4.2 Metallocene Polymerization Catalysts

The classic Ziegler–Natta heterogeneous catalysts initiate polymerization at diverse, somewhat randomly scattered active sites on modified $TiCl_3$ crystals, with the result that polymer chain length is variable even though essentially complete isotacticity can be achieved. Furthermore, these catalysts offer little opportunity to manipulate the degree of stereoregularity and hence the mechanical properties of the product—the Ziegler–Natta polypropylenes are exclusively isotactic.

Largely out of the need to understand the mechanistic details of Ziegler–Natta polymerization in order to take rational steps to improve the performance of Ti-based heterogeneous catalysts, attention has turned to the properties of Group 4 (Ti, Zr, Hf) metallocenes as *homogeneous* polymerization catalysts.[15–17] As noted above, homogeneous catalysts offer the chemist precise knowledge of the nature of the catalytic site, and they also allow the properties and performance of the catalyst to be tailored to meet requirements.

Titanocenes of the type Cp_2TiX_2 consist of two η^5-cyclopentadienide units, bonded to the central Ti^{IV} so that the molecular planes are *not* parallel to each other, with two other *cis* ligands X in the opening of the $TiCp_2$ "jaws." For typical 2-electron donors X, these are 16-electron complexes and can be expected to show some catalytic activity. If the two X ligands can be replaced by an alkyl group and an alkene, bonded in the η^2 (Zeise) fashion, the possibility exists for migration of the alkyl group onto the alkene, forming an extended alkyl ligand, then coordination of another alkene followed by migration of the alkyl ligand onto it, and so on, to build up a long-chain alkane polymer. Thus, the titanocene (or analogous zirconocene) could serve as a model for an isolated Ziegler–Natta Ti site, and, better yet, the molecular geometry could be tinkered with to induce a desired stereoregularity (or lack thereof) in the polymer.

It was discovered in 1957 that Cp_2TiCl_2 can indeed be activated by diethylaluminum chloride $[(C_2H_5)_2AlCl]$ to catalyze polymerization of ethylene; presumably, 14-electron $[Cp_2Ti(C_2H_5)]^+$ or a 16-electron solvent complex such as $[Cp_2Ti(C_2H_5)(THF)]^+$ (THF = tetrahydrofuran) forms, and ethylene coordinates to Ti and then inserts into the Ti–C_2H_5 bond. For many years, however, no progress was made in the homogeneous catalysis of propylene polymerization. It had always been assumed, entirely logically, that water must be rigorously excluded in preparing Ziegler–Natta catalysts or organometallic models thereof, yet it was serendipitously found that

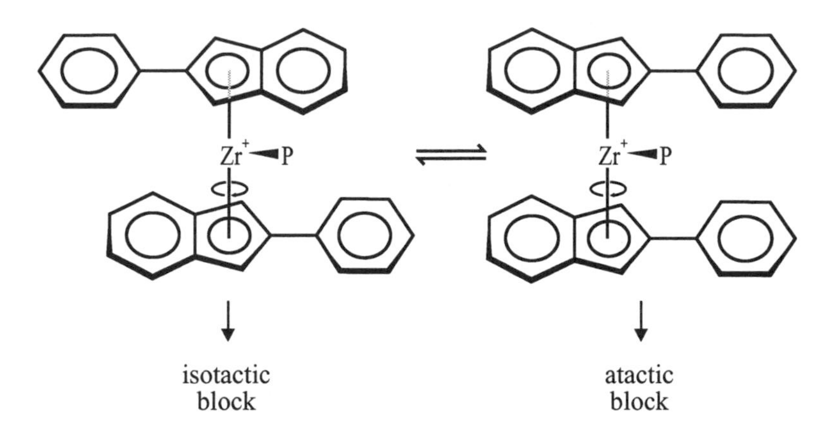

Figure 18.5 Formation of an isotactic–atactic polypropylene copolymer from a bis(phenylindenyl)zirconium complex.

dimethylaluminum chloride or trimethylaluminum do activate Cp_2ZrCl_2 to catalyze polymerization of propylene as well as ethylene *if a little water is present.** It seems that the methylaluminum is partially hydrolyzed to methylaluminoxane (MAO) oligomers:

$$(n + 2)Me_3Al + (n + 1)H_2O \rightarrow Me_2Al(OAlMe)_nOAlMe_2 + (2n + 2)CH_4 \tag{18.20}$$

where n is typically 5–20 and Me represents CH_3. The MAO, in excess over Cp_2ZrCl_2, rapidly converts this to Cp_2ZrMe_2 and then, in a crucial step, evidently abstracts Me^- from the dimethylzirconocene to form $[Cp_2ZrMe]^+[MAOMe]^-$. Now, in the presence of an alkene R—CH=CH_2, a zirconium–methyl–alkene complex forms, the methyl migrates to the alkene, another alkene is picked up, and the process repeats itself a large number of times, producing the polyalkene. An [Al]:[Zr] ratio exceeding 200 is needed to start the process.

A key property of the $[MAOMe]^-$ ion is an inability to bond to the coordinatively unsaturated, 14-electron $[Cp_2ZrMe]^+$ cation, so preserving it for binding the alkene. It appears that any very poorly coordinating anion would serve this purpose in principle, but in practice even anions like tetraphenylborate can be scavenged by the voracious cation. However, tetrakis(pentafluorophenyl)borate will keep the $[Cp_2ZrMe]^+$ ion intact, and the structure of the cation has been authenticated crystallographically in solid salts of that kind. With $[Cp_2ZrMe]^+$ catalysts, polymerization rates of 100–1000 kg polypropylene per hour per mole zirconocene can be achieved, and even faster rates are reached with ethylene. Unfortunately, although

*Zirconocenes are less prone to reduction than are titanocenes, and hence they are less likely to lose catalytic activity.

the molar masses of polyethylene made using MAO-activated zirconocene are on the order of 10^5 to 10^6, the molar masses of the polypropylenes do not exceed about 1000 (i.e., are much lower than with heterogeneous Ziegler–Natta catalysts. On the other hand, the *distribution* of molar masses is much narrower for the polymers produced by homogeneous catalysis.

The stereoregularity of polypropylene can be controlled in detail with purposefully designed homogeneous catalysts. For example, with appropriately engineered catalyst molecular symmetry, η^5-indenylzirconocenes can give isotactic polypropylene, η^5-fluorenylzirconocenes can give the syndiotactic form, and Cp_2Zr complexes in general give the atactic polymer. Coates and Waymouth[17] have show that a bis(phenylindenyl)zirconium complex, in which the two phenylindenyl ligands swivel between atactic and isotactic polypropylene-forming conformations with a frequency a little slower than chain growth, produces a highly stretchable polypropylene elastomer in which atactic and isotactic blocks alternate along the molecule (Fig. 18.5).

Exercises

18.1 (*a*) With the aid of Appendix C, calculate the heat of reaction of liquid diethylzinc ($\Delta H_f^\circ = +10.5$ kJ mol^{-1}) with water vapor to form solid zinc hydroxide ($\Delta H_f^\circ = -642.2$ kJ mol^{-1}).

(*b*) Calculate the heat of reaction of liquid diethylmercury ($\Delta H_f^\circ = +30.1$ kJ mol^{-1}) with water vapor, noting that divalent mercury forms a solid oxide (two forms) but no hydroxide.

(*c*) Comment on the implications of your results, assuming them to be qualitatively representative of the organometallic chemistries of Zn and Hg.
[*Answers:* (*a*) -338.4 kJ mol^{-1}; (*b*) $+36.6$ (yellow HgO) kJ mol^{-1}.]

18.2 Germanium chloride is a colorless fuming liquid (bp 84 °C) that can be prepared simply by distilling a solution of GeO_2 in concentrated aqueous HCl. Suggest a means of making tetraethylgermane (bp 161 °C) from germanium(IV) oxide.

18.3 (*a*) Show that $Fe(CO)_5$, $Ni(CO)_4$, di(η^6-benzene)chromium(0), $(OC)_5Mn{-}Mn(CO)_5$ (note the Mn—Mn single bond), $HCo(CO)_4$, and $(\eta^7\text{-}C_7H_7)Mo(CO)_3{}^+$ conform to the eighteen electron rule.

(*b*) Predict x in $Mn(CO)_xI$, $Co(CO)_xNO$, $(\eta^5\text{-}C_5H_5)Fe(\eta^6\text{-}C_6H_6)^{x+}$, and $Fe(CO)_4{}^{x-}$.

(*c*) Explain why (*i*) the ferricenium ion $Fe(\eta^5\text{-}C_5H_5)_2{}^+$ behaves as a strong oxidizing agent; (*ii*) cobaltocene $Co(\eta^5\text{-}C_5H_5)_2$ acts as a strong reducing agent, in suitable solvents; and (*iii*) nickelocene reacts readily with nitric oxide gas to form (η^5-C_5H_5)Ni(NO).

18.4 Sketch plausible catalytic cycles for (*a*) the OXO [i.e., $HCo(CO)_4$-catalyzed] and (*b*) the Union Carbide hydroformylation reactions.

References

1. C. Elschenbroich and A. Saltzer, "Organometallics: A Concise Introduction," 2nd Ed. VCH, New York, 1992.

2. I. Haiduc and J. J. Zuckerman, "Basic Organometallic Chemistry." de Gruyter, New York, 1985.

3. G. E. Coates, M. L. H. Green, P. Powell, and K. Wade, "Principles of Organometallic Chemistry." Methuen, London, 1968.

4. N. N. Greenwood and A. Earnshaw, "Chemistry of the Elements." Pergamon, Oxford, 1984.

5. F. A. Cotton and G. Wilkinson, "Advanced Inorganic Chemistry," 5th Ed. Wiley (Interscience), New York, 1988.

6. A. Yamamoto, "Organotransition Metal Chemistry." Wiley (Interscience), New York, 1986.

7. C. M. Lukehart, "Fundamental Transition Metal Organometallic Chemistry." Brooks/Cole Publishing, Belmont, California, 1985.

8. J. P. Collman and L. S. Hegedus, "Principles and Applications of Organotransition Metal Chemistry," 2nd. Ed. University Science Books, Mill Valley, California, 1987.

9. M. Bochmann, "Organometallics 1: Complexes with Transition Metal–Carbon σ-Bonds." Oxford Univ. Press, Oxford, 1994; M. Bochmann, "Organometallics 2: Complexes with Transition Metal–Carbon π-Bonds." Oxford Univ. Press, Oxford, 1994.

10. R. F. Heck, "Organotransition Metal Chemistry." Academic Press, New York, 1974.

11. R. H. Crabtree, "The Organometallic Chemistry of the Transition Metals." Wiley, New York, 1988.

12. G. W. Parshall, "Homogeneous Catalysis." Wiley, New York, 1980.

13. C. Masters, "Homogeneous Transition-metal Catalysis—A Gentle Art." Chapman & Hall, London, 1981.

14. A. Nakamura and M. Tsutsui, "Principles and Applications of Homogeneous Catalysis." Wiley (Interscience), New York, 1980.

15. M. Bochmann, Cationic Group 4 metallocene complexes and their role in polymerization catalysis: The chemistry of well defined Ziegler catalysts." *J. Chem. Soc. Dalton Trans.* 255–270 (1996).

16. H. H. Brintzinger, D. Fischer, R. Mülhaupt, B. Rieger and R. M. Waymouth, *Angew. Chem. Int. Ed. Engl.* **34**, 1143–1170 (1995).

17. G. W. Coates and R. M. Waymouth, Oscillating stereocontrol; a strategy for the synthesis of thermoplastic elastomeric polypropylene. *Science*, **267**, 217–219 (1995).

Chapter 19

Some Newer Solid-state Technologies

IN THIS final chapter, some topics introduced earlier are drawn together in the context of some aspects of current materials technology.

19.1 Sol–Gel Science

19.1.1 Gels from Hydrolysis of Metal Aqua Ions

We saw in Section 13.6 that hydrolysis and subsequent polymerization of aqueous metal cations can lead to the precipitation of *gels*. In the case of $Fe(H_2O)_6{}^{3+}$ in mildly acidic solutions, the polymerization sequence of Eqs. 13.25 and 13.26 and Fig. 13.6 first reversibly forms cationic colloidal spherules, 2–4 nm in diameter, with the structure of γ-FeO(OH) [double chains of $Fe(O,OH)_6$ octahedra] on a timescale of about 100 s. These lose H^+ and harden over several hours and then, over several days, form aged polymer rods, then rafts, and ultimately, after several months, needles of solid goethite [α–FeO(OH)].[1,2] Thus, *aging* is an important feature of hydrolytic polymerization.

At higher pH a gel is precipitated immediately, but the submicrometer-sized particles that make up the gel will similarly coarsen and cross-link on aging. Since very small particles have a higher fraction of surface atoms (unsatisfied valences) and hence higher free energies than larger particles, the latter grow at the expense of the former (*Ostwald ripening*), and "necks" between two particles in contact will tend to fill in. The newly formed gel, however, is a voluminous semisolid with a very high water content, and separation of the gel by filtration and subsequent dewatering are frustratingly slow processes. The distinction between a gel and a thick suspension is somewhat arbitrary; the gel state might be said to have been reached when the aging suspension can no longer be poured, or when its viscosity reaches some high value such as 1000 Pa s.

Yet, if the water content can be removed, a very finely particulate, roughly monodisperse* metal hydroxide remains. Furthermore, if hydroxides or basic carbonates of different metals are coprecipitated, the resulting fine, intimately mixed, monodisperse powders can serve as precursors for making mixed-oxide ceramics such as ferrites (MFe_2O_4, where M is a divalent metal), which have important magnetic properties, or the superconducting metal oxides discussed later such as $YBa_2Cu_3O_{7-x}$. Some gels are made as precursors for specialty glasses. A key point is that all this can be done under very mild conditions; it is possible, for example, to make monodisperse ferrites by decomposition of the relevant metal $EDTA^{4-}$ or NTA^{3-} complexes, but hydrothermal conditions are needed. Most intriguingly, with gels that have extensive cross-linking between particles (such as those made from hydrolysis of silicon alkoxides, discussed later), the possibility exists of making a very light, strong, porous solid or *aerogel*—*if* the solvent can be extracted from the gel without causing it to collapse or to crack like drying mud.[1,3–9] We see in Section 19.1.4 how this can be done.

The initial steps in forming gels from aqua ions of a metal M in aqueous solution are *olation* [formation of M—(OH)—M bridges, cf. Eq. 13.26], followed by *oxolation* [loss of H_2O from M—(OH)—M—OH to form an oxo bridge, M—O—M]. The second step is acid catalyzed, since protonation of a terminal —OH group facilitates its removal. The stability of the colloid depends on the mutual repulsions of the electrical double layers surrounding the charged polymer particles, as discussed in Section 14.2, and so the tendency to gel, as well as the nature of the product itself, can be strongly influenced by the nature and concentrations of the counterions and any other electrolytes that may be present. An important parameter is the *point of zero charge* or *pzc*, the pH at which competing protonation and deprotonation of the colloidal particles leave it with no net charge; thus, as a rising pH passes through the pzc, the charge on the colloidal particles reverses from positive to negative. For FeO(OH), the pzc is 6.7; for SiO_2, it is 2.5, and for MgO, 12.

19.1.2 Gels from Hydrolysis of Alkoxides

Most oxide/hydroxide gels of technological interest are made, not from aqua ions, but by hydrolysis of *alkoxides*, $M(OR)_n$ (R = an alkyl group). Furthermore, M is most frequently silicon,[10] rather than a metal, although titanium dioxide gels are made by hydrolysis of $Ti(OC_2H_5)_4$, and alumina gels from $Al(OR)_3$. The initial step in hydrolysis gives an M—OH group,

*The particles of a *monodisperse* solid are all of the same size.

and then, since the alcohol ROH is a better leaving group than H_2O, bridging by "alcoxolation" follows rapidly:

$$\begin{aligned}(RO)_{(n-1)}M{-}OH + RO{-}M(OR)_{(n-1)} &\rightarrow \\ (RO)_{(n-1)}M{-}O{-}M(OR)_{(n-1)} &+ ROH \end{aligned} \tag{19.1}$$

Tetraalkoxysilanes (usually tetramethoxysilane, TMOS, bp 121 °C, or tetraethoxysilane, TEOS, bp 169 °C) for hydrolytic gelling can be made by reaction of $SiCl_4$ (Section 17.8.2) with the relevant alcohol:

$$SiCl_4(l) + 4C_2H_5OH(l) \rightarrow Si(OC_2H_5)_4(l) + 4HCl(g) \tag{19.2}$$

Hydrolysis of tetraalkoxysilanes in pure water is usually incomplete, but it is more effectively carried out in an alcohol–water mixture and can be catalyzed by H^+ or *weak* bases such as ammonia. Polymerization occurs in the range pH 2–7; under neutral conditions, the rate of polymerization is limited by the slowness of hydrolysis, whereas in acidic media hydrolysis is complete before polymerization begins. The nanostructure of the gel (density of particles, fractal dimensions of particle clusters, and degree of cross-linking of particles to form a network) is controlled by these conditions.

19.1.3 Xerogels

Gels dried by conventional methods, known as *xerogels* (xero- = dry), generally have lost the very open structure of the wet gel. The reason for this is that the liquid phase, as it evaporates and shrinks, exerts powerful capillary forces on the pore walls through the surface tension at the meniscus, resulting in collapse or cracking of the large-scale structure of the gel.

As noted earlier, xerogels can form fine, free-flowing, monodisperse powders that are valuable precursors for making inorganic materials. It is also possible, however, to use sol–gel techniques to deposit *thin films* of inorganic solids on suitable substrates. Cracking of the gel on drying is much less prevalent in thin films. These films are of major importance in the fabrication of components for the electronics industry. For example, we noted in Section 4.6 that the ferroelectric properties of lead zirconate titanate (PZT) can be put to use *if* it can be deposited as a very thin film, and indeed suitable coatings can be applied to silicon substrates by sol–gel methods.[11] Thin films of ferromagnetic materials such as nickel ferrite or superconducting ceramics such as $YBa_2Cu_3O_{7-x}$ can similarly be made by sol–gel techniques, which include

(*a*) spraying the sol (or a succession of precursor solutions) onto the substrate;

(*b*) dipping the substrate into the sol and withdrawing it at an appropriate rate; and

(*c*) spinning the sol-wetted substrate (the thickness of the deposited gel layer can be governed by the spin rate).

The deposited gel is then dried in a suitable atmosphere and at an appropriate temperature. For example, a ferroelectric PZT film can be made by dipping a silica substrate into a hydrolyzing mixture of titanium tetraisopropoxide (0.48 mol), zirconium tetra-*n*-propoxide (0.52 mol), and lead(II) 2-ethylhexanoate (1.0 mol), drying the film at 100 °C, firing it at 400 °C, repeating the procedure 20 times, and then finally firing the sample at about 600 °C.

Thin (quarter-wave*) titania–silica antireflection coatings can be applied to optical glass using a bath of aqueous ethanol in which TEOS and titanium tetraethoxide have been allowed to hydrolyze for 1–2 days in the presence of a low concentration of nitric acid. The glass object is dipped into the bath and withdrawn at a rate chosen to give the desired film thickness. The object is then heated in air at a temperature high enough to destroy the residual organic compounds but not high enough to affect the optical quality of the glass (generally <500 °C). The coating is sometimes etched with dilute aqueous HF to produce a gradient in the density of pores in the TiO_2/SiO_2 film, since pore density also affects the refractive index of the film. Conversely, somewhat thicker binary TiO_2/SiO_2 films with a Pd content (Schott Glaswerke's IROX coatings) on architectural glass are used in windows that screen buildings from external radiant heat: the TiO_2 layer improves the reflectivity of the glass, and the Pd absorbs excess sunlight.[3]

Indeed, special purpose lenses themselves can be made by sol–gel methods. By partial leaching of Pb^{2+} over several days from a gel made by hydrolyzing a TMOS/TEOS/lead(II) acetate mixture, it is possible, on drying and sintering the gel, to produce a lens with graded refractive index (a GRIN lens). Such lenses with a radial gradient in refractive index can focus light even though the optical surfaces are flat; in other cases, it is possible to use the GRIN principle to correct for spherical aberration. Sol–gel methods can also be used to make entire components with conventional optical properties but special shapes.

Sol–gel-derived films can also serve to provide protective coverings, flat outer layers for microrough surfaces, or thin dielectric layers. "Spin-on glasses" have been developed for all these applications. Some films can change color under an applied EMF (e.g., TiO_2 turns blue at ~2 V, while

*If the thickness of a transparent film is one quarter of the wavelength of the incident light, a wave of light reflected from the back of the film will be 180° out of phase with the next incoming light wave when they meet at the front surface, and the reflected wave will therefore be cancelled by destructive interference.

brown $\{V_2O_5 + xM^+ + xe^-\}$, where M is an alkali metal, goes to blue $M_xV_2O_5$), and this effect can be utilized in making electrochromic displays.

It is also possible to make ceramic fibers by drawing out viscous sols while polymerization is occurring. For example, although zirconium dioxide fibers are almost impossible to prepare from the melt (mp ~2700 °C), they can be drawn from viscous sols and have been used as alkali-resistant reinforcement fibers for concrete.[1] Silica optical fibers (waveguides) can be made more easily and more economically by sol–gel methods than by chemical vapor deposition, and it is easier to "dope" the fibers with TiO_2, GeO_2, etc., to create an increasing gradient in refractive index toward the fiber surface so as to maximize total internal reflection of a light signal within the fiber.

19.1.4 Aerogels

As noted earlier, conventional drying of gels can lead to gel collapse or the formation of cracks. This can be avoided by dissolving the liquid with a *supercritical fluid* (SCF, Sections 2.4.2 and 8.1.3); then, since there is no gas–liquid distinction (no meniscus) and only minimal solid–fluid surface tension in a SCF, disruptive capillary forces during evaporation are avoided, and a highly porous, extremely light solid *aerogel* is formed. The SCF usually chosen is CO_2, as it is inexpensive, nontoxic, and nonflammable, and has a conveniently low critical point of 31.0 °C and 7.39 MPa. Methanol is sometimes preferred, despite its flammability and less convenient critical parameters (239.4 °C and 8.09 MPa); the surfaces of supercritical methanol-dried silica gels tend to reesterify at the high drying temperature, and so become hydrophobic, whereas the supercritical CO_2-dried aerogels retain surface hydroxyl groups which tend to attract moisture, with the result that the otherwise transparent gel may turn milky or crack over time.

An alternative to SCF drying of silica gels that produces aerogels at ambient pressure has been described.[12] This involves derivatizing the silica surface with trimethylchlorosilane, that is, replacing the —OH groups with —$Si(CH_3)_3$, so that the hydrogen bonding which causes strong capillary action between liquids such as water and alcohols and the silica surface is eliminated. For the present, however, SCF desiccation offers a reliable route to aerogels.

Most aerogels that have been developed are silica based, and these have several special properties[7–9]:

(*a*) Aerogels are the lightest solids known. Densities of aerogels made by acidic hydrolysis of TMOS can range from 800 to only 3 kg m^{-3} (cf. 2635–2660 kg m^{-3} for SiO_2 in the form of natural quartz).[9]

(*b*) Despite the extreme lightness, aerogels have considerable load-bearing capacity and rigidity.

(*c*) Aerogels are excellent thermal insulators, particularly if evacuated.

(*d*) Aerogels are typically transparent, except for a slight blue haze due to scattering of light, and can be used as thermal insulation in windows and solar energy collectors, as well as in lightweight optical devices such as Čerenkov radiation detectors.

(*e*) The velocity of sound in aerogels is very low by the standards of solids ($120\ \mathrm{m\ s^{-1}}$ at $70\ \mathrm{kg\ m^{-3}}$ density, even lower than that of air), and aerogels can be used as acoustic quarter-wave antireflection layers.

These porous solids obviously have very high surface areas, and they lend themselves naturally to service in catalysis. Palladium catalysts supported on alumina aerogels have been used successfully to remove CO and NO from automobile exhausts,[13] and a V_2O_5/TiO_2 aerogel is itself a catalyst for the selective reduction of NO to N_2 and water by gaseous ammonia.[14]

19.2 Materials for Electronics

We introduced the basic principles of semiconduction in Section 5.3. Many semiconductors have the cubic sphalerite (zinc blende) structure (Fig. 4.12), which is the same as the diamond lattice (Fig. 3.1) if all the atoms are the same, as in Si or Ge. Inorganic semiconductors generally have four valence electrons per atom, as is the case for Si and Ge. They can be classified according to the number of valence electrons of the participating atoms[15]:

(*a*) III–V, such as gallium arsenide (GaAs)

(*b*) II–VI, such as cadmium selenide (CdSe) and other combinations of Group 12 and Group 16 ions

(*c*) I–III–VI_2 (e.g., $CuGaSe_2$) and II–IV–V_2 (e.g., $ZnSiP_2$), generally having the *chalcopyrite* ($CuFeS_2$) structure, which is the sphalerite structure with ordered placement of the two kinds of cations.

(*d*) IV–VI, such as PbS, which have more than four valence electrons per atom and generally have the NaCl structure.

As noted in Sections 5.3–5.5, vacancies in the anion or cation array can exist in equilibrium with the vapor of the depleted element. If the enthalpy of formation of a vacancy in the anion array is markedly greater than that of one in the cation sublattice—for example, in ZnTe, where it is about 1 electron volt ($1\ \mathrm{eV} = 96.5\ \mathrm{kJ\ mol^{-1}}$) higher—the heated solid will tend to develop an excess of anions and so will become a p-type semiconductor. The enthalpies of vacancy formation correlate with the anion–cation radius ratios; thus, very large anions such as Te^{2-} matched with relatively small cations such as Zn^{2+} favor "doping" with vapor of the anionic element for p-type behavior. In the case of GaAs, the vacancy formation enthalpies are closely similar for the two sublattices, so doping with either Ga or As from the respective vapors to secure n or p characteristics is equally feasible, whereas n-type behavior can be readily induced in cadmium sulfide.

The dopant atoms, which need not be the same as those of the existing sublattice, create donor or acceptor lattice sites lying close in energy to the inner limits of the conduction or valence bands, respectively (Fig. 5.5). Clearly, impurities are likely to degrade the properties of a semiconductor, so materials for electronics must be of very high purity. Furthermore, care must be taken to minimize introduction of impurities during the construction of an electronic device, for example, when sawing a silicon wafer from a "boule" of purified silicon. Dislocations in the crystal structure (Section 5.2) are useful in that they serve as traps for diffusing impurity atoms, away from the active regions of the semiconductor device; this process is known as "gettering." For example, the back surface of a silicon wafer can be exposed to a laser beam to induce local lattice defects, and then, on annealing the wafer at about 1000 °C, any rapidly diffusing impurities such as transition or alkali metal atoms will be caught on this surface within minutes.

Silicon of electronics-grade purity can be made by Mg reduction of redistilled $SiCl_4$ or, most commonly nowadays, by H_2 reduction of $SiHCl_3$, as described in Section 17.8.2. In either case, the elemental Si is further purified by drawing a spinning boule of solidified Si slowly out of a Si melt in a pure silica crucible, which may be counterrotated to maintain adequate stirring (Czochralski crystal growth).[16] Impurities remain in the melt. Gallium arsenide and indium phosphide (InP) may be similarly purified. Contact with the crucible is the chief source of impurities (notably oxygen) in Czochralski silicon, but contamination can be avoided by the more expensive *float zone growth* technique (zone refining), in which a small, moving zone of a Si rod is melted by passage of a small heater along it: the surface tension of the narrow melt band keeps it in place, and the impurities accumulate in it as it progresses along the rod.

19.2.1 Deposition of Thin Layers

Construction of electronic devices usually involves deposition of thin layers of semiconducting, metallic, and insulating materials onto a suitable substrate (which might be a wafer cut from a Czochralski silicon boule with a diamond saw).[15–24] In some cases, it is possible to grow crystalline layers onto a substrate such that the crystallographic order of the atoms in the film is related to that of the surface of the substrate; this is known as *epitaxial growth*.

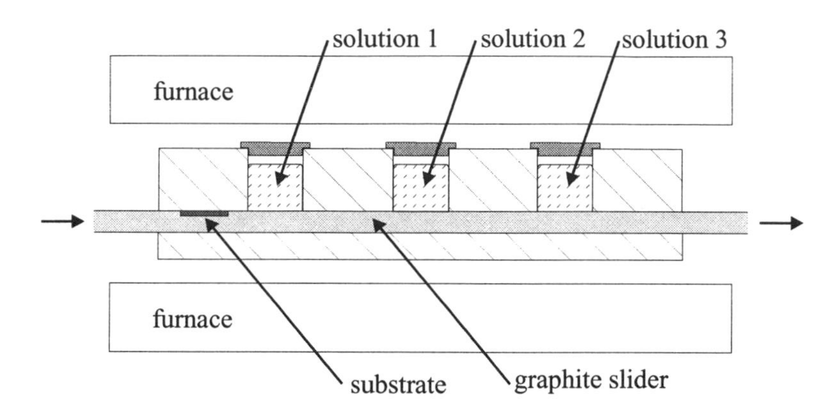

Figure 19.1 Device for liquid phase epitaxial growth of several layers.

Liquid phase epitaxy involves exposure of the solid substrate wafer to an appropriate melt, or to a succession of different melts (or solutions in melts) if it is desired to create several consecutive layers. A simple sliding base device for applying several successive coatings is sketched in Fig. 19.1. This approach is used mainly for III–V semiconductors such as GaAs, but there are sometimes problems with undercooling of the melt (i.e., it may remain metastably liquid well below its normal freezing point). Gallium itself, which is the usual solvent for depositing Al-doped GaAs, refreezes only very slowly even when at 10 °C or more below the remarkably low normal melting point (31 °C).

Vapor-phase deposition of the sputtered or evaporated layer-forming material avoids the undercooling problems associated with liquid phase epitaxy, but it coats everything in the vaporization chamber unselectively. Sputtering is usually done by forming a plasma (ionized gas) in an electrical discharge in the vapor at low pressure.

Chemical vapor deposition (CVD) utilizes thermal decomposition of a gas such as SiH_4, or chemical reaction of a mixture of gases such as $SiCl_4$ with H_2, to deposit a layer of the desired material (in these examples, Si) on a hot ($\leq$1000 °C) substrate surface. This technique deposits layers exclusively on the heated substrate, and in many cases layer growth is epitaxial. Dopants (e.g., phosphorus) can be introduced by admixture of small amounts of a suitable gas (in this example, phosphane, PH_3). Insulating or dielectric layers of SiO_2 can be made by oxidizing deposited Si, but this results in a large volume change in the layer* with associated stresses (cf. the Pilling–Bedworth principle, Section 5.6). It is therefore better to deposit

*The molar volumes of Si and quartz, the densest form of SiO_2, are 12.1 and ~22.8 $cm^3\ mol^{-1}$, respectively; deposited amorphous silica will have an even larger molar volume.

SiO_2 by CVD of a silane–oxygen or, better, SiH_2Cl_2–N_2O mixture. Layers of silicon nitride, Si_3N_4, may be similarly made by CVD of silane–ammonia mixtures at 700 °C:

$$3SiH_4(g) + 4NH_3(g) \rightarrow Si_3N_4(s) + 12H_2 \qquad (19.3)$$

Metal–organic chemical vapor deposition (MOCVD)[20, 21] refers to the use for CVD reagents of compounds selected from the huge range of known organometallics (metal alkyls, metal arene complexes, metal carbonyls, etc.; see Chapter 18), as well as other volatile metal complexes with organic ligands such as chargeless β-diketonates and alkoxides. Mond pioneered this approach at the beginning of the twentieth century, using the pyrolysis (i.e., high temperature thermal decomposition) of $Ni(CO)_4$ (Section 17.4) and related compounds to produce films of nickel and other metals. Pyrolysis is still the easiest MOCVD technique to apply; gallium arsenide, for example, can be deposited on a substrate from a mixture of $Ga(CH_3)_3$ and arsane (formerly called arsine, AsH_3) at 600–700 °C. MOCVD can also be induced photolytically; for example, photochemical decomposition of $(\eta^5\text{-}C_5H_5)Pt(\eta^3\text{-}C_3H_5)$ with 308 nm laser light can be used to deposit a transparent film of platinum on a quartz substrate. Other means of decomposing the metal–organic vapor include plasma-assisted and electron beam MOCVD. The challenge to chemists is production of suitable volatile metal–organic reactants in sufficiently high purities. Spencer[20] presents an extensive tabulation of MOCVD reagents, together with conditions for depositing films on specific substrates.

Sol–gel techniques can be used to produce thin layers, as described in Section 19.1. For example, cadmium sulfide layers for photocells are easily made by spraying an ammoniacal solution of cadmium chloride and thiourea (which hydrolyzes to give sulfide ion) onto a substrate surface and baking the resulting CdS film at up to 500 °C:

$$CdCl_2(aq) + (NH_2)_2CS(aq) + 2H_2O(l) \rightarrow CdS(s) + CO_2(g) + 2NH_4Cl(aq). \qquad (19.4)$$

The solid residue of ammonium chloride left after evaporation of the solvent water decomposes to gaseous HCl and NH_3 during the baking. Antireflective coatings (e.g., of alumina or titania) may also be applied by sol–gel methodology.

19.2.2 Some Simple Electronic Devices

First, let us consider the problem of getting electrical signals into and out of a semiconductor device through metallic contacts. If the contact is with an n-type semiconductor, and if the work function of the metal is smaller than the electron affinity of the semiconductor,* electrons can flow freely in either direction across the contact, which is then referred to as an *ohmic contact.* If, however, the work function of the metal is *larger* than the electron affinity of the conductor, electrons will spontaneously flow out of the semiconductor into the metal, leaving a *depletion zone* in the semiconductor near the contact; this is called a *Schottky junction.*

The simplest electronic devices, such as diodes, light-emitting diodes, lasers, and photocells, have a single p–n junction. If we place, say, a p-type doped Si block in contact with n-typed doped Si block, electrons will normally flow from the n to the p regions but not vice versa. Thus, the p–n diode so created can be fitted with ohmic contacts to function as a rectifier of alternating current. Schottky junctions can act in this way to some degree even without deliberate creation of a p–n junction.

If electron–hole pairs are created in a semiconductor diode by applying a voltage in excess of the band gap and then recombine, light of a frequency corresponding to the band gap may be emitted if alternative deexcitation mechanisms are slower than the process producing luminescence. This qualification rules out Si and Ge as materials for such a *light-emitting diode* (LED); however, GaAs, with a band gap of 1.42 eV, emits near-infrared light at 900 nm, and GaP (band gap 2.24 eV) emits green light at 570 nm. The band gap energy does not necessarily determine the wavelength of the emitted light directly; for example, ZnS has a band gap of 3.7 eV but emits green light because of the inevitable presence of copper ions, and indeed ZnS phosphors are usually doped with Cu ions to take advantage of this effect. Near-infrared emitters such as GaInAsP are of great value in generating optical signals for transmission through silica optical fibers, which are transparent in the region 900–1600 nm.

Conversely, light impinging on a semiconductor may, if the energy of the photons of the light exceeds the band gap, create electron–hole pairs that may become separated across a p–n junction and so generate a voltage (a *photoelectric cell*). In this case, Si serves well in photoelectric cells that collect solar energy, and so do GaAs, Cu_2S, InP, and CdTe. A material with a band gap of about 1.5 eV gives greatest efficiency with sunlight conversion. The photoelectrically active semiconductor is usually covered with a quarter-wave antireflection layer of SiO_2 or TiO_2 to maximize absorption of the light. Such photovoltaic cells now provide reliable electricity supplies from sunlight for unattended telephone repeater units at remote sites, for example, but they are unlikely to replace conventional electrical generating stations on a large scale because of the enormous sunlight collector surfaces required (particularly in high latitudes, where demand is highest when least

*The work function is the energy required to remove an electron from the metal. In this context, the work function and the electron affinity refer to the bulk solids, not to gaseous atoms as in Chapter 2.

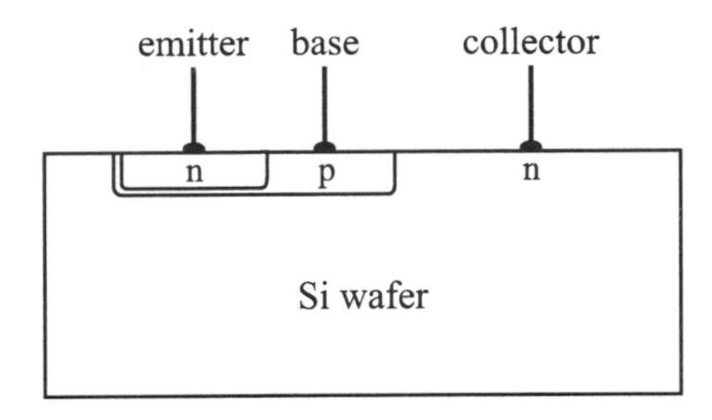

Figure 19.2 Basic structure of an n–p–n bipolar transistor.

sunlight is available).

Transistors are semiconductor devices with at least two p–n junctions. Charge transport within the transistor may be due to movement of *either* electrons *or* holes [a *unipolar* device, such as field-effect transistors (FETs) and charge-coupled devices (CCDs)] or to movement of *both* (a bipolar device). Figure 19.2 shows schematically how a basic n–p–n bipolar transistor is constructed on an n-type Si wafer. The regions under the three contacts are known as the *emitter*, the *base*, and the *collector*, whether the transistor is of the n–p–n type, as shown, or of the alternative p–n–p class. In operation, electrons from the emitter of an n–p–n transistor are injected into the very narrow base region, where they are the minority carriers, and diffuse to the collector, where they are picked up. Their passage through the thin p-type base layer is influenced by signals applied to the base, so that the device can operate as a voltage amplifier or switch.

19.2.3 Construction of Microelectronic Devices

Details of building specific types of microelectronic devices are well described by Grovenor[15]; as an illustration, we consider only the chemical techniques involved in making a metallic contact to a silicon wafer surface. Preparation of a wafer of Si or other material and application of epitaxial layers of semiconducting or insulating materials, if required, were outlined in Section 19.2.1. The construction of shaped features on the wafer is usually done by *photolithography*, or ion- or electron-beam variants thereof.

Suppose that the surface of a Si wafer is covered with a thin film of SiO_2. The oxide surface can be covered with a film of a *photoresist*, meaning a photosensitive material (usually a polymer) that can resist subsequent etching steps. The photoresist is exposed to light (usually ultraviolet) through a mask, or as a projected image, that traces out the shape of the component to be formed on the surface of the device. After exposure, the latent image in the photoresist is developed by subjecting it to an appropriate solvent.

A *positive* photoresist is one that becomes more soluble in the developing fluid as a result of exposure to light, and so dissolves away leaving the desired pattern in the form of undissolved resist; a *negative* resist becomes less soluble on irradiation, so that it is then the illuminated areas that remain.

Organosilanes are particularly effective as positive ultraviolet photoresists (Section 3.5.1). The oxide layer exposed where the resist has been removed may then be etched with an appropriate solution (buffered aqueous HF in the case of a silica layer) to expose the silicon underneath. Application of the desired metal over the whole surface is then followed by removal of the remaining photoresist, taking the unwanted parts of the metal layer with it and leaving the metal contact on the Si that had been exposed by etching. Variations on this approach can be used to attach shaped deposits of other electronically active materials.

19.3 Magnetic Materials and Superconductors

We end this book with an overview of a topic that has immense promise but is still in the laboratory stage of development, namely, *high temperature superconductivity.*[22–24] Since superconductivity is intimately related to magnetic phenomena in solids, a brief consideration of magnetochemistry[25] is prerequisite.

19.3.1 Magnetic Properties

If all of the electrons in a material are paired, the strength of an externally imposed magnetic field will be lower inside a sample of the material than outside it. This is because the magnetic field causes displacement of the moving electrons, which in turn generates an opposing magnetic field (cf. Lenz's law). A freely suspended sample will therefore experience a small force tending to push it out of the applied field (i.e., down the magnetic field gradient). This phenomenon, known as *diamagnetism*, is a property of all materials.

If, on the other hand, there are isolated atomic centers with unpaired electrons in the material such that the spins of the electrons on the separate centers do not interact (e.g., in hydrated $MnSO_4$, which has five unpaired electrons on each Mn^{2+} ion), the magnetic moments due to the electron spins will tend to align independently with an applied field in competition with thermal motion that tends to disrupt the alignment. This phenomenon, called *paramagnetism*, is a stronger interaction than diamagnetism and normally overwhelms the diamagnetism due to the paired electrons in the sample. A paramagnetic sample will therefore tend to be drawn *into* the applied magnetic field (i.e., move *up* the magnetic gradient). Paramagnetism is a characteristic property of many compounds of the transition metals, lanthanides, and actinides, since there are often unpaired electrons in the partly filled *d* or *f* shells in ions of these elements. The magnetism of complexes of these metal ions is a favorite topic of most textbooks of inorganic chemistry,[26] but adequate treatment requires a quantum mechanical development disproportionate to the scope and intent of this book.

Sometimes, potentially paramagnetic materials are not "magnetically dilute," that is, the unpaired electrons on different centers interact with one another, either to reinforce the magnetic moments collectively to produce a very high degree of magnetization (*ferromagnetism*) or to cancel the moments partially (*antiferromagnetism*). Ferromagnetism, as the name suggests, is the intense magnetism commonly associated with metallic iron, and it is technologically by far the most important type of magnetization of materials. Ferromagnetism, being a cooperative phenomenon, remains after the magnetizing field has been removed—in contrast to paramagnetism, which exists within the sample only as long as the magnetizing field is present—but can be canceled by applying an alternating current electromagnetic field that decreases slowly to zero or by heating the sample beyond the *Curie temperature* (766 °C for iron, Section 5.7.1), when thermal agitation overcomes the cooperative forces and normal paramagnetism takes over. In a similar way, antiferromagnetism gives way to normal paramagnetism above the *Néel temperature.* Room temperature antiferromagnetism is displayed by several compounds of transition metals, including NiO, α-FeO(OH), and copper(II) acetate monohydrate (which is actually a dimer with the two Cu^{2+} ions held in close proximity by four bridging acetate ions), but it is of little technological interest at this time.

Most permanent ferromagnets are metallic. Examples include soft iron, cobalt, nickel, alloys of the Alnico genre (such as Alnico I: 12% Al, 20–22% Ni, 5% Co, balance Fe), and also more exotic alloys that give very high intensities of magnetization such as $SmCo_5$ or $Nd_2Fe_{14}B$. A few metal oxides or fluorides are ferromagnetic, but these are relatively rare because the metallic centers must not be too far apart if there is to be cooperative interaction of the unpaired electrons. Nevertheless, ferromagnetic oxides are of major importance in fabricating magnetic disks or tapes. Type I magnetic recording tape is made with a polyester film incorporating fine, needle-shaped crystals of "γ-Fe_2O_3" (Section 5.6), which may be impregnated by sol–gel-type techniques with other magnetic materials such as cobalt ferrite ($CoFe_2O_4$) on the surfaces of the individual iron oxide grains to improve the audio frequency response of the tape. Superior audio performance is obtainable with tapes made with CrO_2, which has the rutile structure (Fig. 4.15; the green normal oxide of chromium, Cr_2O_3, is not ferromagnetic). The different magnetization characteristics of these oxides require different equalization and bias settings on the recorder for the two types of tapes.

19.3.2 Superconductivity

Electrical resistivity of metals generally falls with decreasing temperature. This is because the flowing electrons, as they pass through the lattice of metal cations, collide with one another and with the lattice atoms, and these scattering effects that create the electrical resistance diminish as thermal agitation decreases. For several materials, however, including certain ceramics (metal oxides), alkali metal fullerides (M_3C_{60}, M = K, Rb, or mixed Na, K, Rb, Cs; Section 3.2.2), and other compounds as well as some metals and alloys, the electrical resistivity suddenly *disappears entirely* as the temperature falls below a critical value T_c. This phenomenon, known as *superconductivity*,[22-24, 27-34] is evidently due to pairing up of electrons to form *Cooper pairs* that can pass freely though the metal lattice in an orderly fashion without the scattering interactions. The theory of superconductivity is currently under development, although the Bardeen–Cooper–Schrieffer (BCS) theory of 1957 is still generally considered satisfactory in explaining the phenomenon below about 40 K.

In addition to the zero resistivity, superconducting materials are *perfectly diamagnetic*; in other words, magnetic fields (up to a limiting strength that decreases as the temperature rises toward T_c) cannot penetrate them (the *Meissner effect*). This is a consequence of the mobile, paired state of the electrons. Indeed, it is the demonstration of the Meissner effect, rather than lack of electrical resistivity, that is usually demanded as evidence of superconductive behavior. One entertaining consequence of the Meissner effect is that small but powerful magnets will float (levitate) above the surface of a flat, level superconductor.[30]

Superconductivity was discovered in solid mercury at liquid helium temperature (4.2 K) by H. Kamerlingh Onnes as long ago as 1911; over the next 75 years some other metals, alloys, and ceramics (e.g., Pb, Nb, Nb_3Sn, and $LiTi_2O_4$) were also found to be superconducting at low temperatures, but in no case did T_c exceed 23 K. Niobium–tin and niobium–titanium alloys were developed to wind superconducting electromagnets to generate high fields (10 T or more) for special purposes such as nuclear magnetic resonance spectrometers; once a current is induced in the superconducting coil, it flows indefinitely in the absence of resistance, and very high currents are possible. Unfortunately, these devices require constant refrigeration with liquid helium, so this is expensive and inconvenient technology. If superconductors can be made with T_c above the boiling point (77 K) of liquid nitrogen, which is a relatively inexpensive refrigerant, important new technological opportunities are opened up. For example, transmission of electrical energy over superconducting power lines would generate impressive savings over present methods, which have high resistive losses. Further, Meissner levitation of magnetic passenger trains, for example, would reduce running friction enormously. Better still, of course, would be materials with

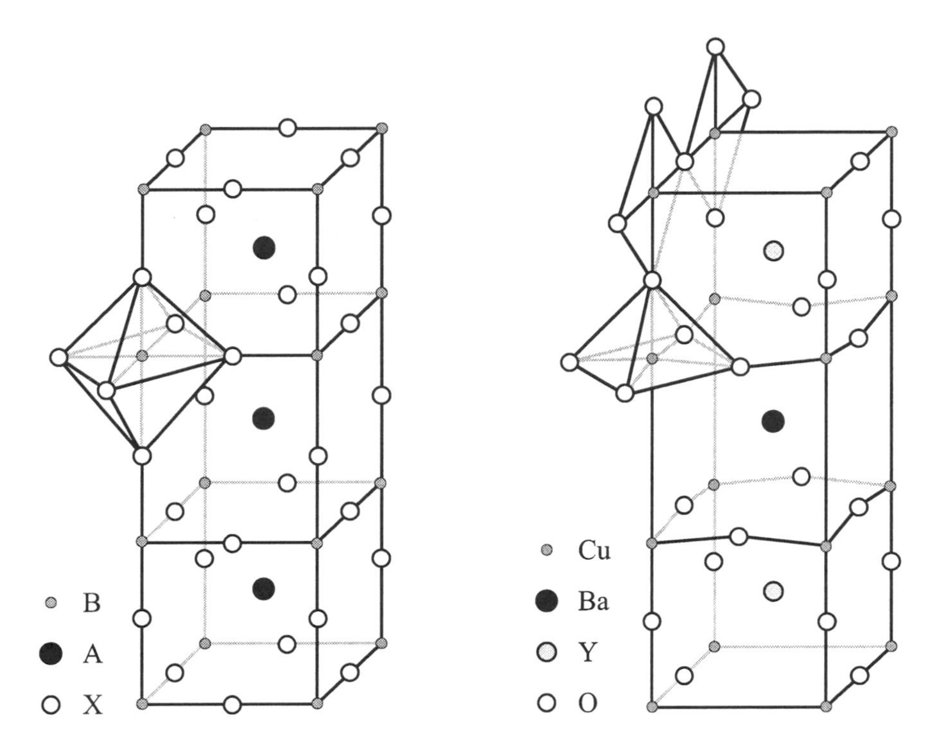

Figure 19.3 Relationship between a triple perovskite ABX_3 unit cell (left) and the $YBa_2Cu_3O_{7-x}$ unit cell (right). Adapted from J. T. S. Irvine, in Ref. 22, p. 287, by permission.

T_c substantially above ambient temperatures.

There was therefore great excitement in 1986 when Bednorz and Müller announced a surprisingly high T_c of 30 K for $La_{2-x}Ba_xCuO_4$, which was improved on in the succeeding months, and then, in the following year, 93 K for $YBa_2Cu_3O_{7-x}$ ("YBCO")—well above the liquid N_2 temperature.[32] A welter of discoveries of high-T_c ceramics followed; by 1993, a T_c of 135 K had been found for $HgBa_2Ca_2Cu_3O_{8+x}$ at atmospheric pressure (rising to 153 K at a pressure of 15 GPa), and subsequently a T_c as high as 250 K has been claimed for a member of the Bi–Sr–Ca–Cu–O ("BSCCO") group of superconductors.[33]

A common feature of all the new ceramic superconductors is that they are *cuprates*, that is, they are complex copper oxides. The structure of YBCO is given in Fig. 19.3, which also shows that it is related to the perovskite structure (Fig. 4.17). Synthesis of YBCO is remarkably easy; appropriate amounts of dry yttrium oxide (Y_2O_3), copper oxide (CuO), and barium carbonate ($BaCO_3$) are ground together into a fine, well-mixed powder and heated at 950 °C to form a black solid that is then cooled, pressed into a disk, sintered at 950 °C, and finally treated with pure oxygen at 500–600 °C in a "sensitizing" step that produces the nonstoichiometric YBCO.[30]

In searching for new superconducting ceramics, there are guidelines that may be followed,[33] some of which are as follows:

(*a*) the structure should include two-dimensional sheets or planes of linked clusters (to date, these have all been composed of Cu–O units);

(*b*) the material should be on the verge of a metal–insulator (Mott) transition, that is, it should be just barely capable of metallic conduction (when not superconducting);

(*c*) on the insulator side of the Mott transition, it should be antiferromagnetic;

(*d*) there should be no extended metal–metal bonds.

The thrust of much current research, however, is not so much in the direction of finding new superconducting compounds; rather, it is aimed at fabrication of known superconducting ceramics, which typically are very brittle, into flexible wires and tapes or other components of practical electrical devices. Furthermore, to be useful, such components must be able to carry very high currents. As noted earlier, high magnetic fields reduce the maximum temperature at which a material can be superconducting, and of course electrical currents themselves generate magnetic fields. In the early years of high-T_c superconductor development, ceramics of the BSCCO family, which have flatter and more regularly shaped grains than YBCO, seemed to be more easily adapted to forming wires and tapes, but YBCO subsequently proved to be workable if an appropriate approach is taken. Thus, in 1995, American and Japanese collaborators succeeded in making a serviceable superconducting tape by depositing a film of yttria-stabilized zirconia (Y_2O_3/ZrO_2, "YSZ") on a flexible nickel alloy strip and growing an epitaxial layer of YBCO on the YSZ layer; a current density of 1,000,000 A cm^{-2} has been sustained at 77 K in this tape in the absence of an externally applied magnetic field, and 500,000 A cm^{-2} in a field of 8 T.[31]

In 1996, a 50 m underground superconducting cable (wound from 6 km of a BSCCO ribbon) and a 150 kW superconducting electric motor were successfully demonstrated. Further development of practical superconductor devices might well depend more on the development of suitable refrigeration technology than on the preparation of new superconducting materials.

Exercises

19.1 Why are strong bases such as NaOH *not* used to catalyze the hydrolysis of alkoxysilanes? Why does polymerization occur in the pH range 2–7?

19.2 Would you expect indium phosphide to lend itself more readily to doping as a p-type or an n-type semiconductor?

19.3 Why does "gettering" trap transition metal and alkali metal impurities without seriously depleting implanted dopants?

19.4 Use the data of Appendix C to answer the following questions. (Assume that ΔC_P° is negligible—but would this affect your answer?) (*a*) At what temperature does CVD of silicon from SiH_4 become thermodynamically feasible? (*b*) Would CVD of solid Si (mp 1410 °C) from $SiCl_4$(g) be feasible under any circumstances, assuming that the liberated chlorine could be tolerated?

19.5 What properties do (*a*) copper ions and (*b*) ions of metals such as yttrium, mercury, and barium have that might conceivably contribute to the formation of a high-T_c ceramic superconductor?

References

1. C. J. Brinker and G. W. Scherer, "Sol–Gel Science." Academic Press, New York, 1990.

2. U. Schwertmann and R. M. Cornell, "Iron Oxides in the Laboratory." VCH, Weinheim, Germany, 1991; C. M. Flynn, Jr., Hydrolysis of iron-(III) salts. *Chem. Rev.* **84**, 31–41 (1984).

3. L. L. Hench and J. K. West, "Chemical Processing of Advanced Materials." Wiley, New York, 1992.

4. L. C. Klein (ed.), "Sol–Gel Technology." Noyes Publications, Park Ridge, New Jersey, 1988.

5. R. W. Jones, "Fundamental Principles of Sol–Gel Technology." Institute of Metals, London, 1989.

6. H. Reuter and M.-T. Brandherm, New dimensions in the sol-gel process. *Angew. Chem. Int. Ed. Engl.* **34**, 1578–1579 (1995).

7. S. J. Teichner, Aerogels–Why they are in vogue. *CHEMTECH* **21**, 372–377 (1991).

8. J. Fricke, Aerogels and their applications. *J. Non-Cryst. Solids* **147–148**, 356–362 (1992).

9. L. W. Hrubesh, Aerogels: The world's lightest solids. *Chem. Ind.* December 17, 824–827 (1990).

10. R. K. Iler, "The Chemistry of Silica." Wiley, New York, 1979; H. Bergna (ed.), "The Colloid Chemistry of Silica." ACS Advances in Chemistry Ser. 234. American Chemical Society, Washington, D.C., 1994.

11. J. F. Scott and C. A. Paz de Araujo, Ferroelectric memories. *Science* **246**, 1400–1405 (1989).

12. S. S. Prakash, C. J. Brinker, A. J. Hurd, and S. M. Rao, Silica aerogel films prepared at ambient pressure by using surface derivatization to induce reversible drying shrinkage. *Nature (London)* **374**, 439–443 (1995).

13. C. Hoang-Van, B. Pommier, R. Harivololna, and P. Pichat, Alumina-based aerogels as carriers for automotive Pd catalysts. *J. Non-Cryst. Solids* **145**, 250–254 (1992) [this issue is devoted to aerogels].

14. M. Schneider, M. Maciejewski, S. Tschudin, A. Wokaun, and A. Baiker, Vanadia–titania aerogels. *J. Catal.* **149**, 326–343 (1994).

15. C. R. M. Grovenor, "Microelectronic Materials." Adam Hilger/IOP Publishing, Bristol, 1989.

16. D. Elwell and H. J. Scheel, "Crystal Growth from High-Temperature Solutions." Academic Press, New York, 1975.

17. D. W. Hess and K. F. Jensen (eds.), "Microelectronics Processing: Chemical Engineering Aspects." ACS Advances in Chemistry Ser. 221. American Chemical Society, Washington, D.C., 1989.

18. D. L. Smith, "Thin Film Deposition: Principles and Practice." McGraw-Hill, New York, 1995.

19. L. V. Interrante, L. A. Casper, and A. B. Ellis (eds.), "Materials Chemistry: An Emerging Discipline." ACS Advances in Chemistry Ser. 245. American Chemical Society, Washington, D.C., 1995.

20. J. T. Spencer, Chemical vapor deposition of metal-containing thin-film materials from metal–organic compounds. *Prog. Inorg. Chem.* **41**, 145–237 (1994).

21. D. J. Cole-Hamilton, Precursors for growing new materials. *Chem. Br.* **26**, 852–856 (1990).

22. D. Thompson (ed.), "Insights into Speciality Inorganic Chemicals." Royal Society of Chemistry, Cambridge, 1995.

23. L. Smart and E. Moore, "Solid State Chemistry." Chapman & Hall, London, 1992.

24. D. W. Bruce and D. O'Hare (eds.), "Inorganic Materials." Wiley, New York, 1992.

25. R. L. Carlin, "Magnetochemistry." Springer-Verlag, Berlin, 1986.

26. D. F. Shriver, P. Atkins, and C. H. Langford, "Inorganic Chemistry," 2nd Ed. Freeman, New York, 1994; F. A. Cotton and G. Wilkinson, "Advanced Inorganic Chemistry," 5th Ed. Wiley, New York, 1988; N. N. Greenwood and A. Earnshaw, "Chemistry of the Elements." Pergamon, London, 1984; M. Gerloch and E. C. Constable, "Transition Metal Chemistry." VCH, New York, 1994.

27. P. P. Edwards and C. N. R. Rao, A new era for chemical superconductors. *Chem. Br.* **30**, 722–726, (1994); and following articles.

28. F. J. Adrian and D. O. Cowan, The new superconductors. *Chem. Eng. News*, December 21, 24–41 (1992).

29. A. W. Sleight, Chemistry of high-temperature superconductors. *Science* **242**, 1519–1527 (1988).

30. A. B. Ellis, Superconductors: Better levitation through chemistry. *J. Chem. Educ.* **64**, 836–841 (1987); F. H. Juergens, A. B. Ellis, G. H. Dieckmann, and R. I. Perkins, Levitating a magnet using a superconductive material. *J. Chem. Educ.* **64**, 851–853 (1987).

31. R. F. Service, Superconductivity turns 10. *Science* **271**, 1804–1806 (1996).

32. K. A. Müller and J. G. Bednorz, The discovery of a class of high-temperature superconductors. *Science* **237**, 1133–1139 (1987).

33. M. Laguës, X. M. Xie, H. Tebbji, X. Z. Xu, V. Mairet, C. Hatterer, C. F. Beuran, and C. Deville-Cavellin, Evidence suggesting superconductivity at 250 K in a sequentially deposited cuprate film. *Science* *262*, 1850–1852 (1993).

34. T. H. Geballe, Paths to higher temperature superconductors. *Science* **259**, 1550–1551 (1993).

Appendix A

Useful Constants

Values recommended by the Committee on Data for Science and Technology (CODATA) Task Force on Fundamental Constants; see E. R. Cohen and B. N. Taylor, *J. Phys. Chem. Ref. Data* **17**, 1795–1803 (1988). Numbers in parentheses represent uncertainty in the last decimal places. For example, the uncertainty in the molar gas constant R is ± 0.000070 J K^{-1} mol^{-1}.

Ice point temperature	$= 273.15$ K exactly
Molar gas constant	$R = 8.314510(70)$ J K^{-1} mol^{-1}
Avogadro constant	$N = 6.0221367(36) \times 10^{23}$ elementary entities per mole
Speed of light in a vacuum	$c = 2.99792458 \times 10^{8}$ m s^{-1} exactly
Planck constant	$h = 6.6260755(40) \times 10^{-34}$ J s^{-1}
Boltzmann constant	$k_B = 1.380658(12) \times 10^{-23}$ J K^{-1}
Charge of the electron	$e = -1.60217733(49) \times 10^{-19}$ A s
Mass of the electron	$m_e = 9.1093897(54) \times 10^{-31}$ kg
Mass of the proton	$m_p = 1.6726231(10) \times 10^{-27}$ kg
Faraday constant	$F = 9.6485309(29) \times 10^{4}$ A s mol^{-1}
Permittivity of a vacuum	$\varepsilon_0 = 8.854187817 \times 10^{-12}$ A^2 s^4 kg^{-1} m^{-3} exactly

Appendix B

The Chemical Elements: Standard Atomic Masses

RECOMMENDATIONS OF the IUPAC Commission on Atomic Weights and Isotopic Abundances, *J. Phys. Chem. Ref. Data* **22**, 1571–1584 (1993), follow. As of November, 1996, the names of elements 104–112 are not agreed on [see *Chem. Eng. News* August 21, 4 (1995)]. Masses are scaled to carbon-12 = 12 atomic mass units exactly. The atomic masses of some elements vary somewhat depending on geological origin, as a result of geochemical isotopic fractionation processes or varying contributions from radioactive decay of other nuclides. For elements that have no stable isotopes, the nominal atomic masses given in square brackets refer to the longest lived known isotope. Names of elements given in parentheses indicate alternative nomenclature or the origin (usually a Latin name) of the chemical symbols.

Element	*Symbol*	*Atomic Number*	*Atomic Mass*
Actinium	Ac	89	[227]
Aluminum	Al	13	26.981539(5)
Americium	Am	95	[243]
Antimony	Sb	51	121.757(3)
(Argentum)	Ag	see Silver	
Argon	Ar	18	39.948(1)
Arsenic	As	33	74.92159(2)
Astatine	At	85	[210]
(Aurum)	Au	see Gold	
Barium	Ba	56	137.327(7)
Berkelium	Bk	97	[247]
Beryllium	Be	4	9.012182(3)
Bismuth	Bi	83	208.98037(3)
Boron	B	5	10.811(5)

Element	*Symbol*	*Atomic Number*	*Atomic Mass*
Bromine	Br	35	79.904(1)
Cadmium	Cd	48	112.411(8)
Calcium	Ca	20	40.078(4)
Californium	Cf	98	[251]
Carbon	C	6	12.011(1)
Cerium	Ce	58	140.115(4)
Cesium	Cs	55	132.90543(5)
Chlorine	Cl	17	35.4527(9)
Chromium	Cr	24	51.9961(6)
Cobalt	Co	27	58.93320(1)
(Columbium)	(Cb)	see Niobium	
Copper	Cu	29	63.546(3)
(Cuprum)	Cu	see Copper	
Curium	Cm	96	[247]
Dysprosium	Dy	66	162.50(3)
Einsteinium	Es	99	[254]
Erbium	Er	68	167.26(3)
Europium	Eu	63	151.965(9)
Fermium	Fm	100	[253]
(Ferrum)	Fe	see Iron	
Fluorine	F	9	18.9984032(9)
Francium	Fr	87	[212]
Gadolinium	Gd	64	157.25(3)
Gallium	Ga	31	69.723(1)
Germanium	Ge	32	72.61(2)
Gold	Au	79	196.96654(3)
Hafnium	Hf	72	178.49(2)
Helium	He	2	4.002602(2)
Holmium	Ho	67	164.93032(3)
(Hydrargyrum)	Hg	see Mercury	
Hydrogen	H	1	1.00794(7)
Indium	In	49	114.818(3)
Iodine	I	53	126.90447(3)
Iridium	Ir	77	192.22(3)
Iron	Fe	26	55.847(3)
(Kalium)	K	see Potassium	
Krypton	Kr	36	83.80(1)
Lanthanum	La	57	138.9055(2)
Lawrencium	Lr	103	[262]
Lead	Pb	82	207.2(1)
Lithium	Li	3	6.941(2)
Lutetium	Lu	71	174.967(1)

Element	*Symbol*	*Atomic Number*	*Atomic Mass*
Magnesium	Mg	12	24.3050(6)
Manganese	Mn	25	54.93805(1)
Mendelevium	Md	101	[258]
Mercury	Hg	80	200.59(2)
Molybdenum	Mo	42	95.94(1)
(Natrium)	Na	see Sodium	
Neodymium	Nd	60	144.24(3)
Neon	Ne	10	20.1797(6)
Neptunium	Np	93	[237]
Nickel	Ni	28	58.6934(2)
Niobium	Nb	41	92.90638(2)
Nitrogen	N	7	14.00674(7)
Nobelium	No	102	[259]
Osmium	Os	76	190.23(3)
Oxygen	O	8	15.9994(3)
Palladium	Pd	46	106.42(1)
Phosphorus	P	15	30.973762(4)
Platinum	Pt	78	195.08(3)
(Plumbum)	Pb	see Lead	
Plutonium	Pu	94	[244]
Polonium	Po	84	[209]
Potassium	K	19	39.0983(1)
Praseodymium	Pr	59	140.90765(3)
Promethium	Pm	61	[145]
Protactinium*	Pa	91	231.03588(2)
Radium	Ra	88	[226]
Radon	Rn	86	[222]
Rhenium	Re	75	186.207(1)
Rhodium	Rh	45	102.90550(3)
Rubidium	Rb	37	85.4678(3)
Ruthenium	Ru	44	101.07(2)
Samarium	Sm	62	150.36(3)
Scandium	Sc	21	44.955910(9)
Selenium	Se	34	78.96(3)
Silicon	Si	14	28.0855(3)
Silver	Ag	47	107.8682(2)
Sodium	Na	11	22.989768(6)
(Stannum)	Sn	see Tin	
(Stibium)	Sb	see Antimony	
Strontium	Sr	38	87.62(1)
Sulfur	S	16	32.066(6)
Tantalum	Ta	73	180.9479(1)

Element	*Symbol*	*Atomic Number*	*Atomic Mass*
Technetium	Tc	43	[98]
Tellurium	Te	52	127.60(3)
Terbium	Tb	65	158.92534(3)
Thallium	Tl	81	204.3833(2)
Thorium*	Th	90	232.0381(1)
Thulium	Tm	69	168.93421(3)
Tin	Sn	50	118.710(7)
Titanium	Ti	22	47.88(3)
Tungsten	W	74	183.84(1)
Uranium*	U	92	238.0289(1)
Vanadium	V	23	50.9415(1)
(Wolfram)	W	see Tungsten	
Xenon	Xe	54	131.29(2)
Ytterbium	Yb	70	173.04(3)
Yttrium	Y	39	88.90585(2)
Zinc	Zn	30	65.39(2)
Zirconium	Zr	40	91.224(2)
		104	[261]
		105	[262]
		106	[263]
		107	[262]
		108	[265]
		109	[266]
		110	[269]
		111	[272]
		112	[277]

*Although this element has no stable isotopes, a characteristic terrestrial isotopic composition can be determined.

Appendix C

Chemical Thermodynamic Data

THE FOLLOWING data for substances mentioned in this book are adapted from D. D. Wagman, W. H. Evans, V. B. Parker, R. H. Schumm, I. Halow, S. M. Bailey, K. L. Churney, and R. L. Nuttall, "The NBS Tables of Chemical Thermodynamic Properties," American Chemical Society/American Institute of Physics, Washington, D.C., 1982. Exceptions are the data for C_{60}, taken from W. V. Steele, R. D. Chirico, N. K. Smith, W. E. Billips, P. R. Elmore, and A. E. Wheeler, *J. Phys. Chem.* **96**, 4731–4733 (1992). The tabulated quantities are the enthalpy of formation, entropy, and heat capacity in SI units for 298.15 K and 100 kPa (1 bar) exactly. For solutes, units are moles per kilogram of solvent (molal scale). The NBS order of entries is used: oxygen first, then hydrogen, followed by descents of the periodic table groupwise, beginning with the noble gases (Group 18) and moving left. As thermodynamic properties of ions in solution cannot be separated experimentally from those of the counterions that must be present, they are conventionally referred to those for H^+(aq); for example, $S°$ for Cl^-(aq) is conventionally taken to be the same as $S°$ for HCl(aq. ions).

Substance	ΔH_f° (*kJ mol*$^{-1}$)	S° (*J K*$^{-1}$ *mol*$^{-1}$)	C_P° (*J K*$^{-1}$ *mol*$^{-1}$)
O(g)	249.170	161.055	21.912
O_2(g)	0	205.138	29.355
O_3(g)	142.7	238.93	39.20
H(g)	217.965	114.713	20.784
H^+(aq)	0	0	0
H_2(g)	0	130.684	28.824
OH(g)	38.95	183.745	29.886

Substance	ΔH_f° (*kJ mol*$^{-1}$)	S° (*J K*$^{-1}$ *mol*$^{-1}$)	C_P° (*J K*$^{-1}$ *mol*$^{-1}$)
OH^-(g)	−143.5	—	—
OH^-(aq)	−229.994	−10.75	−148.5
HO_2(g)	10.5	229.0	34.89
HO_2^-(aq)	−160.33	23.8	—
H_2O(g)	−241.818	188.825	33.577
H_2O(l)	−285.830	69.91	75.291
H_2O_2(aq)	−191.17	143.9	—
F(g)	78.99	158.754	22.744
F^-(g)	−255.39	—	—
F_2(g)	0	202.78	31.30
HF(g)	−271.1	173.779	29.133
HF(aq, undissoc.)	−320.08	88.7	—
Cl(g)	121.679	165.198	21.840
Cl^-(g)	−233.13	—	—
Cl^-(aq)	−167.159	56.5	−136.4
Cl_2(g)	0	223.066	33.907
Cl_2(aq)	−23.4	121	—
ClO(g)	101.84	226.63	31.46
ClO^-(aq)	−107.1	42	—
ClO_2(g)	102.5	256.84	41.97
ClO_2(aq)	74.9	164.8	—
ClO_2^-(aq)	−66.5	101.3	—
ClO_3^-(aq)	−103.97	162.3	—
ClO_4^-(aq)	−129.33	182.0	—
HCl(g)	−92.307	186.908	29.12
HClO(aq, undissoc.)	−120.9	142	—
Br(g)	111.884	175.022	20.786
Br^-(g)	−219.07	—	—
Br^-(aq)	−121.55	82.4	−141.8
Br_2(l)	0	152.231	75.689
Br_2(g)	30.907	245.463	36.02
BrO^-(aq)	−94.1	42	—
BrO_3^-(aq)	−67.07	161.71	—
HBr(g)	−36.40	198.695	29.142
I(g)	106.838	180.791	20.786

Substance	ΔH_f° (kJ mol^{-1})	S° (J K^{-1} mol^{-1})	C_P° (J K^{-1} mol^{-1})
$I_2(s)$	0	116.135	54.438
$I_2(g)$	62.438	260.69	36.90
$I_2(aq)$	22.6	137.2	—
$I^-(aq)$	−55.19	111.3	−142.3
$I_3^-(aq)$	−51.5	239.3	—
$HI(g)$	26.48	206.594	29.158
α-S	0	31.80	22.64
$S^{2-}(aq)$	33.1	−14.6	—
$SO_2(g)$	−296.830	248.22	39.87
$SO_2(aq)$	−322.980	161.9	—
β-$SO_3(s)$	−454.51	70.7	—
$SO_3(g)$	−395.72	256.76	50.67
$SO_3^{2-}(aq)$	−635.5	−29	—
$SO_4^{2-}(aq)$	−909.27	20.1	−293
$S_2O_3^{2-}(aq)$	−648.5	67	—
$S_2O_8^{2-}(aq)$	−1344.7	244.3	—
$S_4O_6^{2-}(aq)$	−1224.2	257.3	−67.8
$HS^-(aq)$	−17.6	62.8	—
$H_2S(g)$	−20.63	205.79	34.23
H_2S(aq, undissoc.)	−39.7	121	—
H_2SO_3(aq, undissoc.)	−608.81	232.2	—
$HSO_3^-(aq)$	−626.22	139.7	—
$HSO_4^-(aq)$	−887.34	131.8	−84
$H_2SO_4(l)$	−813.989	156.904	138.91
$N(g)$	472.704	153.298	20.786
$N_2(g)$	0	191.61	29.125
$N_3^-(aq)$	275.14	107.9	—
$NO(g)$	90.25	210.761	29.844
$NO_2(g)$	33.18	240.06	37.20
$NO_2^-(aq)$	−104.6	123.0	−97.5
$NO_3^-(aq)$	−205.0	146.4	−86.6
$N_2O(g)$	82.05	219.85	38.45
$N_2O_4(g)$	9.16	304.29	77.28
$N_2O_4(l)$	−19.50	209.2	142.7
$NH_3(g)$	−46.11	192.45	35.06
$NH_3(aq)$	−80.29	111.3	—

Substance	ΔH_f° (kJ mol^{-1})	S° (J K^{-1} mol^{-1})	C_P° (J K^{-1} mol^{-1})
NH_4^+(aq)	−132.51	113.4	79.9
N_2H_4(l)	50.63	121.21	98.87
HN_3(l)	264.0	140.6	—
HNO_2(aq)	−119.2	135.6	—
HNO_3(l)	−174.10	155.60	109.87
NH_4NO_2(s)	−256.5	—	—
NH_4NO_3(s)	−365.56	151.08	139.3
NCl_3(l)	230	—	—
NOCl(g)	51.71	261.69	44.69
NH_4Cl(s)	−314.43	94.6	84.1
NH_4ClO_4(s)	−295.3	186.2	—
$(NH_4)_2SO_4$(s)	−1180.85	220.1	187.49
P(s, white)	0	41.09	23.840
P(red, triclinic)	−17.6	22.80	21.21
PO_4^{3-}(aq)	−1277.4	−222	—
P_4O_{10}(s, hexagonal)	−2984.0	228.86	211.71
HPO_4^{2-}(aq)	−1292.14	−33.5	—
$H_2PO_4^-$(aq)	−1296.29	90.4	—
C(graphite)	0	5.740	8.527
C(diamond)	1.895	2.377	6.113
$C(C_{60})$(s)	2.42	—	—
$C(C_{60})$(g)	2.66	—	—
C(g)	716.682	158.096	20.838
CO(g)	−110.525	197.674	29.142
CO_2(g)	−393.509	213.74	37.11
CO_2(aq)	−413.80	117.6	—
CO_3^{2-}(aq)	−677.14	−56.9	—
CH_4(g)	−74.81	186.264	35.309
HCO_3^-(aq)	−691.99	91.2	—
CH_3OH(l)	−238.66	126.8	81.6
CH_3OH(g)	−200.66	239.81	43.89
CF_4(g)	−925	261.61	61.09
CCl_4(l)	−135.44	216.40	131.75
CCl_4(g)	−102.9	309.85	83.30
$COCl_2$(g)	−218.8	283.53	57.66
$CHCl_3$(l)	−134.47	201.7	113.8

Substance	ΔH_f° (kJ mol^{-1})	S° (J K^{-1} mol^{-1})	C_P° (J K^{-1} mol^{-1})
$CF_3Cl(g)$	−695	285.29	66.86
$CF_2Cl_2(g)$	−477	300.77	72.26
$CS_2(l)$	89.70	151.34	75.7
$CS_2(g)$	117.36	237.84	45.40
$COS(g)$	−142.09	231.57	41.51
$CN^-(aq)$	150.6	94.1	—
$HCN(l)$	108.87	112.84	70.63
$HCN(g)$	135.1	201.78	35.86
HCN(aq, undissoc.)	107.1	124.7	—
$CO(NH_2)_2(s)$	−333.51	104.60	93.14
$NH_4CO_2NH_2(s)$	−645.05	133.5	—
$NCS^-(aq)$	76.44	144.3	−40.2
$C_2H_2(g)$	226.73	200.94	43.93
$C_2H_4(g)$	52.26	219.56	43.56
$C_2H_6(g)$	−84.68	229.60	52.63
$\frac{1}{n}(C_2F_4)_n$(Teflon)	−820.5	—	—
$C_2F_4(g)$	−650.6	300.06	80.46
$Si(s)$	0	18.83	20.00
$SiO_2(\alpha$-quartz)	−910.94	41.84	44.43
$SiO_2(\alpha$-cristobalite)	−909.48	42.68	44.18
$SiO_2(\alpha$-tridymite)	−909.06	43.5	44.60
SiO_2(amorphous)	−903.49	46.9	44.4
$H_4SiO_4(aq)$	−1468.6	180	—
$SiH_4(g)$	34.3	204.62	42.84
$SiCl_4(l)$	−687.0	239.7	145.31
$SiCl_4(g)$	−657.01	330.73	90.25
SiC(s, cubic)	−65.3	16.61	26.86
$[(CH_3)_3Si]_2O(l)$	−815.0	433.84	311.37
$[(CH_3)_3Si]_2O(g)$	−777.72	535.06	238.49
Sn(s, white)	0	51.55	26.99
Sn(s, gray)	−2.09	44.14	25.77
$SnO(s)$	−285.8	56.5	44.31
$SnO_2(s)$	−580.7	52.3	52.59
$SnCl_4(l)$	−511.3	258.6	165.3
$SnCl_4(g)$	−471.5	365.8	98.3
$Pb(s)$	0	64.81	26.44

Substance	ΔH_f° (kJ mol^{-1})	S° (J K^{-1} mol^{-1})	C_P° (J K^{-1} mol^{-1})
Pb^{2+}(aq)	−1.7	10.5	—
PbO(s, yellow)	−217.32	68.70	45.77
PbO(s, red)	−218.99	66.5	45.81
PbO_2(s)	−277.4	68.6	64.64
Pb_3O_4(s)	−718.4	211.3	146.9
PbS(s)	−100.4	91.2	49.50
$PbSO_4$(s)	−919.94	148.57	103.207
$Pb(N_3)_2$(s)	478.2	148.1	—
B(s)	0	5.86	11.09
B_2O_3(s)	−1272.77	53.97	62.93
$B(OH)_3$(s)	−1094.33	88.83	81.38
BF_3(g)	−1137.00	254.12	50.46
BF_3NH_3(s)	−1353.9	—	—
Al(s)	0	28.33	24.35
Al(g)	326.4	164.54	21.38
Al^{3+}(g)	5483.17	—	—
Al^{3+}(aq)	−531	−321.7	—
Al_2O(g)	−130	259.35	45.69
Al_2O_3(s, α)	−1675.7	50.92	79.04
Al_2O_3(s, γ)	−1656.9	—	—
$Al_2O_3\cdot H_2O$(boehmite)	−1980.7	96.86	131.25
$Al_2O_3\cdot H_2O$(diaspore)	−1998.91	70.67	106.19
$Al_2O_3\cdot 3H_2O$(gibbsite)	−2586.67	136.90	183.47
$Al_2O_3\cdot 3H_2O$(bayerite)	−2576.5	—	—
$Al(OH)_4^-$(aq)	−1502.5	102.9	—
AlF_3(s)	−1504.1	66.44	75.10
$AlCl_3$(s)	−704.2	110.67	91.84
Al_2Cl_6(g)	−1290.8	490	—
Zn(s)	0	41.63	25.40
Zn(g)	130.729	160.984	20.786
Zn^{2+}(aq)	−153.89	−112.1	46
ZnO(s)	−348.28	43.64	40.25
ZnS(zinc blende)	−205.98	57.7	46.0
ZnS(wurtzite)	−192.63	—	—
$ZnSO_4$(s)	−982.8	110.5	99.2

Substance	ΔH_f° (*kJ mol*$^{-1}$)	S° (*J K*$^{-1}$ *mol*$^{-1}$)	C_P° (*J K*$^{-1}$ *mol*$^{-1}$)
Hg(l)	0	76.02	27.983
Hg(g)	61.317	174.96	20.786
Hg^{2+}(aq)	171.1	−32.2	—
$Hg_2{}^{2+}$(aq)	172.4	84.5	—
HgO(s, red)	−90.83	70.29	44.06
HgO(s, yellow)	−90.46	71.1	—
$HgCl_2$(s)	−224.3	146	—
Hg_2Cl_2(s)	−265.22	192.5	—
HgS(s, red)	−58.2	82.4	48.41
HgS(s, black)	−53.6	88.3	—
Cu(s)	0	33.150	24.435
Cu^+(aq)	71.67	40.6	—
Cu^{2+}(aq)	64.77	−99.6	—
CuO(s)	−157.3	42.63	42.30
Cu_2O(s)	−168.6	93.14	63.64
CuS(s)	−53.1	66.5	47.82
Cu_2S(s)	−79.5	120.9	76.32
$CuSO_4$(s)	−771.36	109	100
$Cu(NH_3)_4{}^{2+}$(aq)	−348.5	273.6	—
Ag(s)	0	42.55	25.351
Ag^+(aq)	105.579	72.68	21.8
Ag_2O(s)	−31.05	121.3	65.86
AgCl(s)	−127.068	96.2	50.79
AgBr(s)	−100.37	107.1	52.38
Ag_2S(s, α)	−32.59	144.01	76.53
Au(s)	0	47.40	25.418
$Au(CN)_2{}^-$(aq)	242.3	172	—
Ni(s)	0	29.87	26.07
Ni^{2+}(aq)	−54.0	−128.9	—
NiO(s)	−239.7	37.99	44.31
$Ni(OH)_2$(s, α)	−529.7	88	—
NiS(s)	−82.0	52.97	47.11
$NiSO_4$(s)	−872.91	92	138
$NiSO_4 \cdot 7H_2O$(s)	−2976.33	378.94	364.59
$Ni(NH_3)_6{}^{2+}$(aq)	−630.1	394.6	—

Substance	ΔH_f° (kJ mol^{-1})	S° (J K^{-1} mol^{-1})	C_P° (J K^{-1} mol^{-1})
$Ni(CO)_4$(l)	−633.0	313.4	204.6
$Ni(CO)_4$(g)	−602.91	410.6	145.18
Co(s)	0	30.04	24.81
Co^{2+}(aq)	−58.2	−113	—
CoO(s)	−237.94	52.97	55.23
$Co(OH)_2$(pink, ppt)	−539.7	79	—
$Co(NH_3)_6{}^{3+}$(aq)	−584.9	146	—
$Co(NH_3)_5Cl^{2+}$(aq)	−628.0	341.4	—
Fe(s)	0	27.28	25.10
Fe^{2+}(g)	2749.93	—	—
Fe^{3+}(g)	5712.8	—	—
Fe^{2+}(aq)	−89.1	−137.7	—
Fe^{3+}(aq)	−48.5	−315.9	—
$FeO_{0.947}$(s, wustite)	−266.27	57.49	48.12
Fe_2O_3(s, hematite)	−824.2	87.40	103.85
Fe_3O_4(s, magnetite)	−1118.4	146.4	143.43
$FeOH^{2+}$(aq)	−290.8	−142	—
$Fe(OH)_2$(s, ppt)	−569.0	88	—
$Fe(OH)_3$(s, ppt)	−823.0	106.7	—
$Fe_{1.000}S$(s)	−100.0	60.29	50.54
FeS_2(s, pyrite)	−178.2	52.93	62.17
Fe_7S_8(s)	−736.4	485.8	398.57
$FeSO_4$(s)	−928.4	107.5	100.58
$Fe(CO)_5$(l)	−774.0	338.1	240.6
$Fe(CO)_5$(g)	−733.9	445.3	—
$Fe(CN)_6{}^{3-}$(aq)	561.9	270.3	—
$Fe(CN)_6{}^{4-}$(aq)	455.6	95	—
Mn(s, α)	0	32.01	26.32
Mn^{2+}(aq)	−220.75	−73.6	50.
MnO(s)	−385.22	59.71	45.44
MnO_2(s)	−520.03	53.05	54.14
$MnO_4{}^-$(aq)	−541.4	191.2	−82.0
Mn_2O_3(s)	−959.0	110.5	107.65
Mn_3O_4(s)	−1387.8	155.6	139.66
$Mn(OH)_2$(s, ppt)	−695.4	99.2	—

Substance	ΔH_f° (*kJ mol*$^{-1}$)	S° (*J K*$^{-1}$ *mol*$^{-1}$)	C_P° (*J K*$^{-1}$ *mol*$^{-1}$)
Cr(s)	0	23.77	23.35
$CrO_4{}^{2-}$(aq)	−881.15	50.21	—
Cr_2O_3(s)	−1139.7	81.2	118.74
$Cr_2O_7{}^{2-}$(aq)	−1490.3	261.9	—
$HCrO_4{}^-$(aq)	−878.2	184.1	—
$FeCr_2O_4$(s)	−1444.7	146.0	133.64
Mo(s)	0	28.66	24.06
MoO_2(s)	−588.94	46.28	55.98
MoO_3(s)	−745.09	77.74	74.98
$MoO_4{}^{2-}$(aq)	−997.9	27.2	—
MoS_2(s)	−235.1	62.59	63.55
W(s)	0	32.64	24.27
WO_3(s)	−842.87	75.90	73.76
V(s)	0	28.91	24.89
VO^{2+}(aq)	−486.6	−133.9	—
$VO_2{}^+$(aq)	−649.8	−42.3	—
$VO_3{}^-$(aq)	−888.3	50	—
V_2O_5(s)	−1550.6	131.0	127.65
Ti(s)	0	30.63	25.02
TiO(s, α)	−519.7	50	39.96
TiO_2(s, anatase)	−939.7	49.92	55.48
TiO_2(s, rutile)	−944.7	50.33	55.02
Ti_2O_3(s)	−1520.9	78.78	97.36
TiH_2(s)	−119.7	29.7	30.1
$TiCl_4$(l)	−804.2	252.34	145.18
$TiCl_4$(g)	−763.2	354.9	95.4
TiN(s)	−338.1	30.25	37.07
TiC(s)	−184.5	24.23	33.64
Zr(s)	0	38.99	25.36
ZrO_2(s, α)	−1100.56	50.38	56.19
ZrH_2(s)	−169.0	35.02	30.96
$ZrCl_4$(s)	−980.52	181.6	119.79
$ZrCl_4$(g)	−870.3	368.3	98.28

Substance	ΔH_f° (*kJ mol*$^{-1}$)	S° (*J K*$^{-1}$ *mol*$^{-1}$)	C_P° (*J K*$^{-1}$ *mol*$^{-1}$)
U(s)	0	50.21	27.665
U^{4+}(aq)	−591.2	−410	—
UO_2(s)	−1084.9	77.03	63.60
$UO_2{}^{2+}$(aq)	−1019.6	−97.5	—
U_3O_8(s, α)	−3574.8	282.59	238.36
UF_4(s)	−1914.2	151.67	116.02
UF_6(g)	−2147.4	377.9	129.62
Be(s)	0	9.50	16.44
BeO(s)	−609.6	14.14	25.52
$BeCl_2$(s)	−490.4	82.68	64.85
Be_2SiO_4(s)	−2149.3	64.31	95.56
Mg(s)	0	32.68	24.89
Mg(g)	147.70	148.650	20.786
Mg^{2+}(g)	2348.504	—	—
Mg^{2+}(aq)	−466.85	−138.1	—
MgO(s, periclase)	−601.70	26.94	37.15
$Mg(OH)_2$(s)	−924.54	63.18	77.03
$MgCl_2$(s)	−641.32	89.62	71.38
$MgCO_3$(s)	−1095.8	65.7	75.52
$MgSiO_3$(s, enstatite)	−1549.0	67.74	81.38
Mg_2SiO_4(forsterite)	−2174.0	95.14	118.49
$Mg_3Si_2O_5(OH)_4$(s, chrysotile)	−4365.6	221.3	273.68
$Mg_3Si_4O_{10}(OH)_2$(talc)	−5922.5	260.7	321.7
$MgAl_2O_4$(spinel)	−2299.9	80.63	116.19
Ca(s)	0	41.42	25.31
Ca(g)	178.2	154.884	20.786
Ca^{2+}(g)	1925.90	—	—
Ca^{2+}(aq)	−542.83	−53.1	—
CaO(s)	−635.09	39.75	42.80
$Ca(OH)_2$(s)	−986.09	83.39	87.49
CaF_2(s)	−1219.6	68.87	67.03
$CaCl_2$(s)	−795.8	104.6	72.59
$CaSO_3 \cdot \frac{1}{2}H_2O$(s)	−1311.7	121.3	—
$CaSO_4$(s, anhydrite)	−1434.11	106.7	99.66
$CaSO_4 \cdot \frac{1}{2}H_2O$(s)	−1576.74	130.5	119.41
$CaSO_4 \cdot 2H_2O$(gypsum)	−2022.63	194.1	186.02

Substance	ΔH_f° (kJ mol^{-1})	S° (J K^{-1} mol^{-1})	C_P° (J K^{-1} mol^{-1})
$Ca_3(PO_4)_2$(s, β)	−4120.8	236.0	227.82
$Ca_5(PO_4)_3OH$(apatite)	−6738.5	390.35	384.95
$Ca_5(PO_4)_3F$(s)	−6872.	387.85	375.95
CaC_2(s)	−59.8	69.96	62.72
$CaCO_3$(calcite)	−1206.92	92.9	81.88
$CaCO_3$(aragonite)	−1207.13	88.7	81.25
$CaTiO_3$(perovskite)	−1660.6	93.64	97.65
$CaMg(CO_3)_2$(dolomite)	−2326.3	155.18	157.53
Ba(s)	0	62.8	28.07
Ba^{2+}(aq)	−537.64	9.6	—
BaO(s)	−553.5	70.42	47.78
$BaSO_4$(s)	−1473.2	132.2	101.75
$BaCO_3$(s)	−1216.3	112.1	85.35
Li(s)	0	29.12	24.77
Li^+(aq)	−278.49	13.4	68.6
Li_2O(s)	−597.94	37.57	54.1
LiOH(s)	−484.93	42.80	49.66
LiCl(s)	−408.61	59.33	47.99
LiI(s)	−270.41	86.78	51.04
Li_2CO_3(s)	−1215.9	90.37	99.12
$LiAlH_4$(s)	−116.3	78.74	83.18
Na(s)	0	51.21	28.24
Na(g)	107.32	153.712	20.786
Na^+(g)	609.358	—	—
Na^+(aq)	−240.12	59.0	46.4
Na_2O(s)	−414.22	75.06	69.12
Na_2O_2(s)	−510.87	95.0	89.24
NaH(s)	−56.275	40.016	36.401
NaOH(s)	−425.609	64.455	59.54
NaF(s)	−573.647	51.46	46.86
NaCl(s)	−411.153	72.13	50.50
$NaClO_3$(s)	−365.774	123.4	—
$NaClO_4$(s)	−383.30	142.3	—
NaBr(s)	−361.062	86.82	51.38
NaI(s)	−287.78	98.53	52.09
Na_2S(s)	−364.8	83.7	—

Substance	ΔH_f° (kJ mol^{-1})	S° (J K^{-1} mol^{-1})	C_P° (J K^{-1} mol^{-1})
Na_2SO_4(s)	−1387.08	149.58	128.20
$NaHSO_4$(s)	−1125.5	113.0	—
NaN_3(s)	21.71	96.86	76.61
$NaNO_2$(s)	−358.65	103.8	—
$NaNO_3$(s)	−467.85	116.52	92.88
Na_3PO_4(s)	−1917.40	173.80	153.47
$Na_5P_3O_{10}$(s)	−4399.1	381.79	327.02
Na_2CO_3(s)	−1130.68	134.98	112.30
$Na_2CO_3 \cdot 10H_2O$(s)	−4081.32	562.7	550.32
$NaHCO_3$(s)	−950.81	101.7	87.61
NaCN(s, cubic)	−87.49	115.60	70.37
Na_2SiO_3(cryst.)	−1554.90	113.85	—
$Na_2B_4O_7 \cdot 10H_2O$(borax)	−6288.6	586.	615.
$NaBH_4$(s)	−188.61	101.29	86.78
Na_3AlF_6(s)	−3301.2	238.5	215.89
$NaAlSi_2O_6 \cdot H_2O$(s, analcite)	−3300.8	234.3	209.91
$NaAlSi_3O_8$(albite)	−3935.1	207.40	205.10
Na_2CrO_4(s)	−1342.2	176.61	142.13
K(s)	0	64.18	29.58
K(g)	89.24	160.336	20.786
K^+(g)	514.26	—	—
K^+(aq)	−252.38	102.5	21.8
KO_2(s)	−284.93	116.7	77.53
K_2O(s)	−361.5	—	—
K_2O_2(s)	−494.1	102.1	—
KOH(s)	−424.764	78.9	64.9
KF(s)	−567.27	66.57	49.04
KCl(s)	−436.747	82.59	51.30
$KClO_3$(s)	−397.73	143.1	100.25
$KClO_4$(s)	−432.75	151.0	112.38
KBr(s)	−393.798	95.90	52.30
KI(s)	−327.900	106.32	52.93
K_2SO_4(s)	−1437.79	175.56	131.46
$K_2S_2O_8$(s)	−1916.1	278.7	213.09
KNO_2(s)	−369.82	152.09	107.40
KNO_3(s)	−494.63	133.05	96.40

Substance	ΔH_f° (*kJ mol*$^{-1}$)	S° (*J K*$^{-1}$ *mol*$^{-1}$)	C_P° (*J K*$^{-1}$ *mol*$^{-1}$)
K_2CO_3(s)	−1151.02	155.52	114.43
$KHCO_3$(s)	−963.2	115.5	—
KCN(s)	−113.0	128.49	66.27
KNCS(s)	−200.16	124.26	88.53
$KAl(SO_4)_2 \cdot 12H_2O$(s)	−6061.8	687.4	651.03
$KAlSi_3O_8$(orthoclase)	−3959.7	232.88	204.51
$KAl_3Si_3O_{10}(OH)_2$(muscovite)	−5984.4	306.3	—
$KMnO_4$(s)	−837.2	171.71	117.57
K_2CrO_4(s)	−1403.7	200.12	145.98
$K_2Cr_2O_7$(s)	−2061.5	291.2	219.24

Appendix D

Standard Electrode Potentials for Aqueous Solutions

FOR MORE comprehensive information, see A. J. Bard, R. Parsons and J. Jordan, "Standard Potentials in Aqueous Solution," Dekker, New York, 1985, from which the following E° values are taken.

Acidic Solutions ($[H^+] = 1.0$ mol kg^{-1})

Half-reaction	$E^\circ(V)$
$Li^+ + e^- \rightleftharpoons Li$	−3.045
$K^+ + e^- \rightleftharpoons K$	−2.925
$Na^+ + e^- \rightleftharpoons Na$	−2.714
$La^{3+} + 3e^- \rightleftharpoons La$	−2.37
$Mg^{2+} + 2e^- \rightleftharpoons Mg$	−2.356
$\frac{1}{2}H_2 + e^- \rightleftharpoons H^-$	−2.25
$Be^{2+} + 2e^- \rightleftharpoons Be$	−1.97
$Zr^{4+} + 4e^- \rightleftharpoons Zr$	−1.70
$Al^{3+} + 3e^- \rightleftharpoons Al$	−1.67
$Ti^{3+} + 3e^- \rightleftharpoons Ti$	−1.21
$Mn^{2+} + 2e^- \rightleftharpoons Mn$	−1.18
$V^{2+} + 2e^- \rightleftharpoons V$	−1.13
$SiO_2(\text{glass}) + 4H^+ + 4e^- \rightleftharpoons Si + 2H_2O$	−0.888
$Zn^{2+} + 2e^- \rightleftharpoons Zn$	−0.763
$U^{4+} + e^- \rightleftharpoons U^{3+}$	−0.52
$Fe^{2+} + 2e^- \rightleftharpoons Fe$	−0.44

Half-reaction	$E°(V)$
$Cr^{3+} + e^- \rightleftharpoons Cr^{2+}$	−0.424
$Cd^{2+} + 2e^- \rightleftharpoons Cd$	−0.403
$PbSO_4 + 2e^- \rightleftharpoons Pb + SO_4{}^{2-}$	−0.351
$Eu^{3+} + e^- \rightleftharpoons Eu^{2+}$	−0.35
$Co^{2+} + 2e^- \rightleftharpoons Co$	−0.277
$H_3PO_4 + 2H^+ + 2e^- \rightleftharpoons H_3PO_3 + H_2O$	−0.276
$Ni^{2+} + 2e^- \rightleftharpoons Ni$	−0.257
$V^{3+} + e^- \rightleftharpoons V^{2+}$	−0.255
$2SO_4{}^{2-} + 4H^+ + 2e^- \rightleftharpoons S_2O_6{}^{2-} + 2H_2O$	−0.253
$N_2 + 5H^+ + 4e^- \rightleftharpoons N_2H_5{}^+$	−0.23
$CO_2 + 2H^+ + 2e^- \rightleftharpoons HCOOH$	−0.16
$AgI + e^- \rightleftharpoons Ag + I^-$	−0.152
$Sn^{2+} + 2e^- \rightleftharpoons Sn$	−0.136
$Pb^{2+} + 2e^- \rightleftharpoons Pb$	−0.125
$2H^+ + 2e^- \rightleftharpoons H_2$	0.000
$HCOOH + 2H^+ + 2e^- \rightleftharpoons HCHO + H_2O$	+0.056
$AgBr + e^- \rightleftharpoons Ag + Br^-$	+0.071
$TiO^{2+} + 2H^+ + e^- \rightleftharpoons Ti^{3+} + H_2O$	+0.100
$S + 2H^+ + 2e^- \rightleftharpoons H_2S$	+0.144
$Sn^{4+} + 2e^- \rightleftharpoons Sn^{2+}$	+0.15
$SO_4{}^{2-} + 4H^+ + 2e^- \rightleftharpoons H_2SO_3 + H_2O$	+0.158
$Cu^{2+} + e^- \rightleftharpoons Cu^+$	+0.159
$AgCl + e^- \rightleftharpoons Ag + Cl^-$	+0.222
$HCHO + 2H^+ + 2e^- \rightleftharpoons CH_3OH$	+0.232
$UO_2{}^{2+} + 4H^+ + 2e^- \rightleftharpoons U^{4+} + 2H_2O$	+0.27
$VO^{2+} + 2H^+ + e^- \rightleftharpoons V^{3+} + H_2O$	+0.337
$Cu^{2+} + 2e^- \rightleftharpoons Cu$	+0.340
$Fe(CN)_6{}^{3-} + e^- \rightleftharpoons Fe(CN)_6{}^{4-}$	+0.361
$2H_2SO_3 + 2H^+ + 4e^- \rightleftharpoons S_2O_3{}^{2-} + 3H_2O$	+0.400
$H_2SO_3 + 4H^+ + 4e^- \rightleftharpoons S + 3H_2O$	+0.500
$4H_2SO_3 + 4H^+ + 6e^- \rightleftharpoons S_4O_6{}^{2-} + 6H_2O$	+0.507
$Cu^+ + e^- \rightleftharpoons Cu$	+0.520
$I_2 + 2e^- \rightleftharpoons 2I^-$	+0.5355
$I_3{}^- + 2e^- \rightleftharpoons 3I^-$	+0.536
$MnO_4{}^- + e^- \rightleftharpoons MnO_4{}^{2-}$	+0.56
$S_2O_6{}^{2-} + 4H^+ + 2e^- \rightleftharpoons 2H_2SO_3$	+0.569

Half-reaction	$E°(V)$
$CH_3OH + 2H^+ + 2e^- \rightleftharpoons CH_4 + H_2O$	+0.59
$HN_3 + 11H^+ + 8e^- \rightleftharpoons 3NH_4{}^+$	+0.695
$O_2 + 2H^+ + 2e^- \rightleftharpoons H_2O_2$	+0.695
$Rh^{3+} + 3e^- \rightleftharpoons Rh$	+0.76
$(NCS)_2 + 2e^- \rightleftharpoons 2NCS^-$	+0.77
$Fe^{3+} + e^- \rightleftharpoons Fe^{2+}$	+0.771
$Hg_2{}^{2+} + 2e^- \rightleftharpoons 2Hg$	+0.796
$Ag^+ + e^- \rightleftharpoons Ag$	+0.799
$2NO_3{}^- + 4H^+ + 2e^- \rightleftharpoons N_2O_4 + 2H_2O$	+0.803
$Hg^{2+} + 2e^- \rightleftharpoons Hg$	+0.911
$NO_3{}^- + 3H^+ + 2e^- \rightleftharpoons HNO_2 + H_2O$	+0.94
$NO_3{}^- + 4H^+ + 3e^- \rightleftharpoons NO + 2H_2O$	+0.957
$HNO_2 + H^+ + e^- \rightleftharpoons NO + H_2O$	+0.996
$N_2O_4 + 4H^+ + 4e^- \rightleftharpoons 2NO + 2H_2O$	+1.039
$Br_2 + 2e^- \rightleftharpoons 2Br^-$	+1.065
$N_2O_4 + 2H^+ + 2e^- \rightleftharpoons 2HNO_2$	+1.07
$H_2O_2 + H^+ + e^- \rightleftharpoons \cdot OH + H_2O$	+1.14
$ClO_4{}^- + 2H^+ + 2e^- \rightleftharpoons ClO_3{}^- + H_2O$	+1.201
$O_2 + 4H^+ + 4e^- \rightleftharpoons 2H_2O$	+1.229
$MnO_2 + 4H^+ + 2e^- \rightleftharpoons Mn^{2+} + 2H_2O$	+1.23
$N_2H_5{}^+ + 3H^+ + 2e^- \rightleftharpoons 2NH_4{}^+$	+1.275
$Cl_2 + 2e^- \rightleftharpoons 2Cl^-$	+1.358
$Cr_2O_7{}^{2-} + 14H^+ + 6e^- \rightleftharpoons 2Cr^{3+} + 7H_2O$	+1.36
$PbO_2 + 4H^+ + 2e^- \rightleftharpoons Pb^{2+} + 2H_2O$	+1.468
$2BrO_3{}^- + 12H^+ + 10e^- \rightleftharpoons Br_2 + 6H_2O$	+1.478
$Mn^{3+} + e^- \rightleftharpoons Mn^{2+}$	+1.51
$Au^{3+} + 3e^- \rightleftharpoons Au$	+1.52
$NiO_2 + 4H^+ + 2e^- \rightleftharpoons Ni^{2+} + 2H_2O$	+1.593
$2HBrO + 2H^+ + 2e^- \rightleftharpoons Br_2 + 2H_2O$	+1.604
$2HClO + 2H^+ + 2e^- \rightleftharpoons Cl_2 + 2H_2O$	+1.630
$PbO_2 + SO_4{}^{2-} + 4H^+ + 2e^- \rightleftharpoons PbSO_4 + 2H_2O$	+1.698
$MnO_4{}^- + 4H^+ + 3e^- \rightleftharpoons MnO_2 + 2H_2O$	+1.70
$Ce^{4+} + e^- \rightleftharpoons Ce^{3+}$	+1.72
$H_2O_2 + 2H^+ + 2e^- \rightleftharpoons 2H_2O$	+1.763
$Au^+ + e^- \rightleftharpoons Au$	+1.83
$Co^{3+} + e^- \rightleftharpoons Co^{2+}$	+1.92

Half-reaction	$E^\circ(V)$
$HN_3 + 3H^+ + 2e^- \rightleftharpoons NH_4^+ + N_2$	+1.96
$S_2O_8^{2-} + 2e^- \rightleftharpoons 2SO_4^{2-}$	+1.96
$O_3 + 2H^+ + 2e^- \rightleftharpoons O_2 + H_2O$	+2.075
$\cdot OH + H^+ + e^- \rightleftharpoons H_2O$	+2.38
$F_2 + 2H^+ + 2e^- \rightleftharpoons 2HF$	+3.053

Basic Solutions ([OH$^-$] = 1.0 mol kg^{-1})

Half-reaction	$E^\circ(V)$
$Ca(OH)_2 + 2e^- \rightleftharpoons Ca + 2OH^-$	−3.026
$Mg(OH)_2 + 2e^- \rightleftharpoons Mg + 2OH^-$	−2.687
$Al(OH)_4^- + 3e^- \rightleftharpoons Al + 4OH^-$	−2.310
$SiO_3^{2-} + 3H_2O + 4e^- \rightleftharpoons Si + 6OH^-$	−1.7
$Mn(OH)_2 + 2e^- \rightleftharpoons Mn + 2OH^-$	−1.56
$2TiO_2 + H_2O + 2e^- \rightleftharpoons Ti_2O_3 + 2OH^-$	−1.38
$Cr(OH)_3 + 3e^- \rightleftharpoons Cr + 3OH^-$	−1.33
$Zn(OH)_4^{2-} + 2e^- \rightleftharpoons Zn + 4OH^-$	−1.285
$Zn(NH_3)_4^{2+} + 2e^- \rightleftharpoons Zn + 4NH_3$	−1.04
$MnO_2 + 2H_2O + 4e^- \rightleftharpoons Mn + 4OH^-$	−0.980
$Cd(CN)_4^{2-} + 2e^- \rightleftharpoons Cd + 4CN^-$	−0.943
$SO_4^{2-} + H_2O + 2e^- \rightleftharpoons SO_3^{2-} + 2OH^-$	−0.94
$2H_2O + 2e^- \rightleftharpoons H_2 + 2OH^-$	−0.828
$HFeO_2^- + H_2O + 2e^- \rightleftharpoons Fe + 3OH^-$	−0.8
$Co(OH)_2 + 2e^- \rightleftharpoons Co + 2OH^-$	−0.733
$CrO_4^{2-} + 4H_2O + 3e^- \rightleftharpoons Cr(OH)_4^- + 4OH^-$	−0.72
$Ni(OH)_2 + 2e^- \rightleftharpoons Ni + 2OH^-$	−0.72
$FeO_2^- + H_2O + e^- \rightleftharpoons HFeO_2^- + OH^-$	−0.69
$2SO_3^{2-} + 3H_2O + 4e^- \rightleftharpoons S_2O_3^{2-} + 6OH^-$	−0.58
$Ni(NH_3)_6^{2+} + 2e^- \rightleftharpoons Ni + 6NH_3$	−0.476
$S + 2e^- \rightleftharpoons S^{2-}$	−0.45
$O_2 + e^- \rightleftharpoons O_2^-$	−0.33
$CuO + H_2O + 2e^- \rightleftharpoons Cu + 2OH^-$	−0.29
$Mn_2O_3 + 3H_2O + 2e^- \rightleftharpoons 2Mn(OH)_2 + 2OH^-$	−0.25
$2CuO + H_2O + 2e^- \rightleftharpoons Cu_2O + 2OH^-$	−0.22

Half-reaction	$E°(V)$
$O_2 + H_2O + 2e^- \rightleftharpoons HO_2^- + OH^-$	−0.065
$MnO_2 + 2H_2O + 2e^- \rightleftharpoons Mn(OH)_2 + 2OH^-$	−0.05
$NO_3^- + H_2O + 2e^- \rightleftharpoons NO_2^- + 2OH^-$	+0.01
$Co(NH_3)_6^{3+} + e^- \rightleftharpoons Co(NH_3)_6^{2+}$	+0.058
$HgO(\text{red form}) + H_2O + 2e^- \rightleftharpoons Hg + 2OH^-$	+0.098
$N_2H_4 + 2H_2O + 2e^- \rightleftharpoons 2NH_3 + 2OH^-$	+0.1
$Co(OH)_3 + e^- \rightleftharpoons Co(OH)_2 + OH^-$	+0.17
$HO_2^- + H_2O + e^- \rightleftharpoons \cdot OH + 2OH^-$	+0.184
$O_2^- + H_2O + e^- \rightleftharpoons HO_2^- + OH^-$	+0.20
$ClO_3^- + H_2O + 2e^- \rightleftharpoons ClO_2^- + 2OH^-$	+0.295
$Ag_2O + H_2O + 2e^- \rightleftharpoons 2Ag + 2OH^-$	+0.342
$Ag(NH_3)_2^+ + e^- \rightleftharpoons Ag + 2NH_3$	+0.373
$ClO_4^- + H_2O + 2e^- \rightleftharpoons ClO_3^- + 2OH^-$	+0.374
$O_2 + 2H_2O + 4e^- \rightleftharpoons 4OH^-$	+0.401
$NiO_2 + 2H_2O + 2e^- \rightleftharpoons Ni(OH)_2 + 2OH^-$	+0.490
$FeO_4^{2-} + 2H_2O + 3e^- \rightleftharpoons FeO_2^- + 4OH^-$	+0.55
$BrO_3^- + 3H_2O + 6e^- \rightleftharpoons Br^- + 6OH^-$	+0.584
$MnO_4^{2-} + 2H_2O + 2e^- \rightleftharpoons MnO_2 + 4OH^-$	+0.62
$ClO_2^- + H_2O + 2e^- \rightleftharpoons ClO^- + 2OH^-$	+0.681
$BrO^- + H_2O + 2e^- \rightleftharpoons Br^- + 2OH^-$	+0.766
$HO_2^- + H_2O + 2e^- \rightleftharpoons 3OH^-$	+0.867
$ClO^- + H_2O + 2e^- \rightleftharpoons Cl^- + 2OH^-$	+0.890
$ClO_2 + e^- \rightleftharpoons ClO_2^-$	+1.041
$O_3 + H_2O + 2e^- \rightleftharpoons O_2 + 2OH^-$	+1.246
$\cdot OH + e^- \rightleftharpoons OH^-$	+1.985

Appendix E

Nomenclature of Coordination Compounds

CONVENTIONS FOR naming inorganic compounds are established by the International Union of Pure and Applied Chemistry (IUPAC) and are reviewed on a continuing basis. The following rules cover only the more commonly encountered complexes.

Ligands

For *neutral ligands*, there are four special names that have survived from the early days of coordination chemistry: *aqua* for coordinated water, *ammine* for ammonia (not to be confused with *amine*, meaning organic compounds RNH_2, R_2NH, etc., which can also act as ligands), *carbonyl* for complexed carbon monoxide, and *nitrosyl* for bound nitric oxide (NO). For all other neutral ligands, the ordinary name of the molecule is used without modification. For *anionic* molecules as ligands, the final "e" of the anion name is replaced with "o", but there are irregular cases involving *some* anions which end in "ide": chloride (Cl^-) becomes chloro (and the other halides likewise); oxide (O^{2-}), hydroxide (OH^-), peroxide ($O_2{}^{2-}$), and superoxide (O^{2-}) become oxo, hydroxo, peroxo and superoxo; cyanide (CN^-) becomes cyano. Most others follow the regular rule:

nitride (N^{3-})	becomes	nitrido
sulfide (S^{2-})		sulfido
azide ($N_3{}^-$)		azido
amide ($NH_2{}^-$)		amido
carbonate ($CO_3{}^{2-}$)		carbonato
nitrate ($NO_3{}^-$)		nitrato
nitrite ($NO_2{}^-$)		nitrito (if bonded through an O)
		nitro (if bonded through N)
thiocyanate (NCS^-)		thiocyanato-*S* (if S-bonded)
		thiocyanato-*N* (if N-bonded)

Complexes

Ligands are named first, starting with any anionic ones, and the name of the metal is followed without a space by the oxidation state in Roman numerals (or the Arabic 0, for zero-valent metal centers) in parentheses. If the complex as a whole is anionic, the metal name is made to end in *-ate*, which replaces endings such as -ium or -um (nickelate, chromate, tantalate) and is followed by the oxidation state. Where the chemical symbol is derived from a Latin name, the anion name is usually also Latinized: cuprate, argentate, aurate, ferrate, stannate, plumbate—but mercurate is an exception.

Numbers of ligands are indicated by Greek prefixes di-, tri-, tetra-, penta-, hexa-, hepta-, octa-, nona- (ennea-), deca-, etc. If, however, the names of the ligands themselves already contain these prefixes (e.g., *di*ethylene*tri*amine), the ligand name is placed in parentheses and the prefix outside becomes bis-, tris-, tetrakis-, pentakis-, hexakis-, etc.

$[Co(NH_3)_5Cl]Cl_2$	chloropentaamminecobalt(III) chloride
$Co(CO)_3NO$	nitrosyltricarbonylcobalt(0)
$K_4[Fe(CN)_6]$	potassium hexacyanoferrate(II)
$Co(en)_3{}^{3+}$	tris(ethylenediamine)cobalt(III) ion

(Strictly speaking, names such as ethylenediamine are not systematic and should not be used, but tradition dies hard.)

Isomers and Bridged Complexes

As explained in Section 13.2, geometric isomers are designated by the prefixes *cis-* and *trans-*, or, for octahedral complexes ML_3X_3, *fac-* (meaning facial—all ligands L adjacent to each other, defining one octahedral face) or *mer-* (for meridional—the three Ls occupy a north–south "meridian").

Bridging ligands are indicated by a prefix μ-.

$$(H_2O)_4Fe(\mu\text{-}OH)_2Fe(OH_2)_4{}^{4+}$$

di(μ-hydroxo)octaaquadiiron(III) ion

Use of the η^n- prefix in organometallic complexes is explained in Chapter 18. The superscript n gives the number of atoms (usually carbon atoms) in a ligand that are directly bonded to the metal center; for example, η^5-cyclopentadienide means that all five carbons in a $C_5H_5{}^-$ ligand are bonded to the metal atom.

Appendix F

Ionic Radii

THE FOLLOWING data are selected mostly from the extensive compilation of R. D. Shannon [*Acta Crystallogr.* **A32**, 751–767 (1976)]. As Shannon emphasizes, ionic radii r are influenced by the coordination number C.N. and, because only the *sums* of anion and cation radii can be measured crystallographically, by the choice of a reference radius. Here, radii are calculated relative to r = 140 pm for O^{2-} (C.N. 6), giving values close to the traditional ionic radii of Goldschmidt, Pauling, and others. (Shannon favors a scale based on r = 126 pm for O^{2-} with C.N. 6, giving radii 14 pm larger for cations and 14 pm smaller for anions, in better accordance with expectations of the actual sizes of atoms and ions.) Apparent radii for atoms in very high oxidation states (e.g., "N^{5+}") are not included because such small, highly charged ions do not exist as such in molecules or crystal lattices. The symbol # indicates that the ion occurs only in environments with different interionic distances; the radius quoted is an approximate average value only. The abbreviation *sq* denotes square planar (as distinct from tetrahedral) coordination, and *ls* indicates the low spin (spin-paired) alternative for ions with electronic configurations nd^4 through nd^7 (all other entries are high spin, where this distinction exists). The ion "Hg^+" exists in compounds as the dimer $Hg_2{}^{2+}$.

Ion	*C.N.*	*r (pm)*	*Ion*	*C.N.*	*r (pm)*	*Ion*	*C.N.*	*r (pm)*
Ag^+	2	67	Bi^{3+}	6	103	Ce^{4+}	6	87
	4	100	Br^-	6	196	Cl^-	6	181
Al^{3+}	4	39	Ca^{2+}	6	100	Co^{2+}	4	58
	6	53.5		8	112		6	74.5
Au^+	6	137	Cd^{2+}	4	78		6*ls*	65
Au^{3+}	4*sq*	68		6	95	Co^{3+}	6	61
Ba^{2+}	6	135	Ce^{3+}	6	101		6*ls*	54.5
	8	142		8	114	Cr^{2+}	6	80#
Be^{2+}	4	27		9	120	Cr^{3+}	6	61.5

Ion	*C.N.*	*r (pm)*	*Ion*	*C.N.*	*r (pm)*	*Ion*	*C.N.*	*r (pm)*
Cs^{+}	6	167	La^{3+}	9	122	Pt^{4+}	6	62.5
	8	174	Li^{+}	4	59	Rh^{3+}	6	66.5
Cu^{+}	2	46		6	76	Rb^{+}	6	152
	4	60	Lu^{3+}	6	86		8	161
Cu^{2+}	4	57		8	98	Ru^{3+}	6	68
	4*sq*	57	Mg^{2+}	4	57	Ru^{4+}	6	62
	6	73#		6	72	S^{2-}	6	184
Eu^{2+}	9	130	Mn^{2+}	4	66	Sc^{3+}	6	74.5
Eu^{3+}	6	95		6	83	Se^{2-}	6	198
	8	107		6*ls*	67	Sn^{2+}	6	96#
	9	112	Mn^{3+}	6	64.5	Sn^{4+}	4	55
F^{-}	2	129		6*ls*	58		6	69
	4	131	Mn^{4+}	6	53	Sr^{2+}	6	118
	6	133	N^{3-}	4	146		8	126
Fe^{2+}	4	63	Na^{+}	4	100	Te^{2-}	6	221
	6	78		6	105	Th^{4+}	8	105
	6*ls*	61		8	118		12	121
Fe^{3+}	4	49	Ni^{2+}	4	55	Ti^{3+}	6	67
	6	64.5		4*sq*	49	Ti^{4+}	4	42
	6*ls*	55		6	69		6	60.5
Ga^{3+}	4	47	Ni^{3+}	6	60	Tl^{+}	6	150
	6	62		6*ls*#	56	Tl^{3+}	6	88.5
Hf^{4+}	6	71	O^{2-}	2	135	U^{3+}	6	103
Hg^{+}	3	97		4	138	U^{4+}	6	89
Hg^{2+}	2	69		6	140	V^{2+}	6	79
	4	96	OH^{-}	2	132	V^{3+}	6	64
	6	102		4	135	V^{4+}	6	58
I^{-}	6	220		6	137	W^{4+}	6	66
In^{3+}	4	62	Pb^{2+}	4	98	Y^{3+}	6	90
	6	80		6	119		8	102
Ir^{3+}	6	68		8	129	Zn^{2+}	4	60
K^{+}	4	137	Pb^{4+}	4	65		6	74
	6	138		6	77.5	Zr^{4+}	4	59
	8	151	Pd^{2+}	4*sq*	64		6	72
La^{3+}	6	103	Pt^{2+}	4*sq*	60		8	84
	8	116		6	80#			

Index

A

B

C

D

E

F

I

J

K

L

M

Q

R

S

U

V

W